Animal Nutrition

Fourth edition

P. McDonald

Formerly Reader in Agricultural Biochemistry, University of Edinburgh and Head of the Department of Agricultural Biochemistry, Edinburgh School of Agriculture

R. A. Edwards

Formerly Head of the Department of Animal Nutrition, Edinburgh School of Agriculture

J. F. D. Greenhalgh

Formerly Professor of Animal Production and Health, University of Aberdeen and Chairman of the Animal Production and Health Group, School of Agriculture, Aberdeen

Longman
Scientific &
Technical

Copublished in the United States with
John Wiley & Sons, Inc., New York

Longman Scientific & Technical,
Longman Group UK Ltd,
Longman House, Burnt Mill, Harlow,
Essex CM20 2JE, England
and Associated Companies throughout the world.

Copublished in the United States with
John Wiley & Sons, Inc., 605 Third Avenue, New York, NY 10158

First published by Oliver & Boyd 1966
Second edition 1973
Reprinted by Longman Group Ltd 1975, 1978
Third edition 1981
Fourth edition 1988
Reprinted 1989, 1990, 1991, 1992

British Library Cataloguing in Publication Data
McDonald, Peter, 1926–
   Animal nutrition.—4th ed.
   1. Animal nutrition
   I. Title   II. Edwards, R. A. (Rhys Alun)
   III. Greenhalgh, J. F. D.
   636.08′52   SF95

ISBN 0-582-40903-9

Library of Congress Cataloging-in-Publication Data
McDonald, Peter, 1926–
   Animal nutrition.

   Bibliography: p.
   Includes index.
   1. Animal nutrition.   2. Feeds.   I. Edwards, R. A.
II. Greenhalgh, J. F. D.   III. Title.
SF95.M38   1988   636.08′52   86–28723
ISBN 0–470–20791–4 (USA only).

Set in Linotron 202 10/13 pt Times
Produced by Longman Singapore Publishers Pte Ltd
Printed in Singapore

Contents

Preface to the fourth edition   vi

1 The animal and its food   1
2 Carbohydrates   8
3 Lipids   26
4 Proteins and nucleic acids   42
5 Vitamins   58
6 Minerals   90
7 Enzymes   117
8 Digestion   130
9 Metabolism   158
10 Evaluation of foods. (A) Digestibility   200
11 Evaluation of foods. (B) Energy content of foods and the partition of food energy within the animal   217
12 Evaluation of foods. (C) Systems for expressing the energy value of foods   244
13 Evaluation of foods. (D) Protein   260
14 Feeding standards for maintenance and growth   284
15 Feeding standards for reproduction and lactation   321
16 Voluntary intake of food   375
17 Grass and forage crops   388
18 Silage   404
19 Hay, artificially dried forages, straws and chaff   416
20 Roots and tubers   430
21 Cereal grains and cereal by-products   438
22 Protein concentrates   455
Appendix—list of tables   484
References   524
Index   525

Preface to the fourth edition

In the twenty-one years since the first edition of this book was published, our scientific understanding of animal nutrition has increased dramatically. In this present edition, although we have kept the same format as in previous editions, we have taken into account recent developments in the subject and brought each chapter up to date. We have maintained the policy introduced in the third edition, of broadening the coverage of the book to include the needs of students outside Britain, and hope that this new edition will be of world-wide interest.

As with earlier editions, we have received many constructive comments from colleagues and friends both at home and abroad and we gratefully acknowledge their help. In particular, we would like to thank Dr Colin A. Morgan for his invaluable advice in preparing the Appendix Tables. We are also grateful to the following individuals and organisations for allowing us to reproduce colour plates of vitamin and mineral deficiencies: Dr W. A. Dewar (Plate 8), Dr M. L. Scott (Plate 3), Dr C. Whitehead (Plate 5), Central Veterinary Laboratory, Weybridge (Plates 1 and 4), F. Hoffmann-La Roche, Switzerland (Plate 2), the Rowett Research Institute, Aberdeen (Plates 6 and 7) and CAB International, Farnham Royal (Plate 6).

P. McDonald R. A. Edwards J. F. D. Greenhalgh

1

The animal and its food

Food is material which, after ingestion by animals, is capable of being digested, absorbed and utilised. In a more general sense we use the term 'food' to describe edible material. Grass and hay, for example, are described as foods, but not all their components are digestible. Where the term 'food' is used in the general sense, as in this book, then those components capable of being utilised by animals are described as *nutrients*.

The diet of farm animals consists of plants and plant products, although some foods of animal origin such as fishmeal and milk are used in limited amounts. Animals depend upon plants for their existence and consequently a study of animal nutrition must necessarily begin with the plant itself.

Plants are able to synthesise complex materials from simple substances such as carbon dioxide from the air, and water and inorganic elements from the soil. By means of photosynthesis, energy from sunlight is trapped and used in these synthetic processes. The greater part of the energy, however, is stored as chemical energy within the plant itself and it is this energy which is used by the animal for the maintenance of life and synthesis of its own body tissues. Plants and animals contain similar types of chemical substances, and we can group these into classes according to constitution, properties and function. The main components of foods, plants and animals are:

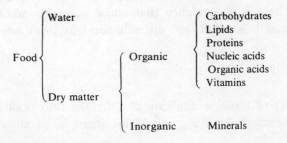

TABLE 1.1   Composition of some plant and animal products (g/kg)

	Water	Carbohydrate	Fat	Protein	Ash
Turnips	910	71	2	10	7
Pasture grass					
(young leafy)	800	100	10	32	24
Wheat grain	130	712	19	122	17
Groundnuts	60	201	449	268	22
Dairy cow	570	2	206	172	50
Blood	820	1	6	164	7
Liver	740	13	65	168	14
Muscle	720	6	43	214	15
Milk (cows)	876	47	36	33	8

WATER

The water content of the animal body varies with age. The newborn animal contains from 750 to 800 g/kg water but this falls to about 500 g/kg in the mature fat animal. It is vital to the life of the organism that the water level in the body be maintained—an animal will die more rapidly if deprived of water than if deprived of food. Water functions in the body as a solvent in which nutrients are transported about the body and in which waste products are excreted. Many of the chemical reactions brought about by enzymes take place in solution and involve hydrolysis. Because of the high specific heat of water, large changes in heat production can take place within the animal with very little alteration in body temperature. Water also has a high latent heat of evaporation, and its evaporation from the lungs and skin gives it a further role in the regulation of body temperature.

The animal obtains its water from three sources: drinking water, water present in its food and metabolic water, this last being formed during metabolism by the oxidation of hydrogen-containing organic nutrients. The water content of food is very variable and, as the figures in Table 1.1 indicate, can range from 60 g/kg in concentrates to over 900 g/kg in some root crops. The water content of growing plants is related to the stage of growth, younger plants containing more water than older plants. In temperate climates drinking water is not usually a problem and animals are provided with a continuous supply. There is no evidence that under normal conditions an excess of drinking water is harmful, and animals normally drink what they require.

DRY MATTER

The dry matter (DM) of foods is conveniently divided into organic and inorganic material, although in living organisms there is no such sharp distinction. Many organic compounds contain mineral elements as structural components. Proteins, for example, contain sulphur, and many lipids and carbohydrates contain phosphorus.

It can be seen from Table 1.1 that the main component of the DM of pasture grass is carbohydrate, and this is true of all plants and many seeds—the oilseeds, such as groundnuts, being exceptional in containing large amounts of protein, and lipid material in the form of fat or oil. In contrast the carbohydrate content of the animal body is very low. One of the main reasons for the difference between plants and animals is that, whereas the cell walls of plants consist of carbohydrate material, mainly cellulose, the walls of animal cells are composed almost entirely of protein. Furthermore, plants store energy largely in the form of carbohydrates such as starch and fructans, whereas an animal's main energy store is in the form of fat.

Fat is the most important lipid present in both plants and animals. The fat content of the animal body is variable and is related to age, the older animal containing a much greater proportion of fat than the young animal. The lipid content of living plants is relatively low, that of pasture grass for example, being 40 to 50 g/kg DM.

In both plants and animals, proteins are the major nitrogen-containing compounds. In plants, where most of the protein is present as enzymes, the concentration is high in the young growing plant and falls as the plant matures. In animals, muscle, skin, hair, feathers, wool and nails contain protein.

Like proteins, nucleic acids are also nitrogen-containing compounds and they play a basic role in the synthesis of proteins in all living organisms. They also carry the genetic information of the living cell.

The organic acids which occur in plants and animals include citric, malic, fumaric, succinic and pyruvic. Although these are normally present in small quantities, they nevertheless play an important role as intermediates in the general metabolism of the cell. Other organic acids occur as fermentation products in the rumen, or in silage, and these include acetic, propionic, butyric and lactic.

Vitamins are present in plants and animals in minute amounts, and many of them are important as components of enzyme systems. An important difference between plants and animals is that, whereas the former can synthesise all the vitamins they require for metabolism, animals cannot, or have very limited powers of synthesis, and are dependent upon an external supply.

The inorganic matter contains all those elements present in plants and animals other than carbon, hydrogen, oxygen and nitrogen. Calcium and phosphorus are the major inorganic components of animals, whereas potassium and silicon are the main inorganic elements in plants.

ANALYSIS OF FOODS

Much of the existing information we have about the composition of foods is based on a system of analysis described as the *proximate analysis of foods*, which was devised about 100 years ago by two German scientists, Henneberg and Stohmann.

Proximate analysis of foods

This system of analysis divides the food into six fractions, as shown in Table 1.2.

The moisture content is determined as the loss in weight which results from drying a known weight of food to constant weight at 100 °C. This method is satisfactory for most foods, but with a few, such as silage, significant losses of volatile material may take place.

The ash content is determined by ignition of a known weight of the food at 500 °C until all carbon has been removed. The residue is the ash and is taken to represent the inorganic constituents of the food. The ash may, however, contain material of organic origin such as sulphur and phosphorus from proteins, and some loss of volatile material in the form of sodium, chloride, potassium, phosphorus and sulphur will take place during ignition. The ash content is thus not truly representative of the inorganic material in the food either qualitatively or quantitatively.

The crude protein content is calculated from the nitrogen content of the food, determined by a modification of a technique originally devised by Kjeldahl over 100 years ago. In this method the food is digested with sulphuric acid, which converts to ammonia all nitrogen present except that in the form of nitrate and nitrite. This ammonia is liberated by adding sodium hydroxide to the digest, distilled off and collected in standard acid, the quantity so collected being determined by titration or by an automated colorimetric method. It is assumed that the nitrogen is derived from protein containing 16 per cent nitrogen, and by multiplying the nitrogen figure by 100/16 or 6.25 an approximate protein value is obtained. This is not 'true protein' since the method determines nitrogen from sources other than protein, and the fraction is therefore designated crude protein.

The ether extract fraction is determined by subjecting the food to a continuous extraction with petroleum ether for a defined period. The residue, after evaporation of the solvent, is the ether extract. As well as true fat it contains waxes, organic acids, alcohols and pigments; designation of the fraction as 'oil' or 'fat' is therefore incorrect.

The carbohydrate of the food is contained in two fractions, the crude fibre and the nitrogen-free extractives. The former is determined by

TABLE 1.2 Components of different fractions in the proximate analysis of foods

Fraction	Components
Moisture	Water (and volatile acids and bases if present).
Ash	Essential elements $\begin{cases} \text{Major: Ca, K, Mg, Na, S, P, Cl.} \\ \text{Trace: Fe, Mn, Cu, Co, I, Zn, Si,} \\ \text{Mo, Se, Cr, F, V, Sn, As, Ni.} \end{cases}$
	Non-essential elements: Ti, Al, B, Pb.
Crude protein	Proteins, amino acids, amines, nitrates, nitrogenous glycosides, glycolipids, B-vitamins, nucleic acids.
Ether extract	Fats, oils, waxes, organic acids, pigments, sterols, vitamins A, D, E, K.
Crude fibre	Cellulose, hemicelluloses, lignin.
Nitrogen-free extractives	Cellulose, hemicelluloses, lignin, sugars, fructans, starch, pectins, organic acids, resins, tannins, pigments, water-soluble vitamins.

subjecting the residual food from ether extraction to successive treatments with boiling acid and alkali of defined concentration; the organic residue is the crude fibre. When the sum of the amounts of moisture, ash, crude protein, ether extract and crude fibre (expressed in g/kg) is subtracted from 1000, the difference is designated the nitrogen-free extractives (NFE). The crude fibre fraction contains cellulose, lignin and hemicelluloses, but not necessarily all of these that are present in the food: a variable proportion of them is contained in the nitrogen-free extractives, depending upon the species and stage of growth of the plant material. The complexity of the nitrogen-free extractives fraction is well illustrated by the constituents shown in Table 1.2. The crude fibre was intended originally to provide a measure of the indigestible part of the food, but quite a large part of it may in fact be digested by ruminant animals. Despite this the figure is valuable because of the correlation existing between it and the digestibility of the food.

Van Soest method of forage analysis

In recent years the proximate analysis procedure has been severely criticised by many nutritionists as being archaic and imprecise, and in many laboratories it has been partially replaced by other analytical procedures. Most criticism has been focused on the crude fibre and nitrogen-free extractives fractions, and alternative procedures using detergents have been proposed by Van Soest (Table 1.3). The neutral-detergent fibre (NDF), which is the residue after extraction with boiling neutral solutions of sodium lauryl sulphate and ethylenediaminetetraacetic acid (EDTA), consists mainly of lignin, cellulose and hemicellulose and can be regarded as a measure of the

TABLE 1.3   Classification of forage fractions using the detergent methods of Van Soest
(After P. J. Van Soest, 1967. *J. Anim. Sci*, **26**, 119)

Fraction	Components
Cell contents (soluble in neutral detergent)	Lipids
	Sugars, organic acids and water-soluble matter
	Pectin, starch
	Non-protein N
	Soluble protein
Cell wall constituents (fibre insoluble in neutral detergent)	
1. Soluble in acid detergent	Hemicelluloses
	Fibre-bound protein
2. Acid-detergent fibre	Cellulose
	Lignin
	Lignified N
	Silica

plant cell wall material. The acid-detergent fibre (ADF) is the residue after refluxing with 0.5 M sulphuric acid and cetyltrimethylammonium bromide, and represents essentially the crude lignin and cellulose fractions of plant material but also includes silica.

The determination of ADF is particularly useful for forages as there is a good statistical correlation between it and digestibility. In the UK the ADF method has been modified slightly, the duration of boiling and acid strength being increased. The term modified acid-detergent fibre (MADF) is used to describe this determination.

Other analytical procedures

If detailed information on individual amino acids, fatty acids, sugars and mineral elements in foods is required, then chromatographic separation and atomic absorption techniques can be used. Many of these methods make use of automated procedures in which the results are processed by computer. More recently, in some laboratories, procedures combining statistical regression techniques with infrared reflectance spectrophotometry, which allow the rapid and non-destructive determination of a wide range of organic components of foods, have been introduced.

FURTHER READING

The Analysis of Agricultural Materials, 1981, Ministry of Agriculture, Fisheries and Food. Ref Book 47. HMSO, London.

The Feeding Stuffs (Sampling and Analysis) Regulations, 1982. HMSO, London.
Association of Official Analytical Chemists, 1980, *Official Methods of Analysis*, 13th
    edn. Washington, DC.
P. J. Van Soest, 1982. *Nutritional Ecology of the Ruminant*. O and B Books,
    Corvallis, Oregon.

2

Carbohydrates

The name carbohydrate is derived from the French *hydrate de carbone* and was originally applied to neutral chemical compounds containing the elements carbon, hydrogen and oxygen, with the last two elements present in the same proportions as in water. Although many carbohydrates have the empirical formula $(CH_2O)_n$ where n is three or more, the above definition is not strictly correct, since some compounds with general properties of the carbohydrates contain phosphorus, nitrogen or sulphur in addition to the elements carbon, hydrogen and oxygen. Furthermore some compounds, e.g. deoxyribose $(C_5H_{10}O_4)$ do not have hydrogen and oxygen in the same ratio as that in water.

A more modern approach is to define carbohydrates as polyhydroxy aldehydes, ketones, alcohols or acids, their simple derivatives, and any compound that may be hydrolysed to these.

CLASSIFICATION OF CARBOHYDRATES

The carbohydrates are usually divided into two major groups, the sugars and the non-sugars (see Table 2.1). The simplest sugars are the monosaccharides which are divided into sub-groups: trioses $(C_3H_6O_3)$, tetroses $(C_4H_8O_4)$, pentoses $(C_5H_{10}O_5)$, hexoses $(C_6H_{12}O_6)$ and heptoses $(C_7H_{14}O_7)$ depending upon the number of carbon atoms present in the molecule. The trioses and tetroses occur as intermediates in the metabolism of other carbohydrates and their importance will be considered in Chapter 9. Monosaccharides may be linked together, with the elimination of one molecule of water at each linkage, to produce di-, tri-, or polysaccharides containing respectively 2, 3 or larger numbers of monosaccharide units or residues.

The term 'sugar' is generally restricted to those carbohydrates containing less than ten monosaccharide residues, while the name oligosaccharides

TABLE 2.1 Classification of carbohydrates

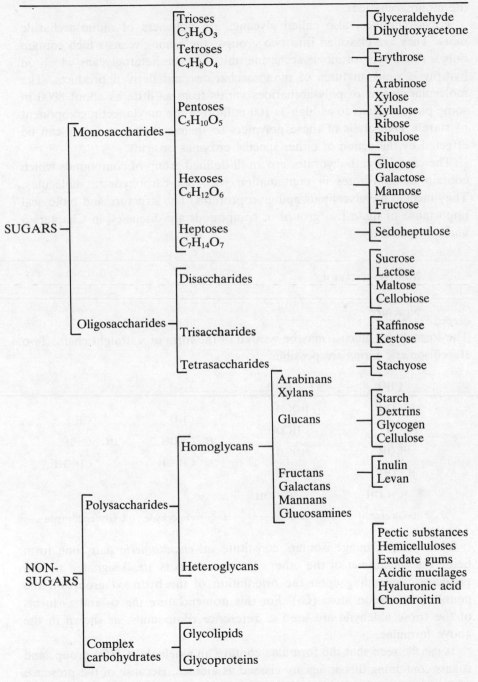

(from Greek *oligos*, a few) is frequently used to include all sugars other than the monosaccharides.

Polysaccharides, also called glycans, are polymers of monosaccharide units. They are classified into two groups, the homoglycans which contain only a single type of monosaccharide unit, and the heteroglycans which on hydrolysis yield mixtures of monosaccharides and derived products. The molecular weight of polysaccharides varies from as little as about 8000 in some plant fructans to as high as 100 million in the amylopectin component of starch. Hydrolysis of these polymers to their constituent sugars can be effected by the action of either specific enzymes or acids.

The complex carbohydrates are an ill-defined group of compounds which contain carbohydrates in combination with non-carbohydrate molecules. They include the glycolipids and glycoproteins. The structure and biological importance of these two groups of compounds are discussed in Chapters 3 and 4 respectively.

MONOSACCHARIDES

Structure

The formula for glucose may be written in the form of a straight chain. Two stereoisomeric forms are possible:

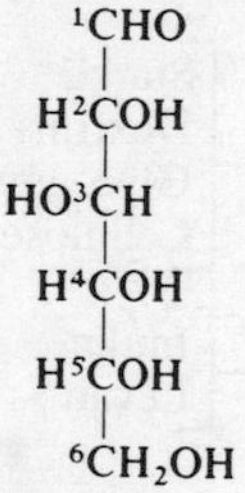

$$
\begin{array}{cccc}
^1CHO & CHO & & \\
| & | & CHO & CHO \\
H^2COH & HOCH & | & | \\
| & | & H-C-OH & HO-C-H \\
HO^3CH & HCOH & | & | \\
| & | & CH_2OH & CH_2OH \\
H^4COH & HOCH & & \\
| & | & & \\
H^5COH & HOCH & & \\
| & | & & \\
^6CH_2OH & CH_2OH & & \\
\text{D-Glucose} & \text{L-Glucose} & \text{D-Glyceraldehyde} & \text{L-Glyceraldehyde}
\end{array}
$$

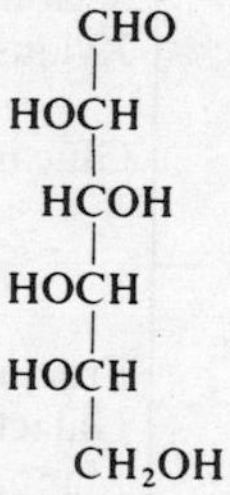

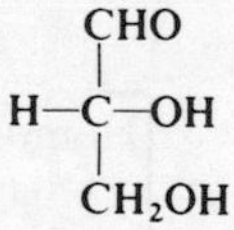

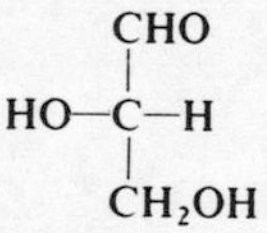

These mirror image isomers constitute an *enantiomeric* pair, one form being the *enantiomer* of the other. The two isomers are designated D- and L-glucose, depending upon the orientation of the hydroxyl group on the penultimate carbon atom (C_5). For this nomenclature the D- and L- forms of the triose aldehyde are used as reference compounds, as shown in the above formulae.

It can be seen that the formulae contain an aldehyde (CHO) group, and sugars containing this group are classed as aldoses. Because of the presence of four asymmetric or chiral carbon atoms in aldohexoses, there are sixteen possible stereoisomeric forms, eight of which are D-sugars and the other eight are mirror-images of them or L-forms. Only a few of these occur naturally; in addition to D-glucose the important ones are D-galactose and D-mannose. The hexose straight-chain formula may contain a ketone group (CO) instead of an aldehyde group. Sugars containing a ketone group are classed as ketoses. Eight stereoisomers are possible, four D-forms and four L-forms. The most important naturally occurring ketohexose is D-fructose:

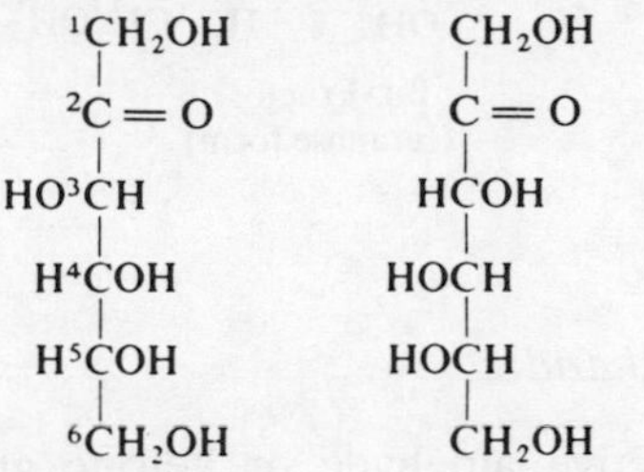

$$
\begin{array}{ll}
^1CH_2OH & CH_2OH \\
^2C=O & C=O \\
HO^3CH & HCOH \\
H^4COH & HOCH \\
H^5COH & HOCH \\
^6CH_2OH & CH_2OH \\
\text{D-Fructose} & \text{L-Fructose}
\end{array}
$$

Sugars can exist in a ring or cyclic form, in the case of D-glucose, a pyranose ring, similar to pyran, which is usually depicted as a regular hexagon.

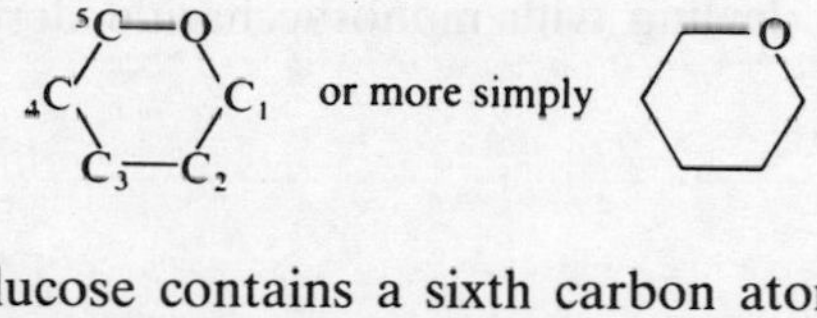

or more simply

D-Glucose contains a sixth carbon atom attached to carbon atom 5 and two forms of this sugar occur known as α- and β-glucose depending upon the configuration of carbon atom 1. Ring closure results in the formation of another chiral carbon which doubles the number of isomeric sugars in any one group. A pair of stereoisomers related to each other, as are α- and β-D-glucose, are said to be *anomers*, and carbon atom 1 is termed the anomeric carbon atom.

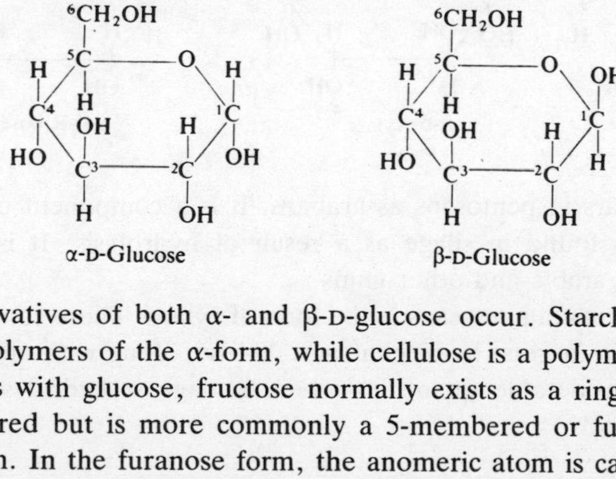

α-D-Glucose   β-D-Glucose

Derivatives of both α- and β-D-glucose occur. Starch and glycogen are both polymers of the α-form, while cellulose is a polymer of β-glucose.

As with glucose, fructose normally exists as a ring, which may be 6-membered but is more commonly a 5-membered or furanose ring similar to furan. In the furanose form, the anomeric atom is carbon atom 2.

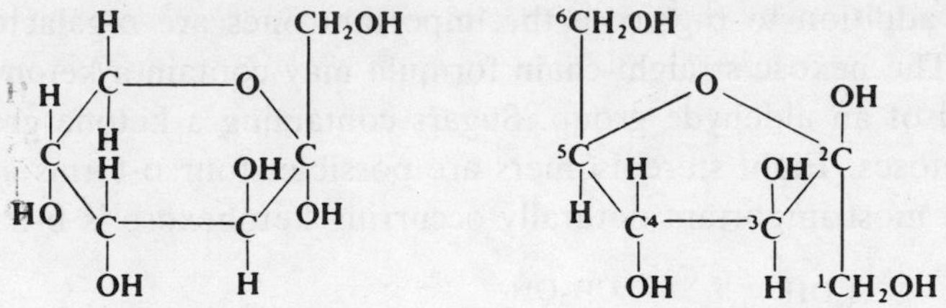

α-D-Fructose
(pyranose form)

β-D-Fructose
(furanose form)

Properties of the monosaccharides

Because of the presence of an active aldehyde or ketone grouping, the monosaccharides act as reducing substances. The reducing properties of these sugars are usually demonstrated by their ability to reduce certain metal ions, notably copper or silver, in alkaline solution. The aldehyde and ketone groups may also be reduced chemically, or enzymatically, to yield the corresponding sugar alcohols. Examples of oxidation and reduction products are given in the section dealing with monosaccharide derivatives (p. 14).

Pentoses

Pentoses have the general formula $C_5H_{10}O_5$. The most important members of this group of simple sugars are the aldoses L-arabinose, D-xylose and D-ribose, and the ketoses D-xylulose and D-ribulose.

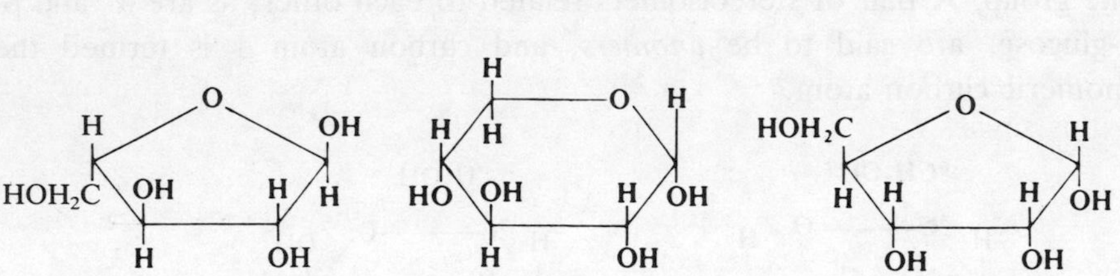

α-L-Arabinose   α-D-Xylose   α-D-Ribose

L-*Arabinose* occurs in pentosans as arabans. It is a component of hemi-celluloses and it is found in silage as a result of hydrolysis. It is also a component of gum arabic and other gums.

D-*Xylose* also occurs in pentosans in the form of xylans. These compounds form the main chain in grass hemicelluloses. Xylose, along with arabinose, is produced in considerable quantities when herbage is hydrolysed with normal sulphuric acid.

D-*Ribose* is present in all living cells as a constituent of ribonucleic acid (RNA), and it is also a component of several vitamins and coenzymes.

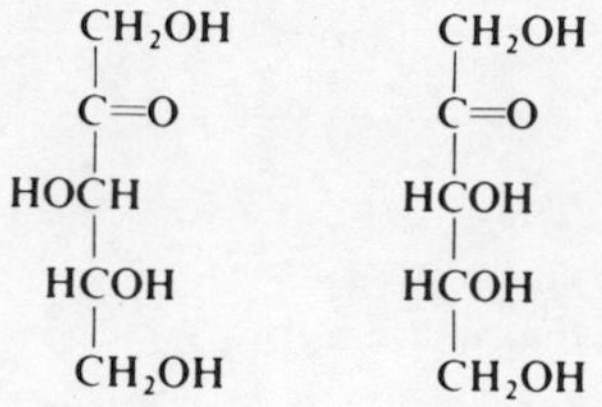

$$
\begin{array}{cc}
CH_2OH & CH_2OH \\
| & | \\
C=O & C=O \\
| & | \\
HOCH & HCOH \\
| & | \\
HCOH & HCOH \\
| & | \\
CH_2OH & CH_2OH \\
\text{D-Xylulose} & \text{D-Ribulose}
\end{array}
$$

The phosphate derivatives of D-*xylulose* and D-*ribulose* occur as intermediates in the pentose phosphate metabolic pathway (see p. 169).

Hexoses

Glucose and fructose are the most important naturally occurring hexose sugars, while mannose and galactose occur in plants in a polymerised form as mannans and galactans.

D-*Glucose*, grape sugar or dextrose, exists in the free state as well as in combined form. The sugar occurs free in plants, fruits, honey, blood, lymph and cerebrospinal fluid, and is the sole or major component of many oligo-saccharides, polysaccharides and glucosides. In the pure state, glucose is a white crystalline solid and like all sugars is soluble in water.

D-*Fructose*, fruit sugar or laevulose, occurs free in green leaves, fruits and honey. It also occurs in the disaccharide sucrose and in fructans. Green leafy crops usually contain appreciable amounts of this sugar both free and in polymerised form. The free sugar is a white crystalline solid and has a sweeter taste than sucrose. The exceptionally sweet taste of honey is due to this sugar.

D-*Mannose* does not occur free in nature but exists in polymerised form as mannan, and also as a component of glycoproteins. Mannans are found widely distributed in yeasts, moulds and bacteria.

D-*Galactose* does not occur free in nature except as a breakdown product during fermentation. It is present as a constituent of the disaccharide lactose, which occurs in milk. Galactose also occurs as a component of the anthocyanin pigments, galactolipids, gums and mucilages.

Heptoses

D-*Sedoheptulose* is an important example of a monosaccharide containing seven carbon atoms.

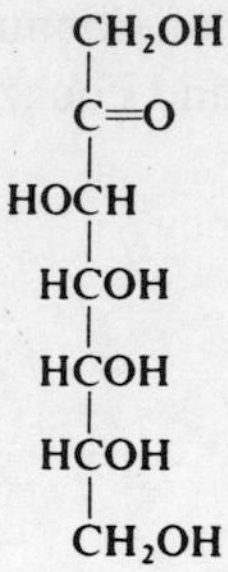
CH₂OH
C=O
HOCH
HCOH
HCOH
HCOH
CH₂OH

D-Sedoheptulose

This heptose occurs, as the phosphate, as an intermediate in the pentose phosphate metabolic pathway (see p. 169).

MONOSACCHARIDE DERIVATIVES

Phosphoric acid esters

The phosphoric acid esters of sugars play an important role in a wide variety of metabolic reactions in living organisms (see Ch. 9). The most commonly occurring derivatives are those formed from glucose, the esterification occurring at either carbon atoms 1 or 6.

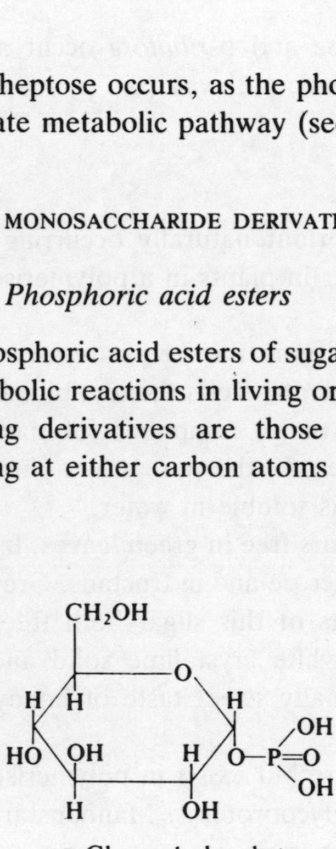
CH₂OH
O
H
H
H
OH
HO
OH
H
O—P=O
OH
H
OH

α-D-Glucose 1-phosphate

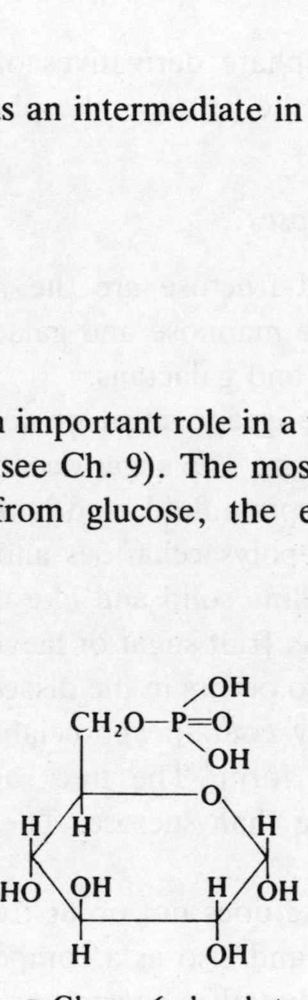
OH
CH₂O—P=O
OH
O
H
H
H
HO
OH
H
OH
H
OH

α-D-Glucose 6-phosphate

Amino sugars

If the hydroxyl group on carbon atom 2 of an aldohexose is replaced by an amino group ($-NH_2$), the resulting compound is an amino sugar. Two such naturally occurring important compounds are D-glucosamine, a major component of chitin, (see p. 23), and D-galactosamine, a component of the polysaccharide of cartilage.

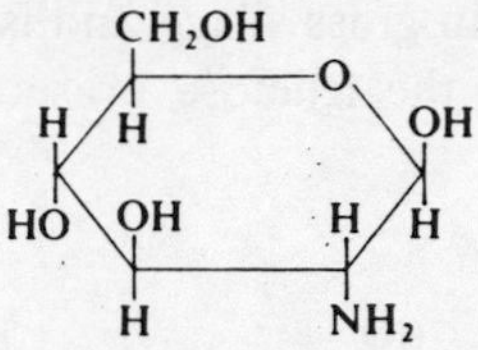

CH₂OH ... β-D-Glucosamine

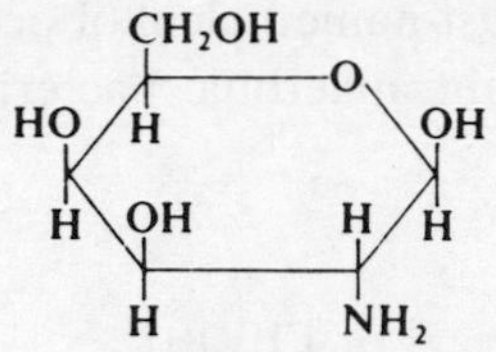

CH₂OH ... β-D-Galactosamine

Deoxy sugars

Replacement of a hydroxyl group by hydrogen yields a deoxy sugar. The derivative of ribose, *deoxyribose* is a component of deoxyribonucleic acid (*DNA*). Similarly deoxy derivatives of the two hexoses, galactose and mannose, occur as *fucose* and *rhamnose* respectively, these being components of certain heteropolysaccharides.

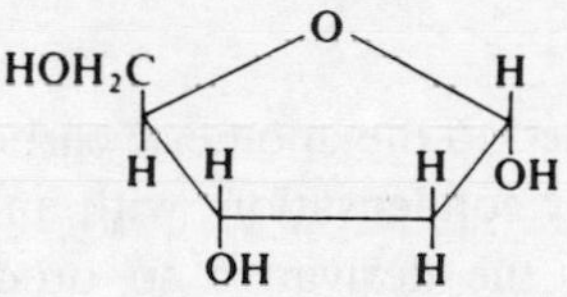

α-D-Deoxyribose

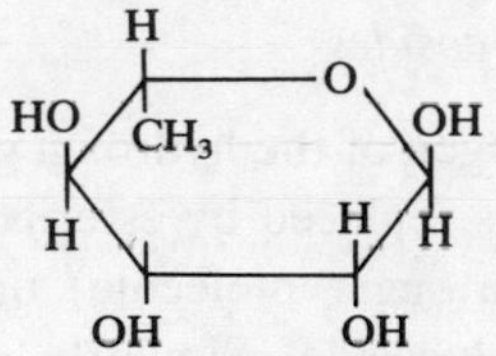

α-L-Rhamnose

Sugar acids

The aldoses can be oxidised to produce a number of acids of which the most important are:

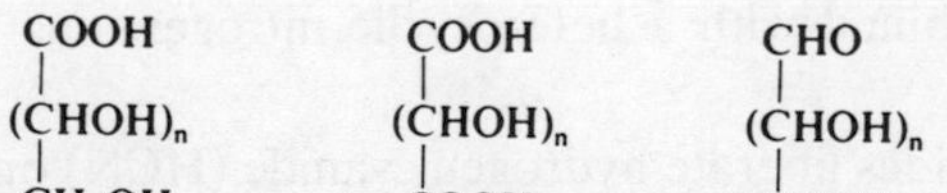

COOH
(CHOH)ₙ
CH₂OH
Aldonic acids

COOH
(CHOH)ₙ
COOH
Aldaric acids

CHO
(CHOH)ₙ
COOH
Uronic acids

In the case of glucose, the derivatives corresponding to these formulae are gluconic, glucaric and glucuronic acids respectively. Of these compounds, the uronic acids, particularly those derived from glucose and galactose, are important components of a number of heteropolysaccharides.

Sugar alcohols

Simple sugars can be reduced to polyhydric alcohols; for example, glucose yields sorbitol, galactose yields dulcitol while both mannose and fructose yield mannitol. This last-named alcohol occurs in grass silage and is formed by the action of certain anaerobic bacteria on the fructose present in the grass.

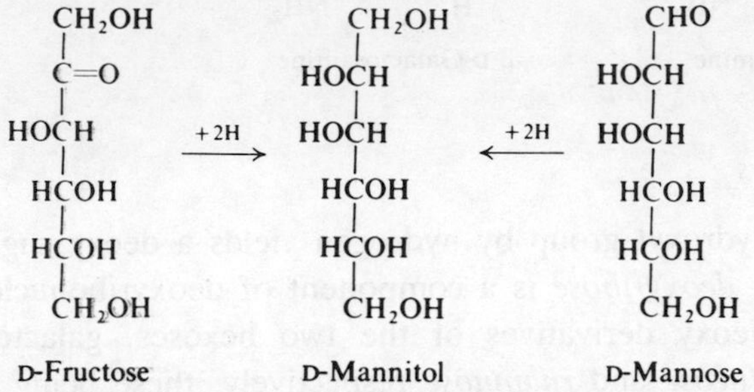

$$
\begin{array}{ccccc}
\text{CH}_2\text{OH} & & \text{CH}_2\text{OH} & & \text{CHO} \\
| & & | & & | \\
\text{C}=\text{O} & & \text{HOCH} & & \text{HOCH} \\
| & & | & & | \\
\text{HOCH} & \xrightarrow{+\,2\text{H}} & \text{HOCH} & \xleftarrow{+\,2\text{H}} & \text{HOCH} \\
| & & | & & | \\
\text{HCOH} & & \text{HCOH} & & \text{HCOH} \\
| & & | & & | \\
\text{HCOH} & & \text{HCOH} & & \text{HCOH} \\
| & & | & & | \\
\text{CH}_2\text{OH} & & \text{CH}_2\text{OH} & & \text{CH}_2\text{OH} \\
\text{D-Fructose} & & \text{D-Mannitol} & & \text{D-Mannose}
\end{array}
$$

Glycosides

If the hydrogen of the hydroxyl group attached to the anomeric carbon atom of glucose is replaced by esterification, or by condensation, with an alcohol (including a sugar molecule) or a phenol, the derivative so produced is termed a glucoside. Similarly galactose forms galactosides, and fructose forms fructosides. The general term glycoside is used collectively to describe these derivatives, and the linkage effected through the anomeric carbon atom is described as a glycosidic bond.

Oligosaccharides and polysaccharides are classed as glycosides, and these compounds yield sugars or sugar derivatives on hydrolysis. Certain naturally occurring glycosides contain non-sugar residues. For example the nucleosides contain a sugar combined with a heterocyclic nitrogeneous base (see Ch. 4.).

The cyanogenetic glycosides liberate hydrogen cyanide (HCN) on hydrolysis, and because of the toxic nature of this compound, plants containing this type of glycoside are potentially dangerous to animals. The glycoside itself is not toxic and must be hydrolysed before poisoning occurs. However, the glycoside is easily broken down to its components by means of an enzyme which is usually present in the plant.

An example of a cyanogenetic glycoside is linamarin (also called phaseolunatin), which occurs in linseed, Java beans and cassava. If wet mashes or gruels containing these foods are given to animals, it is advisable to boil them when mixing in order to inactivate any enzyme present. On hydrolysis linamarin yields glucose, acetone and hydrogen cyanide.

Examples of other cyanogenetic glycosides and their sources are shown in Table 2.2.

TABLE 2.2 Some important naturally occurring cyanogenetic glycosides

Name	Source	Hydrolytic products in addition to glucose and hydrogen cyanide
Linamarin (phaseolunatin)	Linseed (*Linum usitatissimum*) Java beans (*Phaseolus lunatus*) Cassava (*Manihot esculenta*)	Acetone
Vicianin	Seeds of wild vetch (*Vicia angustifolia*)	Arabinose, benzaldehyde
Amygdalin	Bitter almonds, kernels of peach, cherries, plums, apples and fruits of Rosaceae	Benzaldehyde
Dhurrin	Leaves of the great millet (*Sorghum vulgare*)	*p*-hydroxy-benzaldehyde
Lotaustralin	Trefoil (*Lotus australis*) White clover (*Trifolium repens*)	Methylethyl ketone

OLIGOSACCHARIDES

DISACCHARIDES

A large number of disaccharide compounds are theoretically possible, depending upon the monosaccharides present and the manner in which they are linked. The most nutritionally important disaccharides are sucrose, maltose, lactose and cellobiose which on hydrolysis yield two molecules of hexoses:

$$C_{12}H_{22}O_{11} + H_2O \rightarrow 2C_6H_{12}O_6$$

Sucrose consists of one molecule of α-D-glucose and one molecule of β-D-fructose joined together through an oxygen bridge between their respec-

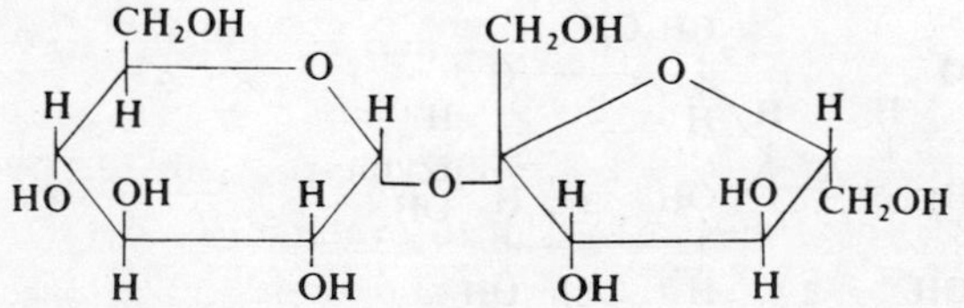

Sucrose

tive anomeric carbon atoms (1,2). As a consequence, sucrose has no active reducing group.

Sucrose is the most ubiquitous and abundantly occurring disaccharide in plants, where it is the main transport form of carbon. This disaccharide is found in high concentration in sugar cane (200 g/kg) and in sugar beet (150–200 g/kg); it is also present in other roots such as mangels and carrots, and occurs in many fruits. Sucrose is easily hydrolysed by the enzyme sucrase (invertase) or by dilute acids. When heated to a temperature of 160 °C it forms barley sugar and at a temperature of 200 °C it forms caramel.

Lactose, or milk sugar, occurs only as a product of the mammary gland. Cow's milk contains 46–48 g/kg. It is not as soluble as sucrose and is less sweet, imparting only a faint sweet taste to milk. Lactose consists of one molecule of β-D-glucose joined to one of β-D-galactose in a β-(1→4)-linkage and has one active reducing group.

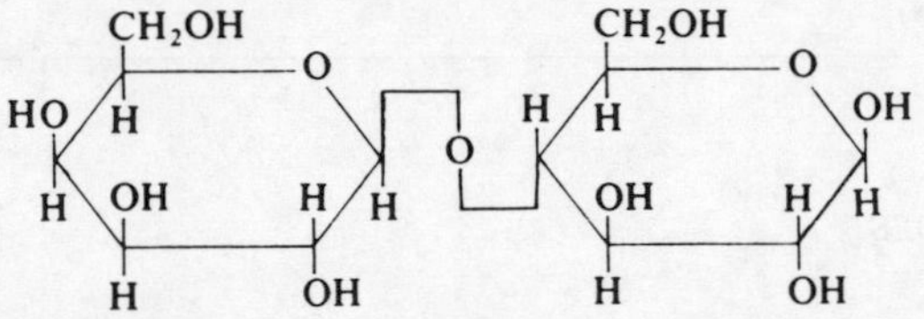

Lactose

Lactose readily undergoes fermentation by a number of organisms, including *Streptoccocus lactis*. This organism is responsible for souring milk by converting the lactose into lactic acid ($CH_3 \cdot CHOH \cdot COOH$). If lactose is heated to 150 °C it turns yellow, and at a temperature of 175 °C the sugar is changed into a brown compound, lactocaramel. On hydrolysis lactose produces one molecule of glucose and one molecule of galactose.

Maltose, or malt sugar, is produced during the hydrolysis of starch and glycogen by dilute acids or enzymes. This sugar is produced from starch, during the germination of barley, by the action of the enzyme amylase. The barley, after controlled germination and drying, is known as malt and is used in the manufacture of beer and Scotch malt whisky. Maltose is water-

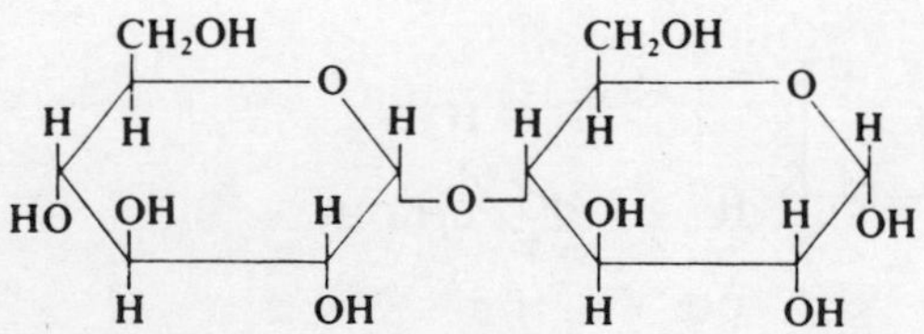

Maltose

soluble, but it is not as sweet as sucrose. Structurally it consists of two α-D-glucose residues linked in the α-1,4 positions and it has one active reducing group.

Cellobiose does not exist naturally as a free sugar, but is the basic repeating unit of cellulose. It is composed of two β-D-glucose residues linked through a β-(1→4)-bond. This linkage cannot be split by mammalian digestive enzymes. It can, however, be split by microbial enzymes. Like maltose, cellobiose has one active reducing group.

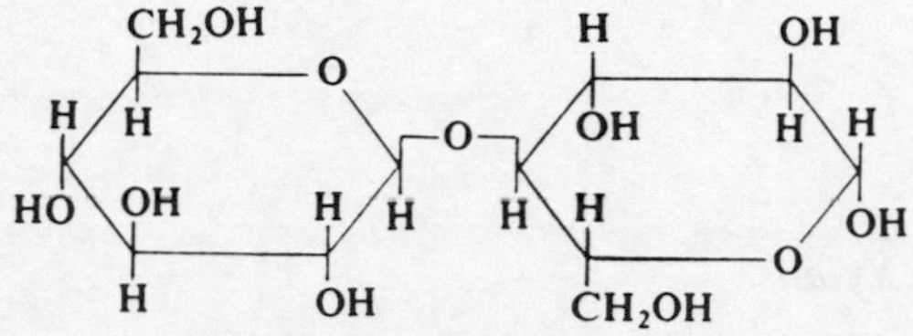

Cellobiose

TRISACCHARIDES

Raffinose and *kestose* are two important naturally occurring trisaccharides. They are both non-reducing and on hydrolysis produce three molecules of hexose sugars.

$$C_{18}H_{32}O_{16} + 2H_2O \rightarrow 3C_6H_{12}O_6$$

Raffinose is the commonest member of the group, occurring almost as widely in plants as sucrose. It exists in small amounts in sugar beet and accumulates in molasses during the commercial preparation of sucrose. Cotton seed contains about 80 g/kg of raffinose. On hydrolysis, this sugar produces glucose, fructose and galactose.

Kestose, and its isomer isokestose, occurs in the vegetative parts and seeds of grasses. These two trisaccharides consist of a fructose residue attached to a sucrose molecule.

TETRASACCHARIDES

Tetrasaccharides are made up of four monosaccharides residues. *Stachyose*, a member of this group, is almost as ubiquitous in higher plants as raffinose and has been isolated from about 165 species. It is a non-reducing sugar and on hydrolysis produces two molecules of galactose, one molecule of glucose and one of fructose.

$$C_{24}H_{42}O_{21} + 3H_2O \rightarrow 4C_6H_{12}O_6$$

POLYSACCHARIDES

HOMOGLYCANS

These carbohydrates are very different from the sugars. The majority are of high molecular weight, being composed of large numbers of pentose or hexose residues. Homoglycans do not give the various sugar reactions characteristic of the aldoses and ketoses. Many of them occur in plants either as reserve food materials such as starch or as structural materials such as cellulose.

Arabinans and Xylans

Arabinans (arabans) and *xylans* are polymers of arabinose and xylose respectively. Although homoglycans based on these two pentoses are known, they are more commonly found in combination with other sugars as constituents of heteroglycans.

Glucans

Starch is a glucan and is present in many plants as a reserve carbohydrate. It is most abundant in seeds, which may contain up to 700 g/kg and in fruits, tubers and roots, which may contain 300 g/kg. Starch occurs naturally in the form of granules, whose size and shape vary in different plants. The granules are built up in concentric layers, and although glucan is the main component of the granules they also contain minor constituents such as protein, fatty acids and phosphorus compounds which may influence their properties.

Starches differ in their chemical composition, and except in rare instances are mixtures of two structurally different polysaccharides, amylose and amylopectin. The proportions of these present in natural starches depend upon the source, although the proportion found in most cereal grains and potatoes is 15–30 per cent amylose and 85–70 per cent amylopectin. The quantity of amylose in starches can be estimated by a characteristic reaction with iodine: amylose produces a deep blue colour while amylopectin solutions produce a blue-violet or purple colour.

A study of the two major fractions of starch has shown that amylose is mainly linear in structure, the α-D-glucose residues being linked between carbon atom 1 of one molecule and carbon atom 4 of the adjacent molecule.

A very small proportion of α-(1→6)-linkages may also be present. Amylopectin has a bush-like structure containing primarily α-(1→4)-linkages, but also an appreciable number of α-(1→6)-linkages.

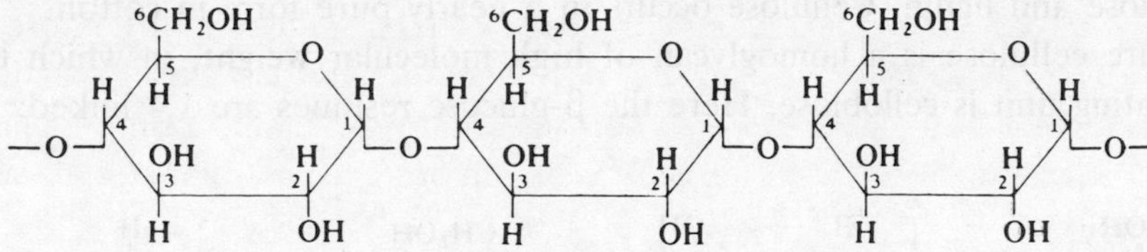

Part of amylose molecule showing 1,4 linkages

Starch granules are insoluble in cold water, but when a suspension in water is heated the granules swell and eventually gelatinise. On gelatinisation, potato starch granules swell greatly and then burst open; cereal starches swell but tend not to burst.

Animals consume large quantities of starch in cereal grains, cereal by-products and tubers.

Glycogen is a term used to describe a group of highly branched polysaccharides isolated from animals or micro-organisms. Glycogens occur in liver, muscle and other animal tissues. They are glucans, analogous to amylopectin in structure, and have been referred to as 'animal starches'. They form colloidal solutions which are dextro-rotatory. Glycogen is the main carbohydrate storage product in the animal body and plays an essential role in energy metabolism.

The molecular weights of glycogen molecules vary considerably depending on the animal species, the type of tissue and the physiological state of the animal. The glycogen of rat liver for example has molecular weights in the range $1\text{--}5 \times 10^8$, whereas that from rat muscle has a rather lower molecular weight of about 5×10^6.

Dextrins are intermediate products of the hydrolysis of starch and glycogen:

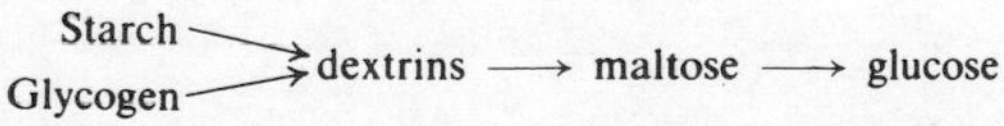

$$\left.\begin{array}{l}\text{Starch}\\\text{Glycogen}\end{array}\right\}\text{dextrins} \longrightarrow \text{maltose} \longrightarrow \text{glucose}$$

Dextrins are soluble in water and produce gum-like solutions. The higher members of these transitional products produce a red colour with iodine, while the lower members do not give a colour. The presence of dextrins gives a characteristic flavour to bread crust, toast and partly charred cereal foods.

Cellulose is a glucan and is the most abundant plant constituent, forming the fundamental structure of plant cell walls. The cell wall material contains other ingredients, and recent evidence suggests that there might be a chemical linkage between cellulose and hemicelluloses as well as between cellulose and lignin. Cellulose occurs in a nearly pure form in cotton.

Pure cellulose is a homoglycan of high molecular weight, in which the repeating unit is cellobiose. Here the β-glucose residues are 1,4 linked:

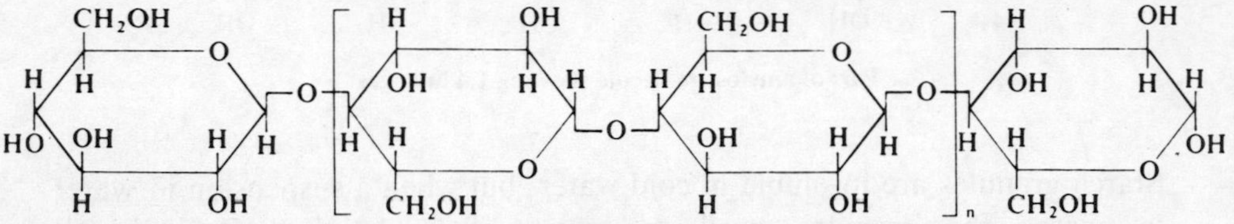

Cellulose

The polymers may have more than 15 000 residues in the chains and hydrogen bonding is extensive, both within and between chains. The result is a compact, tightly bonded fibrous structure possessing great strength.

Fructans

Fructans (formerly called fructosans) occur as reserve material in roots, stems, leaves and seeds of a variety of plants, but particularly in the Compositae and Graminae. These polysaccharides are soluble in cold water and are of a relatively low molecular weight. All known fructans contain β-D-fructose residues joined by 2,6 or 2,1 linkages. They can be divided into three groups: (1) the levan group, characterised by 2,6 linkages; (2) the inulin group, containing 2,1 linkages; and (3) a group of highly branched fructans found, for example, in couch grass (*Agropyron repens*) and in wheat endosperm. This group contains both types of linkages.

Most fructans on hydrolysis yield, in addition to D-fructose, a small amount of D-glucose, which is derived from the terminal sucrose unit in the fructan molecule. The structure of a typical grass fructan is depicted below.

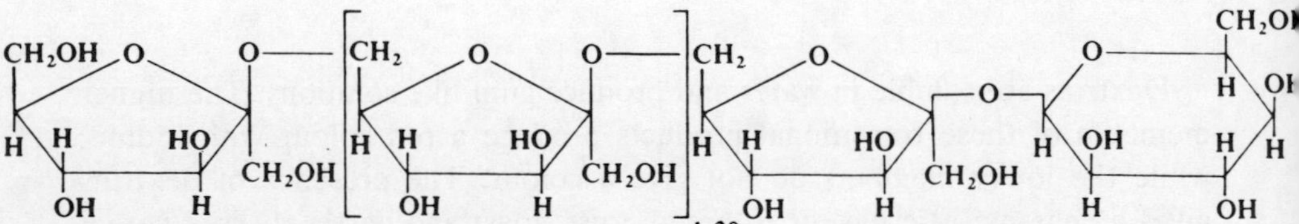

Grass fructan   1-30   (Sucrose residue)

Galactans and mannans

Galactans and mannans are polysaccharides which occur in the cell walls of plants. A mannan is the main component of the cell walls of palm seeds, where it occurs as a food reserve and disappears during germination. A rich source of mannan is the endosperm of nuts from the South American tague tree; the hard endosperm of this nut is known as 'vegetable ivory'. The seeds of many legumes, including clovers, trefoil and lucerne, contain galactans.

Glucosaminans

Chitin is the only known example of a homoglycan containing glucosamine, being a linear polymer of acetyl-D-glucosamine. Chitin is of widespread occurrence in lower animals and is particularly abundant in Crustacea, in fungi and in some green algae. Next to cellulose, it is probably the most abundant polysaccharide of nature.

HETEROGLYCANS

Pectic substances

Pectic substances are a group of closely associated polysaccharides, which are soluble in hot water and occur as constituents of primary cell walls and intercellular regions of higher plants. They are particularly abundant in soft tissues such as the peel of citrus fruits, and sugar beet pulp. Pectin, the main member of this group, consists of a linear chain of D-galacturonic acid units in which varying proportions of the acid groups are present as methyl esters. The chains are interrupted at intervals by the insertion of L-rhamnose residues. Other constituent sugars, e.g. D-galactose, L-arabinose, D-xylose, are attached as side chains. Pectic acid is another member of this class of compounds; it is similar in structure to pectin, but is devoid of ester groups. Pectic substances possess considerable gelling properties and are used commercially in jam making.

Hemicelluloses

Hemicelluloses are defined as alkali-soluble cell wall polysaccharides that are closely associated with cellulose. The name hemicellulose is misleading and implies erroneously that the material is destined for conversion to cellulose. Structurally, hemicelloloses are composed mainly of D-glucose, D-galactose, D-mannose, D-xylose and L-arabinose units joined together in different combinations and in various glycosidic linkages. They may also contain uronic acids.

Hemicelluloses from grasses contain a main chain of xylan made up of β-(1→4)-linked D-xylose units with side-chains containing methylglucuronic acid and frequently glucose, galactose and arabinose.

Exudate gums

Exudate gums are often produced from wounds in plants, although they may arise as natural exudations from bark and leaves. The gums occur naturally as salts, especially of calcium and magnesium, and in some cases a proportion of the hydroxyl groups are esterified, usually as acetates. Gum arabic (acacia gum) has long been a familiar substance; on hydrolysis it yields arabinose, galactose, rhamnose and glucuronic acid.

Acidic mucilages

Acidic mucilages are obtained from the bark, roots, leaves and seeds of a variety of plants. Linseed mucilage is a well known example which produces arabinose, galactose, rhamnose and galacturonic acid on hydrolysis.

Hyaluronic acid and chondroitin

These two polysaccharides have a repeating unit consisting of an amino sugar and D-glucuronic acid. Hyaluronic acid, which contains acetyl-D-glucosamine, is present in the skin, synovial fluid and the umbilical cord. Solutions of this acid are viscous and play an important part in the lubrication of joints.

Chondroitin is chemically similar to hyaluronic acid but contains galactosamine in place of glucosamine. Sulphate esters of chondroitin are major structural components of cartilage, tendons and bones.

LIGNIN

Lignin, which is not a carbohydrate but is closely associated with this group of compounds, confers chemical and biological resistance to the cell wall and mechanical strength to the plant. Strictly speaking the term 'lignin' does not refer to one, well-defined, individual, compound, but is a collective term which embraces a whole series of closely related compounds.

Lignin is a polymer that originates from three derivatives of phenylpropane: coumaryl alcohol, coniferyl alcohol and sinapyl alcohol. The lignin molecule is made up of many phenylpropanoid units associated in a complex cross-linked structure.

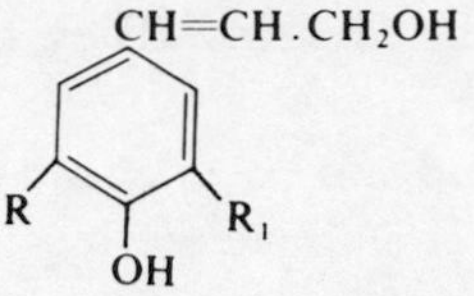

CH=CH.CH₂OH

R   R₁
OH

(1) Coumaryl alcohol, where R = R₁ = H.
(2) Coniferyl alcohol, where R = H, R₁ = OCH₃.
(3) Sinapyl alcohol, where R = R₁ = OCH₃.

This compound is of particular interest in animal nutrition because of its high resistance to chemical degradation. Physical incrustation of plant fibres by lignin renders them inaccessible to enzymes that would normally digest them.

There is evidence that strong chemical bonds exist between lignin and many plant polysaccharides and cell wall proteins which render these compounds unavailable during digestion.

Wood products, mature hays and straws are rich in lignin and consequently are of low digestibility unless treated chemically to break the bonds between lignin and other carbohydrates (see p. 212).

FURTHER READING

G. O. Aspinall (ed.) 1982–85. *The Polysaccharides*, Vol 1–3. Academic Press, New York.

R. W. Bailey, 1973. Structural carbohydrates, in *Chemistry and Biochemistry of Herbage*, Vol. 1, G. W. Butler and R. W. Bailey (eds). Academic Press, New York.

D. J. Candy, 1980. *Biological Functions of Carbohydrates*. Blackie, Glasgow

C. M. Duffus and J. H. Duffus, 1984 *Carbohydrate Metabolism in Plants*. Longman, London.

P. K. Stumpf and E. E. Conn (eds) 1980. *The Biochemistry of Plants* Vol. 3. *Carbohydrates: Structure and Functions*. Academic Press, New York.

R. S. Tipson and D. Horton (eds). *Advances in Carbohydrate Chemistry and Biochemistry* (annual volumes since 1945). Academic Press, New York.

3

Lipids

The lipids, or lipides, are a group of substances found in plant and animal tissues, insoluble in water but soluble in common organic solvents such as benzene, ether and chloroform. They act as electron carriers, as substrate carriers in enzymic reactions, as components of biological membranes and as stores of energy. In the proximate analysis of foods they are included in the ether extract fraction. They may be classified as follows:

TABLE 3.1 Classification of lipids

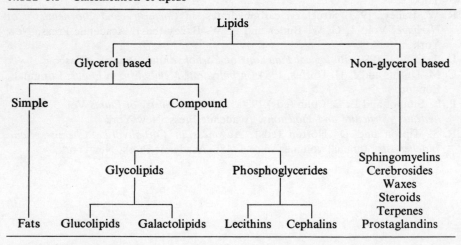

In plants, lipids are of two types, the structural and the storage. The former are present as constituents of various membranes and protective surface layers and make up about 7 per cent of the dry matter of the leaves of higher plants. The surface lipids are mainly waxes with relatively minor contributions from long chain hydrocarbons, fatty acids and cutin. The membrane lipids, present in mitochondria, the endoplasmic reticulum and the plasma membranes are mainly glycolipids (40–50%) and phospholipids. Plant storage lipids occur in fruits and seeds and are predominantly oils.

In animals, lipids are the major form of energy storage, mainly as fat. This may constitute up to 97 per cent of the adipose tissue of obese animals. The structural lipids of animal tissues, mainly phospholipids, constitute between 0.5 and 1 per cent of muscle and adipose tissue, but the concentration in the liver is usually between 2 and 3 per cent. The most important non-glyceride, neutral, lipid fraction of animal tissue is made up of cholesterol and its esters which together make up 0.06 to 0.09 per cent of muscle and adipose tissue.

FATS

Fats and oils are constituents of both plants and animals, and are important sources of stored energy. Both have the same general structure and chemical properties but they have different physical characteristics. The melting points of the oils are such that at ordinary room temperatures they are liquid. The term fat is frequently used in a general sense to include both groups.

Structure of fats

Fats are esters of fatty acids with the trihydric alcohol glycerol, called glycerides or acylglycerols. When all three alcohol groups are esterified with fatty acids the compound is a triacyglycerol (triglyceride):

$$^1CH_2OH \qquad\qquad ^1CH_2.O.CO.R$$
$$^2CHOH + 3R.COOH \longrightarrow {}^2CH.O.CO.R + 3H_2O$$
$$^3CH_2OH \qquad\qquad ^3CH_2.O.CO.R$$

Glycerol + Fatty acid $\longrightarrow$ Triacylglycerol

The glycerol carbon atoms are designated by the numbers 1, 2 and 3 as shown. Di- and monoacyglycerols also occur naturally but in much smaller amounts than the triacyl form.

Triacylglycerols differ in type according to the nature and position of the fatty acid residues. Those with three residues of the same fatty acid are termed simple triacyglycerols as illustrated above. When more than one fatty acid is concerned in the esterification then a mixed triacylglycerol results:

$$CH_2.O.CO.R_1$$
$$CH.O.CO.R_2$$
$$CH_2.O.CO.R_3$$

Mixed triacylglycerol

R_1, R_2, R_3 represent the chains of different fatty acids. Naturally occurring fats and oils are mixtures of such mixed triacyglycerols, although simple types do occur naturally and sometimes are the dominant type present. Laurel oil, for example, contains 31 per cent of the tricylglycerol of lauric acid. Most of the naturally occurring fatty acids contain a single carboxyl group and an unbranched carbon chain, which may be saturated or unsaturated. The unsaturated acids contain either one double bond (monoenoic), two (dienoic), three (trienoic) or many (polyenoic) double bonds. They possess different physical properties from the saturated acids; they have lower melting points and are chemically more reactive. The acids found in normal fats usually have an even number of carbon atoms, which is to be expected in view of their mode of formation (Ch. 9). The carbon atoms of the fatty acids are numbered from the carboxyl group and the position of each double bond Δ is shown by the number of the carbon atom at its carboxyl end, i.e. a Δ^7 acid would have a double bond between carbon atoms 7 and 8.

Oleic acid with 18 carbon atoms and a double bond between carbons 9 and 10 would be represented as $18 : 1\Delta^9$. Similarly, stearic acid with 18 carbons and no double bonds would be written $18 : 0$.

The presence of a double bond in a fatty acid molecule means that the acid can exist in two forms depending upon the spatial arrangement of the hydrogen atoms attached to the carbon atoms of the double bond. Where the hydrogen atoms lie on the same side of the double bond the acid is said to be in the *cis* form, while it is said to be in the *trans* form when they lie on opposite sides. A monoenoic $18 : 1\Delta^9$ acid could be:

$$
\begin{array}{ccc}
H\diagdown\quad\diagup(CH_2)_7.COOH & & H\diagdown\quad\diagup(CH_2)_7.COOH \\
C & & C \\
\|\| & \text{or} & \|\| \\
C & & C \\
H\diagup\quad\diagdown(CH_2)_7.CH_3 & & CH_3.(CH_2)_7\diagup\quad\diagdown H \\
\textit{Cis} & & \textit{Trans}
\end{array}
$$

Most of the naturally occurring fatty acids have the *cis* configuration. Some of the more important fatty acids are shown in Table 3.2. Other acids, containing two carboxyl groups, odd numbers of carbon atoms, or branched chains, have been isolated from natural fats, but they are not considered to be of great importance.

Triacylglycerols are named according to the fatty acids they contain, e.g.

$CH_2.O.CO.C_{17}H_{33}$
|
$CH.O.CO.C_{17}H_{33}$
|
$CH_2.O.CO.C_{17}H_{33}$

Trioleoylglycerol (Triolein)

$CH_2.O.CO.C_{15}H_{31}$
|
$CH.O.CO.C_{17}H_{33}$
|
$CH_2.O.CO.C_{17}H_{35}$

1-palmitoyl 2-oleoyl 3-stearoylglycerol (Palmito-oleostearin)

The fatty acid·composition of triacylglycerols determines their physical nature. Those with a high proportion of low molecular weight (short chain) and unsaturated acids have low melting points. Thus, tristearin is solid at body temperature while triolein is liquid.

TABLE 3.2   Common acids of natural fats
1. *Saturated fatty acids*

Acid	Formula	Melting point, °C
Butyric (butanoic)	C_3H_7COOH	−7.9
Caproic (hexanoic)	$C_5H_{11}COOH$	−3.2
Caprylic (octanoic)	$C_7H_{15}COOH$	16.3
Capric (decanoic)	$C_9H_{19}COOH$	31.2
Lauric (dodecanoic)	$C_{11}H_{23}COOH$	43.9
Myristic (tetradecanoic)	$C_{13}H_{27}COOH$	54.1
Palmitic (hexadecanoic)	$C_{15}H_{31}COOH$	62.7
Stearic (octadecanoic)	$C_{17}H_{35}COOH$	69.6
Arachidic (eicosanoic)	$C_{19}H_{39}COOH$	76.3

2. *Unsaturated fatty acids*

Acid	Formula	Melting point, °C
Palmitoleic (hexadecenoic)	$C_{15}H_{29}COOH$	0
Oleic (octadecenoic)	$C_{17}H_{33}COOH$	13
Linoleic (octadecadienoic)	$C_{17}H_{31}COOH$	− 5
Linolenic (octadecatrienoic)	$C_{17}H_{29}COOH$	−14.5
Arachidonic (eicosatetraenoic)	$C_{19}H_{31}COOH$	−49.5

Essential fatty acids

In 1929 linoleic acid was shown to be effective in preventing the development of certain conditions in rats fed on diets almost devoid of fats. These animals showed a scaly appearance of the skin, and sub-optimal performance in growth, reproduction and lactation; eventually they died. Later, linolenic acid was shown to be partially effective in correcting the deficiency symptoms, and arachidonic acid has been found to be even more effective. The three acids are now referred to as essential fatty acids (EFA). Like other polyunsaturated acids they form part of various membranes and play a part in lipid transport and certain lipoprotein enzymes. In addition they are the source materials for the synthesis of prostaglandins and thromboxanes, hormone-like substances that regulate many cell functions including blood clotting, blood pressure and the immune response. The need for a dietary supply of the acids arises from the inability of animals to introduce double bonds between the ninth carbon atom and the terminal methyl group of the fatty acid chain. This is the result of the absence of the relevant desaturases from animal tissues and linoleic $(18 : 2\Delta^{9,12})$, linolenic $(18 : 3\Delta^{9,12,15})$ and arachidonic $(20 : 4\Delta^{5,8,11,14,})$ acids cannot be synthesised. Once linoleic acid has been absorbed by the animal it can be converted into certain polyunsaturated acids including arachidonic.

Since the original observations were made considerable evidence has accumulated in their support, and to show that the EFA are also required by chicks, pigs, calves and goats. Chicks kept on low fat diets have shown poor growth rates, poor feathering, oedema and high mortality in the first few weeks of life. The evidence in the case of pigs is contradictory. Several experiments have shown skin lesions and poor growth rates in animals maintained on low fat diets. In others only the skin condition occurred and this could be prevented by adding hydrogenated plant oils to the diet. The need for linoleic acid by the pig would thus appear to be in some doubt and it has been claimed that some synthesis of this acid takes place in the pig.

The oil seeds are generally rich sources of linoleic acid, while linseed is a particularly good source of linolenic acid. Pigs and poultry, which normally have considerable quantities of oil seed residues in their diets, will therefore receive an adequate supply of the essential fatty acids.

Ruminants are largely dependent upon grass for their nutritional needs and are thereby supplied with considerable quantities of linoleic and even larger quantities of linolenic acid. Considerable hydrogenation of unsaturated acids takes place in the rumen with a consequent overall reduction in the available EFA (see Ch. 8). Despite this the possibility of ruminants suffering from EFA deficiencies is remote.

It is important to appreciate that excessive amounts of unsaturated fatty acids in the diet may induce a vitamin E shortage and result in the occurrence of deficiency conditions such as muscular dystrophy (see Ch. 5).

Composition of fats

It is frequently important in nutritional investigations to assess the quality of the fat being produced under a certain treatment. Where the effect of the diet is considerable the results may be obvious in a softening or hardening of the fat. Less obvious changes may occur, and for these a more objective assessment is necessary. Differences in fats are a function of their fatty acid composition, since glycerol is common to all fats. The logical method of following changes in fats, therefore, is to measure their fatty acid constitution. Analysis of fats for individual fatty acids has presented great problems in the past, but the introduction in recent years of techniques such as gas chromatography has allowed determinations to be made more easily and accurately. As well as providing a most valuable research tool, gas chromatographic analysis has given detailed quantitative information on the fatty acid constitution of many different fats. This means a more certain identification and characterisation of fats, and provides a more accurate method of detecting and quantitatively estimating the adulteration of a given fat or oil than was previously available.

Some typical values for different fats are given in Table 3.3.

In general, plant and marine oils, especially those of fish, are more highly unsaturated than those of mammalian origin. This is due to the presence of varying amounts of linoleic and linolenic acids, in addition to the unsaturated oleic acid, which is quantitatively the major fatty acid in most natural fats. In mammalian depot fat the proportion of the more unsaturated acids

TABLE 3.3 Fatty acid proportions (mmol/mol) of some common fats and oils

Fatty acids		Butterfat	Lard	Beef tallow	Sperm whale oil	Ground-nut oil	Soya bean oil
Saturated	4:0	90	0	0	0	0	0
	6:0	30	0	0	0	0	0
	8:0	20	0	0	0	0	0
	10:0	40	0	0	1	0	0
	12:0	30	0	0	38	0	0
	14:0	110	10	70	74	0	0
	16:0	230	320	290	94	100	95
	18:0	90	80	210	7	97	37
Unsaturated	$18:1\Delta^9$	260	480	410	325	511	217
	$18:2\Delta^{9,12}$	30	110	20	5	274	571
	$18:3\Delta^{9,12,15}$	3	6	–	98	<1	65

is lower, and there is a higher proportion of high molecular weight saturated acids such as palmitic and stearic acids, with smaller but significant contributions from lauric and myristic acids. For this reason fats like lard, beef and mutton tallow are firm and hard, while the marine mammal, fish and plant oils are softer and are frequently oils in the true sense.

Within individual animals, subcutaneous fats contain a higher proportion of unsaturated acids and are thus softer than deep-body fats. The physical nature of fat varies between animals, marine mammals having softer body fat than land mammals. The reason in both cases is that animal fat has to maintain a degree of malleability at the temperature of the tissue.

Ruminant milk fats are characterised by their high content of low molecular weight fatty acids; these sometimes forming as much as 20 per cent of the total acids present. As a result they are softer than the depot fats of the respective animals but not as soft as fats of vegetable and marine origin, being semi-solid at ordinary temperatures. Milk fats of non-ruminants resemble the depot fat of the particular animal.

In most commercially important, edible plant oils, the dominant fatty acids are oleic, linoleic and linolenic. Coconut oil is an exception in having the saturated C_{12} acid, lauric, as its major acid. Families of plants tend to produce characteristic oils frequently dominated by one unusual fatty acid. Examples are the erucic acid ($C_{22}\Delta^{13}$) of rape seed, the eighteen carbon monoenoic hydroxy acid, ricinoleic, of the castor bean, and vernolic, the eighteen carbon trienoic epoxy acid of the Compositae.

Properties of fats

Hydrolysis. Fats may be hydrolysed by boiling with alkalis, when glycerol and soaps are formed:

$$\begin{array}{llll}
CH_2.O.CO.R & & CH_2OH & \\
| & & | & \\
CO.O.CO.R & + \ 3KOH \longrightarrow & CHOH & + \quad 3R.COOK \\
| & & | & \\
CH_2.O.CO.R & & CH_2OH & \\
\\
\text{Fat} & & \text{Glycerol} & \quad \text{Soap}
\end{array}$$

Such a hydrolysis is known as saponification since it produces soaps, which are sodium or potassium salts of fatty acids. The process may take place naturally under the influence of the enzymes collectively known as the *lipases*. These may have a certain specificity and preferentially catalayse hydrolysis at particular positions in the molecule. Removal of the fatty acid residue attached to acylglycerol carbon atom 2 is more difficult than those at the 1 and 3 positions. Under natural conditions the products of lipolysis are usually mixtures of mono- and diacyglycerols with free fatty acids. Most of these acids are odourless and tasteless, but some of the lower ones, particularly butyric and caproic, have extremely powerful tastes and smells; when such a breakdown takes place in an edible fat it may frequently be rendered completely unacceptable to the consumer. The lipases are mostly derived from bacteria and moulds, which are chiefly responsible for this type of spoilage, commonly referred to as rancidity. Extensive lipolysis of dietary fats takes place in the duodenum, and in their absorption from the small intestine. Lipolysis also precedes hydrogenation of fats in the rumen.

Oxidation. The unsaturated fatty acids readily undergo oxidation, the site of action being the carbon atom adjacent to the double bond. Hydroperoxides are formed:

$$- CH_2 - CH = CH - CH_2 - CH_2 -$$

$$\Big\downarrow \! \nearrow O_2$$

$$- CH - CH = CH - CH_2 - CH_2 -$$
$$\mid$$
$$OOH$$

These break down to give shorter chain products including free radicals which then attack other fatty acids much more readily than the original oxygen. More free radicals are produced with the result that the speed of the oxidation increases exponentially. Eventually the concentration of free radicals becomes such that they react with each other and the reaction is terminated. Such a reaction in which the products catalyse the reaction is described as autocatalytic and the particular reaction described here as autoxidation. The formation of free radicals is catalysed by light and certain metal ions, particularly copper, and the presence of either increases the rate of oxidation dramatically.

The products of oxidation include shorter chain fatty acids, fatty acid polymers, aldehydes, ketones, epoxides and hydrocarbons. The acids and aldehydes are major contributors to the smells and flavours associated with oxidised fat, and which significantly reduce its palatability. The potency of these compounds is typified by deca-2,4 dienal which is detectable in water at concentrations of as little as 1 in 10 000 million.

Oxidation of saturated fatty acids results in the development of a sweet, heavy taste and smell commonly known as ketonic rancidity. This is due to the presence of methyl ketones as a result of oxidation, which may be represented as follows:

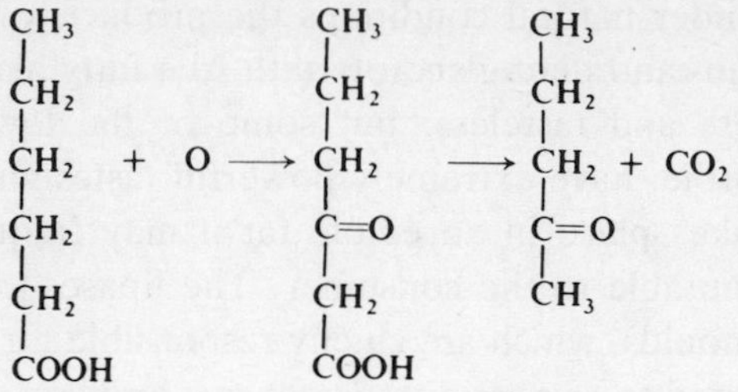

$$
\begin{array}{ccccc}
\text{CH}_3 & & \text{CH}_3 & & \text{CH}_3 \\
| & & | & & | \\
\text{CH}_2 & & \text{CH}_2 & & \text{CH}_2 \\
| & & | & & | \\
\text{CH}_2 & + \text{ O} \longrightarrow & \text{CH}_2 & \longrightarrow & \text{CH}_2 + \text{CO}_2 \\
| & & | & & | \\
\text{CH}_2 & & \text{C}{=}\text{O} & & \text{C}{=}\text{O} \\
| & & | & & | \\
\text{CH}_2 & & \text{CH}_2 & & \text{CH}_3 \\
| & & | & & \\
\text{COOH} & & \text{COOH} & & \\
\text{Caproic acid} & & \text{Pentanone} & &
\end{array}
$$

Similar reactions following mould-induced lipolysis are responsible for the characteristic flavours of various soft and blue cheeses.

Antioxidants. Natural fats possess a certain degree of resistance to oxidation, owing to the presence of compounds termed antioxidants. These prevent the oxidation of unsaturated fats until they themselves have been transformed into inert products. A number of compounds have this anti-oxidant property, including phenols, quinones, tocopherols, gallic acid and gallates. In the United Kingdom propyl, octyl or dodecyl gallate, butylated hydroxyanisole or butylated hydroxytoluene may be added to edible oils in amounts specified in *The Feedingstuffs Regulations* 1982

The most important naturally occurring antioxidant is vitamin E which protects fat by preferential acceptance of free radicals. The possible effects of fat oxidation in diets in which vitamin E levels are marginal, are thus of considerable importance.

Hydrogenation. This is the process whereby hydrogen is added to the double bonds of the unsaturated acids of a fat, converting them to their saturated analogues. Oleic acid, for example, yields stearic acid as shown:

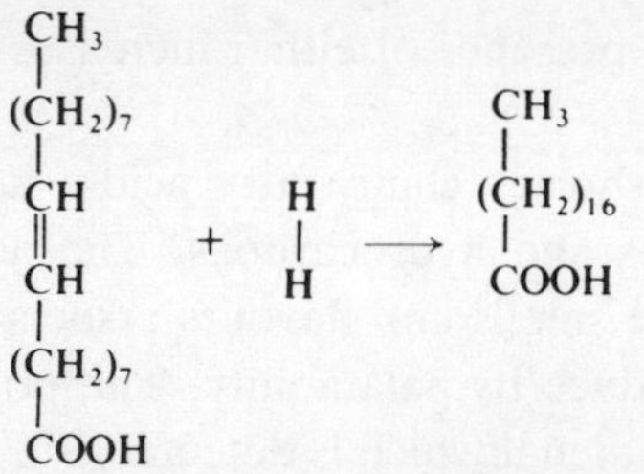

$$
\begin{array}{ccc}
\text{CH}_3 & & \\
| & & \\
(\text{CH}_2)_7 & & \text{CH}_3 \\
| & & | \\
\text{CH} & \quad\text{H} & (\text{CH}_2)_{16} \\
\| & + \quad | \longrightarrow & | \\
\text{CH} & \quad\text{H} & \text{COOH} \\
| & & \\
(\text{CH}_2)_7 & & \\
| & & \\
\text{COOH} & & \\
\text{Oleic acid} & & \text{Stearic acid}
\end{array}
$$

The process is important commercially for producing firm hard fats from vegetable and fish oils in the manufacture of margarine. The hardening results from the higher melting points of the saturated acids. For the rate of reaction to be practicable a catalyst has to be used, usually finely divided nickel. Hardening has the added advantage of improving the keeping quality of the fat, since removal of the double bonds eliminates the chief centres of reactivity in the material.

Dietary fats consumed by ruminants first undergo hydrolysis in the rumen and this is followed by progressive hydrogenation of the unsaturated free fatty acids (mainly 18 : 2 and 18 : 3 acids) to stearic acid. This helps to reconcile the apparent contradiction that whereas their dietary fats are highly unsaturated, the body fats of ruminants are highly saturated.

GLYCOLIPIDS

In these compounds two of the alcohol groups of glycerol are esterified by fatty acids while the other is linked to a sugar residue. The lipids of grasses and clovers, which form the major part of the dietary fat of ruminants, are predominantly (about 60%) galactolipids. Here the sugar is galactose and we have:

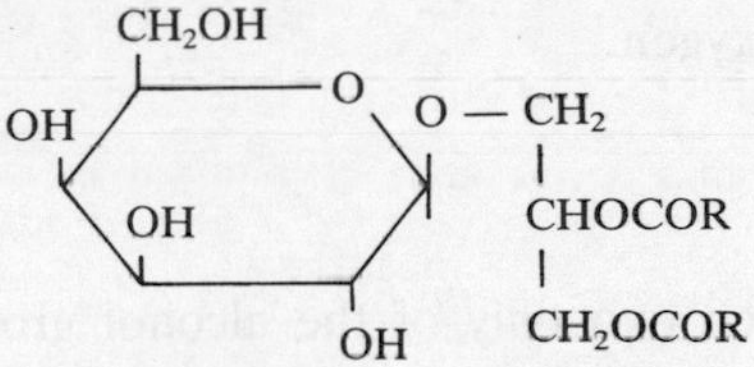

$$CH_2OH$$

$$OH \quad O \quad O - CH_2$$
$$| \quad CHOCOR$$
$$OH \quad | \quad CH_2OCOR$$
$$OH$$

Galactolipid

The galactolipids of grasses are mainly of the monogalactosyl type illustrated above, but smaller quantities of the digalactosyl compounds are also present. These have two galactose residues at the first carbon atom. The fatty acids of the galactolipids of grass are almost entirely (95%) α-linolenic acid with a small contribution (2–3%) from linoleic. Rumen micro-organisms are able to break down the galactolipids to give galactose, fatty acids and glycerol. Preliminary lipolysis appears to be essential so that the galactosyl glycerides can be hydrolysed by the microbial galactosidases.

Glycolipids are present in animal tissues, mainly in the brain and nerve fibres. The glycerol of the plant glycolipids is here replaced as the basic unit by the nitrogenous base sphingosine.

$$OH \quad NH_2$$
$$CH_3.(CH_2)_{12}.CH = CH.CH.CH.CH_2OH$$

Sphingosine

In their simplest form, *the cerebrosides*, the glycolipids have the amino group of the sphingosine linked to the carboxyl group of a long chain fatty acid, and the terminal alcohol group to a sugar residue, usually galactose. The typical structure is illustrated below:

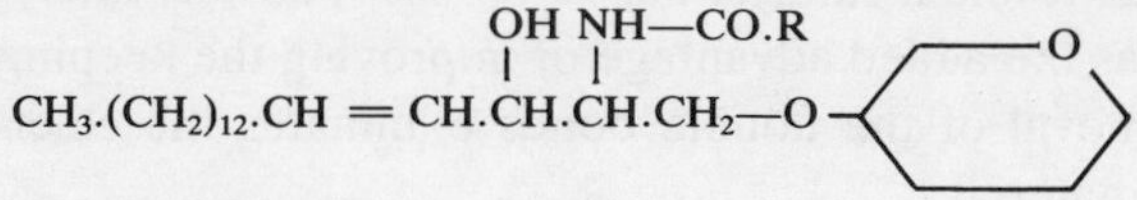

$$\underset{\overset{\displaystyle |}{\text{CH}_3.(\text{CH}_2)_{12}.\text{CH}}}{} = \overset{\overset{\displaystyle \text{OH}}{\displaystyle |}}{\text{CH}}.\overset{\overset{\displaystyle \text{NH—CO.R}}{\displaystyle |}}{\text{CH}}.\text{CH}.\text{CH}_2\text{—O}$$

More complex substances, the gangliosides, are found in the brain. They have the terminal alcohol group linked to a branched chain of sugars with sialic acid as the terminal residue of at least one of the chains.

PHOSPHOLIPIDS

The role of the phospholipids is primarily as constituents of the lipoprotein complexes of biological membranes. They are widely distributed, being particularly abundant in the heart, kidneys and nervous tissues. Myelin of the nerve axons, for example, contains up to 55 per cent of phospholipid. Eggs are one of the best animal sources, while among the plants, soya beans contain relatively large amounts. The phospholipids contain phosphorus in addition to carbon, hydrogen and oxygen.

Phosphoglycerides

These are esters of glycerol in which two only of the alcohol groups are esterified by fatty acids and the third by phosphoric acid. The parent compound of the phosphoglycerides is thus phosphatidic acid, which may be regarded as the simplest phosphoglyceride.

$$\begin{array}{l} \text{CH}_2.\text{O.CO.R}_1 \\ | \\ \text{CH.O.CO.R}_2 \\ | \qquad\quad \text{O} \\ | \qquad\quad \| \\ \text{CH}_2.\text{O.P.OH} \\ \qquad\quad | \\ \qquad\quad \text{OH} \end{array}$$

Phosphatidic acid

Phosphoglycerides are commonly referred to as phosphatides. In the major biologically important compounds, the phosphate group is esterified by one of several alcohols the commonest of which are serine, choline, glycerol, inositol and ethanolamine. The chief fatty acids present are the sixteen carbon saturated and the eighteen carbon saturated and monoenoic, although others with fourteen to twenty-four carbons are also found. The most commonly occurring phosphoglycerides in higher plants and animals are the lecithins and the cephalins.

Lecithins have the phosphoric acid esterified by the nitrogenous base choline and are more correctly termed phosphatidylcholines. A typical example would have the formula:

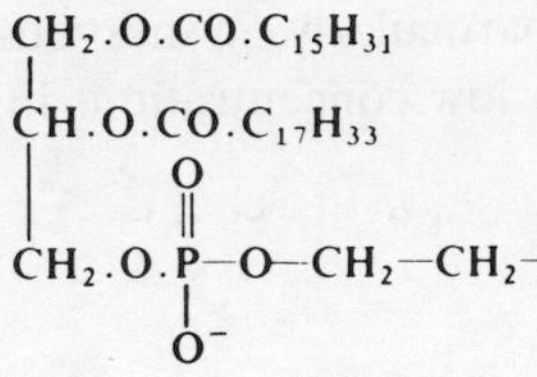

$$CH_2.O.CO.C_{15}H_{31}$$
$$CH.O.CO.C_{17}H_{33}$$
$$\overset{O}{\underset{O^-}{\overset{\parallel}{CH_2.O.P}}}-O-CH_2-CH_2-N^+\equiv(CH_3)_3$$

Lecithin

Cephalins differ from the lecithins in having ethanolamine instead of choline and are correctly termed phosphatidyl-ethanolamines. Ethanolamine has the following formula:

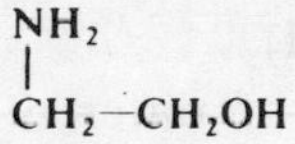

$$NH_2$$
$$CH_2-CH_2OH$$

Phosphoglycerides are white waxy solids which turn brown when exposed to the air owing to oxidation and subsequent polymerisation. When placed in water, phosphoglycerides appear to dissolve, but their true solubility is very low, the apparent solubility being due to the formation of micelles.

Phosphoglycerides are hydrolysed by naturally occurring enzymes, the phospholipases, which specifically cleave certain bonds within the molecule to release fatty acids, the phosphate ester, the alcohol and glycerol. The release of choline, when followed by a further oxidative breakdown, has been considered to be responsible for the development of fishy taints in fats through the release of the trimethylamine group or its oxide; currently these taints are considered to be the result of fat oxidation and not of lecithin breakdown. The phosphoglycerides combine within the same molecule both the hydrophilic (water-loving) phosphate ester groups, and the hydrophobic fatty acid chains. They are therefore surface active and may play an important role as emulsifying agents in biological systems, for example, in the duodenum. Their surface-active nature also explains their function as part of various biological membranes.

Sphingomyelins

Sphingomyelins belong to the large group, the sphingolipids, which have sphingosine instead of glycerol as the parent material. They differ from the cerebrosides in having the terminal hydroxyl group linked to phosphoric acid instead of a sugar residue. The phosphoric acid is esterified by either choline or ethanolamine. The sphingomyelins also have the amino group linked to the carboxyl group of a long chain fatty acid by means of a peptide linkage. Like the lecithins and cephalins they are surface active and are important as components of membranes, particularly in nervous tissue. They are absent from, or present only in very low concentrations in, energy generating tissue.

WAXES

Waxes are simple lipids consisting of a fatty acid combined with a mono-hydric alcohol of high molecular weight. They are usually solid at ordinary temperatures. The fatty acids present in waxes are those found in fats, although acids lower than lauric are very rare while several higher acids like carnaubic ($C_{23}H_{47}COOH$) and melissic ($C_{30}H_{61}COOH$) may also be present. The most common of the alcohols found in waxes are carnaubyl ($C_{24}H_{49}OH$), myricyl ($C_{31}H_{63}OH$) and cetyl ($C_{16}H_{33}OH$).

Natural waxes are usually mixtures of a number of different esters. Beeswax is known to consist of at least five different esters, the main one being myricyl palmitate:

$$C_{15}H_{31}COOH + C_{31}H_{63}OH \longrightarrow C_{15}H_{31}COOC_{31}H_{63} + H_2O$$

Palmitic acid   Myricyl   Myricyl palmitate
                  alcohol

Waxes are widely distributed in plants and animals, where they often have a protective function. In plants water losses due to transpiration are reduced, and in animals, wool and feathers are protected against water by the hydrophobic nature of the wax coating. Among the better-known animal waxes are lanolin, obtained from wool, and spermaceti, a product of marine animals. Unlike fats, waxes are not readily hydrolysed and are unlikely to have any nutritive value. Their presence in foods in large amounts leads to a high ether extract figure and may result in the nutritive value being overestimated.

STEROIDS

The steroids include such biologically important compounds as the sterols, the bile acids, the adrenal hormones and sex hormones. They have a common basic structural unit of a phenanthrene nucleus linked to a cyclo-pentane ring as shown below. The individual compounds differ in the number and positions of their double bonds and in the nature of the side chain at carbon atom 17.

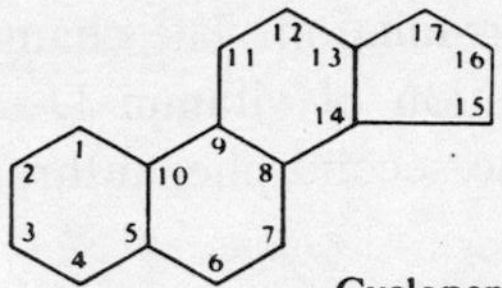

Cyclopentane ring

Sterols

These have 8 to 10 carbon atoms in the side chain, and an alcohol group at carbon atom 3. They may be classified into

    (a) the phytosterols of plant origin,

    (b) the mycosterols of fungal origin,

    (c) the zoosterols of animal origin.

The phytosterols and the mycosterols are not absorbed from the gut and are not found in animal tissues.

Cholesterol is a zoosterol which is quantitatively an important constituent of the brain, where it may form up to 17 per cent of the dry matter. It occurs in smaller amounts in all animal cells and it can be synthesised in the body, but, in spite of this wide distribution and apparent importance, little is known of its actual function. Many of the important sterols of the body may be synthesised from cholesterol, and some authorities consider that its function is to act as a source material in such syntheses. It has attained some prominence in recent years in connection with the condition of atherosclerosis, which involves a thickening of the arterial walls. The thickening is due to deposits containing cholesterol which form on the inside of the arterial walls.

7-Dehydrocholesterol, which is derived from cholesterol, is important as the precursor of vitamin D_3, which is produced when the sterol is exposed to ultraviolet light:

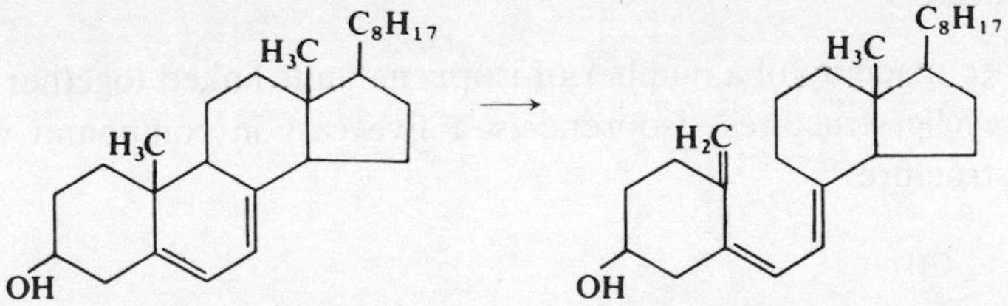

7-Dehydrocholesterol   Cholecalciferol (Vitamin D_3)

This is a good illustration of how relatively small changes in chemical structure may bring about radical changes in physiological activity.

Ergosterol is a phytosterol widely distributed in brown algae, bacteria and higher plants. It is important as the precursor of ergocalciferol or vitamin

D_2, into which it is converted by ultra-violet irradiation. The change is the same as that which takes place in the formation of vitamin D_3 from 7-dehydrocholesterol and involves opening of the second phenanthrene ring.

Bile acids

The bile acids have a five-carbon side chain at carbon 17 terminating in a carboxyl group which is bound by an amide linkage to glycine or taurine:

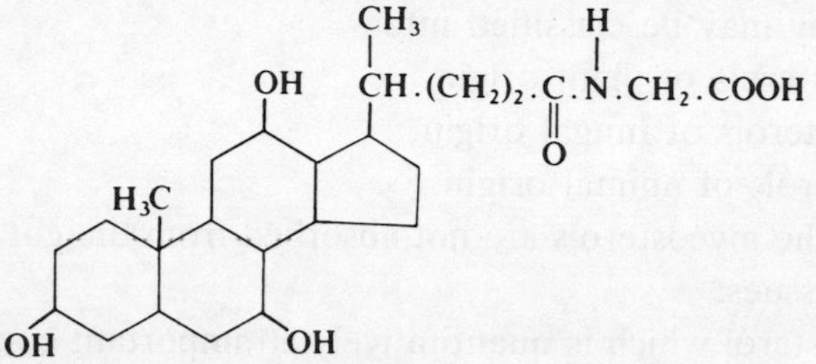

Glycocholic acid

The bile acids are important in the duodenum where they aid the emulsification of fats and the activation of lipase (see p. 135).

Steroid hormones

These include the female sex hormones (oestrogens), the male sex hormones (androgens), and progesterone, as well as cortisol, aldosterone and corticosterone which are produced by the adrenal cortex. The adrenal hormones have an important role in the control of glucose and fat metabolism.

TERPENES

Terpenes are made up of a number of isoprene units linked together to form chains or cyclic structures. Isoprene is a five carbon compound with the following structure:

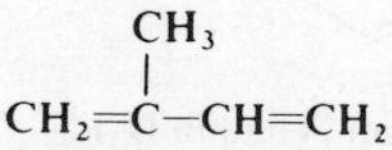

$$CH_2{=}\overset{\displaystyle CH_3}{\overset{|}{C}}{-}CH{=}CH_2$$

Isoprene

Many terpenes found in plants have strong, characteristic odours and flavours and are components of essential oils such as lemon or camphor oil. The word essential is used here to indicate the occurrence of the oils in essences and not to imply that they are required by animals. Among the more important plant terpenes are the phytol moiety of chlorophyll, the carotenoid pigments and the vitamins A, E and K. In animals, some of the coenzymes, including those of the coenzyme Q group, are terpenes.

PROSTAGLANDINS

Prostaglandins are derivatives of certain twenty-carbon unsaturated fatty acids which undergo cyclization to give compounds of which the parent material is prostanoic acid.

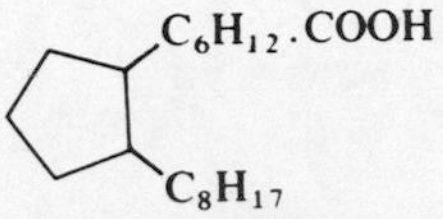

$C_6H_{12}.COOH$

C_8H_{17}

Prostanoic acid

The essential fatty acids act as precursors of the prostaglandins, the two seventeen carbon acids being first changed to the corresponding twenty carbon compounds.

About 14 prostaglandins are known. These differ in their ring substituents and the number of side chain double bonds. They show variable biological activity, stimulating smooth muscle to contract and lowering blood pressure.

The prostaglandins are used for oestrus synchronisation and for therapeutic abortions. The most commonly available is prostaglandin $F_{2\alpha}$ (PGF$_{2\alpha}$)

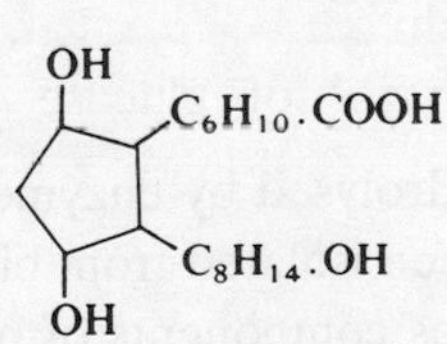

OH

$C_6H_{10}.COOH$

$C_8H_{14}.OH$

OH

OH

Prostaglandin $F_{2\alpha}$

FURTHER READING

G. A. Garton, 1969. Lipid metabolism of farm animals, in *Nutrition of Animals of Agricultural Importance*, D. P. Cuthbertson (ed.). Pergamon Press, Oxford.

M. I. Gurr, 1984. The chemistry and biochemistry of plant fats and their nutritional importance, in *Fats in Animal Nutrition*, J. Wiseman (ed.). Butterworths, London.

M. Enser, 1984. The chemistry, biochemistry and nutritional importance of animal fats, in *Fats in Animal Nutrition*. J. Wiseman (ed.). Butterworths, London.

J. L. Harwood, 1980. Plant acyl lipids: structure, distribution and analysis, in *The Biochemistry of Plants*, Vol. 4. P. K. Stumpf (ed.). Academic Press, London.

4

Proteins and nucleic acids

PROTEINS

Proteins are complex organic compounds of high molecular weight. In common with carbohydrates and fats they contain carbon, hydrogen and oxygen, but in addition they all contain nitrogen and generally sulphur.

Proteins are found in all living cells, where they are intimately connected with all phases of activity that constitute the life of the cell. Each species has its own specific proteins, and a single organism has many different proteins in its cells and tissues. It follows therefore that a large number of proteins occur in nature.

AMINO ACIDS

Amino acids are produced when proteins are hydrolysed by enzymes, acids or alkalis. Although over 200 amino acids have been isolated from biological materials, only 20 of these are commonly found as components of proteins.

Amino acids are characterised by having a basic nitrogenous group, generally an amino group ($-NH_2$), and an acidic carboxyl unit ($-COOH$). Most amino acids occurring naturally in proteins are of the α type, having the amino group attached to the carbon atom adjacent to the carboxyl group, and can be represented by the general formula:

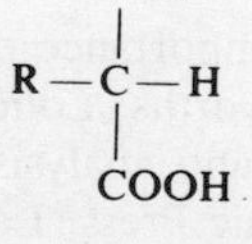

$$
\begin{array}{c}
NH_2 \\
| \\
R - C - H \\
| \\
COOH
\end{array}
$$

The exception is proline which has an imino (NH) instead of an amino group. The nature of the 'R' group, which is referred to as the side-chain, is different in different amino acids. It may simply be a hydrogen atom, as in glycine, or it may be a more complex radical containing, for example, a phenyl group.

The chemical structures of the 20 amino acids commonly found in natural proteins, together with their internationally accepted abbreviations, are shown in Table 4.1.

Special amino acids

Some proteins contain special amino acids which are derivatives of common amino acids. For example collagen, the fibrous protein of connective tissue, contains hydroxyproline and hydroxylysine, which are the hydroxylated derivatives of proline and lysine respectively.

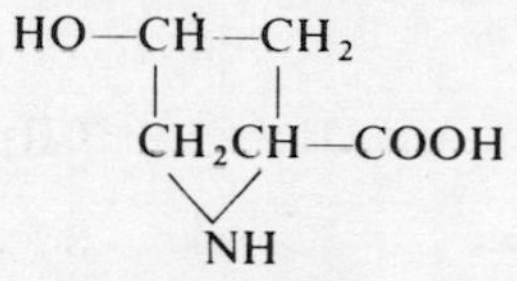

$$HO-CH-CH_2$$
$$\quad\quad | \quad\quad\quad |$$
$$\quad\quad CH_2CH-COOH$$
$$\quad\quad\quad\quad \backslash\!/$$
$$\quad\quad\quad\quad NH$$

Hydroxyproline

$$CH_2NH_2$$
$$|$$
$$CHOH$$
$$|$$
$$CH_2$$
$$|$$
$$CH_2$$
$$|$$
$$NH_2CHCOOH$$

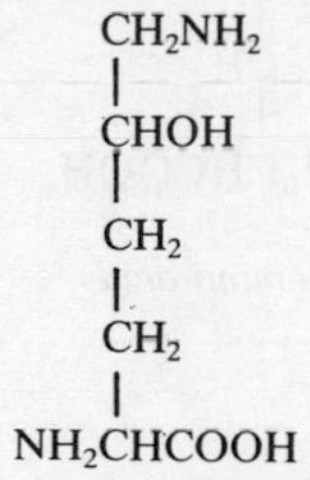

Hydroxylysine

Two iodine derivatives of tyrosine, triiodothyronine and tetraiodothyronine (thyroxine) act as important hormones in the body and are also amino acid components of the protein thyroglobulin (see p. 108).

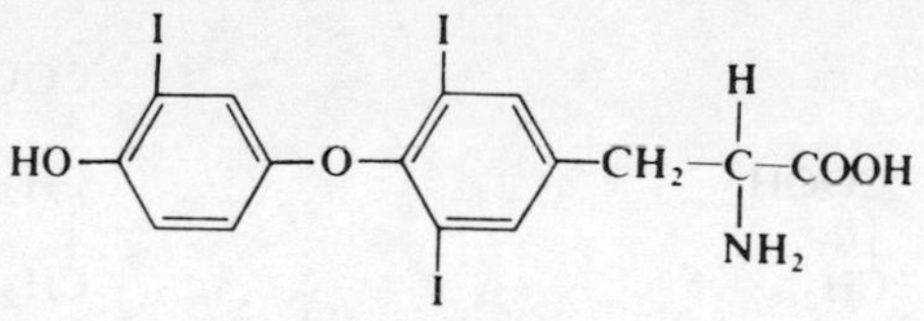

Triiodothyronine

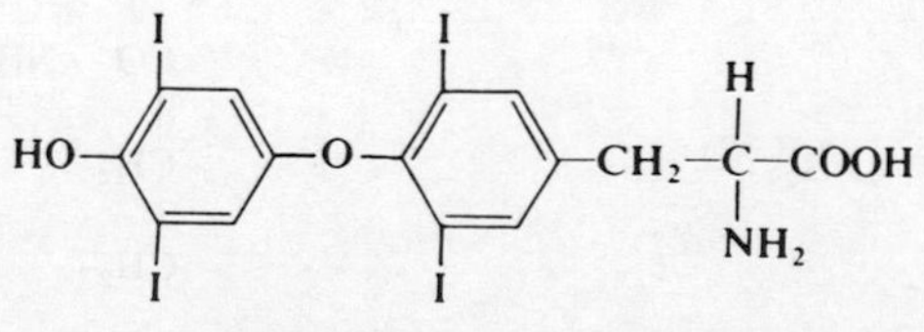

Tetraiodothyronine (thyroxine)

TABLE 4.1   Amino acids commonly found in proteins

1. *Monoamino-monocarboxylic acids*

Glycine NH_2CH_2COOH

Serine
$$\begin{array}{c} CH_2OH \\ | \\ NH_2CHCOOH \end{array}$$

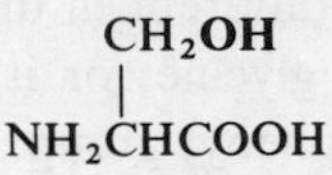

Alanine
$$\begin{array}{c} CH_3 \\ | \\ NH_2CHCOOH \end{array}$$

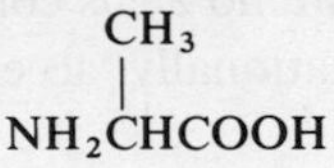

Threonine
$$\begin{array}{c} CH_3 \\ | \\ HCOH \\ | \\ NH_2CHCOOH \end{array}$$

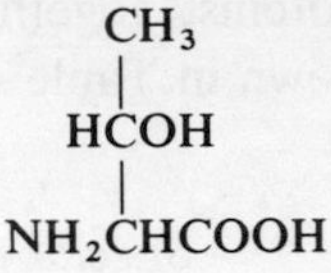

Valine
$$\begin{array}{c} CH_3 \quad CH_3 \\ \diagdown \diagup \\ CH \\ | \\ NH_2CHCOOH \end{array}$$

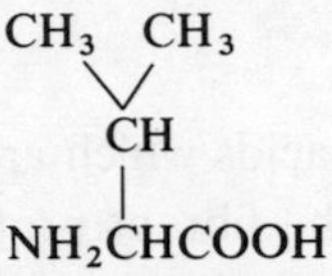

Leucine
$$\begin{array}{c} CH_3 \quad CH_3 \\ \diagdown \diagup \\ CH \\ | \\ CH_2 \\ | \\ NH_2CHCOOH \end{array}$$

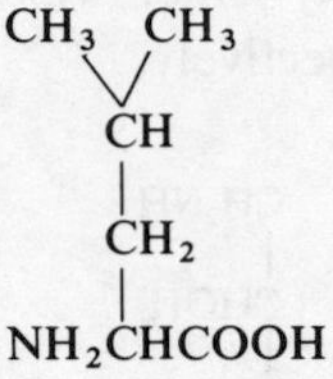

Isoleucine
$$\begin{array}{c} CH_3 \\ | \\ CH_2 \quad CH_3 \\ \diagdown \diagup \\ CH \\ | \\ NH_2CHCOOH \end{array}$$

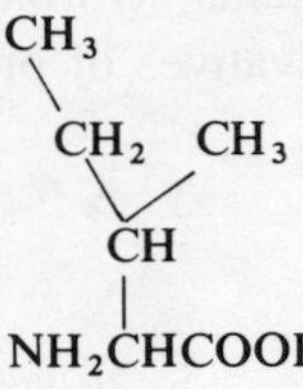

2. *Sulphur-containing amino acids*

Cysteine
$$\begin{array}{c} CH_2SH \\ | \\ NH_2CHCOOH \end{array}$$

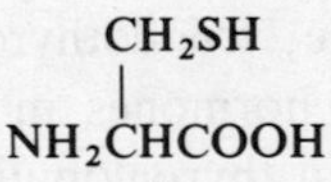

Methionine
$$\begin{array}{c} CH_3 \\ | \\ S \\ | \\ CH_2 \\ | \\ CH_2 \\ | \\ NH_2CHCOOH \end{array}$$

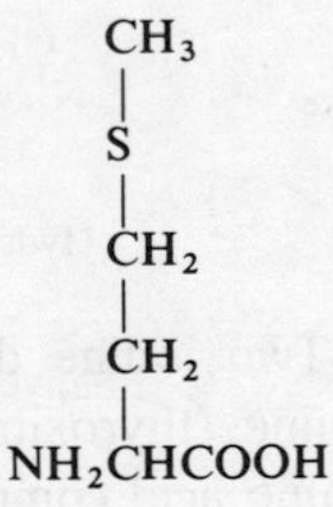

3. *Monoamino-dicarboxylic acids and their amine derivatives*

Aspartic acid
$$\begin{array}{c} COOH \\ | \\ CH_2 \\ | \\ NH_2CHCOOH \end{array}$$

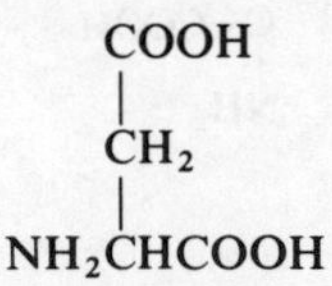

Glutamic acid
$$\begin{array}{c} COOH \\ | \\ CH_2 \\ | \\ CH_2 \\ | \\ NH_2CHCOOH \end{array}$$

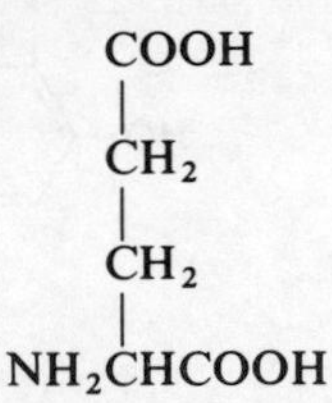

Asparagine
$$\begin{array}{c} CO-NH_2 \\ | \\ CH_2 \\ | \\ NH_2CHCOOH \end{array}$$

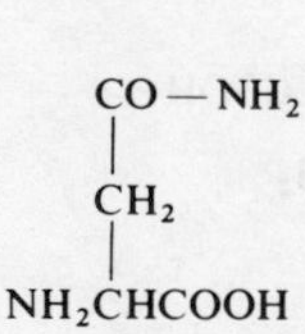

Glutamine
$$\begin{array}{c} CO-NH_2 \\ | \\ CH_2 \\ | \\ CH_2 \\ | \\ NH_2CHCOOH \end{array}$$

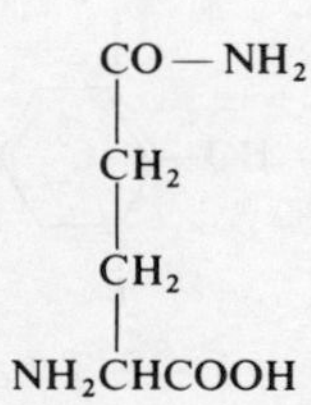

4. Basic amino acids

$$
\begin{array}{c}
CH_2NH_2 \\
| \\
CH_2 \\
| \\
CH_2 \\
| \\
CH_2 \\
\end{array}
$$

Lysine $NH_2CHCOOH$

Histidine $NH_2CHCOOH$

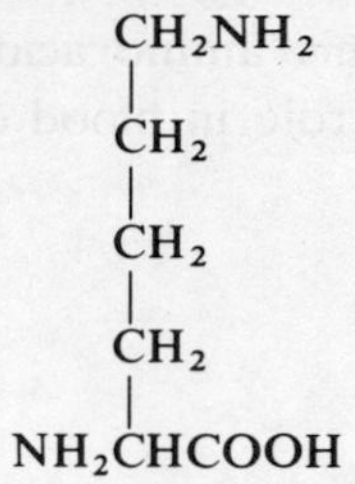

(Histidine ring)

$$
\begin{array}{c}
HC-N \\
\quad \quad \backslash \\
\quad \quad CH \\
\quad / \\
C-NH \\
| \\
CH_2
\end{array}
$$

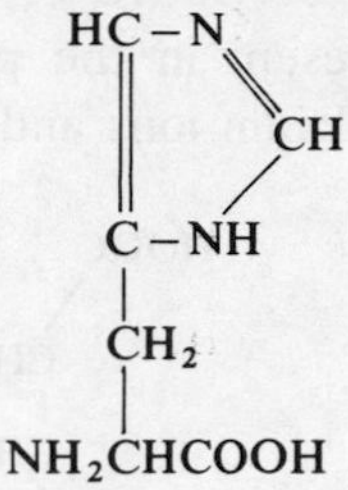

$$
\begin{array}{c}
NH_2 \\
| \\
C=NH \\
| \\
NH \\
| \\
(CH_2)_3 \\
\end{array}
$$

Arginine $NH_2CHCOOH$

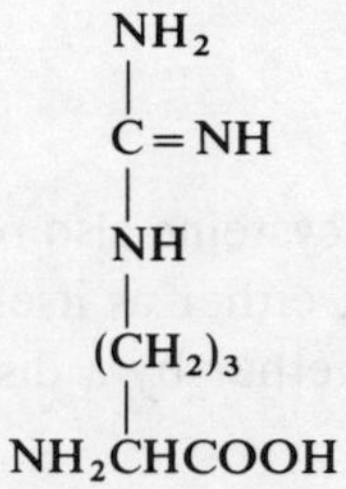

5. Aromatic and heterocyclic amino acids

Phenylalanine $NH_2CHCOOH$

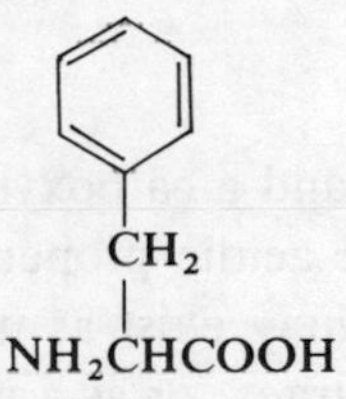

Tyrosine $NH_2CHCOOH$

(Tyrosine with OH)

$$
\begin{array}{c}
CH \\
HC \diagup \quad \diagdown C - C - CH_2 - CH - COOH \\
\quad \quad \quad \quad \| \quad \quad \quad \quad | \\
HC \diagdown \quad \diagup C \quad CH \quad \quad NH_2 \\
\quad CH \quad NH
\end{array}
$$

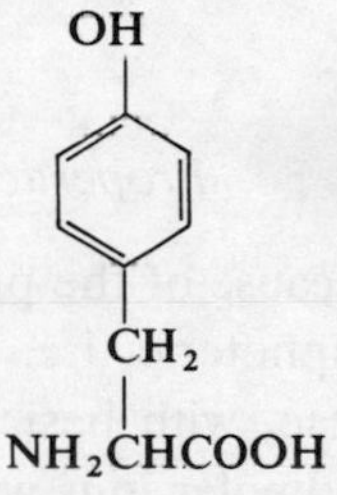

Tryptophan

$$
\begin{array}{c}
CH_2-CH_2 \\
| \quad \quad \quad | \\
CH_2 \quad CH-COOH \\
\diagdown \quad \diagup \\
NH
\end{array}
$$

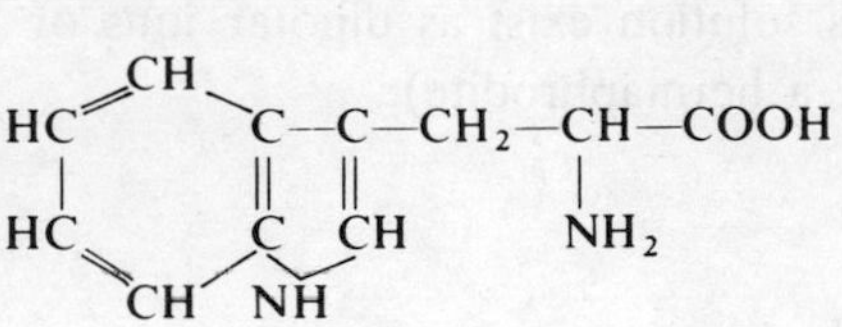

Proline

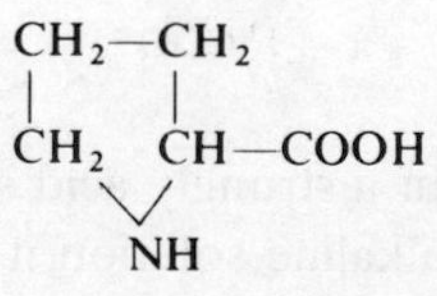

A derivative of glutamic acid, γ-carboxyglutamic acid, is an amino acid present in the protein thrombin. This amino acid is capable of binding calcium ions and plays an important role in blood clotting (see p. 73).

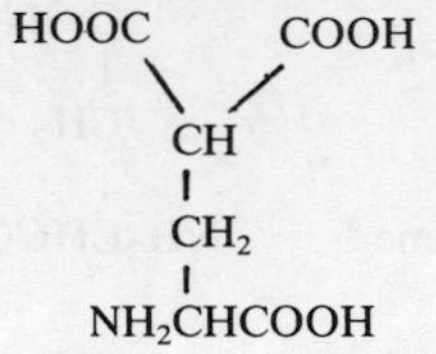

$$HOOC \quad COOH$$
$$\diagdown \diagup$$
$$CH$$
$$|$$
$$CH_2$$
$$|$$
$$NH_2CHCOOH$$

γ-Carboxyglutamic acid

The sulphur-containing amino acid cysteine also requires special mention. It may occur in protein in two forms, either as itself or as cystine in which two cysteine molecules are joined together by a disulphide bridge.

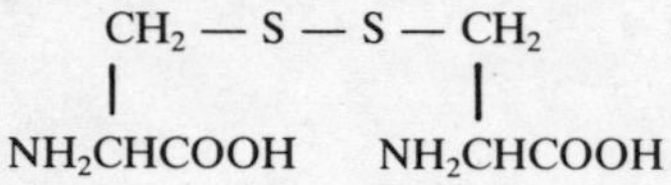

$$CH_2 - S - S - CH_2$$
$$| \qquad\qquad |$$
$$NH_2CHCOOH \quad NH_2CHCOOH$$

Cystine

Properties of amino acids

Because of the presence of an amino and a carboxyl group, amino acids are amphoteric, i.e. have both basic and acidic properties. Molecules such as these, with basic and acidic groups, may exist as uncharged molecules, or as dipolar ions with opposite ionic charges, or as a mixture of these. Amino acids in aqueous solution exist as dipolar ions or 'zwitter ions' (from the German *Zwitter*, a hermaphrodite):

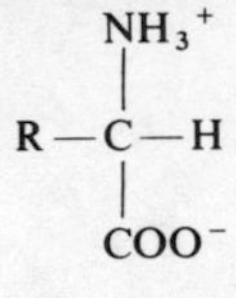

$$NH_3{}^+$$
$$|$$
$$R - C - H$$
$$|$$
$$COO^-$$

In a strongly acid solution an amino acid exists largely as a cation, while in alkaline solution it occurs mainly as an anion. There is a *p*H value for a given amino acid at which it is electrically neutral; this value is known as the *isoelectric point*.

Because of their amphoteric nature acids act as buffers, resisting changes in *p*H. All the α-amino acids except glycine are optically active.

All the amino acids involved in protein structure have an L- configuration of the carbon atom. Configurations are determined by relation to the standard substance D-glyceraldehyde, as described under Carbohydrates (Ch. 2).

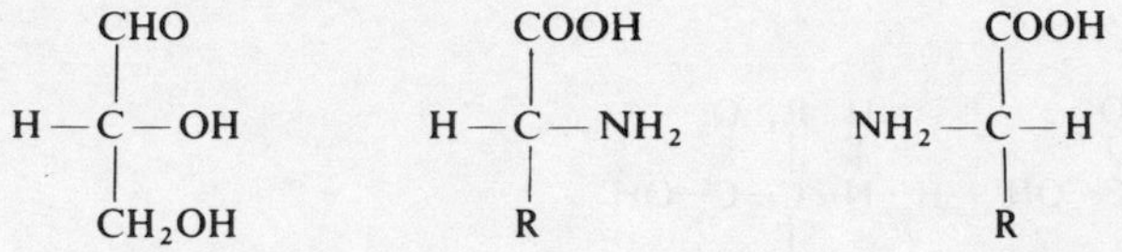

$$
\begin{array}{ccc}
\text{CHO} & \text{COOH} & \text{COOH} \\
| & | & | \\
\text{H}-\text{C}-\text{OH} & \text{H}-\text{C}-\text{NH}_2 & \text{NH}_2-\text{C}-\text{H} \\
| & | & | \\
\text{CH}_2\text{OH} & \text{R} & \text{R} \\
\text{D-Glyceraldehyde} & \text{D-Amino acid} & \text{L-Amino acid}
\end{array}
$$

Indispensable amino acids

Plants and many micro-organisms are able to synthesise proteins from simple nitrogenous compounds such as nitrates. Animals cannot synthesise the amino group, and in order to build up body proteins they must have a dietary source of amino acids. Certain amino acids can be produced from others by a process known as *transamination* (see Ch. 9), but a number cannot be effectively synthesised in the animal body and these are referred to as indispensable or essential amino acids.

Most of the early work in determining the amino acids which could be classed as indispensable was carried out with rats fed on purified diets. The following ten indispensable amino acids are required for growth in the rat:

Arginine	Methionine
Histidine	Phenylalanine
Isoleucine	Threonine
Leucine	Tryptophan
Lysine	Valine

The chick also requires in the diet the 10 amino acids listed above, but in addition needs a dietary source of glycine. The list of indispensable amino acids required by the pig is similar to that of the rat, with the exception of arginine and histidine which can be synthesised by the pig.

In the case of the ruminant, all the indispensable amino acids can be synthesised by the rumen micro-organisms, which theoretically makes this class of animal independent of a dietary source. However, maximum rates of growth or milk production cannot be achieved in the absence of a supply of dietary amino acids in a suitable form (see p. 363).

STRUCTURE OF PROTEINS

For convenience the structure of proteins can be considered under four basic headings.

Primary structure

Proteins are built up from amino acids by means of a linkage between the α-carboxyl of one amino acid and the α-amino group of another acid as shown below.

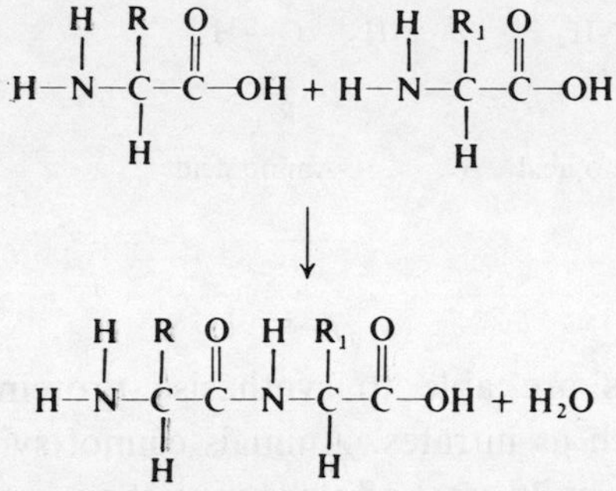

$$H-\underset{\underset{H}{|}}{\overset{\overset{H}{|}}{N}}-\underset{|}{\overset{\overset{R}{|}}{C}}-\overset{\overset{O}{\|}}{C}-OH + H-\underset{\underset{H}{|}}{\overset{\overset{H}{|}}{N}}-\underset{|}{\overset{\overset{R_1}{|}}{C}}-\overset{\overset{O}{\|}}{C}-OH$$

$$\downarrow$$

$$H-\underset{\underset{H}{|}}{\overset{\overset{H}{|}}{N}}-\underset{|}{\overset{\overset{R}{|}}{C}}-\overset{\overset{O}{\|}}{C}-\underset{|}{\overset{\overset{H}{|}}{N}}-\underset{|}{\overset{\overset{R_1}{|}}{C}}-\overset{\overset{O}{\|}}{C}-OH + H_2O$$

This type of linkage is known as the *peptide linkage*; in the example shown a dipeptide has been produced from two amino acids. Large numbers of amino acids can be joined together by this means, with the elimination of one molecule of water at each linkage, to produce polypeptides. The term 'primary structure' refers to the sequence of amino acids along the polypeptide chains of protein.

Secondary structure

The secondary structure of proteins refers to the conformation of the chain of amino acids resulting from the formation of hydrogen bonds between the imido (NH) and carbonyl groups of adjacent amino acids, as shown in Fig. 4.1.

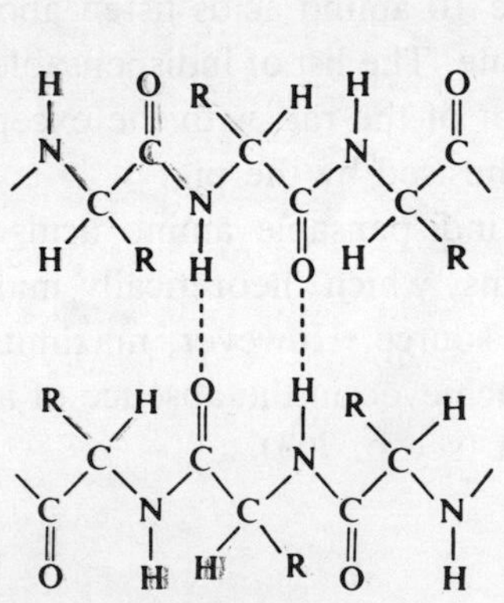

Fig. 4.1.   Configuration of polypeptide chain. Dotted lines represent possible hydrogen bonds.

The secondary structure may be regular, in which case the polypeptide chains exist in the form of an α-helix or a β-pleated sheet, or it may·be irregular and exist, for example, as a random coil.

Tertiary structure

The tertiary structure describes how the chains of the secondary structure further interact through the R groups of the amino acid residues. This interaction causes folding and bending of the polypeptide chain, the specific manner of the folding giving each protein its characteristic biological activity.

Quaternary structure

Proteins possess quaternary structure if they contain more than one polypeptide chain. The forces that stabilise these aggregates are hydrogen bonds and electrostatic or salt bonds formed between residues on the surfaces of the polypeptide chains.

PROPERTIES OF PROTEINS

All proteins have colloidal properties; they differ in their solubility in water, ranging from insoluble keratin to albumins which are highly soluble. Soluble proteins can be precipitated from solution by the addition of certain salts such as sodium chloride or ammonium sulphate. This is a physical effect and the properties of the proteins are not altered. On dilution the proteins can easily be redissolved.

Although the amino and carboxyl groups in the peptide linkage are non-functional to acid-base reactions, all proteins contain a number of free amino and carboxyl groups, either as terminal units or in the side-chain of amino acid residues. Like amino acids, proteins are therefore amphoteric. They exhibit characteristic isoelectric points, and have buffering properties.

All proteins can be *denatured* or changed from their natural state. Denaturation has been more accurately defined by Neurath and coworkers as 'any nonproteolytic modification of the unique structure of a native protein, giving rise to definite changes in chemical, physical or biological properties'. Products of protein hydrolysis are not included under this term. Perhaps the best known example of denaturation is the coagulation of a protein solution, such as egg white, upon heating. Many proteins are heat-coagulable. Apart from heat there are many other agents which can bring about the denaturation of proteins; these include strong acid, alkali, alcohol, acetone, urea, and salts of heavy metals.

The most notable effects of denaturation are the changes in biological properties; for example enzymes are usually inactivated. Changes in solubility and optical activity may also occur. Solutions of proteins are laevorotatory, and denaturation increases the specific rotation.

CLASSIFICATION OF PROTEINS

Proteins may be classified into three main groups according to their shape, solubility and chemical composition.

Fibrous proteins

These proteins are insoluble animal proteins which are very resistant to animal digestive enzymes. They are composed of elongated, filamentous chains which are joined together by cross linkages. This group contains the *collagens* which are the main proteins of connective tissues. Collagens make up about 30 per cent of the total proteins in the mammalian body. The essential amino acid tryptophan is not found in these proteins.

Elastin is the protein found in elastic tissues such as tendons and arteries, while *keratins* are the proteins of hair, nails, wool and hooves. These proteins are very rich in the sulphur-containing amino acid cystine; wool protein, for example, contains about 4 per cent of sulphur (see p. 315).

Globular proteins

This group includes all the enzymes, antigens and those hormones which are proteins. They can be subdivided into *albumins* which are water-soluble and heat-coagulable and which occur in eggs, milk, blood and many plants. The *globulins* are insoluble or sparingly soluble in water and are present in eggs, milk and blood and are the main reserve protein in seeds. Lactoglobulin is a protein of milk; it was at first considered to be a homogeneous protein, but actually consists of two components designated A and B lactoglobulin. Individual cows produce either one or the other, or both, depending upon genetic factors. *Histones* are basic proteins which occur in cell nuclei where they are associated with DNA. They are soluble in salt solutions, are not heat-coagulable, and on hydrolysis yield large quantities of histidine and lysine.

Finally, the *protamines* are basic proteins, of relatively low molecular weight, which are associated with nucleic acids and are found in large quantities in the mature, male germ cells of vertebrates. Protamines are rich in arginine, but contain no tyrosine, tryptophan or sulphur-containing amino acids.

Complex proteins

The complex, or compound, proteins, on hydrolysis, yield non-protein groups, usually called 'prosthetic groups', as well as amino acids. The prosthetic group varies and may be phosphoric acid (phosphoproteins), a carbohydrate or carbohydrate derivative (glycoproteins), a lipid (lipoproteins), a pigment (chromoproteins) or a nucleic acid (nucleoproteins).

Casein of milk, and phosvitin present in egg yolk, are *phosphoproteins* containing phosphoric acid. The term phosphoprotein is usually limited to compounds with a phosphorus-containing radical other than a nucleic acid or phospholipid.

Glycoproteins are complex proteins with one or more heteroglycans as prosthetic groups. In most glycoproteins the heteroglycans contain hexosamine, either glucosamine or galactosamine or both; in addition galactose and mannose may be present. Glycoproteins are components of mucous secretions which act as lubricants in many parts of the body.

Lipoproteins, which are proteins conjugated with lipid (e.g. lecithin, cholesterol) are the main components of cell membranes and play a basic role in lipid transport.

Chromoproteins contain a pigment as the prosthetic group. Examples are haemoglobin which consists of the protein, globin, combined with an iron-containing compound, haem or haematin. Other examples are haemocyanin, cytochrome and flavoproteins. *Nucleoproteins* are compounds of high molecular weight with estimated values ranging from 10 million to several hundred million. The prosthetic group—nucleic acids—will be dealt with in a separate section at the end of this chapter.

NON-PROTEIN NITROGENOUS COMPOUNDS

A considerable variety of nitrogenous compounds, which are not classed as proteins, occur in plants and animals. In plant analysis these compounds have been frequently classed together as non-protein nitrogenous compounds, to distinguish them from 'true proteins' determined in routine chemical analysis. Amino acids form the main part of the non-protein nitrogeneous fraction in plants, and those present in greatest amount include glutamic acid, aspartic acid, alanine, serine, glycine and proline. Other compounds are nitrogenous lipids, amines, amides, purines, pyrimidines, nitrates and alkaloids. In addition many members of the vitamin B complex contain nitrogen in their structure.

It is clearly impossible to deal with these compounds in any detail, and only some of the important ones not previously mentioned will be discussed.

Table 4.2 shows the main non-protein nitrogenous components of two herbage samples.

TABLE 4.2   Composition of non-protein nitrogen (NPN) of two herbages (After W. S. Ferguson and R. A. Terry, 1954. *J. Sci. Fd Agric.*, **5**, 515)

	Perennial ryegrass: *percentage of NPN*	*White clover:* *percentage of NPN*
Amino-N	46.6	49.8
Amide-N	9.7	13.0
Ammonia-N	3.2	2.6
Nitrate-N	7.9	3.9
Purine-N	7.5	6.7
Betaine-N	1.9	1.0
Choline-N	1.8	0.8

Amines

Amines are basic compounds present in small amounts in most plant and animal tissues. Many occur as decomposition products in decaying organic matter and have toxic properties.

A number of micro-organisms are capable of producing amines by decarboxylation of amino acids (Table 4.3). These may be produced in the rumen under certain conditions and may give rise to physiological symptoms; histamine, for example, is an amine formed from the amino acid, histidine, and in cases of anaphylactic shock is found in the blood in relatively large amounts. Silages in which clostridia have dominated the fermentation, usually contain appreciable amounts of amines. (see Ch. 18).

Betaine is a tertiary amine which is formed by the oxidation of choline. Betaine occurs in sugar beet, and the young leaves may contain 25 g/kg; it is this amine which is responsible for the 'fishy' aroma frequently associated

TABLE 4.3   Some important amines and their parent amino acids

Amino acid	Amine
Arginine	Putrescine
Histidine	Histamine
Lysine	Cadaverine
Phenylalanine	Phenylethylamine
Tyrosine	Tyramine
Tryptophan	Tryptamine

with the commercial extraction of sugar from beet. In the animal body betaine may be transformed into trimethylamine, and it is this which gives the fishy taint to milk produced by cows that have been given excessive amounts of sugar beet by-products.

Amides

Asparagine and glutamine are important amide derivatives of the amino acids, aspartic acid and glutamic acid. These two amides are also classed as amino acids (Table 4.1) and occur as components of proteins. They occur as free amides and play an important role in transamination reactions.

Urea is an amide which is the main end-product of nitrogen metabolism in mammals, although it also occurs in many plants and has been detected in wheat, soya bean, potato and cabbage:

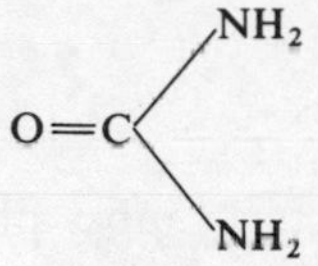

$$O{=}C\begin{cases} NH_2 \\ \\ NH_2 \end{cases}$$

Urea

In man and other primates uric acid is the end-product of purine metabolism and is found in the urine. In subprimate mammals the uric acid is oxidised to *allantoin* before being excreted.

In birds uric acid is the principal end-product of nitrogen metabolism and thus corresponds, in its function, to urea in mammals.

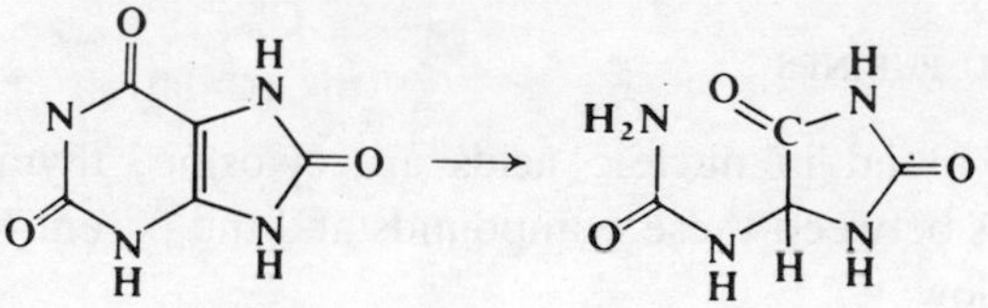

Uric acid   Allantoin

Nitrates

Nitrates may be present in plant materials and, while nitrate itself may not be toxic to animals, it is reduced readily under favourable conditions, as in the rumen, to nitrite, which is toxic. 'Oat hay poisoning' is attributed to the relatively large amounts of nitrate present in green oats.

Quite high levels of nitrate have been reported in herbage given heavy dressings of nitrogenous fertilisers (see Ch. 17).

Alkaloids

These compounds occur only in certain plants, and are of particular interest since many of them have poisonous properties. Their presence is restricted to a few orders in the dicotyledons. A number of the more important alkaloids, with their sources, are listed in Table 4.4.

TABLE 4.4   Some important alkaloids occurring in plants

Name	Source
Coniine	Hemlock
Nicotine	Tobacco
Ricinine	Castor plant seeds
Atropine	Deadly nightshade
Cocaine	Leaves of coca plant
Jacobine	Ragwort
Quinine	Cinchona bark
Strychnine	Seeds of *Nux vomica*
Morphine	Dried latex of opium poppy
Solanine	Unripe potatoes and potato sprouts

NUCLEIC ACIDS

Nucleic acids are high molecular weight compounds which, on hydrolysis, yield a mixture of basic nitrogenous compounds (purines and pyrimidines), a pentose (ribose or deoxyribose) and phosphoric acid. They play a fundamental role in living organisms as a store of genetic information, and they are the means by which this information is utilised in the synthesis of proteins.

PYRIMIDINES AND PURINES

The main pyrimidines found in nucleic acids are cytosine, thymine and uracil. The relationships between these compounds and the parent material, pyrimidine, is given below:

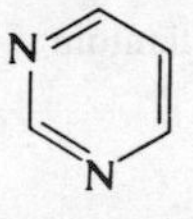

Pyrimidine

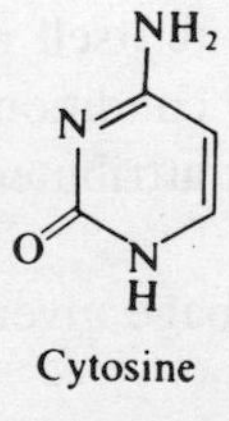

Cytosine

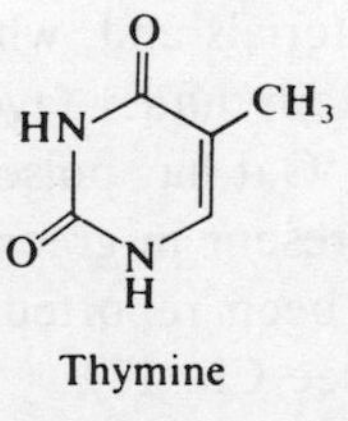

Thymine

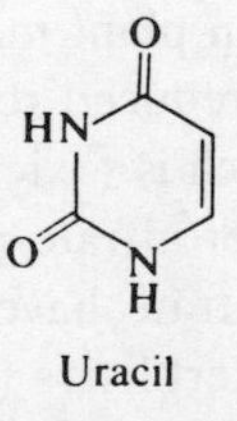

Uracil

Adenine and guanine are the principal purine bases present in nucleic acids.

Purine

Adenine   **Guanine**

The compound formed by linking one of the above nitrogenous compounds to a pentose is termed a *nucleoside*, e.g.

Adenine   D-Ribose   Adenosine

If nucleosides such as adenosine are esterified with phosphoric acid they form *nucleotides*, e.g. Adenosine monophosphate (AMP)

Nucleic acids are polynucleotides of very high molecular weight, generally measured in several millions. Nucleotides containing ribose are termed ribonucleic acids (RNA) while those containing deoxyribose are referred to as deoxyribonucleic acids (DNA).

The nucleotides are arranged in a certain pattern; DNA normally consists of a double-strand spiral or helix (Fig. 4.2). Each strand consists of alternate units of the deoxyribose and phosphate groups. Attached to each sugar group is one of the four bases, cytosine, thymine, adenine or guanine. The bases on the two strands of the spiral are joined in pairs by hydrogen bonds, the thymine on one strand always being paired with the adenine on the other and the cytosine with the guanine. The sequence of bases along these strands carries the genetic information of the living cell (see p. 185). DNA is found in the nuclei of cells as part of the chromosome structure.

There are several distinct types of ribonucleic acids, which are defined in terms of molecular size, base composition and functional properties. They differ from DNA in the nature of their sugar moiety and also in the types

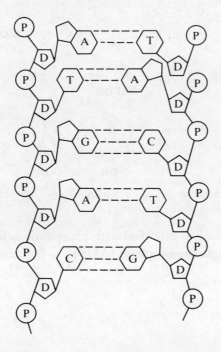

Fig. 4.2. Diagrammatic representation of part of the ladder-like DNA molecule, showing the two strands of alternate phosphate (P) and deoxyribose (D) molecules. The horizontal rods represent the pairs of bases held by hydrogen bonds (represented by dotted lines).

A = Adenine   T = Thymine   C = Cytosine   G = Guanine

of nitrogenous bases present. RNA contains the pyrimidine, uracil, in place of thymine. There is evidence to indicate that unlike DNA, RNA exists in the form of single, folded chains arranged spirally. There are three main forms of RNA, termed messenger RNA, ribosomal RNA and transfer RNA. The functions of these three forms of RNA are dealt with in the protein synthesis section of Chapter 9.

Apart from their importance in the structure of nucleic acids, nucleotides exist free as monomers and many play an important role in cellular metabolism.

Reference has been made previously to the phosphorylation of adenosine to form adenosine monophosphate (AMP). Successive additions of phosphate residues give adenosine diphosphate (ADP) and then the triphosphate (ATP). The importance of ATP in energy transformations is described in Chapter 9.

FURTHER READING

T. E. Creighton, 1984. *Proteins: Structures and Molecular Properties*. W. H. Freeman, Oxford.

R. E. Dickerson and I. Geis, 1983. *Proteins: Structure, Function and Evolution*. Benjamin/Cummings. Menlo Park, Calif.

A. L. Lehninger, 1982. *Principles of Biochemistry*. Worth Publ., New York.

H. Neurath and R. L. Hill (eds) 1982. *The Proteins*. Academic Press, New York.

L. Stryer, 1981. *Biochemistry*. W. H. Freeman, San Francisco.

5

Vitamins

The discovery and isolation of many of the vitamins were originally achieved through work on rats which had been given diets of purified proteins, fats, carbohydrates and inorganic salts. Using this technique, Hopkins in 1912 showed that a synthetic diet of this type was inadequate for the normal growth of rats, but that when a small quantity of milk was added to the diet the animals developed normally. This proved that there was some essential factor, or factors, lacking in the pure diet.

About this time the term 'vitamines', derived from 'vital amines', was coined by Funk to describe these accessory food factors, which he thought contained amino-nitrogen. It is now known that only a few of these substances contain amino-nitrogen and the word has been shortened to vitamins, a term which has been generally accepted as a group name.

Although the discovery of the vitamins dates from the beginning of the twentieth century, the association of certain diseases with dietary deficiencies had been recognised much earlier. In 1753 Lind, a British naval physician, published a treatise on scurvy proving that this disease could be prevented in human beings by including salads and summer fruits in their diet. The action of lemon juice in curing and preventing scurvy had been known, however, since the beginning of the seventeenth century. The use of cod-liver oil in preventing rickets has long been appreciated, and Eijkmann knew at the end of the last century that beri-beri, a disease common in the Far East, could be cured by giving the patients brown rice grain rather than polished rice.

Vitamins are frequently defined as organic compounds which are required in small amounts for normal growth and maintenance of animal life. But this definition ignores the important part that these substances play in plants, and their importance generally in the metabolism of all living organisms.

Vitamins are required by animals in very small amounts compared with other nutrients; for example, the vitamin B₁ (thiamin) requirement of a 50 kg pig is only 3 mg/day. Nevertheless a continuous deficiency in the diet results in disordered metabolism and eventually disease.

Some compounds function as vitamins only after undergoing a chemical change; such compounds, which include β-carotene and certain sterols, are described as provitamins or vitamin precursors.

Many vitamins are destroyed by oxidation, a process speeded up by the action of heat, light and certain metals such as iron. This fact is important since the conditions under which a food is stored will affect the final vitamin potency. Some commercial vitamin preparations are dispersed in wax or gelatin, which act as a protective layer against oxidation.

The system of naming the vitamins by letters of the alphabet was most convenient and was generally accepted before the discovery of their chemical nature. Although this system of nomenclature is still widely used with some vitamins, the modern tendency is to use the chemical name, particularly in describing members of the B complex.

There are at least 15 vitamins which have been accepted as essential food factors, and a few others have been proposed. Only those which are of nutritional importance are dealt with in this chapter.

It is convenient to divide the vitamins into two main groups, the water-

TABLE 5.1 Vitamins important in animal nutrition

Vitamin	Chemical name
Fat-soluble vitamins	
A	retinol
D₂	ergocalciferol
D₃	cholecalciferol
E	tocopherol*
K	phylloquinone†
Water-soluble vitamins	
B complex	
B₁	thiamin
B₂	riboflavin
	nicotinamide
B₆	pyridoxine
	pantothenic acid
	biotin
	folacin
	choline
B₁₂	cyanocobalamin
C	ascorbic acid

* A number of tocopherols have vitamin E activity.
† Several naphthoquinone derivatives possessing vitamin K activity are known.

soluble and the fat-soluble. Table 5.1 lists the important members of these two groups.

Chemical nature

Vitamin A ($C_{20}H_{29}OH$), known chemically as retinol, is an unsaturated monohydric alcohol with the following structural formula:

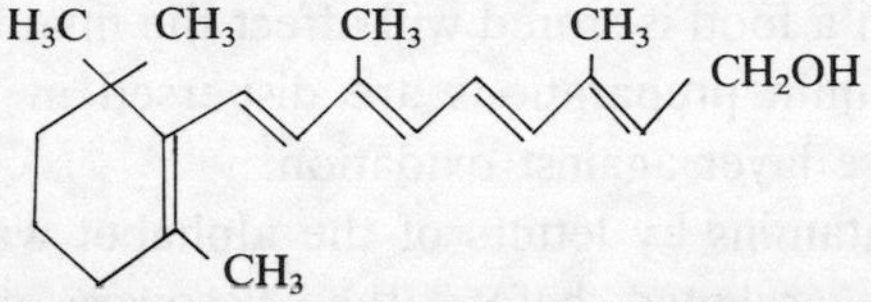

Vitamin A (all-*trans* form)

The vitamin is a pale yellow crystalline solid, insoluble in water but soluble in fat and various fat solvents. It is readily destroyed by oxidation on exposure to air and light. A related compound with the formula $C_{20}H_{27}OH$, originally found in fish, has been designated dehydroretinol or vitamin A_2.

Sources

Vitamin A accumulates in the liver and therefore this organ is likely to be a good source; the amount present varies with species of animal and diet. Table 5.2 shows some typical liver reserves of vitamin A in different species, although these values vary widely within each species.

The oils from livers of certain fish, especially cod and halibut, have long been used as an important dietary source of the vitamin. Egg yolk and milk

TABLE 5.2   Some typical values for liver reserves of vitamin A in different species*
(From T. Moore, 1969. *Fat Soluble Vitamins*, R. A. Morton (ed.) p. 233. Pergamon Press, Oxford)

Species	Vitamin A (μg/g liver)	Species	Vitamin A (μg/g liver)
Pig	30	Horse	180
Cow	45	Hen	270
Rat	75	Codfish	600
Man	90	Halibut	3000
Sheep	180	Polar bear	6000
		Soup-fin-shark	15000

* In every species wide individual variations are to be expected.

fat also are usually rich sources, although the amount in these depends, to a large extent, upon the diets of the animals by which they have been produced.

Vitamin A is manufactured synthetically and can be obtained in a pure form.

Provitamins

Vitamin A does not exist as such in plants, but is present as precursors or provitamins in the form of certain carotenoids which can readily be converted by the animal into the vitamin. At least 80 provitamins are known and these include α-, β- and γ-carotenes, cryptoxanthin, which is present in higher plants, and myxoxanthin which occurs in a blue-green alga. Not all carotenoids are precursors of vitamin A; xanthophylls, for example, which are the main pigments responsible for the orange colour of egg yolk and thought by many to be nutritionally desirable, are hydroxylated carotenoids which are inactive as provitamins. Of the provitamins, β-carotene is the most widely distributed and most active; its structure is shown below:

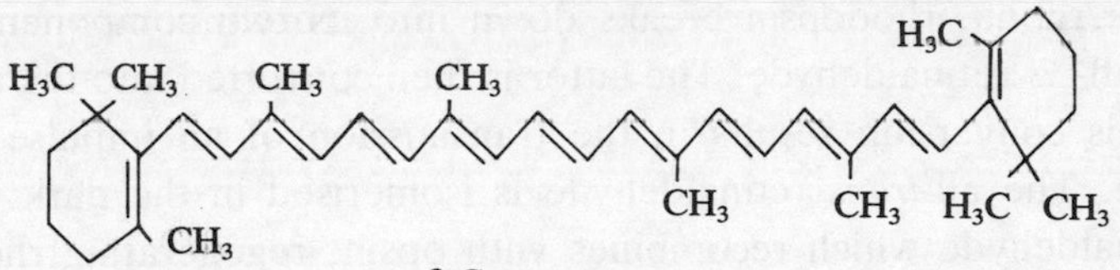

β-Carotene

Pure β-carotene is red in colour, although solutions appear yellowish-orange. All the provitamins are insoluble in water but soluble in fats and fat solvents. Carotenoids are usually accompanied by chlorophyll in plants, though some plant materials such as carrots, tomatoes and certain fungi contain carotenoids but not chlorophyll. Generally, green foods are excellent sources of β-carotene, and in dried crops the degree of greenness is usually a good indication of the β-carotene content. Since carotenes are readily destroyed by oxidation, especially at high temperatures, foods exposed to air and sunlight rapidly lose their vitamin A potency so that large losses can occur during the sundrying of crops.

Apart from yellow maize, most concentrates used in animal feeding are devoid of the provitamins.

Carotenes also occur in certain animal tissues such as the body fat of cattle and horses, but not in sheep or pigs. They are also found in birds' feathers, egg yolk and butterfat.

Conversion of carotene into vitamin A occurs in the intestinal wall and also in the liver. In theory one molecule of β-carotene should form, on hydrolysis, two molecules of vitamin A. The efficiency of conversion is however rarely as great as this; furthermore carotenes are not absorbed from the gut as efficiently as vitamin A. The Agricultural Research Council (ARC) have recommended that the quantity of β-carotene in the diet which is equivalent to 1 μg of retinol should be taken as 11 μg for pigs and 6 μg for ruminants. For poultry, various estimates of conversion efficiencies have been given ranging from 1 to 4. For practical purposes it is suggested that a conversion factor of 3 should be taken. The vitamin A values of foods are often stated in terms of International Units (i.u.), one i.u. of vitamin A being defined as the activity of 0.3 μg of crystalline retinol.

Metabolism

Vitamin A appears to play two different roles in the body according to whether it is acting in the eye or in the general system.

In the eye vitamin A (all-*trans*-retinol) is oxidised to the aldehyde (all-*trans*-retinaldehyde) which is converted into the all-*cis* isomer. The latter then combines with the protein opsin to form rhodopsin (visual purple) which is the photo-receptor for vision at low light intensities. When light falls on the retina, rhodopsin breaks down into its two component parts, opsin and all-*cis* retinaldehyde. The latter is then converted into the all-*trans* isomer. This conversion results in the transmission of an impulse up the optic nerve. The all-*trans* retinaldehyde is isomerised in the dark back to all-*cis* retinaldehyde which recombines with opsin, regenerating rhodopsin and thus continually renewing the light sensitivity of the retina.

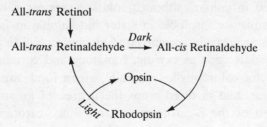

In its second role vitamin A is involved in the formation and protection of epithelial tissues and mucous membranes.

Deficiency symptoms

Ability to see in dim light depends upon the rate of resynthesis of rhodopsin, and where vitamin A is deficient rhodopsin formation is impaired. One of the earliest symptoms of a deficiency of vitamin A in all animals is a lessened ability to see in dim light, commonly known as 'night blindness'.

A physiological manifestation of vitamin A deficiency is an increase in cerebrospinal fluid (CSF) pressure, and a measure of the CSF pressure is often used as a criterion of adequacy of the vitamin.

In adult cattle a mild deficiency of vitamin A is associated with roughened hair and scaly skin. If it is prolonged the eyes are affected, leading to excessive watering, softening and cloudiness of the cornea and development of xerophthalmia, characterised by a drying of the conjunctiva. Constriction of the optic nerve canal may result in blindness in calves. In breeding animals a deficiency may lead to infertility, and in pregnant animals to abortion or to the production of dead, weak or blind calves. Less severe deficiencies may result in calves born with low reserves of the vitamin, and it is imperative that colostrum, rich in antibodies and vitamin A, should be given at birth; otherwise the susceptibility of such animals to infection leads to scours, and if the deficiency is not rectified they frequently die of pneumonia.

In practice severe deficiency symptoms are unlikely to occur in adult animals except after prolonged deprivation. Grazing animals generally obtain more than adequate amounts of provitamin from pasture grass and normally build up liver reserves. If cattle are fed on silage or well preserved hay during the winter months, deficiencies are unlikely to occur. Cases of vitamin A deficiency have been reported among cattle fed indoors on high cereal rations, and under these conditions a high vitamin supplement is recommended.

In ewes, in addition to night blindness, severe cases of deficiency may result in lambs being born weak or dead. A deficiency is not common in sheep, however, because of adequate dietary intakes on pasture.

In pigs, eye disorders such as xerophthalmia and blindness may occur. A deficiency in pregnant animals may result in the production of blind, deformed litters. In less severe cases appetite is impaired and growth retarded. Where pigs are reared out of doors and have access to green food, deficiencies are unlikely to occur except possibly during the winter. Pigs kept indoors on concentrates may not receive adequate amounts in the diet and vitamin A supplements may be required.

In poultry on a diet deficient in vitamin A, the mortality rate is usually high. Early symptoms include retarded growth, weakness, ruffled plumage and a staggering gait. In mature birds, egg production and hatchability are reduced. Since most concentrated foods present in the diets of poultry are low or lacking in vitamin A or its precursors, vitamin A deficiency may be a problem unless precautions are taken. Yellow maize, dried grass or other green food, or alternatively cod or other fish liver oils or vitamin A concentrate, can be added to the diet.

Recent research studies indicate that, in addition to vitamin A, some species may have a dietary requirement for β-carotene *per se*. The ovaries of bovine species are known to contain high concentrations of β-carotene during the luteal phase and it has been postulated that certain fertility disorders in dairy cattle, such as retarded ovulation and early embryonic mortality, may be caused by a deficiency of the provitamin in the diet.

VITAMIN D

Chemical nature

A number of different forms of vitamin D are known, although not all of these are naturally occurring compounds. The two most important forms are ergocalciferol (D_2) and cholecalciferol (D_3). The term D_1 was originally suggested by the earlier workers for an activated sterol which was found later to be impure and to consist mainly of ergocalciferol, which had already been designated D_2. The result of this confusion is that in the group of D-vitamins the term vitamin D_1 has been abolished. The structure of vitamins D_2 adn D_3 are as follows:

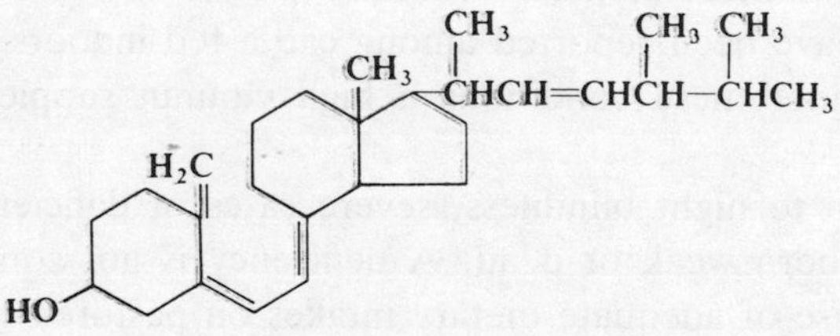

Vitamin D_2 (ergocalciferol)

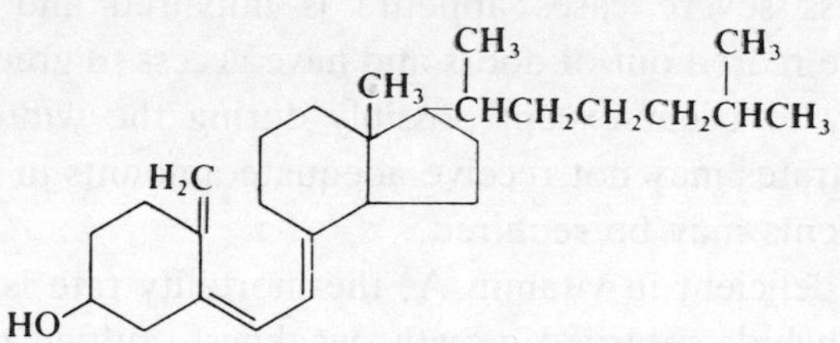

Vitamin D_3 (cholecalciferol)

The D-vitamins are insoluble in water but soluble in fats and fat solvents. The sulphate derivative of vitamin D present in milk is a water soluble form of the vitamin. Both D_2 and D_3 are more stable to oxidation than vitamin A, D_3 being more stable than D_2.

Sources

The D vitamins are limited in distribution. They rarely occur in plants except in sun-dried roughages and the dead leaves of growing plants. In the animal kingdom vitamin D_3 occurs in small amounts in certain tissues, and is abundant only in some fishes. Halibut liver and cod liver oils are rich sources of vitamin D_3. Egg yolk is also a good source, but cows' milk is normally a poor source, although summer milk tends to be richer than winter milk. Colostrum usually contains from 6 to 10 times the amount present in ordinary milk.

Clinical manifestations of avitaminosis D, and other vitamin deficiencies, are frequently treated by injection of the vitamin into the animal.

Provitamins

Reference has been made (Ch. 3) to two sterols, ergosterol and 7-dehydro-cholesterol, as being precursors of vitamins D_2 and D_3 respectively. The provitamins, as such, have no vitamin value and must be converted into calciferols before they are of any use to the animal. For this conversion it is necessary to impart a definite quantity of energy to the sterol molecule, and this can be brought about by the ultra-violet light present in sunlight, by artificially produced radiant energy or by certain kinds of physical treatment. Under natural conditions activation is brought about by irradiation from the sun. The activation occurs most efficiently with light of wavelength between 290 and 315 nm, so that the range capable of vitamin formation is small. The amount of ultra-violet radiation which reaches the earth's surface depends upon latitude and atmospheric conditions; the presence of clouds, smoke and dust reduces the radiation. Ultra-violet radiation is greater in the tropics than in the temperate regions, and the amount reaching the more northern areas in winter may be slight. Since ultra-violet light cannot pass through ordinary window glass, animals housed indoors receive little suitable radiation, if any, for the production of the vitamin. Irradiation is apparently more effective in animals with light-coloured skins. If irradiation is continued for a prolonged period, then the vitamin may itself be altered to compounds which can be toxic.

The chemical transformation occurs in the skin and also in the skin secretions, which are known to contain the precursor. Absorption of the vitamin can take place from the skin, since rickets can be treated successfully by rubbing cod liver oil into the skin.

Vitamin D requirements are often expressed in terms of International Units (iu). One iu of vitamin D is defined as the vitamin D activity of 0.025 μg of crystalline vitamin D_3.

Metabolism

Although the importance of vitamin D in calcium and phosphorus metabolism has long been known, its exact biochemical role is only now being elucidated. Dietary vitamin D_2 or D_3 is absorbed through the small intestine and is transported in the blood to the liver where it is converted into 25-hydroxycholecalciferol. The latter is then transported to the kidney where it is converted into 1,25-dihydroxycholecalciferol, the most biologically active form of the vitamin. This compound is then transported in the blood to the various target tissues, the intestine and the bones. 1,25-Dihydroxycholecalciferol acts in a similar way to a steroid hormone, regulating DNA transcription in the intestinal microvilli, inducing the synthesis of specific messenger RNA (see Ch. 9) which is responsible for the production of calcium binding protein. This protein is involved in the absorption of calcium from the intestinal lumen. The various pathways involved in these transformations are summarised in Fig. 5.1.

The amount of 1,25-dihydroxycholecalciferol produced by the kidney is controlled by the parathyroid hormone. When the level of calcium in the blood is low (hypocalcaemia), the parathyroid gland is stimulated to excrete more parathyroid hormone, which induces the kidney to produce more 1,25-dihydroxycholecalciferol which in turn enhances the intestinal absorption of calcium.

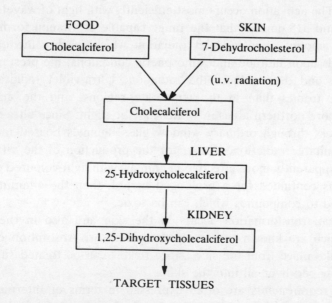

Fig. 5.1. Metabolic pathway showing production of hormonally active form of vitamin D.

In addition to increasing intestinal absorption of calcium, 1,25-dihydroxy-cholecalciferol increases the absorption of phosphorus from the intestine and also enhances calcium and phosphorus reabsorption from the kidney and bone by mechanisms not yet fully understood.

Deficiency symptoms

A deficiency of vitamin D in the young animal results in rickets, a disease of growing bone in which the deposition of calcium and phosphorus is disturbed; as a result the bones are weak and easily broken and the legs may be bowed. In young cattle the symptoms include swollen knees and hocks and arching of the back. In pigs the symptoms are usually enlarged joints, broken bones, stiffness of the joints and occasionally paralysis. The growth rate is generally adversely affected. The term 'rickets' is confined to young growing animals; in older animals vitamin D deficiency causes osteomalacia, in which there is reabsorption of bone already laid down. Osteomalacia due to vitamin D deficiency is not common in farm animals, although a similar condition can occur in pregnant and lactating animals, who require increased amounts of calcium and phosphorus. Rickets and osteomalacia are not specific diseases necessarily caused by vitamin D deficiency, but can be caused by lack of calcium or phosphorus or an imbalance between these two elements.

In poultry, a deficiency of vitamin D causes the bones and beak to become soft and rubbery; growth is usually retarded and the legs may become bowed. Egg production is reduced and eggshell quality deteriorates. Most foods of pigs and poultry, with the possible exception of fish meal, contain little or no vitamin D, and the vitamin is generally supplied to these animals, if reared indoors, in the form of fish liver oils or synthetic preparations.

The need for supplementing the diets of cattle and sheep with vitamin D is probably not so great as that for pigs and poultry. Adult ruminants can receive adequate amounts of the vitamin from hay in the winter months, and from irradiation while grazing. However, since the vitamin D content of hays is extremely variable, it is possible that vitamin D supplementation may be desirable, especially with young growing animals or pregnant animals on winter diets. But there is a considerable lack of information about the vitamin D needs of farm animals under practical conditions.

Vitamins D_2 and D_3 have the same potency for cattle, sheep and pigs, but vitamin D_2 has only about 10 per cent of the potency of D_3 for poultry.

Certain foods such as fresh green cereals and yeast, have been shown to have rachitogenic (rickets-causing) properties for mammals, and raw liver and isolated soya bean protein have a similar effect on poultry. In one study it was shown that in order to overcome the rachitogenic activity present in whole raw soya bean meal, a tenfold increase in vitamin D supplement was necessary. Heating destroys the rachitogenic activity.

VITAMIN E

Chemical nature

Vitamin E is a group name which includes a number of closely related active compounds. Eight naturally occurring forms of the vitamin are known and these can be divided into two groups according to whether the side chain of the molecule is saturated or unsaturated as shown below.

The four saturated vitamins are designated α, β, γ and δ tocopherols and of these the α form is the most biologically active and most widely distributed.

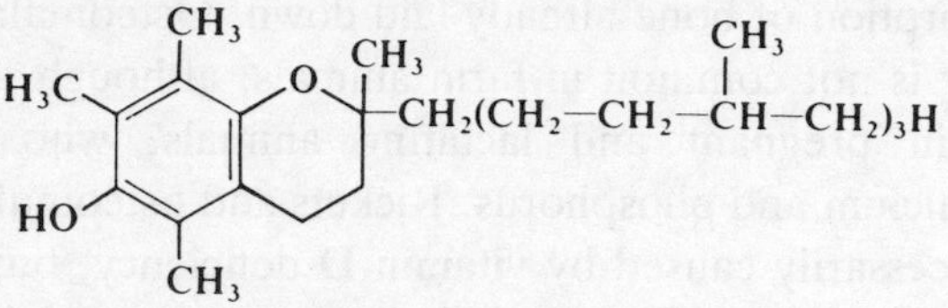

$$H_3C \overset{\displaystyle CH_3}{\underset{\displaystyle CH_3}{\quad}} \quad \overset{\displaystyle CH_3}{\underset{}{O}} \quad \overset{\displaystyle CH_3}{\quad}$$

$$HO\text{---}\quad CH_2(CH_2\text{---}CH_2\text{---}CH\text{---}CH_2)_3H$$

α-Tocopherol

The β, γ and δ forms have only about 56, 16 and 0.5 per cent of the activity of the α form respectively. The unsaturated forms of the vitamin have been designated α, β, γ and δ-tocotrienols and of these only the α form appears to have any significant vitamin E activity, and then only about 16 per cent of its saturated counterpart.

Sources

Vitamin E is very widely distributed in foods. Green fodders are good sources of α-tocopherol, young grass being a better source than mature herbage. The leaves contain 20–30 times as much vitamin E as the stems. Losses during haymaking can be as high as 90 per cent, but losses during ensilage or artificial drying are small.

Cereal grains are also good sources of the vitamin but the tocopherol composition varies with species. Wheat and barley grain resemble grass in containing mainly α-tocopherol, but maize contains, in addition to α-tocopherol, appreciable quantities of γ-tocopherol. During the storage of moist grain in silos, the vitamin E activity can decline markedly. Reduction in the concentration of the vitamin from 9 mg/kg DM to 1 mg/kg DM has been reported in moist barley stored for 12 weeks.

Animal products are relatively poor sources of the vitamin although the amount present is related to the level of vitamin E in the diet. Synthetic α-tocopherol and the acetate are available as commercial preparations.

The vitamin E values of foods are often stated in terms of International Units, one i.u. of vitamin E being defined as the specific activity of 1 mg of synthetic racemic α-tocopherol acetate.

Metabolism

Vitamin E appears to have at least two closely associated metabolic roles: firstly it acts as a non-specific biological antioxidant and secondly it has a more specific action, associated with selenium, in which it protects vital phospholipids from peroxidative damage. Evidence in support of its antioxidant role has been obtained from experiments in which certain synthetic antioxidants, such as methylene blue, can substitute for vitamin E in preventing some, but not all, of the deficiency diseases associated with the vitamin.

In its second role the vitamin acts within the phospholipid membrane of the vital organelles in preventing the formation of peroxides. Those peroxides which do form, in either the presence or absence of vitamin E, are apparently destroyed by the enzyme glutathione peroxidase which contains selenium. The vitamin can therefore be regarded as the first line of defence in preventing peroxide formation, with the selenium-containing enzyme acting as a second line of defence in destroying any peroxides which are formed before they can damage the cell.

More recently, a less well-defined role of vitamin E in the development and function of the immune system has been proposed.

Deficiency symptoms

Early experiments carried out with rats on vitamin E deficient diets showed that the animals failed to reproduce. These experiments led to the vitamin being originally termed the anti-sterility vitamin. Although reproductive failures have been reported in other animals, muscle degeneration (myopathy) is probably the most frequent and, from a diagnostic point of view, the most important manifestation of vitamin E deficiency in domestic animals. A summary of these and other deficiency conditions associated with the vitamin is given in Table 5.3.

Muscular dystrophy occurs in the United Kingdom in suckler herds in which the cows have been wintered on turnips and straw. It is most frequently seen in calves up to 3 months of age, though older animals may also show symptoms which occur most frequently when animals are turned

TABLE 5.3 Nutritional diseases caused by deficiencies of vitamin E and or selenium (From M. L. Scott, in *Feed Energy Sources for Livestock*, Swan and Lewis (eds). Butterworths, 1976, p. 118)

Disease	Experimental animal	Tissue affected	Influenced by poly-unsaturated acids	Prevented by:		
				Vitamin E	*Se*	*Antioxidant*
1. REPRODUCTIVE FAILURE:						
Embryonic degeneration						
Type A	Female rat, hen, turkey	Vascular system of embryo	X	X		X
Type B	Cow, ewe	" "		*	X†	
Sterility	Male rat, guinea pig, hamster, dog, cock	Male gonads	X	X		
Reduced egg production	Hen		X	X		
Decreased hatchability of eggs	Hen		X	X		
2. LIVER, BLOOD, BRAIN, CAPILLARIES, PANCREAS:						
Liver necrosis	Rat, pig, beef calves	Liver	X	X	X	
Fibrosis	Chick, mouse	Pancreas		X	X	
Erythrocyte haemolysis	Rat, chick, man (premature infant)	Erythrocytes	X	X		X

Condition	Species	Tissue affected			
Plasma protein loss	Chick, turkey	Serum albumen		X	
Anaemia	Monkey	Bone marrow	X	X	
Encephalomalacia	Chick	Cerebellum		X	
Exudative diathesis	Chick, turkey	Vascular system	X	X	
Kidney degeneration	Rat, mouse, monkey, mink	Kidney tubular epithelium		X	?
Steatitis (ceroid)	Mink, pig, chick	Adipose tissue	X	X	
Depigmentation	Rat	Incisors	X	X	
Hair and feather loss	Chick, rat, horse, pig	Integument, hair and feathers		X	

3. NUTRITIONAL MYOPATHIES:

Type	Species	Tissue affected			
Type A (nutritional muscular dystrophy)	Rabbit, guinea pig, monkey, duck, mouse, mink	Skeletal muscle		X	?
Type B (white muscle disease)	Lamb, calf, kid	Skeletal and heart muscles	X†	*	
Type C	Turkey	Gizzard, heart	X	*	
Type D	Chicken	Skeletal muscle	‡		

[a] Not effective in diets severely deficient in selenium; [b] When added to diets containing low levels of vitamin E; [c] A low level (0.5%) of linoleic acid necessary to produce dystrophy, higher levels did not increase vitamin E required for prevention

out on to spring pasture. Where the heart muscle is affected, death may be sudden without premonitory signs. Less severe cases may show symptoms of circulatory and respiratory embarrassment on the slightest exertion. Where the skeletal muscles are affected, stiffness, unnatural postures and conformation abnormalities occur. Popular descriptive names for these conditions are 'white muscle disease' and 'stiff lamb disease'. A deficiency of vitamin E may also be induced in calves by the ingestion of diets rich in polyunsaturated fatty acids, such as those containing excessive amounts of cod-liver oil.

Since the quantity of vitamin E present in the tissue of the newborn calf and in the mother's milk is influenced by the mother's diet, it follows that pregnant animals should be given diets containing adequate amounts of the vitamin.

Abnormalities in pigs caused by a vitamin E deficiency are varied; affected pigs show muscular weakness and severe liver damage. 'Fatal syncope' is a vitamin E deficiency condition occurring in pigs and calves in which the heart muscle is affected and sudden death may occur.

Vitamin E deficiency in chicks may lead to a number of distinct diseases: muscular dystrophy, encephalomalacia, or exudative diathesis. In nutritional muscular dystrophy in chicks the main muscles affected are the pectorals although the leg muscles also may be involved. Nutritional encephalomalacia or 'crazy chick disease' is a condition in which the chick is unable to walk or stand, and is accompanied by haemorrhages and necrosis of the brain (Plate 1). Exudative diathesis is a vascular disease of chicks characterised by a generalised oedema of the subcutaneous fatty tissues, associated with an abnormal permeability of the capillary walls.

Vitamin E or synthetic antioxidants are equally effective in preventing nutritional encephalomalacia, while selenium is not important. In the other two diseases affecting chicks, both vitamin E and selenium appear to be involved. Although selenium does not prevent muscular dystropy, its addition to a selenium-low diet decreases the amount of dietary vitamin E needed to prevent the disease. Exudative diathesis is prevented by either vitamin E or selenium.

Selenium itself is a very toxic element and care is required in its use as a dietary additive. The toxic nature of selenium is discussed in the next chapter.

VITAMIN K

Vitamin K was originally discovered in 1935 to be an essential factor in the prevention of haemorrhagic symptoms produced in chicks. The discovery was made by a group of Danish scientists who gave the name 'Koagulation

Factor' to the vitamin, which became shortened to the K factor and eventually to vitamin K.

Chemical nature

A number of compounds are known to have vitamin K activity. The two most important naturally occurring compounds are vitamin K_1 (phylloquinone), found in green plants, and vitamin K_2 (menaquinone), which is a product of bacterial growth. Chemically, both compounds are derivatives of vitamin K_3 (menadione) which is 2-methyl-1,4-naphthoquinone.

All forms of vitamin K are converted in the liver into menaquinone which suggests that this is metabolically the active form of the vitamin.

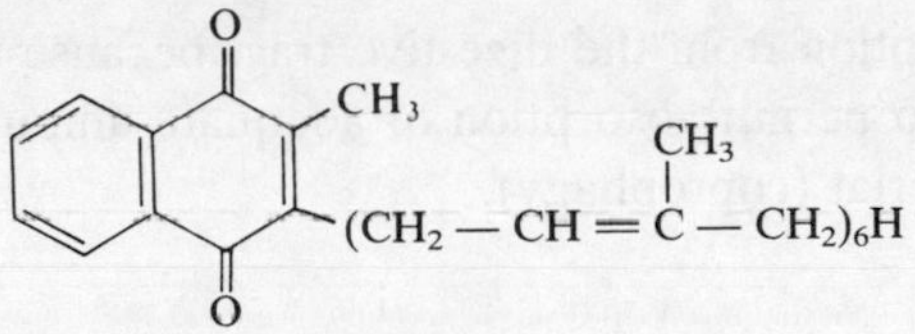

Menaquinone
(2-methyl-3-difarnesyl-1,4-naphthoquinone)

Vitamins K are relatively stable at ordinary temperatures but are rapidly destroyed on exposure to sunlight.

Sources

Vitamin K_1 is present in most green leafy materials, lucerne, cabbage and kale being good sources. The amounts present in foods of animal origin are usually related to the diet, but egg yolk, liver and fish meal are generally good sources. Vitamin K_2 is synthesised by bacteria in the digestive tract.

Metabolism

Vitamin K is necessary for the synthesis of prothrombin in the liver. In the blood-clotting process, prothrombin is the inactive precursor of thrombin, an enzyme that converts the protein fibrinogen in blood plasma into fibrin, the insoluble, fibrous protein that holds blood clots together. Prothrombin normally must bind to calcium ions before it can be activated. If the supply of vitamin K is inadequate, the prothrombin molecule is deficient in γ-carboxyglutamic acid, a specific amino acid responsible for calcium binding.

Deficiency symptoms

Symptoms of vitamin K deficiency have not been reported in ruminants or pigs under normal conditions, and it is generally considered that bacterial synthesis in the digestive tract supplies sufficient vitamin for the animal's needs. A number of micro-organisms are known to synthesise vitamin K, including *Escherichia coli*. A disease of cattle called 'sweet clover disease' is associated with vitamin K in that spoiled sweet clover (*Melilotus albus*) contains a compound, dicoumarol, which lowers the prothrombin content of the blood. The disease can be overcome by administering vitamin K to the animals. For this reason dicoumarol is sometimes referred to as an 'anti-vitamin'.

The symptoms of vitamin K deficiency in chicks are anaemia (Plate 2) and a delayed clotting time of the blood; birds are easily injured and may bleed to death. It is doubtful whether, in birds, microbially synthesised vitamin K is available by direct absorption from the digestive tract because the site of its formation is too distal to permit absorption of adequate amounts except by ingestion of faecal material (coprophagy).

VITAMIN B COMPLEX

The vitamins included under this heading all have the property of being soluble in water, and most of them are components of coenzymes (see Table 5.4).

Unlike the fat-soluble vitamins, members of the vitamin B complex, with the exception of cobalamin, are not stored in the body tissues in appreciable

TABLE 5.4   Some coenzymes and enzyme prosthetic groups involving the B vitamins

Vitamin	Coenzyme or prosthetic group	Enzymic and other function
Thiamin	Thiamin diphosphate (TDP)	Oxidative decarboxylation
Riboflavin	Flavin mononucleotide (FMN)	Hydrogen carrier
Riboflavin	Flavin adenine dinucleotide (FAD)	Hydrogen carrier
Nicotinamide	Nicotinamide adenine dinucleotide (NAD)	Hydrogen carrier
Nicotinamide	Nicotinamide adenine dinucleotide phosphate (NADP)	Hydrogen carrier
Pyridoxal	Pyridoxal phosphate	Transaminases and decarboxylases
Pantothenic acid	Coenzyme A (CoA)	Acyl transfer
Folic acid	Tetrahydrofolic acid	One carbon transfer
Biotin	Biotin	Carbon dioxide transfer
Cobalamin	Adenylcobamide	Group transfer

Plate 1 Encephalomalacia ('crazy chick disease') resulting from a deficiency of vitamin E in the diet. (Reproduced by permission of the Central Veterinary Laboratory, Weybridge.)

Plate 2 Anaemic appearance of fowl resulting from vitamin K deficiency. (Reproduced by permission of Dr H. Brubacher, F. Hoffmann-La-Roche, Switzerland.)

Plate 3   Curled toe paralysis in chick caused by a
deficiency of riboflavin in the diet.
(Reproduced by permission of Professor M. L.
Scott, Cornell University, USA.)

Plate 4   'Goose stepping' (locomotor disturbances,
particularly of the hind legs), a recognised
symptom of pantothenic acid deficiency in the
pig. (Reproduced by permission of the Central
Veterinary Laboratory, Weybridge.)

Plate 5   Different severities of biotin deficiency in chicks, showing
           swellings and cracking of foot pads, inflammations and
           secondary infections. (Reproduced by permission of Dr C.
           Whitehead, Poultry Research Centre, Roslin.)

Plate 6   *Right*: Steer with copper deficiency induced by high dietary
           molybdenum content. (Note typical coat depigmentation.)
           Left: Steer depleted of copper by high dietary iron content,
           but showing no clinical signs of deficiency. (Reproduced by
           permission of the Rowett Research Institute, Aberdeen and of
           CAB International, Farnham Royal, Slough.)

Plate 7 Lambs of the same age and fed on diets differing only in cobalt content. The lamb on the left received a diet containing less than 0.07 mg Co/kg. (Reproduced by permission of the Rowett Research Institute, Aberdeen.)

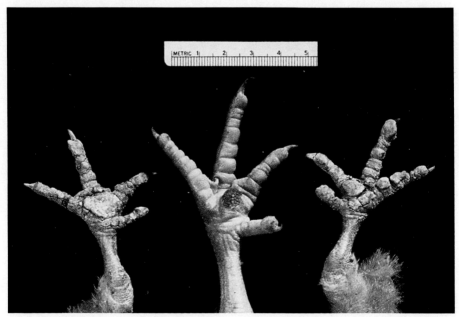

Plate 8 Zinc-deficient chicks showing fissuring and encrustations of the skin of feet compared with normal foot in centre. (Reproduced by permission of Dr W. A. Dewar, Poultry Research Centre, Roslin.)

amounts and a regular exogenous supply is essential. In ruminants, all the vitamins in this group can be synthesised by microbial action in the rumen and generally this will provide satisfactory amounts for normal metabolism in the host and secretion of adequate quantities into milk. However, under certain conditions, deficiencies of thiamin and cobalamin can occur in ruminants.

THIAMIN

Chemical nature

Thiamin (vitamin B_1) is a complex nitrogenous base containing a pyrimidine ring joined to a thiazole ring. Because of the presence of an hydroxyl group at the end of the side chain thiamin can form esters. The main form of thiamin in animal tissues is the diphosphate ester (TDP), formerly known as thiamin pyrophosphate, although the monophosphate (TMP) and the triphosphate (TTP) also occur. The vitamin is very soluble in water and has a characteristic odour and 'meaty' flavour. It is fairly stable in mildly acidic solution but readily decomposes in neutral solutions.

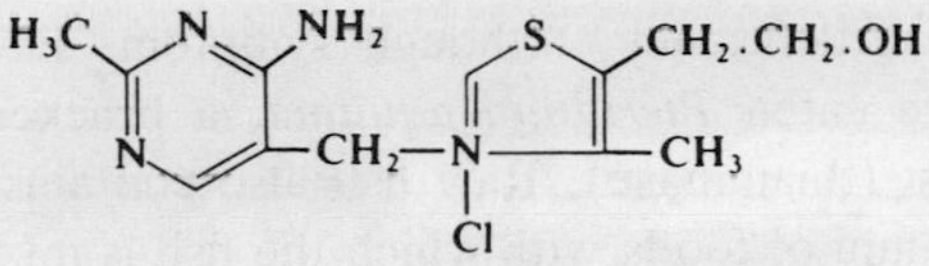

Thiamin chloride

Sources

Thiamin is widely distributed in foods. Brewers' yeast is a rich source. The vitamin is concentrated in the germ of cereal grain and is also present in the aleurone layer. Other good sources include beans, peas and green leafy crops. Animal products rich in thiamin include egg yolk, liver, kidney and pork muscle. The synthetic vitamin is obtainable, usually marketed as the hydrochloride.

Metabolism

Thiamin diphosphate is a coenzyme involved in the oxidative decarboxylation of pyruvate to acetyl coenzyme A, the oxidative decarboxylation of α-ketoglutarate to succinyl coenzyme A in the citric acid cycle, the pentose phosphate pathway and the synthesis of valine in bacteria, yeasts and plants.

Deficiency symptoms

Early symptoms of thiamin deficiency in most species include loss of appetite, emaciation, muscular weakness and a progressive dysfunction of the nervous system. In pigs, appetite and growth are adversely affected and the animals may vomit and have respiratory troubles.

Chicks reared on thiamin-deficient diets have poor appetites and are consequently emaciated. After about ten days they develop polyneuritis, which is characterised by head retraction, nerve degeneration and paralysis.

Many of these deficiency conditions in animals can be explained in terms of the role of TDP in the oxidative decarboxylation of pyruvic acid. On a thiamin-deficient diet animals accumulate pyruvic acid and its reduction product, lactic acid, in their tissues, which leads to muscular weakness. Nerve cells are particularly dependent on the utilisation of carbohydrate and for this reason a deficiency of the vitamin has a particularly serious effect on nervous tissue. Since acetyl coenzyme A is an important metabolite in the synthesis of fatty acids (p. 190), lipogenesis, or fat synthesis, is reduced.

Because thiamin is fairly widely distributed in foods, and in particular because cereal grains are rich sources of the vitamin, pigs and poultry are in practice unlikely to suffer from thiamin deficiency.

Since microbial synthesis occurs in ruminants and in horses these species are unlikely to show thiamin deficiency, although symptoms have been reported in horses that have eaten *Pteridium aquilina*, a bracken which contains a thiamin antagonist (thiaminase). Raw fish also contains thiaminase, which destroys the thiamin of foods with which the fish is mixed. The activity of the thiaminase, however, is destroyed by cooking.

In recent years, cases of thiamin deficiency due to the action of bacterial thiaminases in the alimentary tract have been reported in man, pigs, poultry, cats and dogs. Cerebrocortical necrosis (CCN) in sheep and cattle has also been ascribed to this cause.

RIBOFLAVIN

Chemical nature

Riboflavin (vitamin B_2) consists of a dimethyl-isoalloxazine nucleus combined with ribitol. Its structure is shown on the next page.

It is a yellow, crystalline compound, which has a yellowish-green fluorescence in aqueous solution. Riboflavin is only sparingly soluble in water; it is heat-stable in acid or neutral solutions, but is destroyed by alkali. It is unstable to light, particularly ultra-violet light.

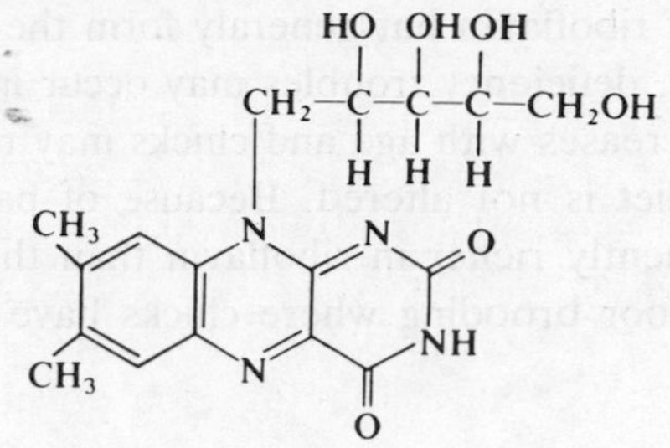

HO OH OH
| | |
CH_2–C–C–C–CH_2OH
| | |
H H H

CH_3

CH_3

N N O

NH

O

Riboflavin

Sources

Riboflavin occurs in all biological materials. The vitamin can be synthesised by all green plants, yeasts, fungi and most bacteria, although the lactobacilli are a notable exception and require an exogenous source. Rich sources are yeast, liver, milk (especially whey), and green leafy crops. Cereal grains are poor sources.

Metabolism

Riboflavin is an important constituent of the flavoproteins. The prosthetic group of these compound proteins contains riboflavin in the form of the phosphate (flavin mononucleotide or FMN) or in a more complex form as flavin adenine dinucleotide (FAD). There are several flavoproteins which function in the animal body; they are all concerned with chemical reactions involving the transport of hydrogen. Their importance in carbohydrate and amino acid metabolism is discussed in Chapter 9.

Deficiency symptoms

In pigs, deficiency symptoms include poor appetite with consequent retardation in growth, vomiting, skin eruptions and eye abnormalities. There is also recent evidence to indicate that riboflavin deficiency can have a detrimental effect on the reproductive system of young sows. Chicks reared on a riboflavin-deficient diet grow slowly and develop 'curled toe paralysis', a specific symptom, caused by peripheral nerve degeneration, in which the chicks walk on their hocks with the toes curled inwards (Plate 3). In breeding hens, a deficiency results in decreased hatchability. Embryonic abnormalities occur, including the characteristic 'clubbed down' condition in which the down feather continues to grow inside the follicle, resulting in a coiled feather.

Since cereals are poor sources of riboflavin but generaly form the major part of the diet of pigs and poultry, deficiency troubles may occur in practice. In poultry the requirement decreases with age and chicks may recover from symptoms even though the diet is not altered. Because of bacterial action, poultry droppings are frequently richer in riboflavin than the diet. This is of great significance with floor brooding where chicks have access to the droppings.

NICOTINAMIDE

Chemical nature

Nicotinamide is the amide derivative of nicotinic acid (pyridine 3-carboxylic acid) and it is the form in which it functions in the body. It has been proposed that the generic name *niacin* should be used to include the acid and its derivatives. The relationship between nicotinic acid, nicotinamide and the amino acid tryptophan, which can act as a precursor is shown below:

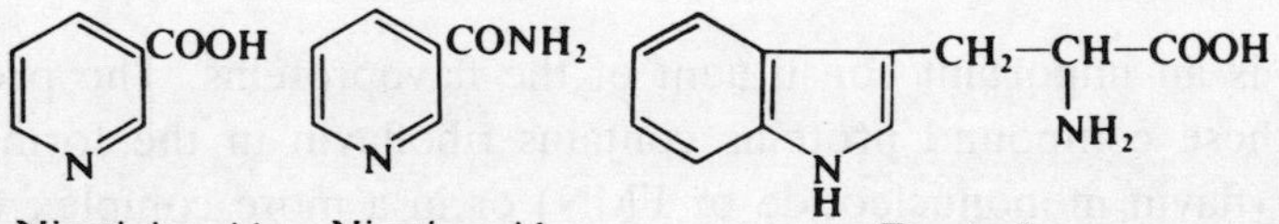

Nicotinic acid   Nicotinamide   Tryptophan

Nicotinamide is a stable vitamin and is not easily destroyed by heat, acids, alkalis, or by oxidation.

Sources

Nicotinic acid can be synthesised from tryptophan in the body tissues, and since animals can convert the acid to the amide-containing coenzyme (see below), it follows that if the diet is adequately supplied with proteins rich in tryptophan, then the dietary requirement for the vitamin itself should be low. However, the efficiency of conversion of tryptophan into nicotinamide is poor, values of about 50 to 1 having been suggested, and it is generally considered that an exogenous source of the vitamin is also necessary. Rich sources of the vitamin are liver, yeast, groundnut and sunflower meals. Although cereal grains contain the vitamin, much of it is present in a bound form which is not readily available to pigs and poultry. Milk and eggs are almost devoid of the vitamin although they contain the precursor tryptophan.

Metabolism

Nicotinamide functions in the animal body as the active group of two important coenzymes, nicotinamide-adenine dinucleotide (NAD) and nicotinamide-adenine dinucleotide phosphate (NADP). These coenzymes, are involved in the mechanism of hydrogen transfer in living cells (see Ch. 9).

Deficiency symptoms

In pigs, deficiency symptoms include poor growth, enteritis and dermatitis. In fowls a deficiency of the vitamin causes bone disorders, dermatitis and inflammation of the mouth and the upper part of the oesophagus.

Most diets contain adequate amounts of nicotinamide or its precursor, tryptophan. Deficiency symptoms are likely in pigs and poultry where diets with a high maize content are used, since maize contains very little of the vitamin or of tryptophan.

VITAMIN B₆

Chemical nature

The vitamin exists in three forms which are interconvertible in the body tissues. The parent substance is known as pyridoxine, the corresponding aldehyde derivative as pyridoxal and the amine as pyridoxamine. The term vitamin B₆ is frequently used to describe the three forms.

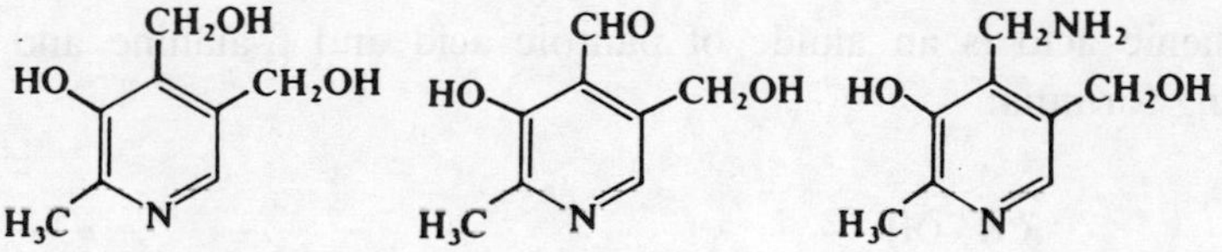

Pyridoxine Pyridoxal Pyridoxamine

The amine and aldehyde derivatives are less stable than pyridoxine and are destroyed by heat.

Sources

The vitamin is widely distributed, and yeast, liver, milk, pulses and cereal grains are rich sources.

Metabolism

Of the three related compounds, the most actively functioning one is pyridoxal in the form of the phosphate. Pyridoxal phosphate plays a central role as a coenzyme in the reactions by which a cell transforms nutrient amino acids into mixtures of amino acids and other nitrogenous compounds required for its own metabolism. These reactions involve the activities of transaminases and decarboxylases (see p. 179), and over 50 pyridoxal phosphate-dependent enzymes have been identified. The vitamin is believed to play a role in the absorption of amino acids from the intestine.

Deficiency symptoms

Because of the numerous enzymes requiring pyridoxal phosphate, a large variety of biochemical lesions are associated with vitamin B_6 deficiency. These lesions are concerned primarily with amino acid metabolism and a deficiency affects the animal's growth rate. Convulsions may also occur, possibly because a reduction in the activity of glutamic acid decarboxylase results in an accumulation of glutamic acid. In addition, pigs exhibit a reduced appetite and may develop anaemia. Chicks on a deficient diet show jerky movements, while in adult birds hatchability and egg production are adversely affected. In practice, vitamin B_6 deficiency is unlikely to occur in farm animals because of the vitamin's wide distribution.

PANTOTHENIC ACID

Chemical nature

Pantothenic acid is an amide of pantoic acid and β-alanine and has the following formula:

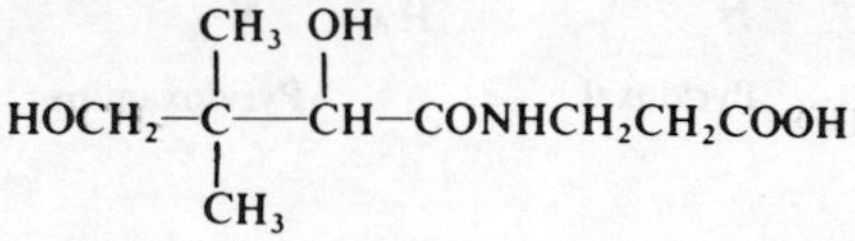

$$HOCH_2 - \underset{\underset{CH_3}{|}}{\overset{\overset{CH_3 \quad OH}{|\quad\;\;\;|}}{C}} - CH - CONHCH_2CH_2COOH$$

Pantothenic acid

Sources

The vitamin is widely distributed, indeed the name is derived from the Greek *pantothen*, 'from everywhere', indicating its ubiquitous distribution. Rich sources are liver, egg yolk, groundnuts, peas, yeast and molasses.

Cereal grains are also good sources of the vitamin. The free acid is unstable and the synthetically prepared calcium pantothenate is the product used commercially.

Metabolism

Pantothenic acid is a constituent of coenzyme A, which is the important coenzyme of acyl transfer. Chemically, coenzyme A is 3-phospho-adenosine-5-diphospho-pantotheine.

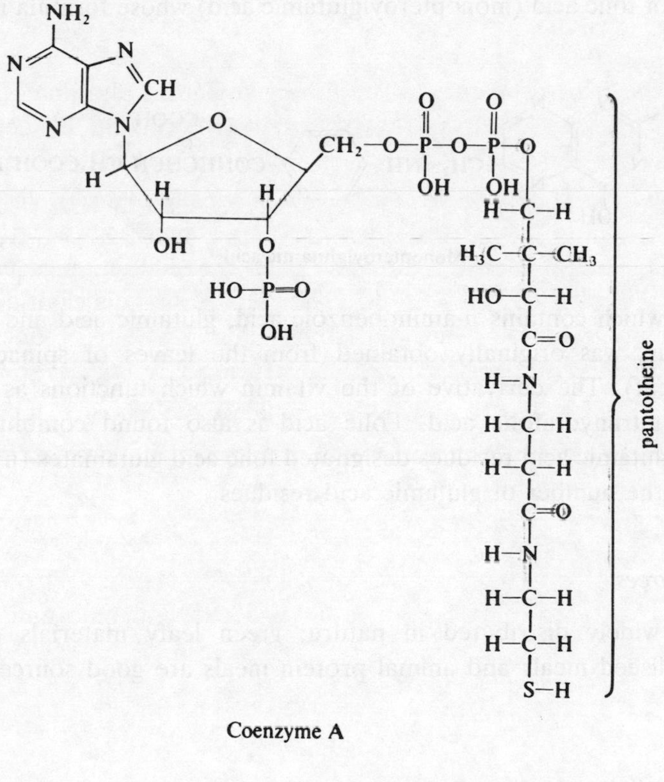

Coenzyme A

The importance of coenzyme A in metabolism is discussed in Chapter 9.

Deficiency symptoms

Deficiency of pantothenic acid in pigs causes slow growth, diarrhoea, loss of hair, scaliness of the skin and a characteristic 'goose-stepping' gait (Plate 4); in severe cases animals are unable to stand. In the chick, growth is retarded and dermatitis occurs. In mature birds, hatchability is reduced. Pantothenic acid, like all the B-complex vitamins, can be synthesised by rumen micro-organisms; *Escherichia coli*, for example, is known to produce this vitamin. Pantothenic acid deficiencies are considered to be rare in practice because of the wide distribution of the vitamin, although deficiency symptoms have been reported in commercial Landrace pig herds.

FOLACIN

Chemical nature

The term 'folacin' is used to describe a number of compounds which are derivatives of folic acid (monopteroylglutamic acid) whose formula is shown below:

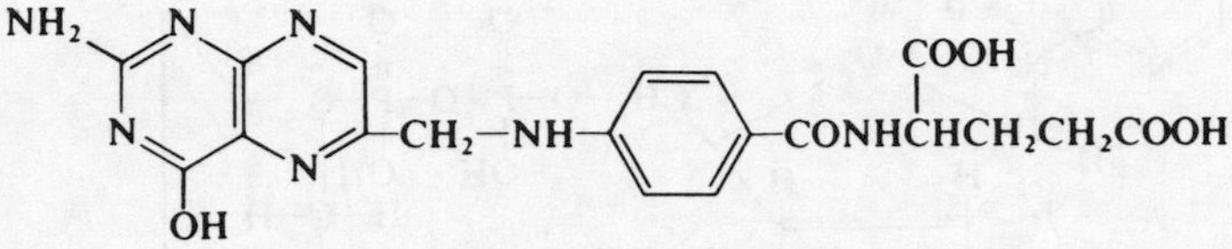

Monopteroylglutamic acid

Folic acid, which contains *p*-aminobenzoic acid, glutamic acid and a pteridine nucleus, was originally obtained from the leaves of spinach (Lat. *folium*, a leaf). The derivative of the vitamin which functions as the coenzyme is tetrahydrafolic acid. Folic acid is also found combined with additional glutamic acid residues designated folic acid glutamates (n), where n indicates the number of glutamic acid residues.

Sources

Folacin is widely distributed in nature; green leafy materials, cereals, extracted oilseed meals and animal protein meals are good sources of the vitamin.

Metabolism

In the form of the coenzyme the vitamin serves as a carrier of various one-carbon groups (e.g. formyl, methyl) that are added to, or removed from, such metabolites as histidine, serine, methionine, and purines.

Deficiency symptoms

A deficiency of folacin in animals is characterised by nutritional anaemia and poor growth. With the exception of young chicks, folacin deficiency symptoms rarely occur in farm animals because of synthesis by intestinal bacteria. Prolonged oral administration of sulpha drugs is known to depress bacterial synthesis of the vitamin and deficiency symptoms may be induced by medication of this kind.

BIOTIN

Chemical nature

Chemically, biotin is 2-keto-3,4-imidazolido-2-tetrahydrothiophene-*n*-valeric acid. Its structure is shown below:

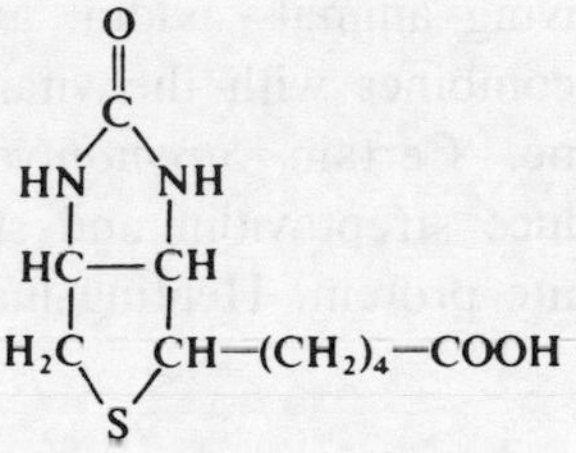

Biotin

Sources

Biotin is widely distributed in foods; liver, yeast, milk and vegetables are rich sources. However, in some foods much of the vitamin may be unavailable. Studies with chicks have shown that the availability of biotin in barley, wheat and some animal protein foods is very low, whereas the biotin in maize and certain oilseed meals, such as soya bean meal, is completely available.

Metabolism

Biotin serves as a prosthetic group of several enzymes which catalyse the transfer of carbon dioxide from one substrate to another. Biotin-dependent enzymes include pyruvate carboxylase, acetyl coenzyme A carboxylase, propionyl coenzyme A carboxylase and methyl crotonyl coenzyme A carboxylase.

Deficiency symptoms

Biotin-dependent enzyme systems are important in carbohydrate metabolism, fatty acid synthesis, protein synthesis, amino acid deamination and nucleic acid metabolism. Because of this wide range of metabolic functions, it is not surprising that a deficiency of the vitamin can result in a large variety of clinical symptoms.

In pigs, biotin deficiency causes foot lesions, alopecia (hair loss) and a dry scaly skin. In growing pigs, both growth rate and food utilisation are adversely affected. In breeding sows, a deficiency of the vitamin can adversely influence reproductive performance.

In poultry, biotin deficiency causes reduced growth, dermatitis, cracked feet (Plate 5), poor feathering and fatty liver and kidney syndrome (FLKS) which is characterised by a lethargic state with death frequently following within a few hours. On autopsy, the liver and kidneys, which are pale and swollen, contain abnormal depositions of lipid.

Biotin deficiency can be induced by giving animals avidin, a protein present in the raw white of eggs, which combines with the vitamin and prevents its absorption from the intestine. Certain *Streptomyces* spp. bacteria present in soil and manure produce streptavidin and stravidin, which have a similar action to the egg white protein. Heating inactivates these antagonist proteins.

CHOLINE

Chemical nature

The chemical structure of choline is given below:

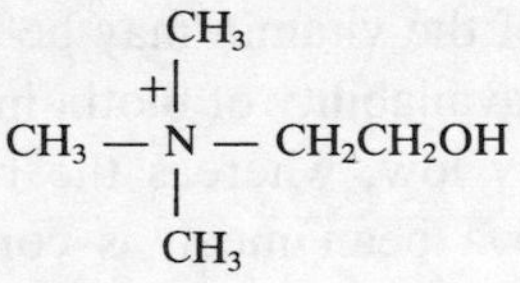

$$CH_3 - \overset{\overset{\displaystyle CH_3}{\overset{\displaystyle +|}{}}}{N} - CH_2CH_2OH$$
$$|$$
$$CH_3$$

Choline

Sources

Green leafy materials, yeast, egg yolk and cereals are rich sources of choline.

Metabolism

Unlike the other B vitamins, choline is not a metabolic catalyst but forms an essential structural component of body tissues. It is a component of lecithins which play a vital role in cellular structure and activity. It also plays an important part in lipid metabolism in the liver by preventing the accumulation of fat in this organ. It serves as a donor of methyl groups in transmethylation reactions and is a component of acetylcholine which is responsible for the transmission of nerve impulses.

Choline can be synthesised in the liver from methionine and the exogenous requirement for this vitamin is, therefore, influenced by the level of methionine in the diet.

Deficiency symptoms

Deficiency symptoms, including slow growth and fatty infiltration of the liver, have been produced in chicks and pigs. Choline is also concerned with the prevention of perosis or slipped tendon in chicks. The choline requirement of animals is unusually large for a vitamin, but in spite of this, deficiency symptoms are not common in farm animals because of its wide distribution and because it can be readily derived from methionine.

VITAMIN B₁₂

Chemical nature

Vitamin B₁₂ has the most complex structure of all the vitamins. The basic

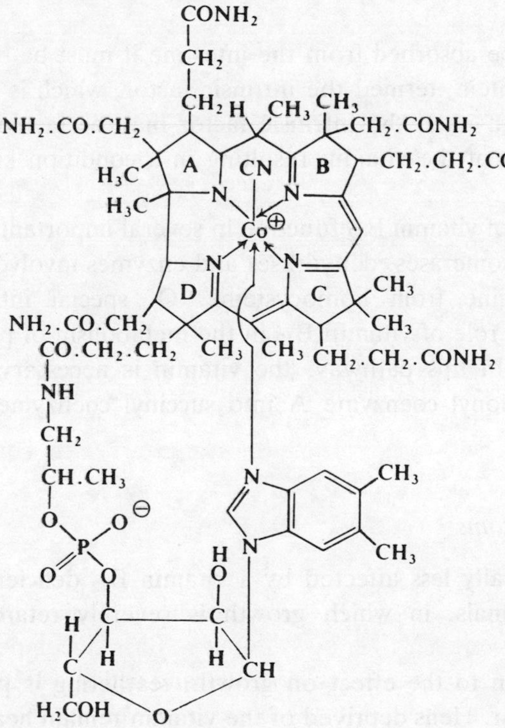

Vitamin B₁₂   (cyanocobalamin)

unit is a corrin nucleus which consists of a ring structure comprising four five-membered rings containing nitrogen. In the active centre of the nucleus is a cobalt atom. A cyano group is usually attached to the cobalt as an artefact of isolation and, as this is the most stable form of the vitamin it is the form in which the vitamin is commercially produced.

In the animal, the cyanide ion is replaced by a variety of ions, e.g. hydroxyl (hydroxycobalamin), methyl (methylcobalamin) and 5'-deoxyadenosyl (5'-deoxyadenosylcobalamin), the last two forms acting as coenzymes in animal metabolism.

Sources

Vitamin B_{12} is considered to be synthesised exclusively by micro-organisms and its presence in foods is thought to be ultimately of microbial origin. The main natural sources of the vitamin are foods of animal origin, liver being a particularly rich source. Its limited occurrence in higher plants is still controversial, since many think that its presence in trace amounts may result from contamination with bacteria or insect remains.

Metabolism

Before vitamin B_{12} can be absorbed from the intestine it must be bound to a highly specific glycoprotein, termed the intrinsic factor, which is secreted by the gastric mucosa. In man, the intrinsic factor may be lacking which leads to poor absorption of the vitamin resulting in a condition known as pernicious anaemia.

The coenzymic forms of vitamin B_{12} function in several important enzyme systems. These include isomerases, dehydrases and enzymes involved in the biosynthesis of methionine from homocysteine. Of special interest in ruminant nutrition is the role of vitamin B_{12} in the metabolism of propionic acid into succinic acid. In this pathway, the vitamin is necessary for the conversion of methylmalonyl coenzyme A into succinyl coenzyme A (see p. 170).

Deficiency symptoms

Adult animals are generally less affected by a vitamin B_{12} deficiency than are young growing animals, in which growth is severely retarded and mortality high.

In poultry, in addition to the effect on growth, feathering is poor and kidney damage may occur. Hens deprived of the vitamin remain healthy but hatchability is adversely affected.

On vitamin B_{12} deficient diets, baby pigs grow poorly and show uncoordination of the hind legs. In older pigs, dermatitis, a rough coat and sub-optimal growth result. Intestinal synthesis of the vitamin occurs in pigs and poultry. Organisms which synthesise vitamin B_{12} have been isolated from poultry excreta and this fact has an important practical bearing on poultry housed with access to litter, where a majority, if not all, of the vitamin requirements can be obtained from the litter.

Vitamin B_{12} and a number of biologically inactive vitamin B_{12} analogues are synthesised by micro-organisms in the rumen and, in spite of poor absorption of the vitamin from the intestine, the ruminant normally obtains an adequate amount of the vitamin from this source. However, if levels of cobalt in the diet are low, a deficiency of the vitamin can arise and cause reduced appetite, emaciation and anaemia (see p. 106). If cobalt levels are adequate, then except with very young ruminant animals, a dietary source of the vitamin is not essential.

OTHER GROWTH FACTORS INCLUDED IN THE VITAMIN B COMPLEX

A number of other chemical substances of an organic nature have been included in the vitamin B complex. These include inositol, orotic acid, lipoic acid, rutin, carnitine and pangamic acid, but it is doubtful if these compounds have much practical significance in the nutrition of farm animals.

Other factors which appear to be of some significance in poultry nutrition include the grass factor and fish factor. The evidence for these has been obtained from growth responses in feeding trials and from hatchability studies.

VITAMIN C

Chemical nature

Vitamin C is chemically known as L-ascorbic acid and has the following formula:

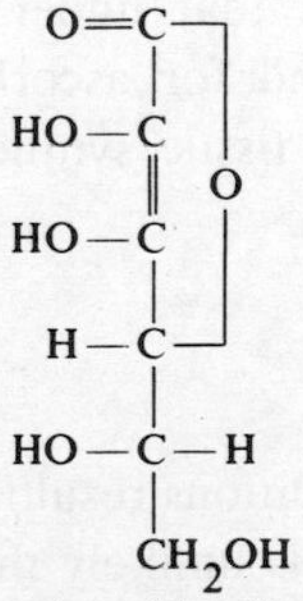

$$
\begin{array}{l}
O=C- \\
HO-C \\
\quad\quad\quad\parallel\quad O \\
HO-C \\
H-C- \\
HO-C-H \\
\quad\quad CH_2OH
\end{array}
$$

L-Ascorbic acid

The vitamin is a colourless, crystalline, water-soluble compound having acidic and strong reducing properties. It is heat-stable in acid solution but is readily decomposed in the presence of alkali. The destruction of the vitamin is accelerated by exposure to light.

Sources

Well-known sources of this vitamin are citrus fruits and green leafy vegetables. Synthetic ascorbic acid is available commercially.

Metabolism

Ascorbic acid and its oxidation product, dehydroascorbic acid, play an important part in various oxidation-reduction mechanisms in living cells. The vitamin is necessary for the maintenance of normal collagen metabolism. It also plays an important role in the transport of iron ions from transferrin, found in the plasma, to ferritin which acts as a store of iron in the bone marrow, liver and spleen. The vitamin is required in the diet of only a few vertebrates—man, other primates, the guinea pig, the red vented bulbul bird and the fruit-eating bat (both native to India) and certain fishes. Some insects and other invertebrates also require a dietary source of vitamin C. Other species synthesise the vitamin from glucose, via glucuronic acid and gulonic acid lactone; the enzyme L-gulonolactone oxidase is required for the synthesis and species requiring ascorbic acid are genetically deficient in this enzyme.

Deficiency symptoms

The classic condition in man arising from a deficiency of vitamin C is scurvy. This is characterised by oedema, emaciation and diarrhoea. Failure in collagen formation results in structural defects in bone, teeth, cartilage, connective tissues and muscles.

Since farm animals can synthesise this vitamin, deficiency symptoms will normally not arise. However, it has been suggested that under certain conditions, e.g. climatic stress in poultry, the demand for ascorbic acid becomes greater than can be provided for by normal tissue synthesis and a dietary supplement may be beneficial.

HYPERVITAMINOSIS

Hypervitaminosis is the name given to pathological conditions resulting from an overdose of vitamins. Under 'natural' conditions it is unlikely that farm animals will receive excessive doses of vitamins, although where synthetic vitamins are added to diets there is always the risk that abnormally large amounts may be ingested if errors are made during mixing. There is experimental evidence that toxic symptoms can occur if animals are given excessive quantities of vitamin A or D.

Clinical signs of hypervitaminosis A in young chicks kept under experimental conditions and given very high doses of vitamin A include loss of appetite, poor growth, diarrhoea, encrustation around the mouth and reddening of the eyelids. In pigs toxic symptoms include rough coat, scaly skin, hyperirritability, haemorrhages over the limbs and abdomen, periodic tremors and death.

Excessive intakes of vitamin D cause abnormally high levels of calcium and phosphorus in the blood, which result in the deposition of calcium salts in the arteries and organs. Symptoms of hypervitaminosis D have been noted in cattle and calves.

Depression in growth and anaemia caused by excessive doses of menadione (vitamin K_3) have been reported.

FURTHER READING

B. M. Babior (ed.) 1975. *Cobalamin*. John Wiley, New York

J. N. Counsell and D. H. Hornig (eds) 1982. *Vitamin C*. Applied Science Publishers, Barking, England.

R. S. Harris *et al.* (eds). *Vitamins and Hormones: Advances in Research and Applications* (annual vols since 1942). Academic Press, New York.

L. J. Machlin (ed.) 1980. *Vitamin E. A Comprehensive Treatise*. Marcel Dekker, New York.

L. J. Machlin (ed.) 1984. *Handbook of Vitamins*. Marcel Dekker, New York.

D. W. Martin, P. A. Mayes and V. W. Rodwell 1983. *Harper's Review of Biochemistry*. Lange Medical Publ., Los Altos, USA.

A. W. Norman. 1979. *Vitamin D*. Academic Press, New York.

6

Minerals

Although most of the naturally occurring mineral elements are found in animal tissues, many of these are thought to be present merely because they are constituents of the animal's food and may not have an essential function in the animal's metabolism. The term 'essential mineral element' is restricted to a mineral element which has been proved to have a metabolic role in the body. Before an element can be classed as essential it is generally considered necessary to prove that purified diets lacking the element cause deficiency symptoms in animals, and that those symptoms can be eradicated or prevented by adding the element to the experimental diet. Most research on mineral nutrition has been carried out in this way, but unfortunately some of the mineral elements required by animals for normal health and growth are needed in such minute amounts that the construction of deficient diets is often difficult to achieve. In such studies it is necessary not only to monitor food and water supplies, but also to ensure that animals do not obtain the element under investigation from cages, troughs, attendants, or dust in the atmosphere.

Until 1950, 13 mineral elements were classified as essential; these comprised the major elements; calcium, phosphorus, potassium, sodium, chlorine, sulphur, magnesium, and the micro- or trace elements: iron, iodine, copper, manganese, zinc and cobalt. In 1953 molybdenum was added, followed in 1957 by selenium and in 1959 by chromium.

In recent years there has been a rapid increase in the number of trace elements known to be essential and in addition to the nine quoted above, the elements fluorine, silicon, vanadium, tin, arsenic and nickel have been added to the list of dietary essentials. It is highly likely that this list is incomplete and it has been suggested that as many as 40 or more elements may have metabolic roles in mammalian tissues. Fortunately, many of these trace elements, especially those of more recent discovery, are required in such minute quantities, or are so widely distributed in animal foods, that deficiencies are likely to be extremely rare under normal practical conditions.

The classification of the essential minerals into major elements and trace elements depends upon their concentration in the animal. Normally trace elements are present in the animal body in a concentration not greater than 50 mg/kg. Those essential mineral elements which are of particular nutritional importance together with their approximate concentrations in the animal body are shown in Table 6.1.

Nearly all the essential mineral elements, both major and trace, are believed to have one or more catalytic functions in the cell. Some mineral elements are firmly bound to the proteins of enzymes, while others are present in prosthetic groups in chelated form. A chelate is a cyclic compound which is formed between an organic molecule and a metallic ion, the latter being held within the organic molecule as if by a claw ('chelate' is derived from the Greek work meaning 'claw'). Examples of naturally occurring chelates are the chlorophylls, cytochromes, haemoglobin and vitamin B_{12}. One of the most potent chelating agents known is the synthetic compound ethylenediaminetetraacetic acid (EDTA) which has the property of forming stable chelates with heavy metals. The addition of chelating agents such as EDTA to poultry diets may in some cases improve the availability of the mineral element. This has been shown to be the case with diets rich in phytic acid where availability of the elements zinc and manganese was increased.

Elements such as sodium, potassium and chlorine have primarily an electrochemical function and are concerned with the maintenance of acid-base balance and the osmotic control of water distribution within the body. Some

TABLE 6.1 Nutritionally important essential mineral elements and their approximate concentration in the animal

Major		Trace	
Element	g/kg	Element	mg/kg
Calcium	15	Iron	20–80
Phosphorus	10	Zinc	10–50
Potassium	2	Copper	1–5
Sodium	1.6	Molybdenum	1–4
Chlorine	1.1	Selenium	1–2
Sulphur	1.5	Iodine	0.3–0.6
Magnesium	0.4	Manganese	0.2–0.5
		Cobalt	0.02–0.1

elements have a structural role, for example calcium and phosphorus are essential components of the skeleton and sulphur is necessary for the synthesis of structural proteins. it is not uncommon for an element to have a number of different roles; for example, magnesium functions catalytically, electrochemically and structurally.

A number of elements have unique functions. Iron is important as a constituent of haem which is an essential part of a number of haemochromagens important in respiration. Cobalt is a component of vitamin B_{12}, and iodine forms part of the hormone thyroxine.

Some elements, for example calcium and molybdenum, may interfere with the absorption and activity of other elements. The interaction of minerals with each other is an important factor in animal nutrition, and an imbalance of mineral elements—as distinct from a simple deficiency—is important in the aetiology of certain nutritional disorders of farm animals. The use of radioactive isotopes in recent years has advanced our knowledge of mineral nutrition, although there are many nutritional diseases associated with minerals whose exact causes are still unknown.

Although we have been considering the essential role of minerals in animal nutrition it is important to realise that many are toxic—causing illness or death—if given to the animal in excessive quantities. This is particularly true of copper, selenium, molybdenum, fluorine, vanadium and arsenic Copper is a cumulative poison, the animal body being unable to excrete it efficiently; small amounts of the element given in excess of the animal's daily needs will, in time, produce toxic symptoms. This also applies to the element fluorine. Supplementation of any diet with minerals should always be carried out with great care and the indiscriminate use of trace elements in particular must be avoided.

MAJOR ELEMENTS

CALCIUM

Calcium is the most abundant mineral element in the animal body. It is an important constituent of the skeleton and teeth, in which about 99 per cent of the total body calcium is found; in addition it is an essential constituent of living cells and tissue fluids. Calcium is essential for the activity of a number of enzyme systems including those necessary for the transmission of nerve impulses and for the contractile properties of muscle. It is also concerned in the coagulation of blood. In blood the element occurs in the plasma. The plasma of mammals usually contains from 80 to 120 mg calcium/l, although that of laying hens contains more (between 300 and 400 mg/l).

Composition of bone

Bone is highly complex in structure, the dry matter consisting of approximately 460 g mineral matter/kg, 360 g protein/kg and 180 g fat/kg. The composition varies, however, according to the age and nutritional status of the animal. Calcium and phosphorus are the two most abundant mineral elements in bone; they are combined in a form similar to that found in the mineral hydroxyapatite $3Ca_3(PO_4)_2.Ca(OH)_2$. Bone ash contains approximately 360 g calcium/kg, 170 g phosphorus/kg and 10 g magnesium/kg.

The skeleton is not a stable unit in the chemical sense, since large amounts of the calcium and phosphorus in bone can be liberated by resorption. This takes place particularly during lactation and egg production, although the exchange of calcium and phosphorus between bones and soft tissue is always a continuous process. Resorption of calcium is controlled by the action of the parathyroid gland. If animals are fed on a low calcium diet, the parathyroid gland is stimulated and the hormone produced causes resorption of bone, liberating calcium to meet the requirements of the animal. Since calcium is combined with phosphorus in bone, the phosphorus is also liberated and excreted by the animal.

The parathyroid hormone also plays an important role in regulating the amount of the calcium absorbed from the intestine by influencing the production of 1,25-dihydroxycholecalciferol, a derivative of vitamin D, which is concerned with the formation of calcium-binding protein (see p. 66).

Deficiency symptoms

If calcium is deficient in the diet of young growing animals, then satisfactory bone formation cannot occur and the condition known as rickets is produced. The symptoms of rickets are misshapen bones, enlargement of the joints, lameness and stiffness. In adult animals calcium deficiency produces osteomalacia, in which the calcium in the bone is withdrawn and not replaced. In osteomalacia the bones become weak and are easily broken. In hens, deficiency symptoms are soft beak and bones, retarded growth and bowed legs; the eggs have thin shells and egg production may be reduced. The symptoms described above for rickets and osteomalacia are not specific for calcium and can also be produced by a deficiency of phosphorus, or an abnormal calcium : phosphorus ratio, or a deficiency of vitamin D (p. 64). It is obvious that a number of factors can be responsible for subnormal calcification.

Milk fever (parturient paresis) is a condition which most commonly occurs in dairy cows shortly after calving. It is characterised by a lowering of the serum calcium level, muscular spasms, and in extreme cases paralysis and unconsciousness. The exact cause of hypocalcaemia associated with milk fever is obscure, but it is generally considered that, with the onset of lactation, the parathyroid gland is unable to respond rapidly enough to increase calcium absorption from the intestine to meet the extra demand. Normal levels of blood calcium can be restored by intravenous injections of calcium gluconate, but this may not always have a permanent effect. It has been shown that avoiding excessive intakes of calcium while maintaining adequate dietary levels of phosphorus during the dry period, reduces the incidence of milk fever. Administration of large doses of vitamin D_3 for a short period prior to parturition has also proved beneficial.

Sources of calcium

Milk and green leafy crops, especially legumes, are good sources of calcium; cereals and roots are poor sources. Animal by-products containing bone, such as fish meal, and meat and bone meal, are excellent sources. Mineral supplements which are frequently given to farm animals, especially lactating animals and laying hens, include ground limestone, steamed bone flour and dicalcium phosphate. If rock calcium phosphate is given to animals it is important to ensure that fluorine is absent, as otherwise this supplement may be toxic.

Calcium : phosphorus ratio

In giving calcium supplements to animals it is important to consider the calcium : phosphorus ratio of the diet, since an abnormal ratio may be as harmful as a deficiency of either element in the diet. The calcium:phosphorus ratio considered most suitable for farm animals other than poultry is generally within the range 1:1 to 2:1, although there is evidence which suggests that ruminants can tolerate rather higher ratios. The proportion of calcium for laying hens is much larger, since they require great amounts of the element for eggshell production. The calcium is usually given to laying hens as ground limestone mixed with the diet, or alternatively calcareous grit may be given *ad libitum*.

PHOSPHORUS

Phosphorus has more known functions in the animal body than any other mineral element. The close association of phosphorus with calcium in bone has already been mentioned. In addition it occurs in phosphoproteins, nucleic acids and phospholipids. The element plays a vital role in energy metabolism in the formation of sugar-phosphates and adenosine di- and tri-phosphates (see Ch. 9). The importance of vitamin D in calcium and phosphorus metabolism has already been discussed in the previous chapter. The phosphorus content of the animal body is rather smaller than the calcium content. Whereas 99 per cent of the calcium found in the body occurs in the bones and teeth, the proportion of the phosphorus in these structures is about 80 per cent of the total.

Deficiency symptoms

Extensive areas of phosphorus-deficient soils occur throughout the world, especially in tropical and subtropical areas, and a deficiency of this element can be regarded as the most widespread and economically important of all the mineral disabilities affecting grazing livestock.

Like calcium, phosphorus is required for bone formation and a deficiency can also cause rickets or osteomalacia. 'Pica' or depraved appetite has been noted in cattle when there is a deficiency of phosphorus in their diet; the affected animals have abnormal appetites and chew wood, bones, rags and other foreign materials. Pica is not specifically a sign of phosphorus deficiency since it may be caused by other factors. Evidence of phosphorus deficiency may be obtained from an analysis of blood serum, which would show a phosphorus content lower than normal. In chronic phosphorus deficiency animals may have stiff joints and muscular weakness.

Low dietary intakes of phosphorus have also been associated with poor fertility, apparent dysfunction of the ovaries causing inhibition, depression or irregularity of oestrus. There are many examples, throughout the world, of phosphorus supplementation increasing fertility in grazing cattle. In cows a deficiency of this element may also reduce milk yield.

Subnormal growth in young animals and low liveweight gains in mature animals are characteristic symptoms of phosphorus deficiency in all species. Phosphorus deficiency is usually more common in cattle than in sheep, as the latter tend to have more selective grazing habits and choose the growing parts of plants which happen to be richer in phosphorus.

Sources of phosphorus

Milk, cereal grains, fish meal and meat products containing bone are good sources of phosphorus; the content in hays and straws is generally very low. Considerable attention has been paid to the availability of phosphorus. Much of the element present in cereal grains is in the form of phytates, which are salts of phytic acid, a phosphoric acid derivative:

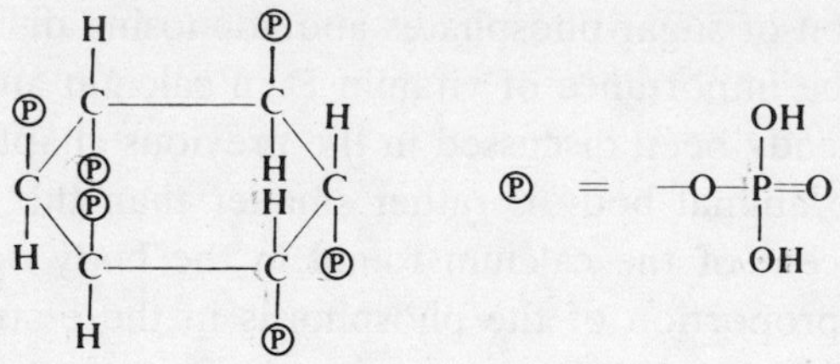

Phytic acid

Insoluble calcium and magnesium phytates occur in cereals and other plant products. Experiments with chicks have shown that the phosphorus of calcium phytate is utilised only 10 per cent as effectively as disodium phosphate. In studies with laying hens, phytate phosphorus was utilised about half as well as dicalcium phosphate. In pigs some of the phytate phosphorus is made available in the stomach by the action of plant phytase enzymes present in the food. It has also been shown with sheep that hydrolysis of phytates by bacterial phytases occurs in the rumen. Phytate phosphorus appears therefore to be utilised by ruminants as readily as other forms of phosphorus, although studies using radioactive isotopes indicate that the availability of phytate phosphorus may range from 0.33 to 0.90.

POTASSIUM

Potassium plays a very important part, along with sodium, chlorine and bicarbonate ions, in the osmotic regulation of the body fluids and in the acid-base balance in the animal. Whereas sodium is the main inorganic cation of extracellular tissue fluids, potassium functions principally as the cation of cells. Potassium plays an important part in nerve and muscle excitability, and is also concerned in carbohydrate metabolism.

Deficiency symptoms

The potassium content of plants is generally very high, that of grass, for example, being frequently above 25 g/kg DM, so that it is normally ingested by animals in larger amounts than any other element. Consequently, potassium deficiency is rare in farm animals kept under natural conditions. There are, however, certain areas in the world where soil potassium levels are naturally low. Such areas occur in Brazil, Panama and Uganda and it is suggested that in these tropical regions, potassium deficiencies may arise in grazing animals especially at the end of the long dry season when potassium levels in the mature herbage are low.

Deficiency symptoms have been produced in chicks given experimental diets low in potassium. They included retarded growth, weakness and tetany, followed by death. Deficiency symptoms, including severe paralysis, have also been recorded for calves given synthetic milk diets low in potassium.

A dietary excess of potassium is normally rapidly excreted from the body, chiefly in the urine. Some research workers believe that high intakes of the element may interfere with the absorption and metabolism of magnesium in the animal, which may be an important factor in the aetiology of hypo-magnesaemic tetany.

SODIUM

Most of the sodium of the animal body is present in the soft tissues and body fluids. Like potassium, sodium is concerned with the acid-base balance and osmotic regulation of the body fluids. Sodium is the chief cation of blood plasma and other extra-cellular fluids of the body. The sodium concentration within the cells is relatively low, the element being replaced largely by potassium and magnesium. Sodium also plays a role in the transmission of nerve impulses and in the absorption of sugars and amino acids from the digestive tract (see p. 139).

Much of the sodium is ingested in the form of sodium chloride (common salt) and it is also mainly in this form that the element is excreted from the body. There is evidence that sodium rather than chlorine is the chief limiting factor in salt-deficient diets of sheep and cows.

Deficiency symptoms

Sodium deficiency in animals occurs in many parts of the world, but especially in the tropical areas of Africa and the arid inland areas of Australia where pastures contain very low concentrations of the element. A deficiency of sodium in the diet leads to a lowering of the osmotic pressure which results in dehydration of the body. Symptoms of sodium deficiency include poor growth and reduced utilisation of digested proteins and energy. In hens, egg production and growth are adversely affected. Rats given experimental diets low in sodium had eye lesions and reproductive disturbances, and eventually died.

Sources of sodium

Most foods of vegetable origin have comparatively low sodium contents; animal products, especially meat meals and foods of marine origin, are richer sources. The commonest mineral supplement given to farm animals is common salt.

CHLORINE

Chlorine is associated with sodium and potassium in acid-base relationships and osmotic regulation. Chlorine also plays an important part in the gastric secretion, where it occurs as hydrochloric acid as well as chloride salts. Chlorine is excreted from the body in the urine and is also lost from the body, along with sodium and potassium, in perspiration.

A dietary deficiency of chlorine may lead to an abnormal increase of the alkali reserve of the blood (alkalosis) caused by an excess of bicarbonate, since inadequate levels of chlorine in the body are partly compensated for by increases in bicarbonate. Experiments with rats on chlorine-deficient diets showed that growth was retarded, but no other symptoms developed.

Sources of chlorine

With the the exception of fish and meat meals, the chlorine content of most foods is comparatively low. The chlorine content of pasture grass varies widely and figures ranging from 3 to 25 g/kg DM have been reported. The main source of this element for most animals is common salt.

Since plants tend to be low in both sodium and chlorine, it is the usual practice to give common salt to herbivores. Unless salt is available deficiencies are likely to occur in both cattle and sheep. Experiments carried out in the USA with dairy cows on salt-deficient diets showed that animals did not exhibit immediate ill effects, but eventually appetite declined, with subsequent loss in weight and lowered milk production. The addition of salt to the diet produced an immediate cure.

Salt is also important in the diet of hens, and it is known to counteract feather picking and cannibalism. Salt is generally given to pigs on vegetable diets, but if fish meal is given the need for added salt is reduced. Swill can also be a rich source of salt, although the product is very variable and can contain excessive amounts. Too much salt in the diet is definitely harmful and causes excessive thirst, muscular weakness and oedema. Salt poisoning is quite common in pigs and poultry, especially where fresh drinking water is limited. When the concentration of salt in the diet of hens exceeds 40 g/kg DM and the supply of drinking water is limited, then death may occur. Hens can tolerate larger amounts of salt if plenty of water is available. Chicks cannot tolerate salt as well as adults, and 20 g/kg DM in the diet should be regarded as the absolute maximum. This value should also not be exceeded in the diets of pigs. Turkey poults are even less tolerant, and 10 g/kg of salt in the diet should not be exceeded.

SULPHUR

Most of the sulphur in the animal body occurs in proteins containing the amino acids cystine, cysteine and methionine. The two vitamins, biotin and thiamin, the hormone, insulin, and the important metabolite, coenzyme A also contain sulphur. Only a small amount of sulphur is present in the body in inorganic form, though sulphates are known to occur in the blood in small quantities. Wool is rich in cystine and contains about 4 per cent of sulphur.

Traditionally, little attention has been paid to the importance of sulphur in animal nutrition since the intake of this element is mainly in the form of protein, and a deficiency of sulphur would indicate a protein deficiency. In recent years, however, with the increasing use of urea as a partial nitrogen replacement for protein nitrogen, it has been realised that the amount of sulphur present in the diet may be the limiting factor for the synthesis in the rumen of cystine, cysteine and methionine. Under these conditions the addition of suphur to urea-containing rations may be beneficial. There is evidence that sulphate (as sodium sulphate) can be used by ruminal micro-organisms more efficiently than elemental sulphur.

It seems likely that in the future sulphur will be of increasing importance in animal nutrition, especially in areas of intensive livestock production where sulphur in soils is not replaced regularly by fertiliser application. Symptoms of sulphur deficiency in crops have been reported from a wide range of geographical locations including the UK, Western Australia, Nigeria, Canada and USA.

Inorganic sulphur seems to be of less practical importance for monogastric animals, although studies with pigs, chicks and poults have indicated that inorganic sulphate can have a sparing effect on their dietary sulphur amino acid requirements.

MAGNESIUM

Magnesium is closely associated with calcium and phosphorus. About 70 per cent of the total magnesium is found in the skeleton, the remainder being distributed in the soft tissues and fluids. Magnesium is the commonest enzyme activator and is particularly important in activating phosphate transferases, decarboxylases and acyl transferases.

Deficiency symptoms

Symptoms due to a simple deficiency of magnesium in the diet have been reported for a number of animals. In rats fed on purified diets the symptoms include increased nervous irritability and convulsions. Experiments carried out on calves reared on low-magnesium milk diets resulted in low serum magnesium levels, depleted bone magnesium, tetany and death. The condition is not uncommon in milk-fed calves about 50–70 days old.

In adult ruminants a condition known as hypomagnesaemic tetany associated with low blood levels of magnesium (hypomagnesaemia) has been recognised since the early thirties. A great deal of attention has been given to this condition in recent years, since it is widespread and the death rate is high.

Hypomagnesaemic tetany has been known under a variety of names including magnesium tetany, lactation tetany and grass staggers, but most of these terms have been discarded because the disease is not always associated with lactation nor with grazing animals. The condition can affect stall-fed dairy cattle, hill cattle, bullocks and cattle at grass as well as sheep. There is some evidence of a breed susceptibility in the United Kingdom where the condition appears to be more common in Ayrshires and least common in Jersey animals. Most cases occur in grazing animals, and, in Europe and North America, the trouble is particularly common in the spring when the animals are turned out on to young, succulent pasture. Because the tetany can develop within a day or two of animals being turned out to graze, the condition has been referred to as the acute form. In this acute type, blood magnesium levels fall so rapidly that the reserves of magnesium within the body cannot be mobilised rapidly enough. In the chronic form of the disease plasma magnesium levels fall over a period of time to low concentrations. This type is not uncommon in suckler herds. Clinical signs of the disease are often brought on by 'stress' factors such as cold, wet and windy weather. In adult animals bone magnesium is not as readily available as the young calf's.

In New Zealand, where cows are pastured throughout the year, hypomagnesaemic tetany occurs most frequently in late winter and early spring. In Australia, a high incidence of the disease has been associated with periods of rapid winter growth of pastures.

The normal magnesium content of blood serum in cattle is within the range of 17 to 40 mg magnesium/l blood serum, but levels below 17 frequently occur without clinical symptoms of disease. Tetany is usually preceded by a fall in blood serum magnesium to about 5 mg/l. Subcutaneous injection of magnesium sulphate, or preferably magnesium lactate, can generally be expected to cure the animal if given early, but in practice this is sometimes difficult. Treatment of this kind is not a permanent cure and oral treatment with magnesium oxide, as described below, should be started immediately. Typical symptoms of tetany are nervousness, tremors, twitching of the facial muscles, staggering gait and convulsions.

The exact cause of hypomagnesaemic tetany in ruminant animals is unknown, although a dietary deficiency of magnesium may be a contributory factor. Some research workers consider the condition to be caused by a cation-anion imbalance in the diet and there is evidence of a positive relationship between tetany and heavy dressings of pasture with nitrogenous and potassic fertilisers.

Recent work using radioactive magnesium in tracer studies indicates that the magnesium present in food is poorly absorbed from the alimentary canal; in some cases only 50 g/kg of the herbage magnesium can be utilised by the ruminant. Why this is so in ruminants is not known. Since adult animals have only very small readily available reserves of body magnesium they are dependent upon a regular dietary supply.

Although the exact cause of hypomagnesaemia is still uncertain, the primary factor would appear to be inadequate absorption of magnesium from the digestive tract. A high degree of success in preventing hypomagnesaemia may be obtained by increasing the magnesium intake. This can be effected by feeding with magnesium-rich mineral mixtures, or alternatively by increasing the magnesium content of pasture by the application of magnesium fertilisers.

Sources of magnesium

Wheat bran, dried yeast and most vegetable protein concentrates, especially cottonseed cake and linseed cake, are good sources of magnesium. Clovers are usually richer in magnesium than grasses, although the magnesium content of forage crops varies widely. The mineral supplement most frequently used is magnesium oxide, which is sold commercially as calcined magnesite. When hypomagnesaemic tetany is likely to occur it is generally considered that about 50 g of magnesium oxide should be given to cows per head per day as a prophylactic measure. The daily prophylactic dose for calves is 7 to 15 g of the oxide, while that for lactating ewes is about 7 g. The mineral supplement can be given mixed with the concentrate ration. Alternatively a mixture of magnesium acetate solution and molasses can be used, which is frequently made available on a free-choice basis from ball feeders placed in the field.

TRACE ELEMENTS

IRON

More than 90 per cent of the iron in the body is combined with proteins, the most important being haemoglobin, which contains about 3.4 g/kg of the element. Iron also occurs in blood serum in a protein called transferrin which is concerned with the transport of iron from one part of the body to another. Ferritin, a protein containing up to 200 g/kg of the element, is present in the spleen, liver, kidney and bone marrow and provides a form of storage for iron. Haemosiderin is a similar storage compound which may contain up to 350 g/kg of the element. Iron is also a component of many enzymes, including cytochromes and certain flavoproteins.

Deficiency symptoms

Since more than half the iron present in the body occurs as haemoglobin, a dietary deficiency of iron would clearly be expected to affect the formation of this compound. The red blood corpuscles contain haemoglobin, and these cells are continually being produced in the bone marrow to replace those red cells destroyed in the animal body as a result of catabolism. Although the haemoglobin molecule is destroyed in the catabolism of these red blood corpuscles, the iron liberated is made use of in the resynthesis of haemoglobin, and because of this the daily requirement of iron by a healthy animal is usually small. If the need for iron increases, as it would after prolonged haemorrhage or during pregnancy, then haemoglobin synthesis may be affected and anaemia will result. Anaemia due to iron deficiency occurs most commonly in rapidly growing sucklings, since the iron content of milk is usually very low; this frequently occurs in pigs housed in pens without access to soil or pasture. The young piglet must retain about 7 mg of iron per day to grow at a normal rate without becoming anaemic; since the sow's milk only provides about 1 mg per day there is an additional requirement of about 6 mg. This should be provided by dosing or injection with iron salts. Giving additional iron to the lactating females does not prevent anaemia occurring in the young piglets, as the iron content of milk is not increased by feeding. Anaemia in piglets is characterised by poor appetite and growth. Breathing becomes laboured and spasmodic—hence the descriptive term 'thumps' for this condition.

Iron deficiency anaemia is not common in lambs and calves because in practice it is unusual to restrict them to a milk diet without supplementary feeding. It does, however, sometimes occur in laying hens, since egg production represents a considerable drain on the body reserves.

Sources of iron

Iron is widely distributed in foods. Good sources of the element are green leafy materials, most leguminous plants and seed coats. Feeds of animal origin, such as meat, blood and fish meals are excellent sources of iron. As mentioned previously milk is a poor source of the element.

It has long been held that the absorption of iron is to a large extent independent of the dietary source, the efficiency of absorption being increased during periods of iron need and decreased during periods of iron overload. The mechanisms whereby the body carries this out are not fully understood. A number of theories have been advanced and one of these, the 'mucosal block' theory, propounded in 1943, is still widely held by many to explain the mechanism. According to this theory, the mucosal cells of the gastro-intestinal tract absorb iron and convert it into ferritin, and when the cells become physiologically saturated with ferritin, further absorption is impeded until the iron is released from ferritin and transferred to plasma. Another more recent theory proposes that a factor in gastric juice that increases absorption is the main regulator of iron uptake.

The adult's need for iron is normally low, as the iron produced from the destruction of haemoglobin is made available for haemoglobin regeneration, only about 10 per cent of the element escaping from this cycle.

Iron toxicity

Iron toxicity is not a common problem in farm animals, but it can result from prolonged oral administration of the element. Chronic iron toxicity results in alimentary disturbances, reduced growth and phosphorus deficiency.

COPPER

Evidence that copper is a dietary essential was obtained in 1924, when experiments with rats showed that copper was necessary for haemoglobin formation. Although copper is not actually a constituent of haemoglobin it is present in certain other plasma proteins such as ceruloplasmin which are concerned with the release of iron from the cells into the plasma. A deficiency of copper impairs the animal's ability to absorb iron, mobilise it from the tissues and utilise it in haemoglobin synthesis. Copper is also a component of other proteins in blood. One of these, erythrocuprein, occurs in erythrocytes where it plays a role in oxygen metabolism. The element is also known to play a vital role in many enzyme systems; for example, it is a component of cytochrome oxidase, which is important in oxidative phosphorylation. The element also occurs in certain pigments, notably turacin, a pigment of feathers. Copper is necessary for the normal pigmentation of hair, fur and wool. It is thought to be present in all body cells, being particularly concentrated in the liver, which acts as the main copper storage organ of the body.

Deficiency symptoms

Since copper performs many functions in the animal body there are a variety of deficiency symptoms. These include anaemia, poor growth, bone disorders, scouring, infertility, depigmentation of hair and wool, gastrointestinal disturbances and lesions in the brain stem and spinal cord. The lesions are associated with muscular incoordination, and occur especially in young lambs. A copper deficiency condition known as 'enzootic ataxia' has been known for some time in Australia. The disorder there is associated with pastures low in copper content (2 to 4 mg/kg DM), and can be prevented by feeding with a copper salt. A similar condition which affects lambs occurs in the United Kingdom and is known as 'swayback'. The signs of swayback range from complete paralysis of the newborn lamb to a swaying staggering gait which affects in particular, the hind limbs. The condition can occur in two forms, one congenital, in which the signs are apparent at birth, and the other, in which the onset of the clinical disease is delayed for several weeks. The congenital form of the condition is irreversible and can only be prevented by ensuring that the ewe receives an adequate level of copper in her diet. Delayed swayback can be prevented or retarded in copper-deficient lambs by parenteral injection of small doses of copper complexes.

Although the dietary level of copper is an important factor in the aetiology of swayback, the condition does not appear to be caused invariably by a simple dietary deficiency of the element. Swayback has been reported to occur on pastures apparently normal or even high (7–15 mg/kg DM) in copper content. One important factor is that the efficiency of absorption of dietary copper is very variable. For example there is about a tenfold variation in the efficiency with which Scottish Blackface ewes absorb copper from autumn pasture (0.012) and from leafy brassicas (0.132). It is also known that genetic factors influence the concentration of copper in the blood, liver and brain of the sheep and that the incidence of swayback can be affected by genotype.

Copper plays an important role in the production of 'crimp' in wool. The element is present in an enzyme which is responsible for the disulphide bridge in two adjacent cysteine molecules. In the absence of the enzyme the protein molecules of the wool do not form this bridge and the wool, which lacks crimp, is referred to as 'stringy' or 'steely' (see p. 316).

Nutritional anaemia resulting from copper deficiency has been produced experimentally in young pigs by diets very low in the element and this type of anaemia could easily arise in such animals fed solely on milk. In older animals copper deficiency is unlikely to occur and copper supplementation of practical rations is generally considered unnecessary. There are, however, certain areas in the world where copper deficiency in cattle occurs. A condition in Australia, known locally as 'falling disease' was found to be related to a progressive degeneration of the myocardium of animals grazing on copper-deficient pastures.

Copper-molybdenum-sulphur interrelations

Certain pastures on calcareous soils in parts of England and Wales have been known for over a hundred years to be associated with a condition in cattle described as 'teart', characterised by unthriftiness and scouring (Plate 6). A similar disorder occurs on reclaimed peat lands in New Zealand, where it is known as 'peat scours'. Molybdenum levels in teart pasture are of the order of 20 to 100 mg/kg DM compared with 0.5 to 3.0 mg/kg DM in normal pastures, and teart was originally regarded as being a straight-forward molybdenosis. In the late 1930s, however, it was demonstrated that feeding with copper sulphate controlled the scouring, and hence a molybdenum-copper relationship was established.

It is now known that the effect of molybdenum is complex, and it is considered that the element exerts its limiting effect on copper retention in the animal only in the presence of sulphur. A mechanism which explains this interaction has recently been suggested. Sulphide is formed by ruminal micro-organisms from dietary sulphate or organic sulphur compounds; the sulphide then reacts with molybdate to form thiomolybdate which in turn combines with copper to form an insoluble copper thiomolybdate ($CuMoS_4$) thereby limiting the absorption of dietary copper. In addition it is considered likely that if thiomolybdate is formed in excess, it may be absorbed from the digestive tract and exert a systemic effect on copper metabolism in the animal.

Sources of copper

Copper is widely distributed in foods and under normal conditions the diet of farm animals is likely to contain adequate amounts. The copper content of crops is related to some extent to the soil copper level, but other factors such as drainage conditions and the herbage species affect the copper content. Seeds and seed by-products are usually rich in copper, but straws contain little. The normal copper content of pasture ranges from about 4 to 8 mg/kg DM. The copper content of milk is low, and hence it is customary when dosing young animals, especially piglets, with an iron salt to include a trace of copper sulphate.

Copper toxicity

It has long been known that copper salts given in excess to animals are toxic. Continuous ingestion of copper in excess of nutritional requirements leads to an accumulation of the element in the body tissues, especially in the liver. Copper can be regarded as a cumulative poison, so that considerable care is required in administering copper salts to animals. The tolerance to copper varies considerably between species. Pigs are highly tolerant and cattle relatively so. On the other hand, sheep are particularly susceptible and chronic copper poisoning has been encountered in housed sheep on concentrate diets containing 40 mg/kg of copper. With diets containing as little as 20–30 mg/kg there is a gradual accumulation of copper in the livers of sheep until the danger level of about 1000 mg/kg (fat-free, DM) is reached. Poisoning has been known to occur in areas where the herbage contains copper of the order of 10 to 20 mg/kg DM and low levels of molybdenum. Chronic copper poisoning results in necrosis of the liver cells, jaundice, loss of appetite and death from hepatic coma. It has been suggested that there may be a genetic variation in animals' susceptibility to copper poisoning. It is unwise to administer copper supplements to sheep unless deficiency conditions are liable to occur—many cases of death due to copper poisoning caused by the indiscriminate use of copper-fortified diets have been reported. Chronic copper poisoning in sheep has occurred under natural conditions in parts of Australia where the copper content of the pasture is high.

COBALT

A number of disorders of cattle and sheep, characterised by emaciation, anaemia and listlessness, have been recognised for many years and have been described as 'pining', 'salt sick', 'bush sickness' and 'wasting disease'. These disorders occur in Europe, Australia, New Zealand and the USA. In the United Kingdom 'pining pastures' occur in many counties and are particularly common in the border counties of England and Scotland.

As early as 1807 Hogg, the Ettrick shepherd, recognised 'pining' or 'vinquish' as being a dietary upset. Pining is associated with a dietary deficiency of cobalt caused by low-concentrations of the element in the soil and pastures. Pining can be prevented in these areas by feeding with small amounts of cobalt.

The physiological function of cobalt was only discovered when vitamin B_{12} was isolated and was shown to contain the element. Cobalt is required by micro-organisms in the rumen for the synthesis of vitamin B_{12}, and if the element is deficient in the diet then the vitamin cannot be produced in the rumen in amounts sufficient to satisfy the animal's requirements, and symptoms of pining occur. Pining is therefore regarded as being due to a deficiency of vitamin B_{12}. There is evidence for this, since injections of vitamin B_{12} into the blood alleviate the condition, whereas cobalt injections have little beneficial effect. Although vitamin B_{12} therapy will prevent pining occurring in ruminant animals, it is more convenient and cheaper in cobalt-deficient areas to supplement the diet with the element, allowing the micro-organisms in the rumen to synthesise the vitamin for subsequent absorption by the host.

When ruminants are confined to cobalt-deficient pastures it may be several months before any manifestations of pine occur because of body reserves of vitamin B_{12} in the liver and kidneys. When these are depleted there is a gradual decrease in appetite with consequent loss of weight followed by muscular wasting, pica, severe anaemia and eventually death. If the deficiency is less severe then a vague unthriftiness, difficult to diagnose, may be the only sign. Deficiency symptoms are likely to occur where levels of cobalt in the pastures are below 0.1 mg/kg DM. Under grazing conditions, lambs are the most sensitive to cobalt deficiency (Plate 7), followed by mature sheep, calves and mature cattle in that order.

Ruminants have a higher requirement for the element than non-ruminants because some of the element is wasted in microbial synthesis of organic compounds with no physiological activity in the host's tissues. Furthermore, vitamin B_{12} is poorly absorbed from the digestive tract of ruminants, the availability in some cases being as low as 0.03. The ruminant has an additional requirement for the vitamin because of its involvement in the metabolism of propionic acid (see p. 170), an important acid absorbed from the rumen.

There is evidence that the intestinal micro-organisms in non-ruminants also can synthesise vitamin B_{12} although in pigs and poultry this synthesis may be insufficient to meet their requirements. It is common practice to include in pig and poultry diets some animal protein food rich in vitamin B_{12} in preference to including a cobalt salt.

Apart from the importance of cobalt as a component of vitamin B_{12}, the element is believed to have other functions in the animal body as an activating ion in certain enzyme reactions.

Sources of cobalt

Most foods contain traces of cobalt. Normal pastures have a cobalt content within the range 100 to 250 μg/kg DM.

Cobalt deficiency in ruminants can be prevented by dosing the animals with cobalt sulphate, although this form of treatment has to be repeated at short intervals, or by allowing the animals access to cobalt-containing salt licks. A continuous supply from a single dose can be obtained by giving a cobalt bullet containing 900 g cobaltic oxide/kg; the bullet remains in the recticulum and slowly releases the element over a long period of time. Some of this cobalt is not utilised by the animal and is excreted, and this of course has the effect of improving the cobalt concentration of the pasture. Alternatively, deficient pastures can be top-dressed with small amounts of cobalt sulphate (about 2 kg/ha).

Cobalt toxicity

Although an excess of cobalt can be toxic to animals, there is a wide margin of safety between the nutritional requirement and the toxic level. Cobalt toxicosis is extremely unlikely to occur under practical farming conditions. Unlike copper, cobalt is poorly retained by the body tissues and an excess of the element is soon excreted. The toxic level of cobalt for cattle is 1 mg cobalt/kg body weight daily. Sheep are less susceptible to cobalt toxicosis than cattle and have been shown to tolerate levels up to 3.5 mg/kg.

IODINE

The concentration of iodine present in the animal body is very small and in the adult is usually less than 600 μg/kg. Although the element is distributed throughout the tissues and secretions its only known role is in the synthesis of the two hormones, triiodothyronine and tetraiodothyronine (thyroxine) produced in the thyroid gland (see p. 43).

The element also occurs in the gland as monoiodotyrosine and diiodotyrosine which are intermediates in the biosynthesis of the hormones from the amino acid tyrosine. The two hormones are stored in the thyroid gland as components of the protein thyroglobulin which releases the hormones into the blood capillaries as required.

The thyroid hormones accelerate reactions in most organs and tissues in the body, thus increasing the basal metabolic rate, accelerating growth, and increasing the oxygen consumption of the whole organism.

Deficiency symptoms

When the diet contains insufficient amounts of iodine the production of thyroxine is decreased. The main indication of such a deficiency is an enlargement of the thyroid gland, termed endemic goitre. The thyroid being situated in the neck, the deficiency condition in farm animals manifests itself as a swelling of the neck, 'big neck'. Reproductive failure is one of the most outstanding consequences of reduced thyroid function; breeding animals deficient in iodine give birth to hairless, weak or dead young.

A dietary deficiency of iodine is not the sole cause of goitre: it is known that certain foods contain goitrogenic compounds and cause goitre in animals if given in large amounts. These foods include most members of the *Brassica* family, especially kale, cabbage and rape, and also soya beans, linseed, peas and groundnuts. Goitrogens have been reported in milk of cows fed on goitrogenic plants. A goitrogen present in swede seeds has been identified as L-5-vinyl-2-oxazolidine-2-thione (goitrin) which acts by inhibiting the iodination of tyrosine. Thiocyanate, which may also be present in members of the *Brassica* family, is known to be goitrogenic and may be produced in the tissues from a cyanogenetic glycoside present in some foods. Goitrogenic activity of the thiocyanate type is prevented by supplying adequate iodine in the diet.

Sources of iodine

Iodine occurs in traces in most foods and is present mainly as inorganic iodide in which form it is absorbed from the digestive tract. The richest sources of this element are foods of marine origin, and values as high as 6 g/kg DM have been reported for some seaweeds; fish meal is also a rich source of the element. The iodine content of land plants is related to the amount of iodine present in the soil, and consequently wide variations occur in similar crops grown in different areas.

In areas where goitre is endemic, precautions are generally taken by supplementing the diet with the element, usually in the form of iodised salt. This contains the element either as sodium or potassium iodide or as sodium iodate.

Iodine toxicity

The minimum toxic dietary level of iodine for calves of 80–112 kg bodyweight has been shown to be about 50 mg/kg, although some experimental animals have been adversely affected at lower levels. Symptoms of toxicity include depressions in weight gain and feed intake. In studies with laying hens, diets with iodine contents of 312–5 000 mg/kg DM resulted in cessation of egg production within the first week at the higher level and reduced egg production at the lower level. The fertility of the eggs produced was not affected but early embryonic death, reduced hatchability and delayed hatching resulted. Pigs seem to be more tolerant of excess iodine and the minimum toxic level is considered to lie between 400 and 800 mg/kg.

MANGANESE

The amount of manganese present in the animal body is extremely small. Most tissues contain traces of the element, the highest concentrations occurring in the bones, liver, kidney, pancreas and pituitary gland. Manganese is important in the animal body as an enzyme activator and resembles magnesium in activating a number of phosphate transferases and decarboxylases.

Deficiency symptoms

Manganese deficiency has been found in ruminants, pigs and poultry. The effects of acute deficiency are similar in all species and include retarded growth, skeletal abnormalities, ataxia of the newborn and reproductive failure. Deficiencies of this element in grazing ruminants are likely to be rare although the reproductive performance of grazing Dorset Horn ewes in Australia was improved by giving manganese over two consecutive years. Low manganese diets for cows and goats have been reported to cause depressed or delayed oestrus and conception, as well as increased abortion. Manganese is an important element in the diet of young chicks, a deficiency leading to perosis or 'slipped tendon', a malformation of the leg bones. It is not however, the only factor involved in the aetiology of this condition, as perosis in young birds may be aggravated by high dietary intakes of calcium and phosphorus. Manganese deficiency in breeding birds reduces hatchability and shell thickness, and causes head retraction in chicks. In pigs lameness is a symptom.

Sources of manganese

The element is widely distributed in foods, and most pastures contain 40–200 mg/kg DM. The manganese content of pastures, however, can vary over a much wider range and in acid conditions may be as high as 500–600 mg/kg DM. Seeds and seed products contain moderate amounts, except for maize which is low in the element. Yeast and most foods of animal origin are also poor sources of manganese. Rich sources are rice bran and wheat offals. Most green foods contain adequate amounts.

Manganese toxicity

There is a wide margin of safety between the toxic dose of manganese and the normal level in foods. Levels as high as 1 g/kg DM in the diet have been given to hens without evidence of toxicity. Growing pigs are less tolerant, levels of 0.5 g/kg DM having been shown to depress appetite and retard growth.

ZINC

Zinc has been found in every tissue in the animal body. The element tends to accumulate in the bones rather than the liver, which is the main storage organ of many of the other trace elements. High concentrations have been found in the skin, hair and wool of animals. Several enzymes in the animal body are known to contain zinc; these include carbonic anhydrase, pancreatic carboxypeptidase, lactate dehydrogenase, alkaline phosphatase and thymidine kinase. In addition zinc is an activator of several enzyme systems.

Deficiency symptoms

Zinc deficiency in pigs is characterised by subnormal growth, depressed appetite, poor feed conversion and parakeratosis. The latter is a reddening of the skin followed by eruptions which develop into scabs. A deficiency of this element is particularly liable to occur in young, intensively housed pigs fed *ad libitum* on a dry diet, though a similar diet given wet may not cause the condition. It is aggravated by high calcium levels in the diet and reduced by decreased calcium and increased phosphorus levels. Gross signs of zinc deficiency in chicks are retarded growth, foot abnormalities (Plate 8), 'frizzled' feathers, parakeratosis and a bone abnormality referred to as the 'swollen hock syndrome'.

Symptoms of zinc deficiency in calves include inflammation of the nose and mouth, stiffness of the joints, swollen feet and parakeratosis. The response of severely zinc-deficient calves to supplemental zinc is rapid and dramatic. Improvements in skin condition are usually noted within 2 to 3 days.

Manifestations of zinc deficiency, responsive to zinc therapy, have been observed in growing and mature cattle in parts of Guyana, Greece, Australia and Scandinavia. As levels in the pasture are apparently comparable with those of other areas the deficiency is believed to be conditioned by some factor in the herbage or general environment.

Sources of zinc

The element is fairly widely distributed. Yeast is a rich source, and zinc is concentrated in the bran and germ of cereal grains. Animal protein by-products such as meat meal and fish meal are usually richer sources of the element than plant protein supplements.

Zinc toxicity

Although cases of zinc poisoning have been reported, most animals have a high tolerance for this element. Excessive amounts of zinc in the diet are known to depress food consumption and may induce copper deficiency.

MOLYBDENUM

The first·indication of an essential metabolic role for molybdenum was obtained in 1953 when it was discovered that xanthine oxidase, important in purine metabolism, was a metallo-enzyme containing molybdenum. Subsequently the element was shown to be a component of two other enzymes, aldehyde oxidase and sulphite oxidase. The biological functions of molybdenum, apart from its reactions with copper previously mentioned (p. 105), are concerned with the formation and activities of these three enzymes. In addition to being a component of xanthine oxidase, molybdenum participates in the reaction of the enzyme with cytochrome C and also facilitates the reduction of cytochrome C by aldehyde oxidase.

Deficiency symptoms

In early studies with rats, low-molybdenum diets resulted in reduced levels of xanthine oxidase, but did not affect growth nor purine metabolism. Similar molydenum-deficient diets have been given to chicks without adverse effect but when tungstate (a molybdenum antagonist) was added, growth was reduced and the chick's ability to oxidise xanthine to uric acid was impaired. These effects were prevented by the addition of molybdenum to the diet. A significant growth response has been obtained in young lambs by the addition of molybdate to a semi-purified diet low in the element. It has been suggested that this growth effect could have arisen indirectly by stimulation of cellulose breakdown by ruminal micro-organisms. Molybdenum deficiency has not been observed under natural conditions in any species.

Molybdenum toxicity

The toxic role of molybdenum in the condition known as 'teart' is described under the element copper (p. 105). All cattle are susceptible to molybdenosis, with milking cows and young animals suffering most. Sheep are less affected and horses are not affected on teart pastures. Scouring and weight loss are the dominant manifestations of the toxicity.

SELENIUM

The nutritional importance of selenium became evident in the 1950s when it was shown that most myopathies in sheep and cattle, and exudative diathesis in chicks could be prevented by supplementing the diet with the element or vitamin E (p. 69). A biochemical role of selenium in the animal body was demonstrated in 1973 when it was discovered that the element was a component of glutathione peroxidase, an enzyme which catalyses the removal of hydrogen peroxide.

In parts of Australia and New Zealand a condition known as 'ill thrift' occurs in lambs, beef cattle and dairy cows at pasture. The clinical signs include loss of weight and sometimes death. Ill thrift can be prevented by selenium treatment with, in some instances, dramatic increases in growth and wool yield. Similar responses with sheep have also been noted in experiments carried out in Scotland, Canada and USA. In hens selenium deficiency reduces hatchability and egg production.

Selenium toxicity

The level of selenium in foods of plant origin is extremely variable depending mainly on the soil conditions under which they are grown. Normal levels of the element in pasture are usually between 100 and 300 μg/kg DM. Some species of plants which grow in seleniferous areas, contain very high levels of selenium. One such plant *Astragalus racemosus*, grown in Wyoming, was reported to contain 14 g selenium/kg DM, while the legume *Neptunia amplexicaulis* grown on a selenized soil in Queensland contained over 4 g/kg DM of the element. Localised seleniferous areas have also been identified in Ireland, Israel, South Africa and the USSR. Selenium is a highly toxic element and a concentration in a dry diet of 5 mg/kg or 500 μg/kg in milk or water may be potentially dangerous to farm animals. Alkali disease and blind staggers are localised names for chronic diseases of animals grazing certain seleniferous areas in the USA. Symptoms include dullness, stiffness of the joints, loss of hair from mane or tail and hoof deformities. Acute poisoning, which results in death from respiratory failure, can arise from sudden exposure to high selenium intakes.

FLUORINE

Fluorine was added to the list of nutritionally essential elements in 1972 when it was shown that growth rates in rats were increased by as much as 30 per cent when fluorine-free diets were supplemented with up to 7.5 mg/kg DM of the element as potassium fluoride. Addition of the trace element to the diet was also found to improve the deposition of incisor pigment. The importance of fluorine in the prevention of dental caries had, however, been well established before 1972.

Most plants have a limited capacity to absorb fluorine from the soil and normal levels in pasture range from about 2 to 16 mg/kg DM. Cereals and other grains usually contain only about 1 to 3 mg/kg DM of the element.

Fluorine is a very toxic element and a level in the diet above 20 mg/kg DM causes fluorosis in which the teeth become pitted and worn until the pulp cavities are exposed; they also become sensitive to cold water. Appetite declines, growth is slow and bone and joint abnormalities also occur. The commonest sources of danger from this element are fluoride-containing water and herbage contaminated by dust from industrial pollution. The chronic fluorosis in cattle, sheep and horses, which occurs in parts of North Africa and is known locally as darmous, is caused by contamination of water and herbage with rock phosphate dusts which contain fluoride.

SILICON

Rats previously fed on specially purified foods showed increased growth rates from the addition to the diet of 500 mg/kg of silicon (as sodium metasilicate).

Similar results have also been obtained with chicks. Silicon is essential for growth and skeletal development in these two species. The element is believed to function as a biological cross-linking agent, possibly as an ether derivative of silicic acid of the type R_1—O—Si—O—R_2. Such bridges are important in the strength, structure and resilience of connective tissue. In silicon deficient rats and chicks, bone abnormalities occur because of a reduction in mucopolysaccharide synthesis in the formation of cartilage. Silicon is also thought to be involved in other processes involving mucopolysaccharides such as the growth and maintenance of the arterial wall and the skin.

Silicon is so widely distributed in the environment and in foods that it is difficult to foresee any deficiencies of this element arising under practical conditions. Whole grasses and cereals may contain as much as 14 to 19 g Si/kg DM, with levels of up to 28 g/kg DM in some range grasses.

Silicon toxicity (silicosis), has long been known as an illness of miners caused by the inhalation of silical particles into the lungs. Under some conditions, part of the silicon present in urine is deposited in the kidney, bladder or urethra to form calculi or uroliths. Silica urolithiasis occurs in grazing wethers in Western Australia and in grazing steers in Western Canada and North Western parts of the USA. Excessive silica in feeds, for example in rice straw, is known to depress organic matter digestibility.

CHROMIUM

Chromium was first shown to be essential for normal glucose utilisation in rats. Mice and rats fed on diets composed of cereals and skim milk and containing 100 μg chromium/kg wet weight were subsequently shown to grow faster if given a supplement of chromium acetate. Chromium may also play a role in lipid and protein synthesis and in serum cholesterol homeostasis. Whether the element is likely to have any practical significance in the nutrition of farm animals remains to be seen. Chromium is not a particularly toxic element and a wide margin of safety exists between the normal amounts ingested and those likely to produce deleterious effects. Levels of 50 mg Cr/kg DM in the diet have caused growth depression and liver and kidney damage in rats.

VANADIUM

Vanadium deficiency has been demonstrated in rats and chicks. Deficiency symptoms included impaired growth and reproduction, and disturbed lipid metabolism. In chicks consuming diets containing less than 10 μg vanadium/kg DM, growth of wing and tail feathers was significantly reduced. Subsequent studies with chicks demonstrated a significant growth response when diets containing 30 μg/kg DM of the element were increased in vanadium content to 3 mg/kg DM. There is little information about the vanadium content of foods but levels ranging from 30 to 110 μg/kg DM have been reported for ryegrass. Herring meal appears to be a relatively rich source containing about 2.7 mg/kg DM. Vanadium is a relatively toxic element. When diets containing 30 mg/kg DM of the element were given to chicks, growth rate was depressed; at levels of 200 mg/kg DM the mortality rate was high.

NICKEL

Physiological symptoms of nickel deficiency have been produced in chicks, rats and pigs kept under laboratory conditions. Chicks given a diet containing the element in a concentration of less than 40 μg/kg DM developed skin pigmentation changes, dermatitis and swollen hocks. It is thought that nickel may play a role in nucleic acid metabolism. Normal levels of the element in pasture are 0.5 to 3.5 mg/kg DM while wheat grain contains 300 to 600 μg/kg DM. Nickel is a relatively non-toxic element, is poorly absorbed from the digestive tract and does not normally present a serious health hazard.

TIN

In 1970 it was reported that a significant growth effect in rats maintained on purified amino acid diets in a trace element-free environment was obtained when the diets were supplemented with tin. These studies suggested that tin was an essential trace element for mammals. The element is normally present in foods in amounts less than 1 mg/kg DM, the values in pasture herbage grown in Scotland, for example, being of the order of 300 to 400 μg/kg DM. The nutritional importance of this element has yet to be determined. Tin is poorly absorbed from the digestive tract and ingested tin has a low toxicity.

ARSENIC

Arsenic is widely distributed throughout the tissues and fluids of the body but is concentrated particularly in the skin, nails and hair. Recently it has been shown that the element has a physiological role in the rat. Animals given an arsenic deficient diet had rough coats and slower growth rates than control animals given a supplement of arsenic. The toxicity of the element is well known; symptoms include nausea, vomiting, diarrhoea and severe abdominal pain.

FURTHER READING

J. M. Gawthorne, J. McC. Howell and C. L. White (eds) 1982. *Trace Element Metabolism in Man and Animals*. Springer-Verlag, Berlin.

V. I. Georgievskii, B. N. Annenkov and V. I. Samokhin, 1982. *Mineral Nutrition of Animals*. Butterworths, London.

National Research Council, 1980. *Mineral Tolerances of Domestic Animals*. National Academy of Sciences, Washington, DC.

E. J. Underwood, 1977. *Trace Elements in Human and Animal Nutrition*. Academic Press, New York.

E. J. Underwood, 1981. *The Mineral Nutrition of Livestock*. Commonwealth Agricultural Bureaux, Slough.

7

Enzymes

The existence of living things involves a continuous series of chemical changes. Thus green plants elaborate chemical compounds such as carbon dioxide and water into complex compounds like sugars, starch and proteins, and in doing so fix and store energy. Subsequently these compounds are broken down, by the plants themselves or by animals which consume the plants, and the stored energy is utilised. The complicated reactions involved in these processes are reversible, and when not associated with living organisms are often very slow; extremes of temperature and/or pressure would be required to increase their velocity to practicable levels. In living organisms such conditions do not exist. Yet the storage and release of energy in such organisms must take place quickly when required, and this necessitates a high velocity of the reactions involved. The required velocity is attained through the activity of numerous catalysts present in the organisms. A catalyst in the classical chemical sense is a substance which affects the velocity of a chemical reaction without appearing in the final products; characteristically the catalyst remains unchanged in mass upon completion of the reaction. The catalysts elaborated and used by living organisms are organic in nature and are known as enzymes. They are capable of increasing rates of reaction by a factor of at least 10^6. At the same time it must be appreciated that the reactions which they catalyse are reversible and that the equilibrium is not changed. Both the forward and reverse reactions are accelerated equally and the concentrations of the reactants and products at equilibrium are not changed. Equilibrum is, however, achieved more quickly. Each living cell contains hundreds of enzymes and can function effectively only if the action of these enzymes is suitably coordinated.

CATALYTIC ACTION

Typical of enzyme-catalysed reactions are the hydrolyses of various substances such as fats and proteins which are essential for the normal functioning of the organism. A fat may be broken down to glycerides (acylglycerols) and fatty acids under the influence of a lipase:

$$
\begin{array}{llll}
CH_2.O.CO.R_1 & & CH_2OH & \\
| & & | & \\
CH.O.CO.R_2 & + \ 2H_2O \rightleftharpoons & CH.O.CO.R_2 + R_3.COOH + R_1.COOH \\
| & & | & \\
CH_2.O.CO.R_3 & & CH_2OH & \\
\text{Triglyceride} & & \text{Monoglyceride} & \text{Fatty acids} \\
\\
\text{(Triacylglycerol)} & & \text{(Monoacylglycerol)} &
\end{array}
$$

Such a reaction, which in the laboratory requires the action of strong alkalis at high temperatures, or superheated steam, takes place in the body at ordinary temperatures and without hydrolysing reagents. Similarly, peptidases split proteins by hydrolysis of the peptide linkage between the constituent amino acids. Enzymes of this type are classed as hydrolysing enzymes, or *hydrolases*.

Another large group are the transferring enzymes, or *transferases*, which catalyse the transfer of groups such as acetyl, amino and phosphate from one molecule to another. For example, in the formation of citrate from oxalacetate during the release of energy in the body, addition of an acetyl group takes place in the presence of citrate synthetase:

$$
\begin{array}{llllll}
CO.COO^- & CH_3 & & CH_2.COO^- & & \\
| & + & | & +H_2O \rightleftharpoons & | & + \ HSCoA \\
CH_2.COO^- & COS.CoA & & COH.COO^- & & \\
& & & | & & \\
& & & CH_2.COO^- & & \\
\text{Oxalacetate} & \text{Acetyl CoA} & & \text{Citrate} & & \text{Coenzyme A}
\end{array}
$$

The *oxidoreductases* catalyse the transfer of hydrogen, oxygen or ele trons from one molecule to another. Lactate is oxidised to pyruvate in ' presence of lactate dehydrogenase:

$$
\begin{array}{llllll}
CH_3 & & & CH_3 & & \\
| & & & | & & \\
CHOH & + & NAD^+ \rightleftharpoons & C{=}O & + NADH(+H^+) \\
| & & & | & & \\
COO^- & & & COO^- & & \\
\text{Lactate} & \text{Nicotinamide} & & \text{Pyruvate} & \text{Reduced NAD} \\
& \text{adenine} & & & \\
& \text{dinucleotide} & & &
\end{array}
$$

The *lyases* are enzymes which catalyse non-hydrolytic decompositions such as decarboxylation and deamination. Pyruvate decarboxylase, for example, catalyses the conversion of a 2-oxo acid to an aldehyde and carbon dioxide.

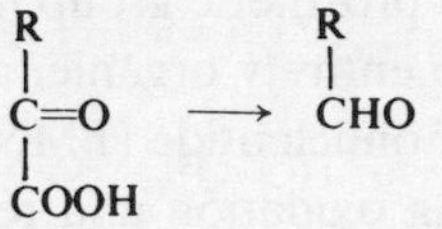

$$
\begin{array}{ccc}
R & & R \\
| & & | \\
C{=}O & \longrightarrow & CHO \quad +CO_2 \\
| & & \\
COOH & &
\end{array}
$$

Certain syntheses in the body involve the breakdown of ATP or similar triphosphate, which provides the energy for the reaction. These reactions are catalysed by enzymes known as *ligases*. The production of acetyl CoA from acetate by acetyl coenzyme A synthetase is typical:

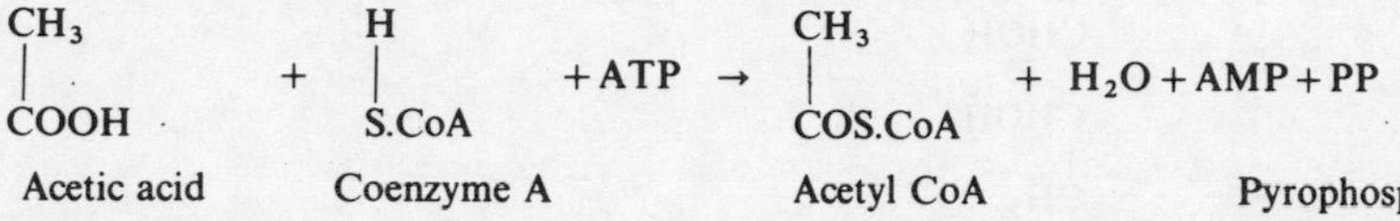

$$
\begin{array}{cccccc}
CH_3 & & H & & CH_3 & \\
| & + & | & +ATP \rightarrow & | & + H_2O+AMP+PP \\
COOH & & S.CoA & & COS.CoA & \\
\text{Acetic acid} & & \text{Coenzyme A} & & \text{Acetyl CoA} & \text{Pyrophosphate}
\end{array}
$$

The *isomerases* are a group of enzymes which catalyse intramolecular rearrangement. Typical of this class of enzymes are the epimerases such as uridine diphosphate glucose 4-epimerase. This catalyses the change of configuration at the fourth carbon atom of glucose, and galactose is produced (see Ch. 9).

NATURE OF ENZYMES

Many enzymes have now been isolated in the pure state and their structure has been elucidated. All have proved to be complex proteins of high molecular weight. Many of the proteins need special groups to aid their activity, especially in the case of the transfer enzymes. An example is the cytochromes; these are important in certain oxidation reactions during which they accept electrons from a reduced substance which is consequently oxidised. The cytochromes are haem proteins, and as such contain the general grouping:

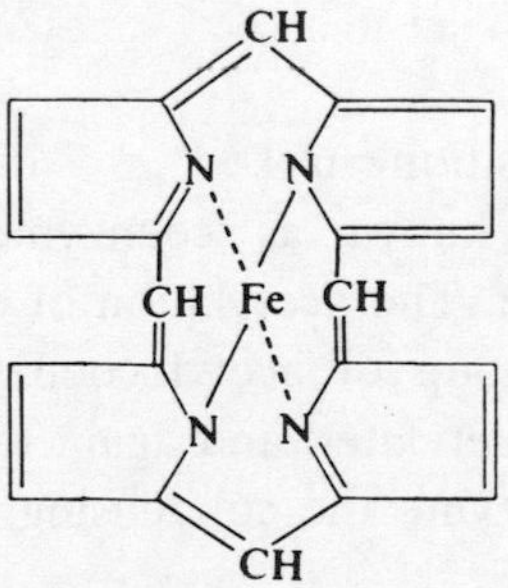

Without this iron-containing haem group the enzyme would be inactive, since electron exchange takes place at the iron atom. Such active groups are known as 'prosthetic' groups when they are actually part of the enzyme molecule; the remainder of the molecule is then known as the apo-enzyme, and the whole molecule as the holo-enzyme. The prosthetic group need not necessarily contain an inorganic entity but may be entirely organic, as in the case of the flavoproteins, which contain flavin mononucleotide (FMN) as the active group. Exchange of hydrogen atoms during oxidation and reduction takes place at the positions marked*.

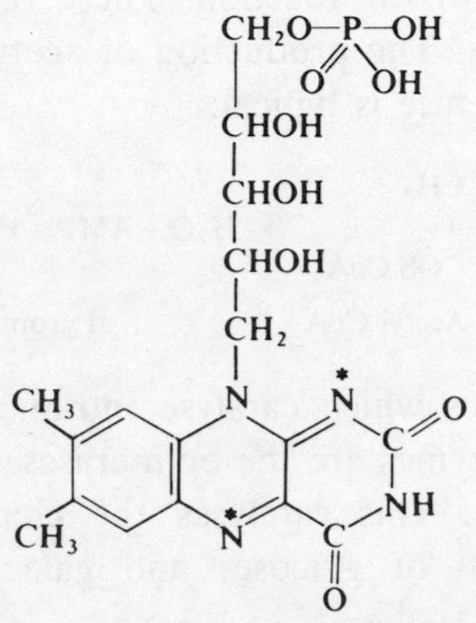

$$CH_2O—P—OH$$
$$O\quad OH$$
$$CHOH$$
$$CHOH$$
$$CHOH$$
$$CH_2$$

Flavin mononucleotide

Sometimes the enzyme needs for its activity a group which is a separate entity. In the oxidation of lactate to pyruvate the hydrogen released must be removed. This is done by a substance nicotinamide-adenine dinucleotide (NAD) or nicotinamide-adenine dinucleotide phosphate (NADP), which contain the nicotinamide grouping:

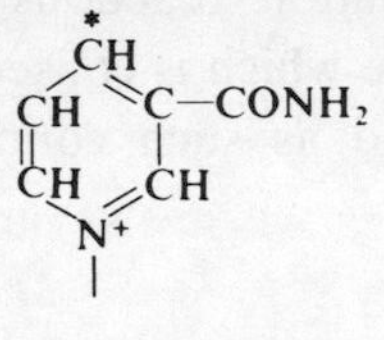

$$CH$$
$$CH\quad C—CONH_2$$
$$CH\quad CH$$
$$N^+$$

The hydrogen exchange takes place at the position marked*.

·Such substances as NAD and NADP are known as 'coenzymes', and several enzymes may have the same coenzyme. The acetylation of oxalacetate depends upon donation of an acetyl group of acetyl coenzyme A, which reverts to coenzyme A; this can be reacetylated and again take part in an acetylation. In the absence of the coenzyme the condensing enzyme would be quite inactive. Many enzymes require the presence of metal ions in order to 'activate' the enzyme. Such metal ions are known as activators. Arginase, for example, needs manganese ions for its action.

Some enzymes are present in an inactive form which is changed to the active state at the place and time when it is required. Thus trypsinogen is synthesised in the pancreas, transported to the small intestine and there changed to the active digestive enzyme trypsin. This kind of mechanism confers considerable control over the siting and timing of enzyme activity. The inactive precursors of proteolytic enzymes are known as *zymogens*.

Small groups may be attached to enzymes by bonds other than peptide bonds, and thus render them active or inactive. This phenomenon is referred to as *covalent modification* and can be reversed by removal of the group by hydrolysis.

MECHANISM OF ENZYME ACTION

Enzyme action involves the formation of a complex between the enzyme and the substrate, the substance to be acted upon. The complex then under-goes breakdown yielding the products and the unchanged enzyme:

$$E + S \rightleftarrows ES \rightleftarrows E + P$$

In the absence of the enzyme the reaction $S \rightleftarrows P$ would proceed through a transition state with free energy greater than that of either S or P. The difference between the free energy of S and that of the transition state is the activation energy of the reaction. The rate of the reaction is proportional to the number of substrate molecules having free energy greater than the activation energy. The formation of the enzyme-substrate complex creates a new reaction pathway and greatly reduces activation energy. More substrate molecules will have free energy in excess of this and the rate of reaction will consequently be accelerated. The situation is illustrated in Fig. 7.1.

The complexes are formed between the substrate, or substrates, and relatively few active centres on the enzyme. The bonding may be ionic, typified by:

$$E \quad \begin{array}{c} COO^- \quad {}^+H_3N.R \\ \\ COO^- \quad {}^+H_3N.R \end{array}$$

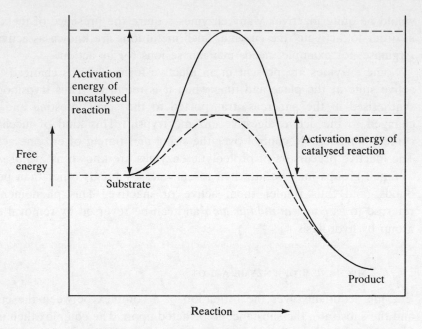

Fig. 7.1. Mechanism of enzyme catalysis.

or by hydrogen bonds:

$$\text{E} \quad \text{C} = \text{O} \, \text{-----} \, \text{H}\text{—N} \Big\langle$$

or covalent linkage typified by sulphydryl groups on the enzyme:

$$\text{E} \text{—S—C—R} \atop \text{OH}$$

or by van der Waals bonds. These arise as the result of non-specific attractive forces between atoms when the distance between them is from 3 to 4 Å. The bonds are weaker than the ionic or hydrogen, but may be very important because of the large number which may be formed when conditions are favourable.

The type of mechanism envisaged for enzyme catalysis is illustrated by that suggested for the chymotrypsin-catalysed hydrolysis of an ester as shown.

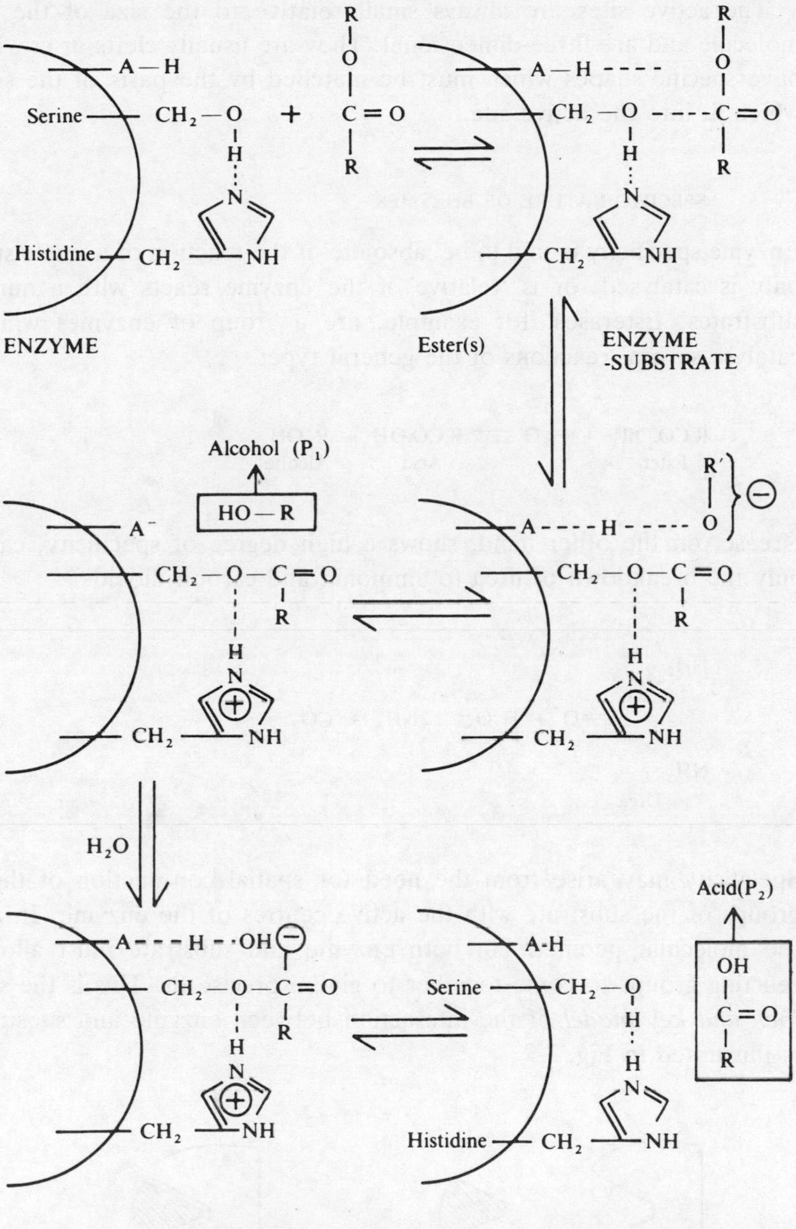

After F. H. Westheimer, 1962. *Advances in Enzymology*, **24**, 464.

In this example the active sites of the enzyme are the alcohol group of serine and the imidazole nitrogen of histidine. The enzyme-substrate complex has a lowered activation energy and an increased reaction rate results. How the lowering occurs is not clear. One suggestion has been that placement at the active sites of the enzyme results in a distortion of the substrate molecule. Certain bonds are thus placed under stress and are easily ruptured.

The active sites are always small relative to the size of the enzyme molecule and are three-dimensional. They are usually clefts or crevices and have specific shapes which must be matched by the parts of the substrate which fit into the active site.

SPECIFIC NATURE OF ENZYMES

Enzyme specificity is said to be 'absolute' if the reaction of a single substrate only is catalysed, or is 'relative' if the enzyme reacts with a number of substrates. Esterases, for example, are a group of enzymes which may catalyse several reactions of the general type:

$$R.CO.OR_1 + H_2O \rightleftarrows R.CO.OH + R_1OH$$
$$\text{Ester} \qquad\qquad \text{Acid} \qquad \text{Alcohol}$$

Urease, on the other hand, shows a high degree of specificity, catalysing only the breakdown of urea to ammonia and carbon dioxide:

$$
\begin{array}{c}
NH_2 \\
\diagdown \\
\quad\; C{=}O + H_2O \rightleftarrows 2NH_3 + CO_2 \\
\diagup \\
NH_2 \\
\text{Urea}
\end{array}
$$

Specificity may arise from the need for spatial conjunction of the active groups of the substrate with the active centres of the enzyme. In addition the molecular geometry of both enzyme and substrate must allow these reacting groups to come together to give a precise fit. This is the so-called *lock and key model* of the interaction between enzyme and substrate and is illustrated in Fig. 7.2.

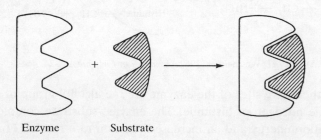

Enzyme   Substrate

Fig. 7.2.   Lock and key model of enzyme-substrate complex formation.

Very slight modifications to the non-reacting groups in either enzyme or substrate may be sufficient to hinder, or prevent, the fit and the catalytic action of the enzyme is lost. In some cases the conformation of the enzyme may not be absolute but may change in response to the presence of the substrate to allow the formation of the enzyme-substrate complex. This is usually referred to as the *induced fit* model, illustrated in Fig. 7.3.

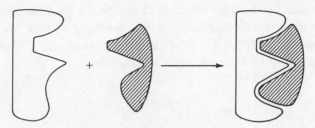

Fig. 7.3.   Induced-fit model of enzyme-substrate complex formation.

In yet other instances, where more than one substrate is involved, union of substrate 2 and enzyme may not take place until, and unless, substrate 1 and the enzyme have united.

FACTORS AFFECTING ENZYME ACTIVITY

Substrate concentration

In a system where the enzyme is in excess and the concentration remains constant, then an increase in substrate concentration increases reaction velocity. This is due to increased utilisation of the available active centres of the enzyme. If the substrate concentration continues to increase, utilisation of the active centres becomes maximal and there is no further increase in reaction rate. In fact continued increases in substrate concentration may lead to a reduction in reaction rate owing to an incomplete linkage of enzyme and substrate as a result of competition for the active centres by the excess substrate molecules.

Enzyme concentration

In a system where the substrate is present in excess, an increase in enzyme concentration gives a straight-line response in reaction velocity owing to the provision of additional active centres for the formation of enzyme-substrate complexes. Further increase in the enzyme concentration may result in some limiting factor, such as availability of coenzyme, becoming operative. Enzymes are rarely saturated with substrate under physiological conditions.

Inhibitors

The catalytic action of enzymes may be inhibited by substances which prevent the formation of a normal enzyme-substrate complex. Such inhibition may be reversible. One of the best known examples of enzyme inhibition is provided by the sulphonamide drugs. In the bacteria controlled by these drugs a vital metabolic process is the synthesis of folic acid from *para*-aminobenzoic acid (PABA). The similarity between PABA and sulphanilamide released by the sulphonamides is obvious:

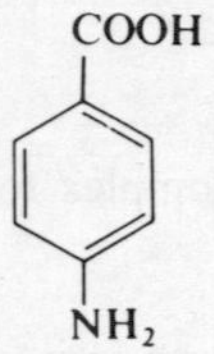

COOH SO_2NH_2

Para-aminobenzoic acid 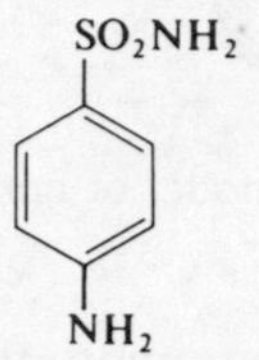   Sulphanilamide

The sulphanilamide forms a complex with the relevant enzyme, thus preventing normal enzyme-substrate combination and the synthesis of folic acid. Addition of excess PABA overcomes the inhibition since the formation of the sulphanilamide-enzyme complex is reversible. The extent of the inhibition, then, depends entirely upon the relative concentrations of the true substrate and the inhibitor. Such inhibition, which depends upon competition with the substrate for the active sites of the enzyme is termed *competitive inhibition*. In other cases the inhibitor combines with the enzyme-substrate complex to give an inactive enzyme-substrate-inhibitor complex, which cannot undergo further reaction to give the usual product. This is referred to as *uncompetitive inhibition*. *Non-competitive inhibition* involves combination of the inhibitor with the enzyme or the enzyme-substrate complex, to give inactive complexes. The inhibitors bind to sites, on the enzymes, other than the active sites. This results in deformation of the enzyme molecule so that formation of the enzyme-substrate complex is slower than normal, or the formation of the final products from the enzyme-substrate complex is slower than normal.

Some enzymes undergo irreversible inactivation. This involves reaction of the inhibitor with a functional group of the enzyme, resulting in a loss of its catalytic activity. Typical of this type of inhibition is the action of organophosphorus nerve poisons in inactivating acetylcholinesterase which is essential for the transfer of nerve impulses. These inhibitors are potent insecticides. The inhibition depends upon reaction of the organophosphorus compound with an active serine residue in the acetylcholinesterase as shown:

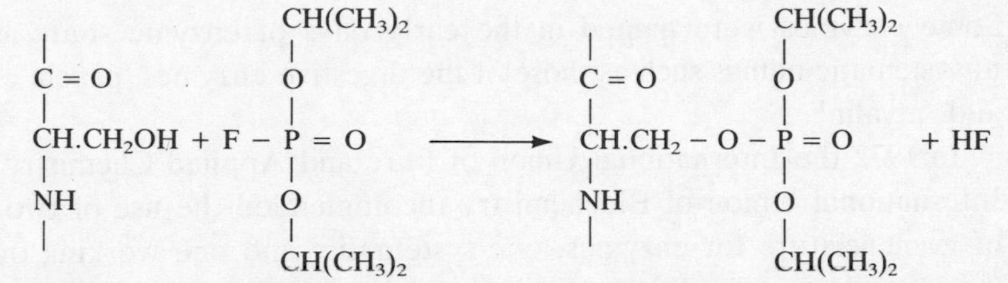

$$
\begin{array}{ccc}
& CH(CH_3)_2 & \\
| & | & \\
C = O & O & \\
| & | & \\
CH.CH_2OH + F - P = O & \longrightarrow & CH.CH_2 - O - P = O \quad + HF \\
| & | & \\
NH & O & \\
| & | & \\
& CH(CH_3)_2 &
\end{array}
$$

Enzyme	Diisopropyl	Inactive enzyme-inhibitor
	phosphofluoridate	complex

Temperature

As in other chemical reactions, the efficiency of enzyme-catalysed reactions is increased by raising the temperature. Very approximately, the speed of reaction is doubled for each increase of 10°C. As the temperature rises, however, a complicating factor comes into play, in that denaturation of the enzyme protein begins. This is a molecular rearrangement which causes a loss of the active centres of the enzyme surface and results in a loss of efficiency. Above 50°C the destruction of the enzyme becomes more rapid, and all enzymes are destroyed when heated to 100°C. The time for which the enzyme is subjected to a given temperature affects the extent of efficiency loss. As might be expected, all enzymes have a temperature range within which they are most efficient.

Acidity

Hydrogen-ion concentration also has an important effect on the efficiency of enzyme action. Many enzymes are most effective in the region of pH 6 to 7, which is that of the cell. Extra-cellular enzymes show maximum activity in the acid or the alkaline pH range, but in any case the actual range of pH within which the enzyme works is only about 2.5 to 3.0 units; outside this range the activity drops off very rapidly. The reduction in efficiency caused by changes in the pH is due to changes in the degree of ionisation of the substrate and enzyme. Where the linkage between the active centres is electrostatic, this affects the facility with which the intermediate complex is formed and thus the efficiency of enzyme action. In addition highly acidic or alkaline conditions bring about a denaturation and subsequent loss of enzymatic activity.

NOMENCLATURE OF ENZYMES

Some enzymes were named in the early days of enzyme study and have unsystematic names such as those of the digestive enzymes 'pepsin', 'trypsin' and 'ptyalin'.

In 1972 the International Union of Pure and Applied Chemistry and the International Union of Biochemistry recommended the use of two systems of nomenclature for enzymes, one systematic, and one working or recommended. The systematic name, formed in accordance with definite rules, was to show the action of the enzyme as exactly as possible, thus identifying it precisely. The recommended name was not necessarily very systematic but was short enough for convenient use. In addition code numbers were assigned to enzymes according to the following scheme:

  (i) the first number shows to which of the six main classes the enzyme belongs
  (ii) the second number shows the sub-class
  (iii) the third figure gives the sub-sub-class
  (iv) the fourth figure identifies the enzyme

The following are examples of this system:

Reaction	Recommended name	Systematic name	Code number
L-Lactate + NAD$^+$ → pyruvate + NADH + H$^+$	Lactate dehydrogenase	L-Lactate: NAD$^+$ oxidoreductase	1.1.1.27
Hydrolysis of terminal, non-reducing 1,4-linked α-D-glucose residues with release of α-glucose e.g. maltose → glucose	α-Glucosidase	α-D-Glucoside glucohydrolase	3.2.1.20
L-Glutamate → 4-aminobutyrate + CO$_2$	Glutamate decarboxylase	L-Glutamate l-carboxy-lyase	4.1.1.15

FURTHER READING

Enzyme Nomenclature, 1973. *Recommendations (1972) of the International Union of Pure and Applied Chemistry and the International Union of Biochemistry*. American Elsevier, New York.

A. L. Lehninger, 1982. *Principles of Biochemistry*, 1st edn. Worth Publ., New York.

A. White, P. Handler and E. L. Smith, 1978. *Principles of Biochemistry*, 6th edn. McGraw-Hill, New York.

L. Stryer, 1981. *Biochemistry*, 2nd edn. W. H. Freeman San Francisco.

8

Digestion

Many of the organic components of food are in the form of large insoluble molecules which have to be broken down into simpler compounds before they can pass through the mucous membrane of the alimentary canal into the blood and lymph. The breaking-down process is termed 'digestion', the passage of the digested nutrients through the mucous membrane 'absorption'.

The processes important in digestion may be grouped into mechanical, chemical and microbial. The mechanical activities are mastication and the muscular contractions of the alimentary canal. The main chemical action is brought about by enzymes secreted by the animal in the various digestive juices, though it is possible that plant enzymes present in unprocessed foods may in some instances play a minor role in food digestion. Microbial digestion of food, also enzymic, is brought about by the action of bacteria and protozoa, micro-organisms which are of special significance in ruminant digestion. In simple-stomached animals microbial activity occurs in the large intestine.

DIGESTION IN MONOGASTRIC MAMMALS

The alimentary canal

The various parts of the alimentary canal of the pig, which will be used as the reference animal, are shown in Fig. 8.1. The digestive tract can be considered as a tube extending from mouth to anus, lined with mucous membrane, whose function is the ingestion, comminution, digestion and absorption of food, and the elimination of solid waste material. The various parts are mouth, pharynx, oesophagus, stomach, small and large intestine.

Pig

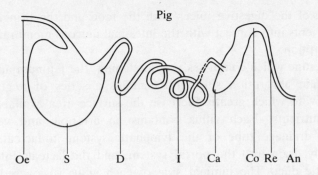

Oe   S   D   I   Ca   Co Re An

Fowl

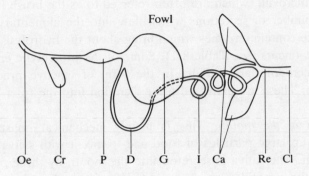

Oe   Cr   P   D   G   I Ca   Re Cl

Cow

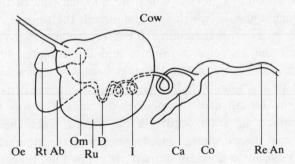

Om D
Oe Rt Ab   Ru   I   Ca   Co   Re An

Fig. 8.1. Diagrammatic representation of the digestive tracts of different farm animals.
An = Anus.   Ab = Abomasum.   Ca = Caecum.   Cl = Cloaca.
Co = Colon.   Cr = Crop.   D = Duodenum.   G = Gizzard.   I = Ileum.
Oe = Oesophagus.   Om = Omasum.   P = Proventriculus.
Re = Rectum.   Rt = Reticulum.   Ru = Rumen.   S = Stomach.

The movement of the intestinal contents along the tract is produced by peristaltic waves, which are contractions of the circular muscle of the intestinal wall. Several different kinds of movement of the intestinal wall are recognised, the functions of these being the transport of materials along the tract, the mixing of the digestive juices with the food and the bringing of the digested nutrients into contact with the intestinal mucous membrane for subsequent absorption.

The small intestine which comprises the duodenum, the jejunum and the ileum, is the main absorption site, and contains a series of finger-like projections, the villi, which greatly increase the surface area available for absorption of nutrients. Each villus contains an arteriole and venule, together with a drainage tube of the lymphatic system, a lacteal. The venules ultimately drain into the portal system, and the lacteals into the thoracic lymphatic duct. The luminal side of each villus is covered with projections, the microvilli, which are often referred to as the brush border.

There are a number of secretions which flow into the alimentary canal and many of these contain enzymes which bring about the hydrolysis of the various food components (see Table 8.1). Some of the proteolytic enzymes present in the secretions are initially in the form of inactive precursors termed zymogens. These are activated after secretion into the tract.

Digestion in the mouth. This is mainly mechanical, mastication helping to break up large particles of food and to mix it with saliva which acts as a lubricant. The saliva is secreted into the mouth by three pairs of salivary glands: the parotids, which are sited in front of each ear; the submandibular (submaxillary) glands, which lie on each side of the lower jaw; and the sublingual glands, which are underneath the tongue. Saliva is about 99 per cent water, the remaining 1 per cent consisting of mucin, inorganic salts, and the enzymes α-amylase and the complex lysozyme. Some animals, such as the horse, cat and dog, lack salivary α-amylase, while the saliva of other species, man included, has strong α-amylase activity. The enzyme is present in the saliva of the pig but the activity is low and it is doubtful whether much digestion occurs in the mouth, since the food is quickly swallowed and passed along the oesophagus to the stomach where the pH is unfavourable for α-amylase activity. It is possible, however, that some digestion of starch by the enzyme can occur in the stomach, since the food mass is not immediately mixed intimately with the gastric juice. The pH of pig's saliva is about 7.3, which is only slightly above the value regarded as optimal for α-amylase activity. This enzyme hydrolyses the α-$(1{\rightarrow}4)$-glucan links in polysaccharides containing three or more α-$(1{\rightarrow}4)$-linked D-glucose units. The enzyme therefore acts on starch, glycogen and related polysaccharides and oligosaccharides. When amylose, which contains exclusively α-$(1{\rightarrow}4)$-glucosidic bonds (p. 20) is attacked by α-amylase, random cleavages of these bonds give rise to a mixture of glucose and maltose. Amylopectin, on the other hand, contains in addition to α-$(1{\rightarrow}4)$-glucosidic bonds a number of branched α-$(1{\rightarrow}6)$-glucosidic bonds which are

TABLE 8.1 Main digestive enzymes

Recommended name	Trivial name	Systematic name	Number	Source	Substrate
A. Enzymes hydrolysing peptide links					
Pepsin		—	3.4.23	Gastric mucosa	Proteins and peptides
Chymosin	Rennin	—	3.4.23.4	Gastric mucosa (young calves)	Proteins and peptides
Trypsin	—	—	3.4.21.4	Pancreas	Proteins and peptides
Chymotrypsin	—	—	3.4.21.1	Pancreas	Proteins and peptides
Carboxypeptidase A	Carboxypeptidase	Peptidyl-L-amino-acid hydrolase	3.4.12.2	Small intestine	Peptides
Carboxypeptidase B	Protaminase	Peptidyl-L-lysine (-L-arginine) hydrolase	3.4.12.3	Small intestine	Peptides
Aminopeptidases	—	α-Aminoacyl-peptide hydrolases	3.4.11	Small intestine	Peptides
Dipeptidases	—	Dipeptide hydrolases	3.4.13	Small intestine	Dipeptides
B. Enzymes hydrolysing glycoside links					
α-Amylase	Diastase	1, 4-α-D-Glucan glucanohydrolase	3.2.1.1	Saliva, pancreas	Starch, glycogen, dextrin
α-Glucosidase	Maltase	α-D-Glucoside glucohydrolase	3.2.1.20	Small intestine	Maltose
Oligo-1, 6-glucosidase	Isomaltase	Dextrin 6-α-glucanohydrolase	3.2.1.10	Small intestine	Dextrins
β-Galactosidase	Lactase	β-D-Galactoside galactohydrolase	3.2.1.23	Small intestine	Lactose
β-Fructofuranosidase	Sucrase	β-D-Fructofuranoside fructohydrolase	3.2.1.26	Small intestine	Sucrose
C. Enzymes acting on ester links					
Triacylglycerol lipase	Lipase	Triacylglycerol acyl-hydrolase	3.1.1.3	Pancreas	Triacylglycerols
Cholesterol esterase		Sterol-ester hydrolase	3.1.1.13	Pancreas and small intestine	Cholesterol esters
Phospholipase A$_2$	Lecithinase A	Phosphatide 2-acyl-hydrolase	3.1.1.4	Pancreas and small intestine	Lecithins and Cephalins
Lysophospholipase	Lysolecithinase	Lysolecithin acyl-hydrolase	3.1.1.5	Small intestine	Lysolecithin
Deoxyribonuclease	DNAase	Deoxyribonucleate 5'-oligonucleotidohydrolase	3.1.4.5	Pancreas and small intestine	DNA
Ribonuclease I	RNAase	Ribonucleate 3'-pyrimidino-oligonucleotidohydrolase	3.1.4.22	Pancreas and small intestine	RNA
Nucleosidase	—	N-Ribosyl-purine ribohydrolase	3.2.2.21	Small intestine	Nucleosides
Phosphatases	—		3.1.3	Small intestine	Orthophosphoric acid esters

not attacked by α-amylase, and the products include a mixture of branched and unbranched oligosaccharides (termed 'limit dextrins') in which α-(1→6)-bonds are abundant.

The enzyme lysozyme has been detected in many tissues and body fluids. It is capable of hydrolysing the β-(1→4)-*N*-acetyl-glucosaminidic linkage of the repeating disaccharide unit in the polysaccharides of the cell walls of many different species of bacteria, thereby killing and dissolving them.

Digestion in the stomach. The stomach of the adult pig has a capacity of about 8 litres, and consists of a simple compartment which functions not only as an organ for the digestion of food but also for storage. Viewed from the exterior, the stomach can be seen to be divided into the cardia (entrance), fundus and pylorus (terminus), the cardia and pylorus being sphincters controlling the passage of food through the stomach. The inner surface of the stomach is increased in area by an infolding of the epithelium, and contains a variety of secretory cells which collectively secrete the gastric juice. A number of factors are concerned in the stimulation of the glands to secrete the gastric juice, among them being the presence of food in the stomach and the hormone, gastrin which is produced in the pylorus. The gastric juice consists mainly of water, with pepsinogens, inorganic salts, mucous, hydrochloric acid and the intrinsic factor important for the efficient absorption of vitamin B_{12}. Pepsinogens are the inactive forms of pepsins which hydrolyse proteins. The acid concentration of the gastric juice varies with the diet but is generally about 0.1 M, which is sufficient to lower the *p*H to 2.0. The acid activates the pepsinogens, converting them into pepsins by removing low molecular weight peptides from each precursor molecule. Four pepsins have been found in the pig which have optimum activity at two different *p*H levels, 2.0 and 3.5. Pepsins preferentially attack those peptide bonds adjacent to aromatic amino acids, e.g. phenylalanine, tryptophan and tyrosine, but also have a significant action on linkages involving glutamic acid and cysteine. Pepsins also have a strong clotting action on milk. Rennin, an enzyme which occurs in the gastric juice of the calf and the young piglet, resembles pepsins in its activity. The products of protein digestion in the stomach are mainly polypeptides of variable chain length, and a few amino acids.

Digestion in the small intestine. The partially digested food leaving the stomach enters the small intestine, where it is mixed with secretions from the duodenum, liver and pancreas. The duodenal (Brunner's) glands produce an alkaline secretion which enters the duodenum through ducts situated between the villi. This secretion acts as a lubricant and also protects the duodenal wall from the hydrochloric acid entering from the stomach.

Bile is secreted by the liver and passes to the duodenum through the bile duct. It contains the sodium and potassium salts of bile acids, chiefly glycocholic and taurocholic (see p. 40), the bile pigments biliverdin and bilirubin, cholesterol and mucin. In all farm animals, except the horse, bile is stored in the gall bladder until required. The bile salts play an important part in digestion by activating pancreatic lipase and emulsifying fats.

The pancreatic juice is secreted by the pancreas, a gland which lies in the duodenal loop and opens into the duodenum through the pancreatic duct. A number of factors induce the pancreas to secrete its juice into the duodenum. When acid enters the duodenum, the hormone *secretin* is liberated from the epithelium of the small intestine into the blood. When it reaches the pancreatic circulation it stimulates the pancreatic cells to secrete a watery fluid containing a high concentration of bicarbonate ions, but very little enzyme. Another hormone, cholecystokinin (pancreozymin), is also liberated from the mucosa when peptides and other digestive products reach the duodenum. This hormone stimulates the secretion into the pancreatic juice of proenzymes and enzymes such as trypsinogen, chymotrypsinogen, procarboxypeptidases A and B, proelastase, α-amylase, lipase, lecithinases and nucleases. Unlike pepsin, these enzymes have *p*H optima around 7 to 9. The inactive zymogen, trypsinogen, is converted to the active trypsin by enterokinase, an enzyme liberated from the duodenal mucosa. This activation is also catalysed by trypsin itself, thus constituting an autocatalytic reaction. The activation process results in the liberation of a hexapeptide from the amino terminal end of trypsinogen. Trypsin is very specific and only acts upon peptide linkages involving the carboxyl groups of lysine and arginine. Trypsin also converts chymotrypsinogen into the active enzyme chymotrypsin which has a specificity towards peptide bonds involving the carboxyl groups of tyrosine, tryptophan, phenylalanine and leucine. Procarboxypeptidases are converted by trypsin into the proteolytic enzymes carboxypeptidases which attack the peptides from the end of the chain, splitting off the terminal amino acid which has a free α-carboxyl group. Such an enzyme is classified as an exopeptidase, as distinct from trypsin and chymotrypsin which attack peptide bonds in the interior of the molecule and which are known as endopeptidases.

Pancreatic α-amylase is similar in function to the salivary amylase and attacks the α-$(1\rightarrow4)$-glucan links in starch and glycogen.

The breakdown of fats is achieved by pancreatic lipase. This enzyme does not completely hydrolyse the triacylglycerol and the action stops at the monoacylglycerol stage.

Dietary fat leaves the stomach in the form of relatively large globules which are difficult to hydrolyse rapidly. Fat hydrolysis is helped by emulsification, which is brought about by the action of bile salts. These bile salts

α' $CH_2OOC.R_1$ $CH_2OOC.R_1$ CH_2OH
| $R_3.COOH$ | $R_1.COOH$ |
β $CHOOC.R_2$ ────↗────→ $CHOOC.R_2$ ────↗────→ $CHOOC.R_2$
| | |
α $CH_2OOC.R_3$ CH_2OH CH_2OH
     Triacylglycerol   Diacylglycerol   Monoacylglycerol

are detergents or amphipaths, the sterol nucleus being lipid soluble and the hydroxyl groups and ionised conjugate of glycine or taurine being water soluble (see p. 40). These amphipaths, in addition to being emulsifying agents also have the property of being able to aggregate together to form micelles. Although triacylglycerols are insoluble in these micelles, mono-acylglycerols and most fatty acids dissolve forming mixed micelles. Some fatty acids e.g. stearic, are not readily soluble in pure bile salt micelles but dissolve in mixed micelles.

Lecithinase A is an enzyme which hydrolyses the bond linking the fatty acid to the β-hydroxyl group of lecithin (see p. 37). The product formed from this hydrolysis, lysolecithin, is further hydrolysed by lysolecithinase (lecithinase B) to form glycerolphosphocholine and a fatty acid. Cholesterol esterase catalyses the splitting of cholesterol esters.

The nucleic acids DNA and RNA (see p. 55) are hydrolysed by the polynucleotidases, deoxyribonuclease (DNase) and ribonuclease (RNase) respectively. These enzymes catalyse the cleavage of the ester bonds between the sugar and phosphoric acid in the nucleic acids. The end prod-ucts are the component nucleotides. Nucleosidases attack the linkage between the sugar and the nitrogenous bases, liberating the free purines and pyrimidines. Phosphatases complete the hydrolysis by separating the ortho-phosphoric acid from the ribose or deoxyribose.

The hydrolysis of oligosaccharides to monosaccharides and of small peptides to amino acids is brought about by enzymes associated with the intestinal villi. Only a small proportion of hydrolysis occurs intraluminally, arising from enzymes present in aged cells discarded from the intestinal mucosa. Most of the enzymatic hydrolysis occurs at the luminal surface of the epithelial cells, although some peptides are absorbed by the cells before being broken down by enzymes present in the cytoplasm.

Enzymes produced by the villi are sucrase, which converts sucrose to glucose and fructose; maltase, which breaks down maltose to two molecules of glucose; lactase which hydrolyses lactose to one molecule of glucose and one of galactose; and oligo-1,6-glucosidase, which attacks the α-(1→6)-links in limit dextrins. Aminopeptidases act on the peptide bond adjacent to the free amino group of simple peptides, while dipeptidases complete the break-down of dipeptides to amino acids.

Digestion in the large intestine. The main site of absorption of digested nutrients is the small intestine, so that by the time the food material has reached the entrance to the colon most of the hydrolysed nutrients will have been absorbed. With normal diets there is always a certain amount of material which is resistant to the action of the enzymes secreted into the alimentary canal. Cellulose and many of the hemicelluloses are not attacked by any of the enzymes present in the digestive secretions of the pig. Lignin is known to be completely unaffected and is thus indigestible. It is also conceivable that lignified tissues may trap proteins and carbohydrates and protect them from the action of digestive enzymes. The glands of the large intestine are mainly mucous glands which do not produce enzymes, and digestion in the large intestine is therefore brought about by enzymes which have been carried down in the food from the upper part of the tract, or occurs as a result of microbial activity.

Extensive microbial activity occurs in the large intestine, especially the caecum. The bacteria present include lactobacilli, streptococci, coliforms, bacteroides, clostridia and yeasts. These organisms metabolise both dietary and endogenous residues, resulting in the formation of a number of products including indole, skatole, phenol, hydrogen sulphide, amines, ammonia and the volatile fatty acids, acetic, propionic and butyric. In the pig it is known from digestibility studies that the cellulose of certain foods is broken down to a limited extent; this microbial breakdown occurs only in the large intestine. The digestion of cellulose and other higher polysaccharides is nevertheless small compared with that taking place in the horse and ruminants, which have a digestive system adapted to deal with fibrous foods. The products of this microbial breakdown of polysaccharides are not sugars but mainly the volatile fatty acids, listed above.

Bacterial action in the large intestine may have a beneficial effect owing to the synthesis of some of the B vitamins, which may be absorbed and utilised by the host. Synthesis of most of the vitamins in the digestive tract of the pig is, however, insufficient to meet the daily requirements and a dietary source is needed.

The waste material, or faeces, voided from the large intestine via the anus consists of water, undigested food residues, digestive secretions, epithelial cells from the tract, inorganic salts, bacteria and products of microbial decomposition.

Digestion in the young pig

From birth until about the age of five weeks the concentration and activity of many digestive secretions in the young pig are different from those in the adult animal. During the first few days after birth the intestine is permeable

Table 8.2   Weight of disaccharide hydrolysed per kg bodyweight per hour by small intestine enzymes in young pigs

	Lactose (g)	Sucrose (g)	Maltose (g)
Newborn	5.9	0.06	0.3
Five weeks	0.8	1.3	2.5

to native proteins. In the young pig, as in other farm animals, this is essential for the transfer of γ-globulins (antibodies) via the mother's milk to the new-born animal. The ability of the young pig to absorb these proteins declines rapidly, and is low by 24 hours *post partum*.

At birth the activity of pepsin, α-amylase, maltase and sucrase is low, while lactase activity is initially high and decreases as the animal matures. Table 8.2 shows the activity of some of the important carbohydrases in the young pig.

These differences in enzyme activities are of special significance where piglets are reared on early weaning diets. If young pigs are weaned at 14 days of age, their diet, especially regarding the types of carbohydrates, should be different from that for older animals weaned at 5–6 weeks. Early weaning mixtures usually include a high proportion of dried milk products containing lactose.

Digestion in the fowl

The digestive tract of the fowl differs in a number of respects from that of the pig (see Fig. 8.1). In the fowl the lips and cheeks are replaced by the beak, the teeth being absent. The crop or diverticulum of the oesophagus is a pear-shaped sac whose main function is to act as a reservoir for holding food, although some microbial activity occurs there and results in the formation of lactic and acetic acids. The oesophagus terminates at the proventriculus or glandular stomach which leads into the gizzard, a muscular organ which undergoes rhythmic contractions and grinds the food with moisture into a smooth paste. The gizzard, which has no counterpart in the pig, although it is often compared to the pyloric part of the mammalian stomach, leads into the duodenum, which encloses the pancreas as in mammals. The pancreatic and bile ducts open into the intestine at the termination of the duodenum. Where the small intestine joins the large intestine there are two large blind sacs known as the caeca. The large intestine is relatively short and terminates in the cloaca, from which urine and faeces are excreted together.

The enzymes present in the digestive secretions of the fowl are similar to those in the secretions of mammals, although lactase has not been detected. Salivary α-amylase is known to occur in the fowl, and the action of this enzyme on starch continues in the crop.

The presence of grit in the gizzard, although not essential, has been shown to increase the breakdown of whole grains by about 10 per cent.

The pancreatic juice of fowls contains the same enzymes as the mammalian secretion, and the digestion of proteins, fats and carbohydrates in the small intestine is believed to be similar to that occurring in the pig. The intestinal mucosa produces mucin, α-amylase, maltase, sucrase and proteolytic enzymes.

The caeca, which function mainly as absorptive organs, are not essential to the fowl, since surgical removal causes no harmful effects. Recent experiments with adult fowls indicate that the cellulose present in cereal grains is not broken down by microbial activity to any extent during its passage through the digestive tract, although some hemicellulose breakdown occurs.

Absorption of digested nutrients

The main organ in the monogastric mammal for the absorption of dietary nutrients is the small intestine. This part of the tract is specially adapted for absorption because its inner surface area is increased by folding and the presence of villi. Absorption of a nutrient from the lumen of the intestine can take place by *passive transport*, involving simple diffusion, provided there is a high concentration of the nutrient outside the cell and a low concentration inside. Absorption by this method is generally a relatively slow process. An alternative and faster process is by *active transport* which involves the use of a specific carrier. Absorption of sugars and amino acids is mainly by this method. The carrier has two specific binding sites and the organic nutrient is attached to one of these while the other site picks up sodium. The loaded carrier moves across the intestinal membrane and deposits the organic nutrient and the sodium inside the cell. The empty carrier then returns back across the membrane free to pick up more nutrients. The sodium which enters the cell is actively pumped back into the lumen where it is available for attachment to another carrier site. A number of different carriers are thought to exist although some may carry more than one nutrient; for example xylose can be bound by the same carrier as glucose. Little is known about the nature of the carriers although one such compound which transports D-glucose has recently been isolated from the microvilli of the hamster and has been shown to be a protein with a molecular weight of 55 000.

A third method of absorption is by *pinocytosis* (cell drinking) in which cells have the capacity to engulf large molecules in solution or suspension. Such a process is particularly important in many newborn suckled mammals, in which immunoglobulins present in colostrum are absorbed intact.

Carbohydrates. The digestion of carbohydrates by enzymes secreted by the pig and other monogastric animals results in the production of monosaccharides. The formation of these simple sugars from disaccharides takes place on the surface of the microvillus membrane. The aldoses, such as glucose, are actively transported across the cell after attachment to the specific carrier and carried by the portal blood systems to the liver. The mechanism for the absorption of ketoses is unclear, although the existence of a carrier for fructose, which is sodium independent, has been postulated. The rates of absorption of various sugars differ. At equal concentrations, galactose, glucose, fructose, mannose, xylose or arabinose are absorbed in decreasing order of magnitude.

Fats. Fats, after digestion, are present in the small intestine in the form of mixed micelles. The exact pattern of absorption of these micelles is not entirely clear, but it is suggested that the lipid components are absorbed into the mucosal cells of the jejunum by passive diffusion, while the bile salts are absorbed by an active process in the ileum. Following absorption there is a resynthesis of triacylglycerols into chylomicrons (minute fat droplets), which then pass into the lacteals of the villi, enter the thoracic duct and join the general circulation. Medium and short chain fatty acids, such as those occurring in butterfat, do not require bile salts nor micelle formation as they can be absorbed very rapidly from the lumen of the intestine directly into the portal blood stream. The entry of these fatty acids is sodium dependent and takes place against a concentration gradient by active transport. In fowls the lymphatic system is negligible and most of the fat is transported in the portal blood as low-density lipoproteins.

Proteins. The products of protein digestion in the lumen of the intestine are free amino acids and small (oligo-) peptides. The latter enter the epithelial cells of the small intestine where they are hydrolysed by specific di- and tri-peptidases. The amino acids, which pass into the portal blood and thence to the liver, are absorbed from the small intestine by an active transport mechanism which in most cases is sodium dependent. In the case of glycine, proline and lysine, the sodium molecule is unnecessary. Several systems have been described for amino acid transfer and these can be classified into four main groups. One is concerned with the transfer of neutral amino acids, whereas separate carriers transfer dicarboxylic and basic amino acids. In addition, a fourth system is involved in the movement of the imino acids and glycine. These mechanisms, however, are not completely rigid and some amino acids can be transferred by more than one system.

Reference has already been made to the absorption of intact proteins, such as immunoglobulins, in the newborn animal by pinocytosis.

Minerals. Absorption of mineral elements is either by simple diffusion or by carrier-mediated transport. The exact mechanisms for all minerals have not been established, but the absorption of calcium, for example, is regulated by 1,25-dehydroxycholecalciferol (see p. 66). Low alimentary *p*H favours calcium absorption but it is inhibited by a number of dietary factors such as the presence of oxalates and phytates (see p. 96). An excess of either calcium or phosphorus interferes with the absorption of the other. The absorption of calcium is also influenced by the requirements of the animal. For example, the absorption of calcium from the digestive tract of laying hens is much greater when shell formation is in progress than when the shell gland is inactive.

The absorption of iron is to a large extent independent of the dietary source. The animal has difficulty in excreting iron from the body in any quantity, and therefore a method exists of regulating the iron absorption to prevent excessive amounts entering the body (see p. 103). In adults the absorption of the element is generally low, but after severe bleeding and during pregnancy the requirement for iron is increased, so that absorption of the element is also increased. Anaemia due to iron deficiency may, however, develop on iron-low diets. Experiments carried out with dogs have shown that the absorption of iron by anaemic animals may be 20 times as great as that by normal healthy dogs.

Zinc resembles iron in being poorly absorbed from the alimentary canal. Calcium is believed to inhibit the absorption of zinc.

Iodine is present mainly as inorganic iodide in plants, whereas in foods of animal origin it exists partly in an organic form. It is thought that the iodine in organic combination is less well absorbed than the inorganic form.

Vitamins. The fat-soluble vitamins A,D,E and K pass through the intestinal mucosa mainly by simple diffusion. Within the cells they may combine with proteins and enter the general circulation as lipoproteins.

Vitamin A is more readily absorbed from the digestive tract than its precursor carotene, although it is thought that vitamin A esters must first be hydrolysed by an esterase to the alcohol form before being absorbed. Phytosterols are poorly absorbed, and it is generally considered that unless ergosterol has been irradiated to vitamin D_2 before ingestion it cannot be absorbed from the tract in any quantity.

Water-soluble vitamins are believed to be absorbed both by simple diffusion and by carrier-mediated transport which is sodium dependent. The importance of a carrier glycoprotein (intrinsic factor) for the absorption of vitamin B_{12} has already been stressed in an earlier chapter (p. 86).

DIGESTION IN THE RUMINANT

The stomach of the ruminant is divided into four compartments (see Fig. 8.1). In the young suckling the first two compartments, the rumen and its continuation the reticulum, are relatively undeveloped, and milk, on reaching the stomach, is channelled by a tube-like fold of tissue, known as the oesophageal or reticular groove, directly to the third and fourth compartments, the omasum and abomasum. As the calf or lamb begins to eat solid food the first two compartments (often considered together as the reticulo-rumen) enlarge greatly, until in the adult they comprise 85 per cent of the total capacity of the stomach. In the adult, the oesophageal groove does not function under normal feeding conditions, and both food and water pass into the reticulo-rumen. However, the reflex closure of the groove to form a channel can be stimulated even in adults, particularly if they are allowed to drink from a teat.

The food is diluted with copious amounts of saliva, firstly during eating and again during rumination: typical quantities of saliva produced per day are 150 l in cattle and 10 l in sheep. Rumen contents contain 850–930 g water/kg on average, but they often exist in two phases: a lower liquid phase, in which the finer food particles are suspended, and a drier upper layer of coarser solid material. The breakdown of food is accomplished partly by physical and partly by chemical means. The contents of the rumen are continually mixed by the rhythmic contractions of its walls, and during rumination material at the anterior end is drawn back into the oesophagus and returned by a wave of contraction to the mouth. Any liquid is rapidly swallowed again, but coarser material is thoroughly chewed before being returned to the rumen. The major factor inducing the animal to ruminate is probably the tactile stimulation of the epithelium of the anterior rumen; some diets, notably those containing little or no coarse roughage, may fail to provide sufficient stimulation for rumination. The time spent by the animal in rumination depends on the fibre content of the food. In grazing cattle it is commonly about 8 hours per day, or about equal to the time spent in grazing. Each bolus of food regurgitated is chewed 40–50 times and thus receives a much more thorough mastication than during eating.

The chemical breakdown of food in the reticulo-rumen is brought about by enzymes secreted, not by the animal itself, but by micro-organisms. The reticulo-rumen provides a continuous culture system for anaerobic bacteria and protozoa (and also some fungi). Food and water enter the rumen and the former is partially fermented to yield principally volatile fatty acids, microbial cells and the gases methane and carbon dioxide. The gases are lost by eructation (belching) and the volatile fatty acids are mainly absorbed through the rumen wall. The microbial cells, together with undegraded food components, pass to the abomasum and small intestine; there they are digested by enyzmes secreted by the host animal, and the products of digestion are absorbed.

Like other continuous culture systems, the rumen requires a number of homeostatic mechanisms. The acids produced by fermentation are theoretically capable of reducing the pH of rumen liquor to 2.5–3.0, but under normal conditions the pH is maintained at 5.5–6.5. Phosphate and bicarbonate contained in the saliva act as buffers; in addition, the rapid absorption of the acids (and also of ammonia—see below) helps to stabilise the pH. The osmotic pressure of rumen contents is kept near that of the blood by the flux of ions between them. Oxygen entering with the food is quickly used up, and anaerobiosis is maintained. The temperature of rumen liquor remains close to that of the animal (38–42°C). Finally, the undigested components of the food, together with soluble nutrients and bacteria, are eventually removed from the rumen by the passage of digesta through the reticulo-omasal orifice.

Rumen micro-organisms

The bacteria number 10^9–10^{10} per ml of rumen contents. Over 60 species have been identified, and for descriptions of them the reader is referred to the works listed at the end of this chapter. Most are non spore forming anaerobes. Table 8.3 lists a number of the more important species and indicates the substrate they utilise and the products of the fermentation. This information is based on studies of isolated species *in vitro* and is not completely applicable *in vivo*. For example, it appears from Table 8.3 that succinic acid is an important end product, but in practice this is converted into propionic acid by other bacteria such as *Selenomonas ruminantium* (Fig. 8.3) such interactions between micro-organisms are an important feature of rumen fermentation. A further point is that the activities of a given species of bacteria may vary from one strain of that species to another. The total numbers of bacteria, and the relative population of individual species, vary with the animal's diet; for example, diets rich in concentrate foods promote high total counts and encourage the proliferation of lactobacilli.

Protozoa are present in much smaller numbers (10^6 ml) than bacteria but, being larger, may equal the latter in total mass. In adult animals, most of

TABLE 8.3   Typical rumen bacteria, their energy sources and fermentation products *in vitro*

Species	Description	Typical energy source	Typical fermentation products*						Alternative energy sources
			Acetic	Propionic	Butyric	Lactic	Succinic	Formic	
Bacteroides succinogenes	Gram negative rods	Cellulose	+				+	+	Glucose, (starch)
Ruminococcus flavefaciens	Catalase negative streptococci with yellow colonies	Cellulose	+				+	+	Xylan
Ruminococcus albus	Single or paired cocci	Cellobiose	+				+	+	Xylan
Streptococcus bovis	Gram positive, short chains of cocci; capsulated	Starch				+			Glucose
Bacteroides ruminicola	Gram negative, oval or rod	Glucose	+				+	+	Xylan, starch
Megasphaera elsdenii	Large cocci, paired or in chains	Lactate	+	+	+				Glucose, glycerol

* Excluding gases

the protozoa are ciliates belonging to two families. The Isotrichidae commonly called the holotrichs, are ovoid organisms covered with cilia; they include the genera *Isotricha and Dasytricha*. The Ophryoscolecidae or oligotrichs, include many species varying considerably in size, shape and appearance; they include the genera *Entodinium, Diplodinium, Epidinium* and *Ophryoscolex*. The oligotrichs can ingest food particles and can utilise both simple and complex carbohydrates, including cellulose. The holotrichs, on the other hand, do not generally ingest food particles and cannot utilise cellulose.

A normal rumen flora (bacteria) and fauna (protozoa) is established quite early in life—as early as six weeks of age in calves. If ruminants are born into, and reared in a special germ-free environment (so-called gnotobiotic animals) they may be inoculated with one or more selected species of bacteria for experimental studies. Animals reared in a normal environment, but kept away from other ruminants, do not develop a protozoal population and may likewise be used for the study of selected protozoal species. Protozoa are easily killed by a low rumen pH, and are generally absent from animals fed on diets which promote—even transiently—a low pH; all-concentrate diets, which are fermented rapidly, come into this category.

Ruminants without rumen protozoa are apparently normal and healthy, and the benefit they gain from these organisms is therefore questionable. Protozoa take up soluble sugars from the rumen liquor (thus retarding their fermentation), and they attack cell walls. They also ingest and digest bacteria, in effect remodelling bacterial protein to protozoal protein. About 25 per cent of the microbial protein digested by the host animal is likely to be protozoal protein. However, protozoa have the undesirable characteristic of being retained in the rumen, where they may 'lock up' protein and prevent its use by the host animal; with low-protein diets this can be a distinct disadvantage. With better quality diets, several comparisons of faunated and defaunated ruminants have shown the former to grow faster.

As the microbial mass synthesized in the rumen provides about 20 per cent of the nutrients absorbed by the host animal, the composition of the micro-organisms is important. The bacterial dry matter contains about 100 g N/kg, but only 80 per cent of this is in the form of amino acids, the remaining 20 per cent being present as nucleic acid N. Moreover, some of the amino acids are contained in the peptidoglycan of the cell wall membrane, and are not digested by the host animal.

The diet of the ruminant contains considerable quantities of cellulose, hemicelluloses, starch and water-soluble carbohydrates mostly in the form of fructans. Thus in young pasture herbage, which is frequently the sole food of the ruminant, each kilogram of dry matter may contain about 400 g cellulose and hemicelluloses, and 200 g of water-soluble carbohydrates. In mature herbage, and in hay and straw, the proportion of cellulose and hemicelluloses is much higher and that of water-soluble carbohydrates is much lower. The β-linked carbohydrates are associated with lignin, which may comprise 20–120 g/kg DM. All the carbohydrates, but not lignin, are attacked by the rumen micro-organisms.

The breakdown of carbohydrates in the rumen may be divided into two stages, the first of which is the digestion of complex carbohydrates to simple sugars. This is brought about by extracellular microbial enzymes and is thus analogous to the digestion of carbohydrates in non-ruminants. Cellulose is decomposed by one or more β-1,3-glucosidases to cellobiose which is then converted either to glucose or, through the action of a phosphorylase, to glucose-1-phosphate. Starch and dextrins are first converted by amylases to maltose and isomaltose and then by maltases, maltose phosphorylases or 1,6-glucosidases to glucose or glucose-1-phosphate. Fructans are hydrolysed by enzymes attacking 2,1 and 2,6 linkages to give fructose, which may also be produced—together with glucose—by the digestion of sucrose (see Fig. 8.2).

Pentoses are the major product of hemicellulose breakdown which is brought about by enzymes attacking the β-1,4 linkages to give xylose and uronic acids. The latter are then converted to xylose. Uronic acids are also produced from pectins, which are first hydrolysed to pectic acid and methanol by pectinesterase. The pectic acid is then attacked by polygalacturonidases to give galacturonic acids, which in turn yield xylose. Xylose may also be produced from hydrolysis of the xylans which may form a significant part of the dry matter of grasses.

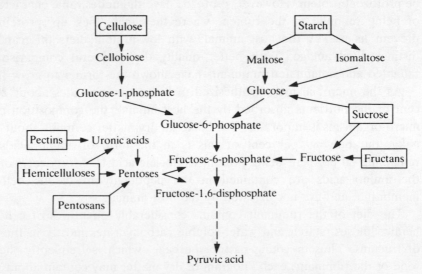

Fig. 8.2.   Conversion of carbohydrates to pyruvate in the rumen.

The simple sugars produced in the first stage of carbohydrate digestion in the rumen are rarely detectable in the rumen liquor because they are immediately taken up and metabolised intra-cellularly by the micro-organisms. For this second stage, the pathways involved are in many respects similar to those followed in the metabolism of carbohydrates by the animal itself, and are thus discussed in the next chapter. The main end products of the metabolism of carbohydrates by rumen micro-organisms are acetic, propionic and butyric acids, and carbon dioxide and methane (see Fig. 8.3). Pyruvic, succinic and lactic acids are important intermediates, and lactic acid can sometimes be detected in rumen liquor. Additional fatty acids are also formed, generally in small quantities, by deamination of amino acids in the rumen; these are isobutyric from valine, valeric from proline, 2-methylbutyric from isoleucine and 3-methylbutyric from leucine. The total concentration of volatile fatty acids in rumen liquor varies between 2 and 15 g/l according to the animal's diet and the time that has elapsed since the previous meal. The relative proportions of the acids also vary, and some typical figures are shown in Table 8.4.

The predominant acid is acetic, and roughage diets high in cellulose give rise to acid mixtures particularly high in acetic acid. As the proportion of concentrates in the diet is increased, the proportion of acetic acid falls and that of propionic rises. With all-concentrate diets the proportion of propionic acid may even exceed that of acetic. However, even with this type of diet, acetic acid predominates if the rumen ciliate protozoa survive in large numbers (see Table 8.4). The grinding (and pelleting) of diets consisting of roughage alone does not have a large effect on the proportions of the volatile fatty acids produced. But if roughage forms only a part of the diet, grinding and pelleting it causes a marked change in the relative proportions of acetic and propionic acids.

The total weight of acids produced may be as high as 4 kg per day in cows. Much of the acid produced is absorbed directly from the rumen, reticulum and omasum, although some may pass through the abomasum and be absorbed in the small intestine. In addition some of the products of carbohydrate digestion in the rumen are used by bacteria and protozoa to form their own cellular polysaccharides, but the amounts passing to the small intestine are probably small and hardly significant.

The rate of gas production in the rumen is most rapid immediately after a meal and in the cow may exceed 30 l/hour. The typical composition of rumen gas is carbon dioxide, 40 per cent; methane, 30–40 per cent; hydrogen, 5 per cent; together with small and varying proportions of oxygen and nitrogen (from ingested air). Carbon dioxide is produced partly as a by-product of fermentation and partly by the reaction of organic acids with the bicarbonate present in the saliva. The basic reaction by which methane is

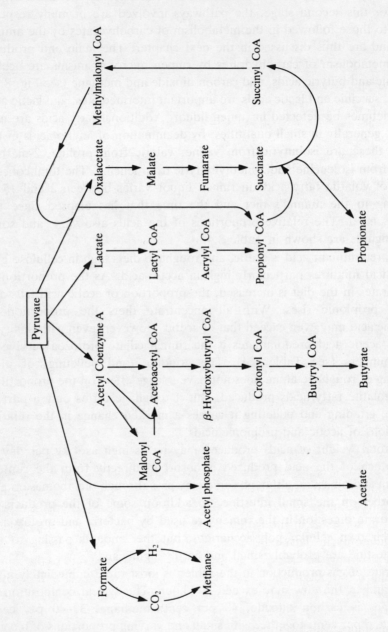

Fig. 8.3.   Conversion of pyruvate to volatile fatty acids in the rumen.

TABLE 8.4   Volatile fatty acids (VFA) in the rumen liquor of cattle or sheep fed on various diets

Animal	Diet	Total VFA (m moles/litre)	Individual VFA (molar proportions)			
			Acetic	Propionic	Butyric	Others
Sheep	Young ryegrass herbage	107	0.60	0.24	0.12	0.04
Cattle	Mature ryegrass herbage	137	0.64	0.22	0.11	0.03
Cattle	Grass silage	108	0.74	0.17	0.07	0.03
Sheep	Chopped lucerne hay	113	0.63	0.23	0.10	0.04
	Ground lucerne hay	105	0.65	0.19	0.11	0.05
Cattle	Long hay (0.4), concentrates (0.6),	96	0.61	0.18	0.13	0.08
	Pelleted hay (0.4), concentrates (0.6)	140	0.50	0.30	0.11	0.09
Sheep	Hay : concentrate					
	1 : 0	97	0.66	0.22	0.09	0.03
	0.8 : 0.2	80	0.61	0.25	0.11	0.03
	0.6 : 0.4	87	0.61	0.23	0.13	0.02
	0.4 : 0.6	76	0.52	0.34	0.12	0.03
	0.2 : 0.8	70	0.40	0.40	0.15	0.05
Cattle	Barley (no ciliate protozoa in rumen)	146	0.48	0.28	0.14	0.10
Cattle	Barley (ciliates present in rumen)	105	0.62	0.14	0.18	0.06

formed is the reduction of carbon dioxide by hydrogen, some of which may be derived from formate. Methanogenesis, however, is a complicated process which involves folic acid and vitamin B_{12}. About 4.5 g of methane is formed for every 100 g of carbohydrate digested, and the ruminant loses about 7 per cent of its food energy as methane (see Ch. 11).

Most of the gas produced is lost by eructation; if gas accumulates it causes the condition known as bloat, in which the distension of the rumen may be so great as to result in the collapse and death of the animal. Bloat occurs most commonly in dairy cows grazing on young, clover-rich herbage, and is due not so much to excessive gas production as to the failure of the animal to eructate. Frequently the gas is trapped in the rumen in a foam, whose formation may be promoted by substances present in the clover. It is also possible that the reflex controlling eructation is inhibited by a physiologically active substance which is present in the food or formed during fermentation. Bloat is a particularly serious problem on the clover-rich pastures of New Zealand, where it is prevented by dosing the cows or spraying the pasture with anti-foaming agents such as vegetable oils. Another form of bloat, termed 'feed lot bloat' occurs in cattle fattened intensively on diets containing much concentrate and little roughage.

The extent to which cellulose is digested in the rumen depends particularly on the degree of lignification of the plant material. Lignin, and also the related substance cutin, is resistant to attack by anaerobic bacteria, probably because of its low oxygen content and its condensed structure (which inhibits hydrolysis). Lignin appears to hinder the breakdown of the cellulose with which it is associated. Thus, in young pasture grass containing only 50 g lignin/kg DM, 80 per cent of the cellulose may be digested, but in older herbage with 100 g lignin/kg the proportion of cellulose digested may be less than 60 per cent. Ruminant diets based on cereals may contain as much as 500 g/kg of starch (and sugars), of which over 90 per cent is fermented in the rumen and the rest digested in the small intestine. This fermentation is rapid, and the resulting fall in the pH of rumen liquor inhibits cellulose-fermenting organisms and thus depresses the breakdown of cellulose.

The breakdown of cellulose and other resistant polysaccharides is undoubtedly the most important digestive process taking place in the rumen. Besides contributing to the energy supply of the ruminant, it ensures that other nutrients which might escape digestion are exposed to enzyme action. Although the main factor in the process is the presence of micro-organisms in the rumen, there are other factors of importance. The great size of the rumen—its contents normally contribute 10–20 per cent of the liveweight of ruminants—allows food to accumulate and ensures that sufficient time is allowed for the rather slow breakdown of cellulose. In addition, the movements of the reticulo-rumen and the act of rumination play a part by breaking up the food and exposing it to attack by micro-organisms.

Digestion of protein

Food proteins are hydrolysed to peptides and amino acids by rumen microorganisms, but some amino acids are degraded further, to organic acids, ammonia and carbon dioxide. An example of the deamination of amino acids is provided by valine which, as mentioned above, is converted to isobutyric acid. Thus the branched-chain acids found in rumen liquor are derived from amino acids. The ammonia produced, together with some small peptides and free amino acids, is utilised by the rumen organisms to synthesise microbial proteins. When the organisms are carried through to the abomasum and small intestine their cell proteins are digested and absorbed. An important feature of the formation of microbial protein is that bacteria are capable of synthesising indispensable as well as dispensable amino acids, thus rendering their host independent of dietary supplies of the former.

The extent to which dietary protein is degraded to ammonia in the rumen, and conversely the extent to which it escapes rumen degradation and is subsequently digested in the small intestine, will be discussed in later chapters (10 and 13). At this point it is sufficient to emphasise that with most diets, the greater part (and sometimes all) of the protein reaching the ruminant's small intestine will be microbial protein of reasonably constant composition. The lesser part will be undegraded food protein, which will vary in amino acid composition according to the nature of the diet.

The ammonia in rumen liquor is the key intermediate in the microbial degradation and synthesis of protein. If the diet is deficient in protein, or if the protein resists degradation, the concentration of rumen ammonia will be low (c 50 mg/l) and the growth of rumen organisms will be slow; in

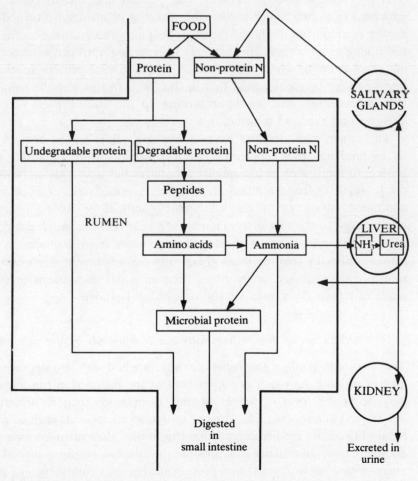

Fig. 8.4. Digestion and metabolism of nitrogenous compounds in the rumen.

consequence, the breakdown of carbohydrates will be retarded. On the other hand if protein degradation proceeds more rapidly than synthesis, ammonia will accumulate in rumen liquor and the optimum concentration will be exceeded. When this happens, ammonia is absorbed into the blood, carried to the liver and converted to urea (see Fig. 8.4). Some of this urea may be returned to the rumen via the saliva, and also directly through the rumen wall, but the greater part is excreted in the urine and thus wasted.

Estimates of the optimum concentration of ammonia in rumen liquor vary widely, from 85 to over 300 mg/l. Rather than expressing the optimum as the concentration in rumen liquor, it would probably be more realistic to relate ammonia to fermentable organic matter, since it is known that for each kg of organic matter fermented, approximately 30 g N is taken up by rumen bacteria as protein and nucleic acid.

If the food is poorly supplied with protein and the concentration of ammonia in rumen liquor is low, the quantity of nitrogen returned to the rumen as urea from the blood (see Fig. 8.4) may exceed that absorbed from the rumen as ammonia. This net gain in 'recycled' nitrogen is converted to microbial protein, which means that the quantity of protein reaching the intestine may be greater than that in the food. In this way the ruminant is able to conserve nitrogen by returning to the rumen urea that would otherwise be excreted in urine.

The rumen microbes thus have a 'levelling' effect on the protein supply of the ruminant; they supplement, both quantitatively and qualitatively, the protein of foods such as low-quality roughages but have a deleterious effect on protein-rich concentrates. It is now an established practice to take additional advantage of the synthesising abilities of rumen bacteria by adding urea to the diet of ruminants (see below). A more recent development is the protection of good-quality proteins from degradation in the rumen, either by treating them chemically (with formalin, for example) to reduce their solubility, or by giving them in liquid suspensions that can be made to bypass the rumen via the oesophageal groove.

Utilisation of non-protein nitrogen compounds by the ruminant

The ammonia pool in the rumen is not supplied only by degradation of dietary protein. As much as 30 per cent of the nitrogen in ruminant foods may be in the form of simple organic compounds such as amino acids, amides and amines (see Ch. 4) or of inorganic compounds such as nitrates. Most of these are readily degraded in the rumen, their nitrogen entering the ammonia pool. In practice it is possible to capitalise on the ability of rumen micro-organisms to convert non-protein nitrogenous compounds to protein, by adding such compounds to the diet. The substance most commonly employed is urea, but various derivatives of urea, and even ammonium salts, may also be used.

Urea entering the rumen is rapidly hydrolysed to ammonia by bacterial urease, and the rumen ammonia concentration is therefore liable to rise considerably. For this ammonia to be efficiently incorporated in microbial protein, two conditions must be met. Firstly, the initial ammonia concentration must be below the optimum (otherwise the ammonia 'peak' produced will simply be absorbed and lost from the animal as described above), and secondly the micro-organisms must have a readily available source of energy for protein synthesis. Feeding practices intended to meet these conditions include mixing urea with other foods (to prolong the period over which it is ingested and deaminated). Such foods should be low in rumen-degradable protein, and high in readily fermentable carbohydrate.

It is important to avoid accidental over-consumption of urea, since the subsequent rapid absorption of ammonia from the rumen can overtax the ability of the liver to re-convert it to urea, hence causing the ammonia concentration of peripheral blood to reach toxic levels.

Derivatives of urea have been used for animal feeding with the intention of retarding the release of ammonia. Biuret is less rapidly hydrolysed than urea but requires a period of several weeks for rumen micro-organisms to adapt to it. However, neither biuret nor isobutylidene diurea (IBDU) nor urea-starch compounds have consistently proved superior to urea itself.

An additional non-protein nitrogenous compound which can be utilised by rumen bacteria, and hence by the ruminant, is uric acid. This is present in high concentration in poultry excreta, and these are sometimes dried for inclusion in diets for ruminants.

The practical significance of these non-protein nitrogenous substances as potential protein sources is discussed in Chapter 22.

Digestion of lipids

The triacylglycerols present in the foods consumed by ruminants contain a high proportion of residues of the C_{18} poly-unsaturated acids, linoleic and linolenic. These triacylglycerols are to a large extent hydrolysed in the rumen by bacterial lipases, as are phospholipids. Once they are released from ester combination, the unsaturated fatty acids are hydrogenated by bacteria, yielding first a monoenoic acid and ultimately, stearic acid. Both linoleic and linolenic acid have all *cis* double bonds, but before they are hydrogenated one double bond in each is converted to the *trans* configuration; thus, *trans* acids can be detected in rumen contents. The rumen micro-organisms also synthesise considerable quantities of lipids, which contain some unusual fatty acids (such as those containing branched side chains) and these acids are eventually incorporated in the milk and body fats of ruminants.

The capacity of rumen micro-organisms to digest lipids is strictly limited. The lipid content of ruminant diets is normally low (i.e. < 50 g/kg) and if it is increased above 100 g/kg the activities of the rumen microbes are reduced. The fermentation of carbohydrates is retarded, and food intake falls.

Unlike their short-chain counterparts, long-chain fatty acids are not absorbed directly from the rumen. When they reach the small intestine they are mainly saturated and unesterified, but some—in the bacterial lipids— are esterified. Monoacylglycerols, which play an important role in the formation of mixed micelles in non-ruminants, are not present. The formation of micelles in the gut of the ruminant, and hence the absorption of long-chain acids, is dependent on the phospholipids present in bile.

As mentioned earlier (p. 31) it is possible in non-ruminants to vary the fatty acid composition of body fats by altering the composition of dietary fats. In ruminants, this is normally not the case, and the predominating fatty acid of ruminant depot fats is the stearic acid resulting from hydrogenation in the rumen. However, it is possible to treat dietary lipids in such a way that they are protected from attack in the rumen but remain susceptible to enzymic hydrolysis and absorption in the small intestine. If such lipids contain unsaturated acids, they will modify the composition of body (and milk) fats.

Synthesis of vitamins

The synthesis by rumen micro-organisms of all members of the vitamin B complex and of vitamin K has already been mentioned (see Ch. 5). In ruminants receiving foods well supplied with B vitamins the amounts synthesised are relatively small, but they increase if the vitamin intake in the diet decreases. The adult ruminant is therefore independent of a dietary source of these vitamins, but it should be remembered that adequate synthesis of vitamin B_{12} will take place only if there is sufficient cobalt in the diet.

Dynamics of digestion in the ruminant

Food and water entering the rumen may leave it in different forms and by different routes. The liquid phase of rumen contents may be envisaged as a tank of fixed volume, and liquid or suspended material entering the tank causes a corresponding volume of liquid to flow out through the reticulo-omasal orifice. This liquid carries with it the soluble food constituents that have escaped fermentation, suspended fine particles of food, bacteria, and the volatile fatty acids that have not been absorbed through the rumen wall. The passage of liquid through the rumen can be expressed as the *dilution rate*, which is defined as the fraction of the liquid volume which leaves the rumen per unit time. Dilution rate can be measured by injecting a dose of soluble but unabsorbable marker, such as the high molecular weight polymer polyethylene glycol, into the rumen and recording the subsequent decline in its concentration in rumen liquor. In sheep, dilution rates are typically 0.03 to 0.15 per hour, but higher in cattle (up to 0.2 per hour). Dilution rates are higher with roughage diets than with those containing concentrates, and they increase as intake increases. Increasing the dilution rate, for example by adding salts to the diet and so increasing water intake, can change the bacterial population of the rumen and modify digestion. Thus increasing the dilution rate often reduces cellulolysis and increases the proportion of propionic acid in the volatile fatty acids; it may also increase the quantity of microbial protein synthesised per unit of organic matter fermented. If a faster dilution rate removes more fine particles, the digestion of fibre in the rumen will be reduced, but this may be corrected by fermentation in the hind gut.

Material entering the rumen as large particles spends longer in the rumen than small particles and soluble nutrients, because—as explained above (p. 142)—large particles must be broken down by rumination and microbial attack before they can leave the rumen. The passage of particulate matter through the tract can be followed by staining some of the food particles with a dye, or marking them with a chemical 'tracer'. The indigestible residues of these particles are recovered in faeces and counted, and their average *retention time* can be calculated. Foods with highly lignified cell walls, such as straws, have long retention times (typically 50–80 hours), whereas more readily digested foods, like immature pasture herbage or concentrates, have short retention times (30–50 hours). Foods which are slowly digested (and hence have long retention times) reduce the throughput of the digestive tract and thus reduce the food consumption of ruminants. (see Ch. 16). Grinding such feeds (i.e. reducing their particle size) increases throughput and hence intake, but is likely to reduce the efficiency of digestion because micro-organisms have insufficient time in which to degrade cell walls (see Ch. 10).

ALTERNATIVE SITES OF MICROBIAL DIGESTION

Its great size, and its location at the anterior end of the gut, make the rumen pre-eminent as a digestive organ for bulky foods. Large quantities of food can be stored rapidly for later mastication and fermentation; cell contents are released at an early stage; the main products of fermentation have ample opportunity to be absorbed in the remainder of the tract. These advantages of rumen digestion are, however, diminished by the disadvantage of all food constituents being exposed to fermentation. If fermentation is delayed until food reaches the large intestine, this disadvantage is overcome, but some of the benefits of the rumen are lost.

The parts of the large intestine which are capable of sustaining a significant microbial population are the colon and the caecum (Fig. 8.1). The caecum is blind-ended, and is duplicated in the fowl; in some animals the walls of both caecum and colon are sacculated. The digestive capacity of these organs depends on their volume relative to the rest of the tract, as this determines the time for which food residues may be delayed for fermentation. The substrate for the large intestine differs from that entering the rumen, because most of the more readily digestible nutrients will have been removed, and also some endogenous materials (such as mucopolysaccharides and enzymes) will have been added. However, as described earlier (pp. 137), microbial digestion in the large intestine is similar to that occurring in the rumen. Volatile fatty acids are produced and absorbed; methane and other gases are present. Proteins and non-protein sources of nitrogen (such as urea from the bloodstream) are re-formed into microbial proteins; in some cases, but not invariably, these undergo proteolysis to amino acids, which may be absorbed. Water-soluble vitamins are synthesised and inorganic elements and water are reabsorbed. But in general, hind-gut fermentation is less effective than rumen digestion, because digesta are not held for sufficient time and because many of the products of digestion (particularly amino acids and vitamins) are not absorbed. Some species of animal overcome the last problem by practising coprophagy (the consumption of faeces). The rabbit has perfected this practice by producing two types of faeces, the normal hard pellets (which are not eaten) and the soft faeces or caecotrophes, which contain well-fermented material from the caecum, and which are consumed.

Ruminants have a substantial capacity for hind-gut fermentation, and this is used to good effect when the diet or the level of feeding cause much fibrous material to reach the caecum (see Table 10.2). The horse relies on its enlarged colon for the digestion of the forages it consumes. In the pig, the hind gut is less enlarged and forages are poorly digested. Nevertheless, the pig can digest as much as half of the cellulose and hemicellulose of cereal grains and their by-products. If starch grains escape digestion in the small intestine, as happens when pigs are fed on uncooked potatoes, these will also be fermented.

Although poultry have two caeca and a colon in which to ferment food residues, they gain little or nothing from hind-gut fermentation when fed on their usual concentrate diets. Indeed, it has been suggested that in poultry the intestinal microflora is more of a handicap than an advantage, as 'germ-free' birds (i.e. reared in isolation and with no micro-organisms) tend to grow larger than normal birds.

FURTHER READING

D. C. Church, 1976. *Digestive Physiology and Nutrition of Ruminants*. Vol. 1, *Digestive Physiology*, 2nd edn. O & E Books, Corvallis, Oregon, USA.

B. M. Freeman (ed.) 1983. *Physiology and Biochemistry of the Domestic Fowl*. Academic Press, London

R. E. Hungate, 1966 *The Rumen and its Microbes*. Academic Press, New York

D. E. Kidder and M. J. Manners, 1978. *Digestion in the Pig*. Scientechnica, Bristol.

A. G. Low and I. G. Partridge (eds) *Current Concepts of Digestion and Absorption in Pigs*. National Institute for Research in Dairying, Reading.

D. W. Martin, P. A. Mayes and V. W. Rodwell, 1983. *Harpers Review of Biochemistry*. Lange Medical Publ., Los Altos, USA.

L. P. Milligan, W. L. Grovum and A. Dobson, 1986. *Control of Digestion and Metabolism in Ruminants*. (Proceedings of the Sixth International Symposium on Ruminant Physiology, Banff, Canada, 1984). Prentice-Hall, Eaglewood Cliffs, New Jersey, USA (see also other volumes in this series).

P. A. Sandford, 1982. *Digestive System Physiology*, Edward Arnold, London.

M. J. Swenson (ed.) 1977. *Dukes Physiology of Domestic Animals*, 9th edn. Cornell Univ. Press, Ithaca, USA.

P. J. Van Soest, 1982. *Nutritional Ecology of the Ruminant*. O & B Books, Corvallis, Oregon, USA.

9

Metabolism

Metabolism is the name given to the sequence, or succession, of chemical processes that take place in the living organism. Some of the processes involve the degradation of complex compounds to simpler materials and are designated by the general term *catabolism. Anabolism* describes those metabolic processes in which complex compounds are synthesised from simpler substances. Waste products arise as a result of metabolism and these have to be chemically transformed and ultimately excreted; the reactions necessary for such transformations form part of the general metabolism. As a result of the various metabolic processes energy is made available for mechanical work, and for chemical work such as the synthesis of carbohydrates, proteins and lipids. Figure 9.1 summarises the sources of the major metabolites available to the body, and their subsequent metabolism.

The starting points of metabolism are the substances produced by the digestion of food. For all practical purposes we may regard the endproducts of carbohydrate digestion in the simple-stomached animal as glucose, with very small amounts of galactose and fructose. These are absorbed into the portal blood and carried to the liver. In ruminant animals the major part of the dietary carbohydrate is broken down in the rumen to acetic, propionic and butyric acids, with small amounts of branched chain and higher volatile acids. Butyric acid is changed in its passage across the rumen wall and passes into the portal blood as β-hydroxybutyric acid (BHBA). Acetic acid and propionic acid pass almost unchanged across the rumen wall into the portal blood and are carried, together with the BHBA, to the liver. Acetic acid and BHBA pass from the liver, via the systemic blood, to various organs and tissues where they are used as sources of energy and fatty acids. Propionic acid is converted to glucose in the liver and joins the liver glucose pool. This may be converted partly into glycogen and stored, or to fatty

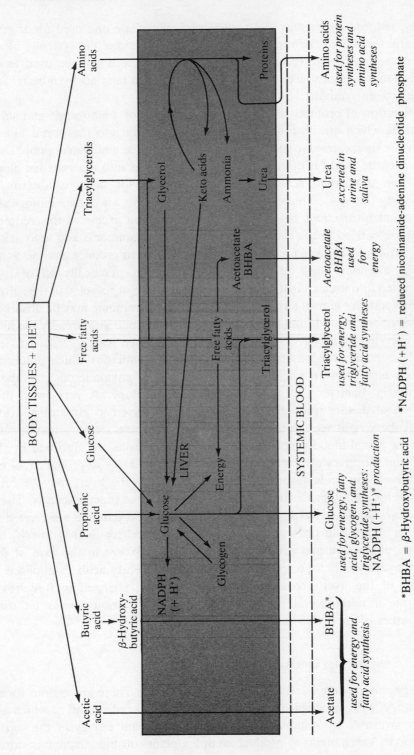

Fig. 9.1. Sources and fates of major body metabolites.

*BHBA = β-Hydroxybutyric acid   *NADPH (+H⁺) = reduced nicotinamide-adenine dinucleotide phosphate

acids, reduced coenzymes and L-glycerol-3-phosphate and used for triacyl-glycerol synthesis. The remainder of the glucose enters the systemic blood supply and is carried to various body tissues where it may be used as an energy source, as a source of reduced coenzymes in fatty acid synthesis, and for glycogen synthesis.

Digestion of proteins results in the production of amino acids and small peptides which are absorbed via the intestinal villi into the portal blood. They are then carried to the liver where they join the amino acid pool. They may then be used for protein synthesis or may pass into the systemic blood and join the amino acids produced as a result of tissue catabolism in providing the raw material for synthesis of proteins and other biologically important nitrogenous compounds. Amino acids in excess of this require-ment are carried to the liver and broken down to ammonia and keto acids. The latter may be used for amino acid synthesis or to produce energy. Some of the ammonia may be used in amination but practically all of it is converted into urea and either excreted in the urine or recycled in the saliva. In the ruminant animal a considerable amount of ammonia may be absorbed into the portal blood from the rumen, and transformed into urea by the liver and excreted or recycled via the saliva or the rumen wall.

Most dietary lipids enter the lacteals as chylomicrons which enter the venous blood vessels via the thoracic duct. The chylomicrons are about 500 nm in diameter with a thin lipoprotein envelope. A very small pro-portion of dietary triacylglycerols may be hydrolysed to glycerol and low molecular weight acids in the digestive tract, and these are absorbed directly into the portal blood supply. Circulating chylomicrons are absorbed by the liver and the triacylglycerols hydrolysed. The fatty acids so produced, along with free fatty acids absorbed from the blood by the liver, may be cata-bolised for energy production or used for synthesis of triacylglycerols. These then re-enter the blood supply in the form of lipoprotein and are carried to various organs and tissues where they may be used for lipid synthesis, for energy production and for fatty acid synthesis. Except in the case of the liver, hydrolysis is a prerequisite for absorption. Fatty acids catabolised in excess of the liver's requirement for energy are changed to β-hydroxy-butyrate and acetoacetate which are transported to various tissues and used as sources of energy.

ENERGY METABOLISM

Energy may be defined as the capacity to do work. There are various forms of energy, such as chemical, thermal, electrical and radiant, all of which are interconvertible by suitable means. Thus the radiant energy of the sun is utilised by green plants to produce complex plant constituents, and is stored as such. The plants are consumed by animals and the constituents broken down, releasing energy which is used by the animal for mechanical work, transport, maintenance of the integrity of cell membranes, for synthesis and for providing heat under adverse climatic conditions.

Traditionally, heat units have been used to represent the various forms of energy involved in metabolism, since all the forms are convertible into heat. The basic unit used was the thermochemical calorie (cal), based on the calorific value of benzoic acid as the reference standard. Usually the kilocalorie (= 1000 cal) or the megacalorie (= 1 000 000 cal) were used in practice since the calorie was inconveniently small. The International Union of Nutritional Sciences and the Nomenclature Committee of the International Union of Physiological Sciences have suggested the joule (J) as the unit of energy for use in nutritional, metabolism and physiological studies. This suggestion has been widely adopted in a number of countries and is followed in this book. The joule is defined as 1 newton per metre, and $4.184J \equiv 1$ cal. The conversion of nutritional data to joules necessitates their multiplication by this factor. By analogy with previous practice the units which will be employed henceforth will be the kilojoule (kJ) and the megajoule (MJ).

The chemical reactions taking place in the animal body may be *endergonic* and involve a gain of free energy by the system, or they may result in a loss of free energy from the system, in which case they are termed *exergonic*. Most of the synthetic reactions of the body are endergonic and require a supply of energy so that they may take place. This energy can be obtained from exergonic catabolic changes. Before the energy released by the exergonic reactions can be utilised for syntheses and other vital body processes, a link between the two must be established. This is brought about by mediating compounds which take part in both processes, picking up energy from one and transferring it to the other.

The most important mediating compound in the body is adenosine triphosphate (ATP). Adenosine is formed from the purine base, adenine, and the sugar, D-ribose. Phosphorylation of the hydroxyl group at carbon atom 5 of the sugar gives adenosine monophosphate (AMP) (see Ch. 4); successive additions of phosphate residues give adenosine diphosphate (ADP), and then the triphosphate. The formation of these last two phosphate bonds requires a considerable amount of energy, which may be obtained directly by reaction of AMP or ADP with an energy-rich material. Thus, in carbohydrate breakdown, one of the steps is the change of phosphoenolpyruvate to pyruvate, which results in one molecule of ATP being produced from ADP.

Where production of ATP from ADP takes place directly during a reaction as in this case, the process is known as *substrate level phosphorylation*.

$$
\begin{array}{c}
\underset{\displaystyle \underset{\text{COO}^-}{|}}{\overset{\displaystyle \overset{\text{CH}_2}{||}}{\text{C}}} - \text{O} - \underset{\displaystyle \underset{\text{O}^-}{|}}{\overset{\displaystyle \overset{\text{O}}{||}}{\text{P}}} - \text{O}^-
\end{array}
\;+\; \text{ADP} \;\xrightarrow{\text{Mg}^{2+}}\;
\begin{array}{c}
\underset{\displaystyle \underset{\text{COO}^-}{|}}{\overset{\displaystyle \overset{\text{CH}_3}{|}}{\text{C}}} = \text{O}
\end{array}
\;+\; \boxed{\text{ATP}}
$$

Phosphoenolpyruvate   Pyruvate

Alternatively ATP may be produced indirectly. Most biological oxidations involve the removal of hydrogen from a substrate, but the final combination with oxygen to form water occurs only at the end of a series of reactions. A typical example is the removal of hydrogen coupled to nicotinamide adenine dinucleotide (NAD^+), as illustrated in Fig. 9.2 for the oxidation of isocitrate to α-ketoglutarate. In this example hydrogen removed from isocitrate is accepted by NAD^+ and is then passed to the flavin coenzyme. This donates two electrons to ubiquinone and two protons ($2H^+$) are formed. The electrons are then transferred via the sequential cytochromes to cytochrome a_3 which is capable of transferring the electrons to oxygen. The negatively charged oxygen finally unites with the two protons, yielding water. During the operation of this pathway ATP is produced from ADP and inorganic phosphate, the process being called *oxidative phosphorylation*. It is confined to the mitochondria and to the reduced NAD^+ produced within them. The mechanism of oxidative phosphorylation is not yet clear but considerations of energy release indicate that production of ATP takes place at the transfer of hydrogen from reduced NAD^+ to FAD, at the transfer of electrons from cytochrome b to c_1, and from cytochrome a_3 to oxygen. The series of reactions may be represented as follows:

$$
\text{NADH}(+\,\text{H}^+) + \tfrac{1}{2}\text{O}_2 + 3\text{ADP} + 3\,\text{P}_i \longrightarrow \text{NAD}^+ + \boxed{3\ \text{ATP}} + \text{H}_2\text{O}
$$

Oxidation of each mole of reduced NAD^+ yields three moles of ATP from ADP and inorganic phosphate.

The energy fixed as ATP may be used for doing mechanical work during the performance of essential life processes in maintaining the animal. Both contraction and relaxation of muscle involve reactions which require a supply of energy which is provided by breakdown of ATP to ADP and inorganic phosphate. The energy fixed in ATP may also be used to drive reactions during which the terminal phosphate group is donated to a large variety of acceptor molecules. Among these is D-glucose:

$$
\boxed{\text{ATP}} + \text{D-glucose} \rightleftharpoons \text{ADP} + \text{D-glucose phosphate}
$$

In this way the glucose is energised for subsequent biosynthetic reactions.

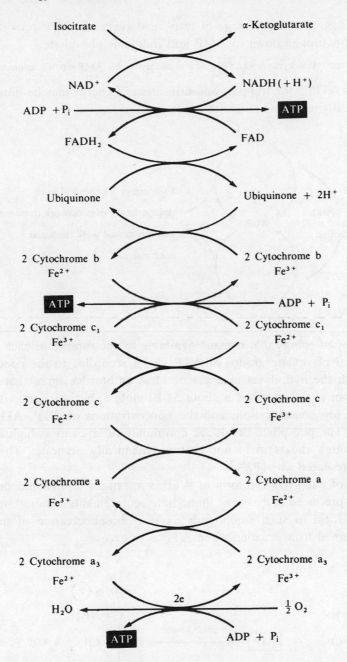

Fig. 9.2. Oxidative phosphorylation system.

In others, such as the first stage of fatty acid synthesis, ATP provides the energy and is broken down to AMP and inorganic phosphate:

$$\text{Acetate} + \text{Coenzyme A} + \boxed{\text{ATP}} \rightleftharpoons \text{Acetyl coenzyme A} + \text{AMP} + \text{pyrophosphate}$$

The role of ATP in the trapping and utilisation of energy may be illustrated diagramatically as follows:

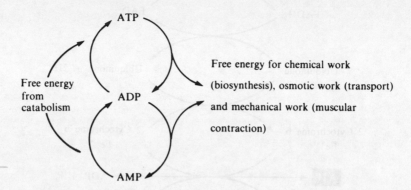

The quantity of energy that is made available by the rupture of each of the two terminal phosphate groups of ATP varies according to the conditions under which the hydrolyses take place. Most authorities agree that under the conditions in intact cells it is about 52 kJ mole^{-1}, but will vary with pH, magnesium ion concentration, and the concentrations of ATP, ADP and phosphate. The phosphate bonds are commonly referred to as high energy bonds although the term is not thermodynamically accurate. They are usually represented as $\sim\!\!\textcircled{P}$.

Fixation of energy in the form of ATP is a transitory phenomenon, and any energy produced in excess of immediate requirements is stored in more permanent form in such compounds as the phosphocreatine of muscle, which is formed from creatine when ATP is in excess:

$$
\begin{array}{lcr}
\begin{array}{l}
\text{NH}_2 \\
| \\
\text{C} = \text{NH}_2{}^+ \\
| \\
\text{N} - \text{CH}_3 \\
| \\
\text{CH}_2 \\
| \\
\text{COO}^-
\end{array}
+ \boxed{\text{ATP}}
\xrightleftharpoons{\textit{Creatine kinase}}
\begin{array}{l}
\text{NH} \sim \textcircled{P} \\
| \\
\text{C} = \text{NH}_2^+ \\
| \\
\text{N} - \text{CH}_3 \\
| \\
\text{CH}_2 \\
| \\
\text{COO}^-
\end{array}
+ \text{ADP}
\end{array}
$$

Creatine   Phosphocreatine

Then when the supply of ATP is insufficient to meet the demands for energy more ATP is produced from phosphocreatine by the reverse reaction.

Even materials like phosphocreatine are only minor, temporary stores of energy—most energy is stored in the body as depot fat, together with small quantities of carbohydrate in the form of glycogen. In addition, protein may be used to provide energy under certain circumstances.

As well as this stored energy, the body also derives energy directly from nutrients absorbed from the digestive tract. One of the chief of these is glucose.

Glucose as an energy source

The major pathway whereby glucose is metabolised to give energy has two stages. The first stage, known as glycolysis, can occur under anaerobic conditions and results in the production of pyruvate. The sequence of reactions from glucose (Fig. 9.3) is often referred to as the Embden-Meyerhof pathway. Two moles of ATP are used in the initial phosphorylations of steps 1 and 3, and the fructose diphosphate so formed then breaks down to yield two moles of glyceraldehyde-3-phosphate. Subsequently one mole of ATP is produced directly at each of steps 8 and 11. Four moles of ATP are thus produced from one mole of glucose. Since two moles of ATP are used up, the net production of ATP from ADP is two moles per mole of glucose. Under aerobic conditions the reduced NAD^+ produced at step 7 may be oxidised via the oxidative phosphorylation pathway, with the production of three moles of ATP per mole of reduced coenzyme. Under aerobic conditions, therefore, glycolysis yields eight moles of ATP per mole of glucose.

Under aerobic conditions too, the pyruvate is oxidised to carbon dioxide and water with further production of energy. The first step in this process is the oxidative decarboxylation of pyruvate in the presence of thiamin diphosphate:

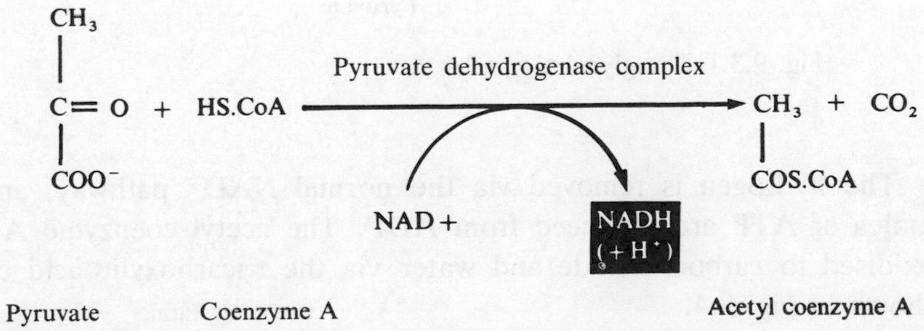

$$
\begin{array}{ccc}
CH_3 & & CH_3 \\
| & & | \\
C = O \ + \ HS.CoA \xrightarrow{\text{Pyruvate dehydrogenase complex}} & CH_3 \ + \ CO_2 \\
| & NAD+ \qquad NADH(+H^{\cdot}) & | \\
COO^- & & COS.CoA \\
\text{Pyruvate} \qquad \text{Coenzyme A} & & \text{Acetyl coenzyme A}
\end{array}
$$

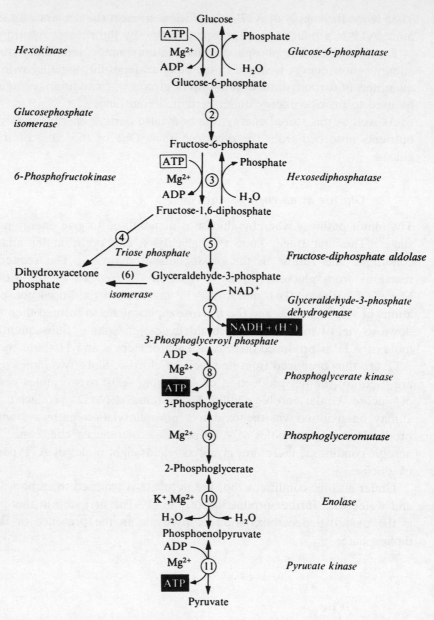

Fig. 9.3. The glycolytic pathway.

The hydrogen is removed via the normal NAD$^+$ pathway, and three moles of ATP are produced from ADP. The acetyl coenzyme A is then oxidised to carbon dioxide and water via the tricarboxylic acid cycle, as shown in Fig. 9.4.

This involves four dehydrogenations, three of which are NAD$^+$ linked and one FAD linked, resulting in 11 moles of ATP being formed from ADP. In addition one mole of ATP arises directly with the change of succinyl coenzyme A to succinic acid. The oxidation of each mole of pyruvate thus yields 15 moles of ATP. The total ATP production from the oxidation of one mole of glucose is then:

	Moles ATP
1 mole of glucose to 2 moles of pyruvate	8
2 moles of pyruvate to CO_2 and water	30
Total per mole of glucose	38

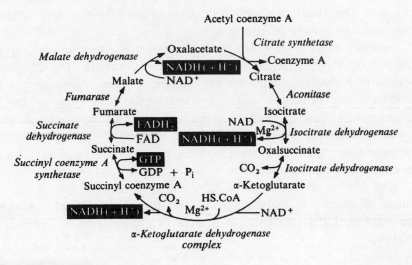

Fig. 9.4. The tricarboxylic acid cycle.

The capture of energy represented by the formation of 38 high energy phosphate bonds may be calculated as $38 \times 52 = 1976$ kJ mole^{-1} of glucose, the total free energy content of which is 2870 kJ. The efficiency of free energy capture by the body is thus $1976/2870 = 0.69$. Such calculations assume perfect coupling of reactions and normal environmental cell conditions.

Under anaerobic conditions, oxygen is not available for the oxidation of the reduced NAD$^+$ by oxidative phosphorylation. In order to allow release of a small amount of energy, by a continuing breakdown of glucose to pyruvate, reduced NAD$^+$ must be converted to the oxidised form. If not, step 7 of Fig. 9.3 will not take place and energy production will be blocked. Oxidation of reduced NAD$^+$ may be achieved under such conditions by the formation of lactate from pyruvate in the presence of lactate dehydrogenase:

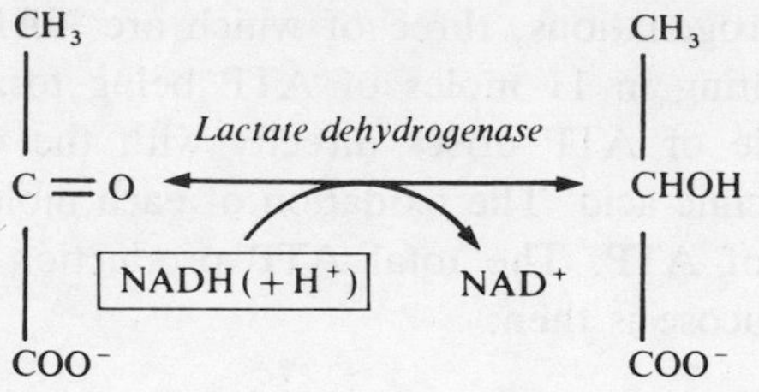

$$
\begin{array}{ccc}
CH_3 & & CH_3 \\
| & & | \\
C=O & \xleftrightarrow{\textit{Lactate dehydrogenase}} & CHOH \\
| & \boxed{NADH(+H^+)} \quad NAD^+ & | \\
COO^- & & COO^-
\end{array}
$$

Pyruvate   Lactate

A net production of two moles of ATP is then possible from each mole of glucose under such adverse conditions as prolonged intense muscular activity. The lactate so produced is transported to the liver where it is converted to glucose or glycogen.

Another pathway by which glucose is metabolised within the body is that known variously as the pentose phosphate pathway, the phosphogluconate oxidative pathway, and the hexose phosphate shunt. Although the system encompassing glycolysis and the tricarboxylic acid cycle is the major pathway of glucose metabolism in the body, the pentose phosphate pathway is of considerable importance in the cytoplasm of the cells of the liver, adipose tissue and the lactating mammary glands. The steps of this pathway are shown in Fig. 9.5.

The net result of this series of reactions is the removal of one carbon atom of glucose as carbon dioxide and the production of two moles of reduced $NADP^+$. Oxidation of a mole of glucose may be represented as follows:

$$\text{Glucose-6-phosphate} + 12\,NADP^+ \longrightarrow 6CO_2 + PO_4^{3-} + \boxed{12\,NADPH(+H^+)}$$

Unlike reduced NAD^+, reduced $NADP^+$ does not undergo oxidative phosphorylation to produce ATP, and the main function of the pentose phosphate pathway is to provide reduced $NADP^+$ for tissues which have a specific demand for it, particularly those actively synthesising fatty acids. Reduced $NADP^+$ can be converted to reduced NAD^+ and thus serve indirectly as a source of ATP.

Glycogen as an energy source

The release of energy in a utilisable form from glycogen necessitates its breakdown to glucose which is then degraded as previously described. Breakdown of glycogen within cells takes place through the action of inorganic phosphate and glycogen phosphorylase. This enzyme catalyses the rupture of the 1,4-glycosidic linkages and degradation begins at the non-

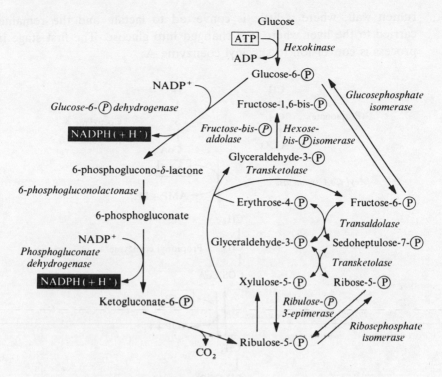

Fig. 9.5. The pentose phosphate pathway.

reducing end of the chain. Glucose-1-phosphate molecules are released successively until a branch point is approached. A rearrangement of the molecule then takes place in the presence of oligotransferase and a limit dextrin with a terminal 1,6 linked glucose is produced. Free glucose is then released by attack at the 1,6 linkage by an oligo-1,6-glucosidase, and glucose-1-phosphate is produced by further activity of the phosphorylase. The net effect of glycogen breakdown is the production of glucose-1-phosphate plus a little glucose. Glucose-1-phosphate is converted, by phospho-glucomutase, to glucose-6-phosphate which enters the Embden-Meyerhof or pentose phosphate pathway, as does the residual glucose. The production of glucose-6-phosphate from glycogen does not involve expenditure of ATP so that energy production from glycogen is slightly more efficient than it is from glucose.

Propionic acid as an energy source

In ruminant animals considerable amounts of propionate are produced from carbohydrate breakdown in the rumen. The acid then passes across the rumen wall, where a little is converted to lactate and the remainder is carried to the liver where it is changed into glucose. The first stage in this process is conversion to succinyl coenzyme A:

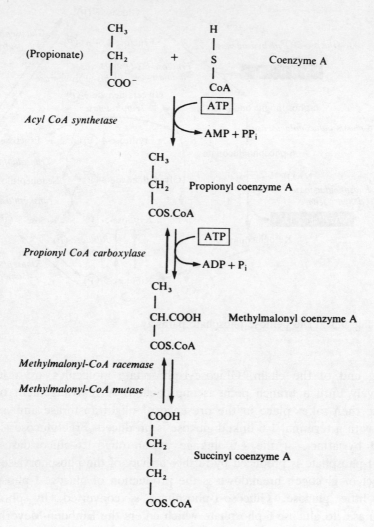

This then enters the tricarboxylic acid cycle and is converted to malate (Fig. 9.4) and the equivalent of three moles of ATP are produced. The malate diffuses into the cytoplasm where it is converted to oxalacetate and then phosphoenolpyruvate as shown below.

The phosphoenolpyruvate may then be converted to fructose diphosphate by reversal of steps 10,9,8,7 and 5 in the glycolytic sequence shown in Fig. 9.3 This is then converted to fructose-6-phosphate by hexose-diphosphatase and then to glucose-6-phosphate by the reverse of step 2 and finally

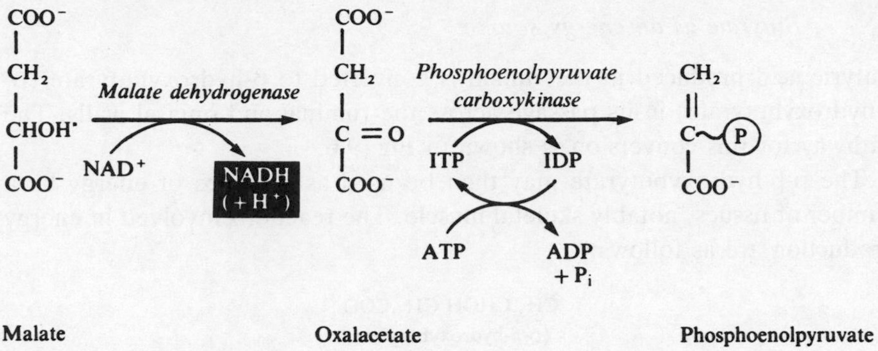

$$
\begin{array}{ccc}
\text{COO}^- & \text{COO}^- & \text{CH}_2 \\
| & | & \\
\text{CH}_2 & \text{CH}_2 & \\
| & | & \\
\text{CHOH}^{\cdot} & \text{C}=\text{O} & \\
| & | & \\
\text{COO}^- & \text{COO}^- & \text{COO}^-
\end{array}
$$

Malate dehydrogenase → ; *Phosphoenolpyruvate carboxykinase* →

NAD^+ → $NADH\,(+H^+)$

ITP IDP

ATP $ADP + P_i$

Malate   Oxalacetate   Phosphoenolpyruvate

to glucose by glucose-6-phosphatase. The glucose may be used eventually to provide energy and we may prepare an energy balance sheet as follows:

	Moles ATP +	Moles ATP −
2 moles propionate to 2 moles succinyl coenzyme A		6
2 moles succinyl coenzyme A to 2 moles malate	6	
2 moles malate to 2 moles phosphoenolpyruvate	6	2
2 moles phosphoenolpyruvate to 1 mole glucose		8
1 mole glucose to carbon dioxide and water	38	
Total	50	16
Net gain of ATP	34	

There is thus a net gain of 17 moles of ATP per mole of propionic acid. Small amounts of propionate are present in the peripheral blood supply. They may arise owing to incomplete removal by the liver, or from oxidation of fatty acids with an odd number of carbon atoms. Such propionate could conceivably be used directly for energy production. The pathway would be the same as that described as far as phosphoenolpyruvate. This would then follow the Embden-Meyerhof pathway via pyruvate, acetyl coenzyme A and the tricarboxylic acid cycle. The balance sheet for this process is:

	Moles ATP +	Moles ATP −
1 mole propionate to 1 mole succinyl coenzyme A		3
1 mole succinyl coenzyme A to 1 mole malate	3	
1 mole malate to 1 mole phosphoenolpyruvate	3	1
1 mole phosphoenolpyruvate to 1 mole acetyl coenzyme A	4	
1 mole acetyl coenzyme A to carbon dioxide and water	12	
Total	22	4
Net gain of ATP	18	

This pathway is thus marginally more efficient than that via glucose.

Butyrate as an energy source

Butyric acid produced in the rumen is converted to β-hydroxybutyrate (D-3-hydroxybutyrate) in its passage across the ruminal and omasal walls. The pathway for this conversion is shown in Fig. 9.6.

The D-β-hydroxybutyrate may then be used as a source of energy by a number of tissues, notably skeletal muscle. The reactions involved in energy production are as follows:

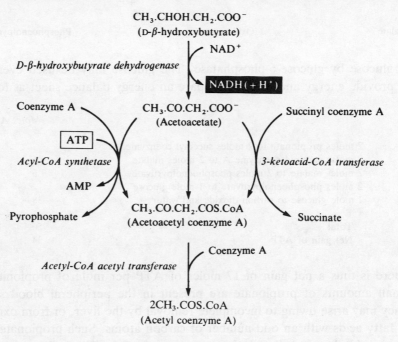

The acetyl coenzyme A is then metabolised via the tricarboxylic acid cycle. We may calculate the energy produced from butyrate by the synthetase pathway as follows:

	Moles ATP +	Moles ATP −
1 mole butyrate to 1 mole β-hydroxybutyrate	5	5
1 mole β-hydroxybutyrate to 2 moles acetyl CoA	3	2
2 moles acetyl CoA to carbon dioxide and water	24	
Total	32	7
Net gain of ATP per mole butyric acid	25	

If the change of acetoacetate to acetoacetyl coenzyme A takes place via the succinyl coenzyme A pathway then there is a saving of two moles of ATP and the net gain per mole of butyric acid is equivalent to 27 high energy phosphate bonds.

Acetic acid as an energy source

Acetic acid is the major product of carbohydrate digestion in the ruminant and is the only volatile fatty acid present in the peripheral blood in significant amounts. It is used by a wide variety of tissues as a source of energy. The initial reaction in this case is conversion of acetate to acetyl coenzyme A in the presence of acetyl coenzyme A synthetase.

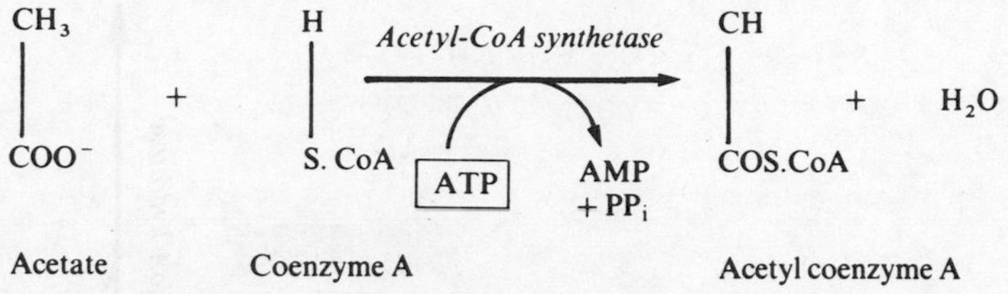

$$
\begin{array}{ccc}
CH_3 & H & CH \\
| & | & \xrightarrow{\text{\textit{Acetyl-CoA synthetase}}} & | \\
COO^- & S.\,CoA & COS.CoA
\end{array}
$$

ATP → AMP + PP$_i$

Acetate   Coenzyme A   Acetyl coenzyme A + H$_2$O

The acetyl coenzyme A is then oxidised via the tricarboxylic acid cycle yielding 12 moles of ATP per mole. Since two high-energy phosphate bonds are used in the initial synthetase-mediated reaction the net yield of ATP is 10 moles per mole of acetate.

Fat as an energy source

The store of triacylglycerol in the body is mobilised to provide energy by the action of the lipases, which catalyse the production of glycerol and fatty acids. The glycerol is glycogenic and enters the glycolytic pathway (see Fig. 9.3) as dihydroxyacetone phosphate, produced as shown in the following reactions:

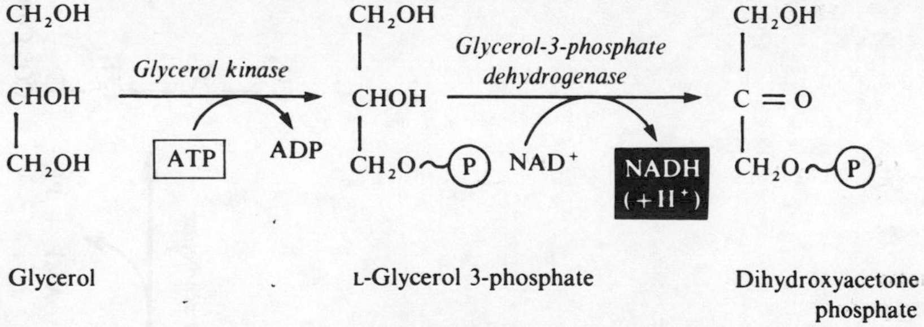

$$
\begin{array}{ccc}
CH_2OH & CH_2OH & CH_2OH \\
| & | & | \\
CHOH & CHOH & C=O \\
| & | & | \\
CH_2OH & CH_2O\sim\!P & CH_2O\sim\!P
\end{array}
$$

Glycerol kinase   ATP → ADP   Glycerol-3-phosphate dehydrogenase   NAD$^+$ → NADH (+H$^+$)

Glycerol   L-Glycerol 3-phosphate   Dihydroxyacetone phosphate

Glucose may then be produced by the reverse of the aldolase reaction to give fructose-1,6-diphosphate which is then converted to glucose by the action of hexose diphosphatase, glucosephosphate isomerase and glucose-6-

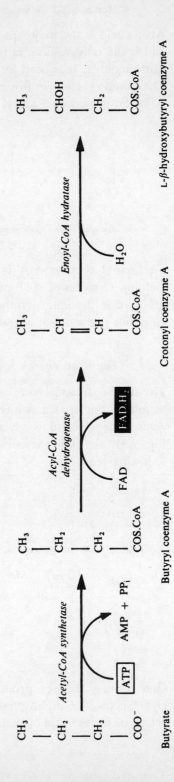

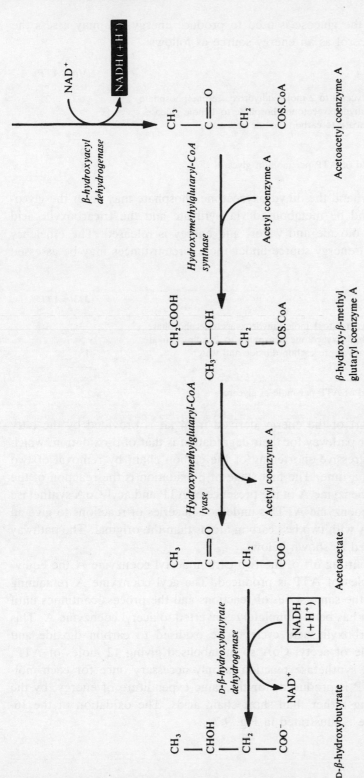

Fig. 9.6. Pathway for production of β-hydroxybutyrate from butyrate.

phosphatase. If the glucose is used to produce energy, we may assess the efficiency of glycerol as an energy source as follows:

	Moles ATP	
	+	−
2 moles glycerol to 2 moles dihydroxyacetone phosphate	6	2
2 moles dihydroxyacetone phosphate to 1 mole glucose		
1 mole glucose to carbon dioxide and water	38	
Total	44	2
Net yield of ATP per mole of glycerol	21	

On the other hand the dihydroxyacetone phosphate may enter the glycolytic pathway and be metabolised via pyruvate and the tricarboxylic acid cycle to carbon dioxide and water, and energy is released. The efficiency of glycerol as an energy source under these circumstances may be assessed as follows:

	Moles ATP	
	+	−
1 mole glycerol to 1 mole dihydroxyacetone phosphate	3	1
1 mole dihydroxyacetone phosphate to 1 mole pyruvate	5	
1 mole pyruvate to carbon dioxide and water	15	
Total	23	1
Net yield of ATP per mole of glycerol	22	

The major part of the energy derived from fat is provided by the fatty acids. The major pathway for their degradation is that of β-oxidation, which results in a progressive shortening of the carbon chain by removal of two carbon atoms at a time. The first stage of β-oxidation is the reaction of the fatty acid with coenzyme A in the presence of ATP and acyl-CoA synthetase to give an acyl coenzyme A. This undergoes a series of reactions to give an acyl coenzyme A with two less carbon atoms than the original. The pathway may be illustrated as shown below.

During the splitting off of the two carbon acetyl coenzyme A the equivalent of five moles of ATP is produced. The acyl coenzyme A remaining then undergoes the same series of reactions and the process continues until the carbon chain has been completely converted to acetyl coenzyme A. This enters the tricarboxylic acid cycle to be oxidised to carbon dioxide and water, each mole of acetyl CoA so metabolised giving 12 moles of ATP. Since the initial synthetase reaction is only necessary once for each molecule, more ATP is produced, for the same expenditure of energy, by the oxidation of long rather than short chain acids. The oxidation of the 16-carbon palmitate is illustrated in Fig. 9.7.

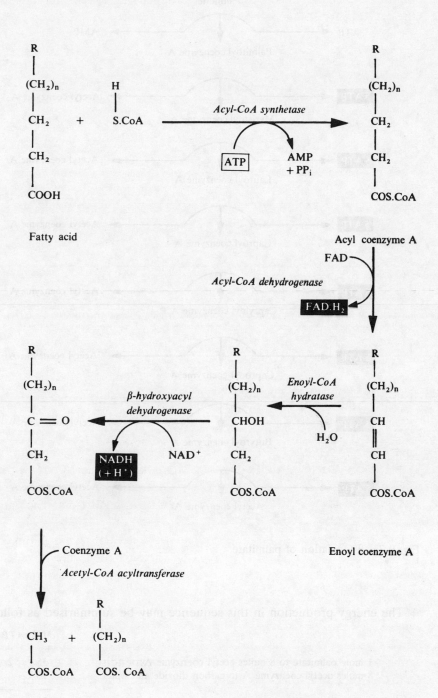

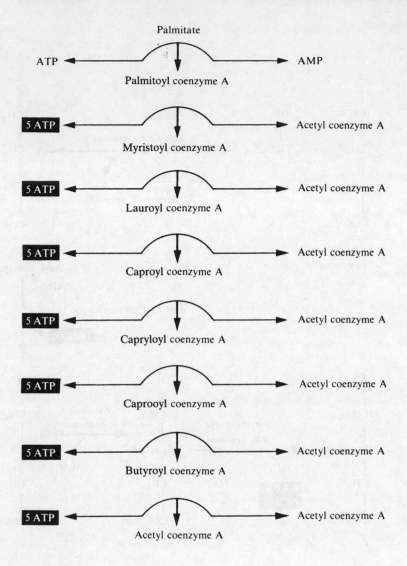

Fig. 9.7. β-Oxidation of palmitate.

The energy production in this sequence may be summarised as follows:

	Moles ATP +	Moles ATP −
1 mole palmitate to 8 moles acetyl coenzyme A	35	2
8 moles acetyl coenzyme A to carbon dioxide and water	96	
Total	131	2
Net gain of ATP per mole of palmitate	129	

Amino acids as an energy source

When amino acids are available in excess of the animal's requirements, or when the animal is forced to catabolise body tissues to maintain essential body processes, amino acids may be broken down to provide energy. Amino acid degradation takes place mainly in the liver although the kidney shows considerable activity, unlike muscular tissue which is relatively inactive.

The first stage in the oxidative degradation of amino acids is the removal of the amino group by one or other of two main pathways, oxidative deamination or transamination. In transamination the amino group is transferred to the α-carbon atom of a keto-acid, usually α-ketoglutarate, resulting in the production of another keto-acid and glutamate. The reactions are catalysed by enzymes known as aminotransferases. The reaction for aspartate may be represented as follows:

$$
\begin{array}{cc}
\underset{\text{Aspartate}}{\overset{\displaystyle \text{NH}_3^-}{\underset{|}{\text{H}-\text{C}-\text{COO}^-}}\atop{|}\atop{\text{CH}_2\text{COO}^-}} +
\underset{\alpha\text{-Ketoglutarate}}{\overset{\displaystyle \text{O}}{\overset{||}{\text{C.COO}^-}}\atop{|}\atop{\text{CH}_2}\atop{|}\atop{\text{CH}_2.\text{COO}^-}}
\underset{\text{transferase}}{\overset{\textit{Aspartate amino}}{\rightleftharpoons}}
\underset{\text{Oxalacetate}}{\overset{\displaystyle \text{O}}{\overset{||}{\text{C.COO}^-}}\atop{|}\atop{\text{CH}_2.\text{COO}^-}} +
\underset{\text{Glutamate}}{\overset{\displaystyle \text{NH}_3^+}{\underset{|}{\text{H}-\text{C}-\text{COO}^-}}\atop{|}\atop{\text{CH}_2}\atop{|}\atop{\text{CH}_2.\text{COO}^-}}
\end{array}
$$

The glutamate so formed, as well as that which becomes available from the digestive tract and from protein breakdown in tissues, may undergo oxidative deamination in the presence of glutamate dehydrogenase:

$$
\underset{\text{Glutamate}}{\overset{\displaystyle \text{NH}_3^+}{\underset{|}{\text{H}-\text{C}-\text{COO}^-}}\atop{|}\atop{\text{CH}_2}\atop{|}\atop{\text{CH}_2.\text{COO}^-}} + \text{H}_2\text{O} + \text{NAD}^+
\xrightarrow{\underset{\textit{dehydrogenase}}{\textit{Glutamate}}}
\underset{\alpha\text{-Ketoglutarate}}{\overset{\displaystyle \text{O}}{\overset{||}{\text{C.COO}^-}}\atop{|}\atop{\text{CH}_2}\atop{|}\atop{\text{CH}_2.\text{COO}^-}} + \text{NH}_4^+ + \boxed{\text{NADH} (+\text{H}^+)}
$$

The α-ketoglutarate may then be used in further transaminations while the reduced coenzyme is oxidised by oxidative phosphorylation (p. 163). Flavin linked D- and L-amino acid oxidases exist which catalyse the production of keto acids and ammonia, but these are of only minor importance.

The final product of amino acid degradation is acetyl coenzyme A, which is then processed via the tricarboxylic acid cycle to yield energy. The acetyl

CoA may be produced directly as in the case of tryptophan and leucine, via pyruvate (alanine, glycine, serine, threonine and cysteine) or via acetoacetyl CoA (phenylalanine, tyrosine, leucine, lysine and tryptophan). Other amino acids are degraded by pathways of varying complexity to give products such as α-ketoglutarate, oxalacetate, fumarate and succinyl CoA which enter the tricarboxylic acid cycle and yield acetyl CoA via phosphenolpyruvate (p. 167).

One of the consequences of amino acid catabolism is the production of ammonia which is highly toxic. Some of this may be used in amination for amino acid synthesis in the body. In this case ammonia reacts with α-keto-glutarate to give glutamate which, is then used for the syntheses. The reaction is the reverse of the oxidative deamination except that NADP$^+$ takes the place of NAD$^+$. Most is excreted from the body, as urea in mammals and uric acid in birds.

The formation of urea which takes place in the liver involves two stages, both of which require an energy supply in the form of ATP. The first is the formation of carbamoyl phosphate from carbon dioxide and ammonia in the presence of carbamoyl phosphate synthetase:

$$CO_2 + NH^+_4 + H_2O \xrightarrow{\text{Carbamoyl phosphate synthetase}} NH_2 - \overset{\overset{O}{\|}}{C} - O - \overset{\overset{O}{\|}}{\underset{\underset{O^-}{|}}{P}} - O^-$$

$$\boxed{2\,ATP} \qquad \begin{array}{c} 2ADP \\ + 2P_i \end{array}$$

Carbamoyl phosphate

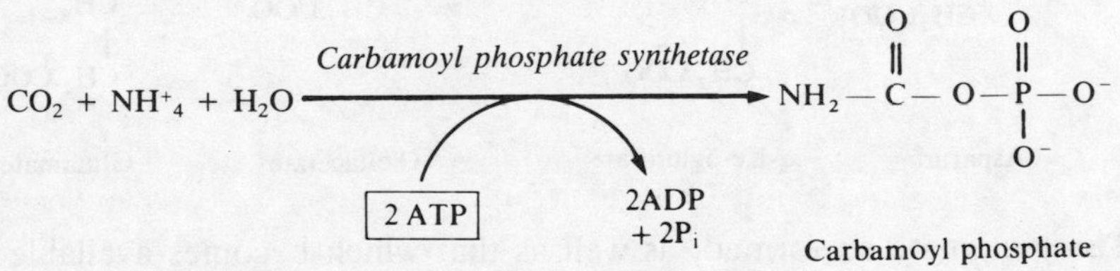

The carbamoyl phosphate then reacts with ornithine to start a cycle of reactions resulting in the production of urea as follows:

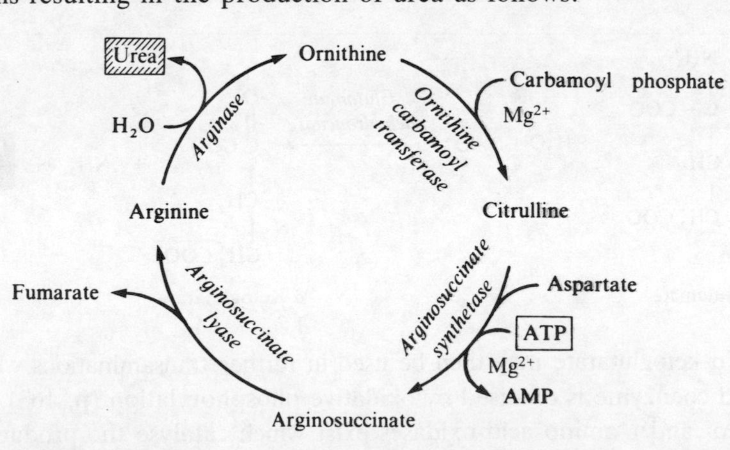

This involves the introduction of another amino group as aspartate and its elimination as urea, and the production of fumarate which joins the tricarboxylic acid pool.

Ammonia absorbed directly from the rumen undergoes the same series of reactions. Most of the urea is excreted, but a certain amount, depending upon the nitrogen status of the animal, is recycled via the saliva and directly across the rumen wall.

Uric acid production involves incorporation of ammonia into glutamine by reaction with glutamate.

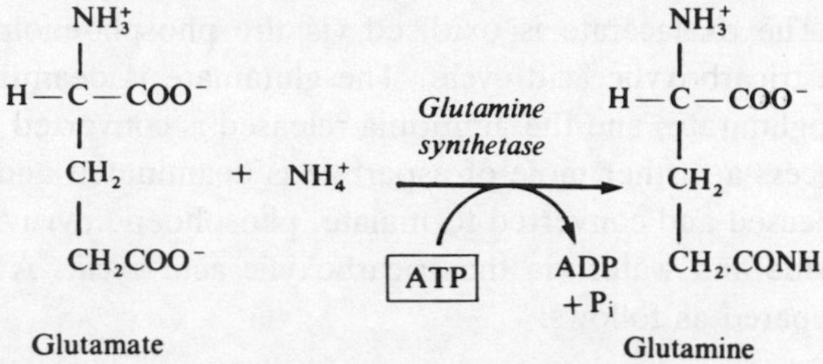

Glutamate ... Glutamine

Glutamine is then involved in a series of reactions with ribose-5-phosphate, glycine and aspartate to give inosinic acid which contains a purine nucleus. This series of reactions may be represented as follows:

$$\text{2 Glutamine} + \text{Glycine} + \text{Aspartate} \xrightarrow{\boxed{\text{8ATP}}} \text{Inosinic acid} + \text{Fumarate}$$

The ribose-5-phosphate residue is then removed giving hypoxanthine, which undergoes two NAD$^+$ linked oxidations to give xanthine and then uric acid.

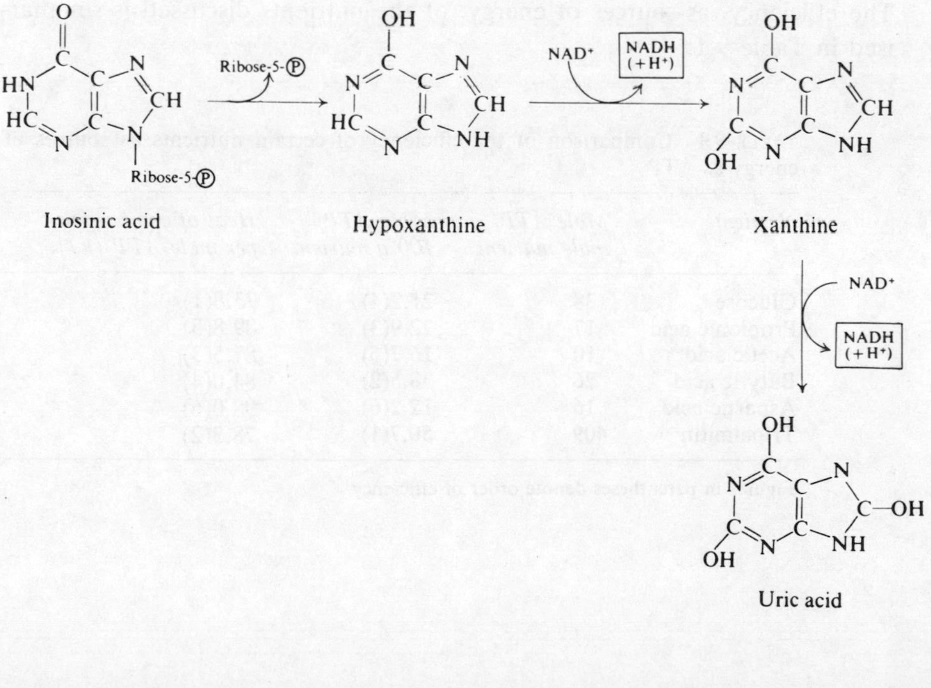

Inosinic acid ... Hypoxanthine ... Xanthine

Uric acid

Elimination of two moles of ammonia results in a net loss of 4 moles ATP and in addition 2 moles of glutamate, 1 mole glycine and 1 mole aspartate are used up and 1 mole fumarate is produced.

In assessing the efficiency of energy production from amino acids the energy needed for urea synthesis must be set against that obtained by oxidation of the carbon skeleton of the acid. If we take aspartate as an example, this is first converted to oxalacetate and glutamate by reaction with α-ketoglutarate. The oxalacetate is oxidised via the phosphoenolpyruvate pathway and the tricarboxylic acid cycle. The glutamate is deaminated to regenerate α-ketoglutarate, and the ammonia released is converted to urea, during which process a further mole of aspartate is deaminated and a mole of fumarate is released and converted to malate, phosphoenolpyruvate, and then carbon dioxide and water via the tricarboxylic acid cycle. A balance sheet may be prepared as follows:

	Moles ATP +	Moles ATP −
1 mole aspartate to glutmate + oxalacetate	0	0
1 mole oxalacetate to CO_2 and H_2O	16	1
1 mole glutamate to α-ketoglutarate	3	0
1 mole ammonia + 1 mole aspartate to urea and fumarate	0	4
1 mole fumarate to CO_2 and H_2O	19	1
Total	38	6
Net gain from 2 moles aspartate	32	

The efficiency, as sources of energy, of the nutrients discussed is summarised in Table 9.1:

TABLE 9.1   Comparison of the efficiency of certain nutrients as sources of energy as ATP

Nutrient	Mole ATP/ mole nutrient	Mole ATP/ 100 g nutrient	Heat of combustion per mole ATP (kJ)
Glucose	38	21.2(4)	73.8(1)
Propionic acid	17	22.9(3)	89.8(5)
Acetic acid	10	16.7(5)	87.5(3)
Butyric acid	26	38.5(2)	84.0(4)
Aspartic acid	16	12.2(6)	98.0(6)
Tripalmitin	409	50.7(1)	78.3(2)

Figures in parentheses denote order of efficiency.

N.B. The average energy cost of producing a mole of ATP is 85.2 kJ and the energy obtained per $\sim\!\!\boxed{P}$ expended is 52 kJ.

PROTEIN SYNTHESIS

Proteins are synthesised from amino acids, which become available either as the end products of digestion or as the result of synthetic processes within the body.

Direct amination may take place as in the case of α-ketoglutarate which yields glutamate:

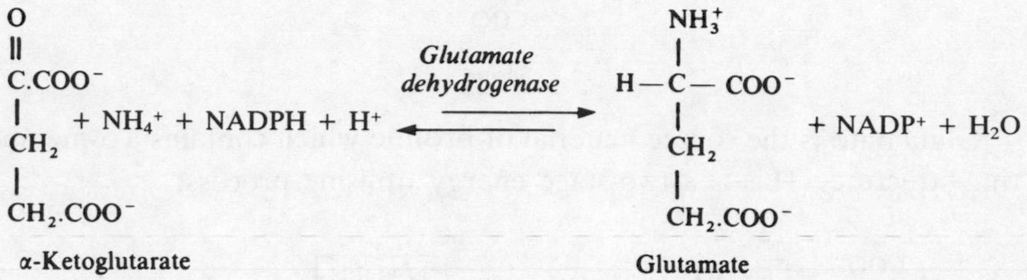

$$
\begin{array}{c}
\text{O} \\
\parallel \\
\text{C.COO}^- \\
\mid \\
\text{CH}_2 \quad + \text{NH}_4^+ + \text{NADPH} + \text{H}^+ \\
\mid \\
\text{CH}_2.\text{COO}^-
\end{array}
\quad
\underset{\text{dehydrogenase}}{\overset{\textit{Glutamate}}{\rightleftharpoons}}
\quad
\begin{array}{c}
\text{NH}_3^+ \\
\mid \\
\text{H}-\text{C}-\text{COO}^- \\
\mid \\
\text{CH}_2 \quad + \text{NADP}^+ + \text{H}_2\text{O} \\
\mid \\
\text{CH}_2.\text{COO}^-
\end{array}
$$

α-Ketoglutarate   Glutamate

The glutamate may undergo further amination to give glutamine but more importantly may undergo transamination reactions with various keto-acids to give amino acids, e.g.

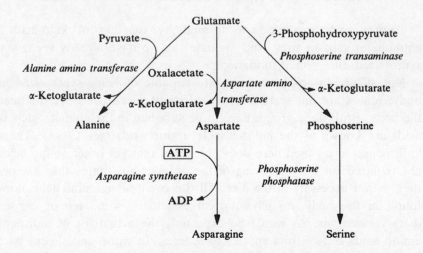

Amino acids other than glutamate may undergo such transaminations to produce new amino acids. Thus, both alanine and glycine react with phospho-hydroxypyruvate to give serine:

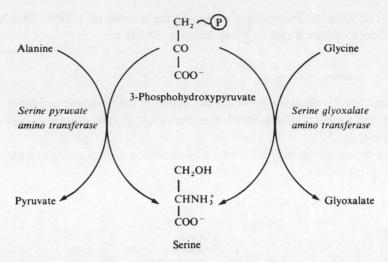

CH₂ ~ P
CO
COO⁻
3-Phosphohydroxypyruvate
Alanine
Glycine
Serine pyruvate amino transferase
Serine glyoxalate amino transferase
Pyruvate
Glyoxalate
CH₂OH
CHNH₃⁺
COO⁻
Serine

Glutamate is the source material of proline which contains a 5-membered ring structure. This is a two-stage energy utilising process:

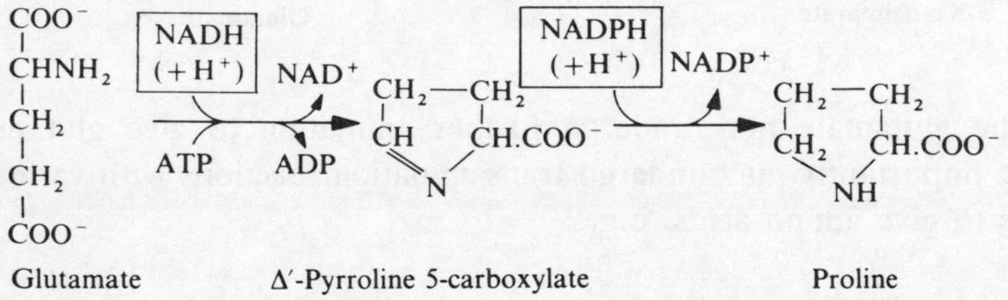

COO⁻
CHNH₂
CH₂
CH₂
COO⁻
Glutamate
NADH (+H⁺)
NAD⁺
ATP
ADP
CH₂—CH₂
CH
CH.COO⁻
N
Δ′-Pyrroline 5-carboxylate
NADPH (+H⁺)
NADP⁺
CH₂—CH₂
CH₂
CH.COO⁻
NH
Proline

Amino acids may also be formed by reaction of keto-acids with ammonium salts or urea, and arginine, as we have already seen, may be synthesised during urea formation.

Not all amino acids, however, are capable of being synthesised in the body; others are not synthesised at sufficient speed to satisfy the needs of the body. Both these groups have to be supplied to the animal. Such amino acids are known as the indispensable amino acids (see Ch. 4). The word 'indispensable' as used here does not mean that yet other amino acids are not required for the well-being of the animal, but simply that a supply of them is not necessary in the diet. All the twenty-five amino acids normally found in the body are physiological essentials; some ten or eleven are dietary essentials. As would be expected, the actual list of indispensable amino acids differs from species to species. In cattle and sheep, bacterial synthesis of amino acids in the rumen renders the inclusion of any specific amino acids in the diet unnecessary (see Ch. 13) except under conditions of intensive production such as high milk production or when small animals are making high weight gains.

The amino acids absorbed into the blood stream from the gut are transferred into the cells. This requires a supply of energy, since the concentration of amino acids in the cell may be up to one hundred times that in the blood and transfer into the cell has to take place against a very considerable concentration gradient. A continuous exchange takes place between the blood and cellular amino acids, but not between the free amino acids and those of the tissue proteins. The tissue proteins themselves undergo breakdown and resynthesis, their stability varying with different tissues. Thus liver protein has a half-life of seven days while collagen is so stable as to be considered almost completely inert.

The process of protein synthesis may be divided conveniently into four stages namely, activation of individual amino acids, initiation of peptide chain formation, chain elongation and termination.

Activation

The first step is enzymatic and requires the presence of ATP to give complexes as follows:

Amino acid + $\boxed{\text{ATP}}$ + Enzyme $\longrightarrow$ Amino-acyl-AMP-Enzyme + pyrophosphate

The amino-acyl group is then coupled to a molecule of transfer RNA (tRNA):

Amino-acyl-AMP-Enzyme + tRNA $\longrightarrow$ Amino-acyl-tRNA + AMP + Enzyme

Both reactions are catalysed by a single amino acyl synthetase, Mg^{2+}— dependent and specific for the amino acid and the tRNA. The synthetases discriminate between the twenty naturally occurring amino acids, but the specificity is not absolute.

The tRNA molecule is composed of a strand of nucleotides (see Ch. 4) which exhibits considerable folding, stabilised by hydrogen bonding. At one end of the chain is a final base arrangement of

——C——C——A——OH

i.e. cytidine-cytidine-adenosine. The amino acid is attached to the ribose of the terminal adenosine. Frequently, the other end of the chain terminates in a guanine nucleotide. We may visualise a typical tRNA molecule as shown below. There is at least one tRNA for each amino acid, but only one amino acid for a given tRNA. Since the terminal regions of the various tRNA species are so similar their specificity must reside within some arrangement in the interior of the molecules. This is thought to be a sequence of three bases, the nature and arrangement of which are specific for a particular amino acid.

When the amino acid has been coupled to the tRNA it is carried to one

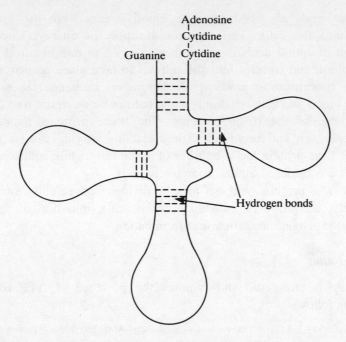

of the sites of protein synthesis, the ribosomes. These form part of the structures known as polysomes in which several ribosomes are linked by a strand of messenger RNA (mRNA). It is the sequence of bases on this mRNA strand, originally transcribed from the nuclear DNA, which dictates the amino acid sequence in the primary structure of the protein to be synthesised. A particular amino acid will be placed at the mRNA surface at a point having a specific arrangement of three bases, i.e. there is a base triplet code, known as the codon, for a given amino acid. The tRNA carrying the specific amino acid for a particular codon will have a complementary arrangement of three bases known as an anti-codon. There are 64 possible base triplet combinations and 61 of these have been shown to code for the 20 amino acids involved in the process of protein synthesis. There is thus more than one codon per amino acid and the code is said to be 'degenerate'. The codons for the amino acids have been elucidated and examples are given in Table 9.2.

The ribosomes of bacteria and algae consist of two parts designated 50 S (S = Svedberg unit) and 30 S according to their sedimentation characteristics in the ultra-centrifuge. In higher organisms the ribosomes consist of a small (40 S) and a large (60 S) subunit. During protein synthesis the growing peptide chain is attached to the large component while the next amino acid to be added, is attached, as its tRNA complex, to the smaller. Figure 9.8 depicts the sequence of events at the ribosome during initiation and elongation.

TABLE 9.2 Examples of known codons in mRNA

Codon	Amino acid
UUU	Phenylalanine
UUA	Leucine
UCC	Serine
UCA	Serine
CCC	Proline
CGA	Arginine
AGC	Serine
AGA	Arginine

U = uracil: C = cytosine: A = adenine: G = guanine

Initiation of peptide chain formation

This involves the formation of a complex between the smaller ribosome component the messenger RNA and a tRNA with formyl methionine or methionine as the attached amino acid (AA1). The larger ribosomal component then becomes attached to this complex to give the situation pictured in phase 1 of Fig. 9.8. The insertion of the amino acid residue requires the expenditure of the energy of one high energy phosphate bond as GTP.

Elongation

The requisite amino acid, AA2, is then placed at codon n + 1 by its specific tRNA as shown in phase 2. This process requires the expenditure of the energy of one high energy bond in GTP. A peptide bond is then formed between AA1 and AA2 and the tRNA for formyl methionine is simultaneously ejected (phase 3). The ribosome and mRNA then move relative to one another, and codon n + 1 is placed at the position previously occupied by codon n while codon n + 2 moves into that previously occupied by n + 1, as shown in phase 4. This translocation requires the expenditure of a further high energy bond in GTP. The process is then repeated with AA3 being placed at n + 2 followed by formation of a peptide bond and movement of the ribosome to expose codon n + 3. This continues until the chain is complete, with each movement requiring the expenditure of one high energy bond from GTP.

Termination

Chain elongation continues until a codon is reached which does not code for any amino acid i.e. UAA, UAG, or UGA. It then ceases and the

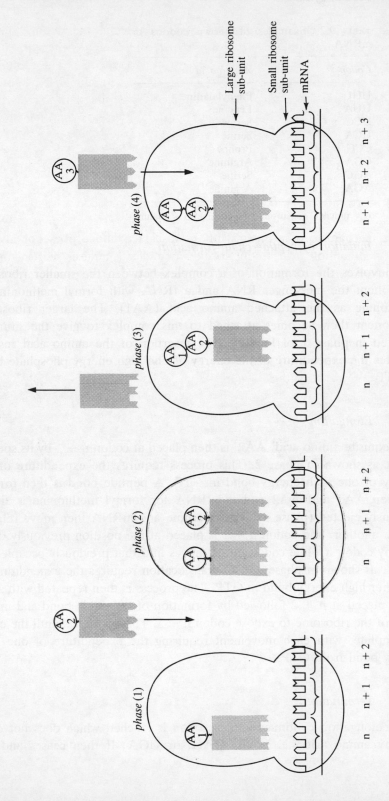

Fig. 9.8. Diagrammatic representation of the sequence of events occurring in the ribosome during the process of polypeptide synthesis. (After J. D. Watson, 1965. *Molecular Biology of the Gene*. W. A. Benjamin Inc., New York).

formed peptide chain is liberated and the formyl methionine residue removed enzymatically since it is not involved in protein structure.

The polypeptide is the primary structure of the protein. The chain then becomes arranged in a secondary spiral form stabilised by hydrogen bonding. The tertiary structure involves extensive coiling and folding of the chain and is stabilised by hydrogen bonding, salt linkages and sulphur bridges. The quaternary structure involves polymerisation of these basic tertiary units (see Ch. 4).

The mRNA forms a small proportion of the cell's RNA and has but a transient existence. In some micro-organisms it may function as a template in synthesis only ten or twenty times, but in mammalian tissues its active life may be much longer; in some cases it may persist for several days.

The mechanism of protein synthesis discussed above does not involve addition of amino acids to preformed peptides; synthesis starts with an amino acid, and a polypeptide chain is synthesised by the successive addition of single amino acids. Unless all the amino acids required to synthesise the peptide are present at the right time, no synthesis will take place and the amino acids which are present are removed and may be catabolised. Considerable wastage of amino acids may thus take place if an incomplete mixture is presented for synthesis.

During protein synthesis energy is provided by hydrolysis of ATP and GTP, the production of each mole requiring the expenditure of 85.2 kJ (Table 9.1) by the body. If we make certain assumptions, an estimate of the energetic efficiency of protein synthesis may be made. Let us assume that the average gram molecular weight of amino acids in a given protein is 100. In such a protein the number of amino acids is large, say n, and the number of peptide bonds will be n—1 (but can be taken as n for all practical purposes), i.e. each amino acid requires the formation of one peptide bond. We may now construct an energy balance sheet as follows:

	Energy expended (kJ)	*Energy stored (kJ)*
100 g amino acid	2437	
2 mole ATP (activation)	170.4	
1 mole GTP (initiation)	85.2	
1 mole GTP (elongation)	85.2	
100 g protein		2437
	2777.8	2437

$$\text{Energetic efficiency} = \frac{2437}{2777.8} = 0.88$$

FAT SYNTHESIS

The glycerides (triacylglycerols) of the depot fat may be derived from the glycerides of the blood, or may be synthesised in the body from fatty acyl CoAs and L-glycerol-3-phosphate.

Fatty acyl-CoA synthesis

It is generally considered that there are two systems of fatty acid synthesis. The first is a cytoplasmic system which is highly active and results finally in the production of palmitic acid from acetyl coenzyme A. The second is a mitochondrial system and results in the elongation of existing fatty acids by two-carbon addition by means of acetyl coenzyme A. The latter is produced in the mitochondria by the oxidative decarboxylation of pyruvate (p. 165), by oxidative degradation of certain amino acids or by β-oxidation of long chain fatty acids. The acetyl CoA is changed to citrate, by reaction with oxalacetate, and this diffuses into the cytoplasm where the acetyl CoA is regenerated.

$$\text{Citrate} + \text{Coenzyme A} \xrightarrow[\text{ATP} \quad \text{ADP}]{\textit{ATP-citrate lyase}} \text{Acetyl coenzyme A} + \text{Oxalacetate}$$

Acetyl CoA may also pass into the cytoplasm as a complex with carnitine.

In the ruminant animal acetate becomes available by direct absorption from the gut and this may form acetyl coenzyme A in the presence of acetyl-CoA synthetase, the energy being supplied by hydrolysis of ATP to AMP as shown on page 173. This is the major source of acetyl CoA in ruminants, in which ATP-citrate lyase activity is greatly reduced.

Cytoplasmic synthesis of fatty acids

This system is active in liver, kidney, brain, lungs, mammary gland and adipose tissues. The requirements of the system are reduced $NADP^+$, ATP, carbon dioxide and manganese ions (Mn^{2+}). The first stage is the transformation of acetyl coenzyme A to malonyl coenzyme A:

$$\begin{array}{c} CH_3 \\ | \\ COS.CoA \end{array} + \boxed{ATP} + H_2O + CO_2 \xrightarrow[Mn^{2+}]{\textit{Acetyl CoA carboxylase}} \begin{array}{c} COOH \\ | \\ CH_2 \\ | \\ COS.CoA \end{array} + ADP + P_i$$

The malonyl coenzyme A then reacts with acyl carrier protein (ACP) in the presence of [Acyl-carrier-protein] malonyl transferase to give the malonyl-ACP complex. Acetyl coenzyme A is then coupled with ACP in the presence of [Acyl-carrier-protein] acetyl transferase and this reacts with the malonyl-ACP, the chain length being increased by two carbon atoms to give butyryl-ACP complex. The reactions involved are as shown:

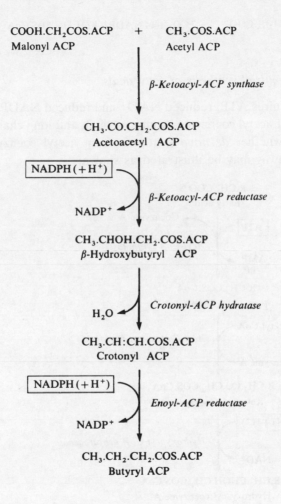

$$COOH.CH_2COS.ACP \quad + \quad CH_3.COS.ACP$$

Malonyl ACP   Acetyl ACP

β-Ketoacyl-ACP synthase

$$CH_3.CO.CH_2.COS.ACP$$
Acetoacetyl ACP

$NADPH(+H^+)$

β-Ketoacyl-ACP reductase

$NADP^+$

$$CH_3.CHOH.CH_2.COS.ACP$$
β-Hydroxybutyryl ACP

H_2O *Crotonyl-ACP hydratase*

$$CH_3.CH:CH.COS.ACP$$
Crotonyl ACP

$NADPH(+H^+)$

Enoyl-ACP reductase

$NADP^+$

$$CH_3.CH_2.CH_2.COS.ACP$$
Butyryl ACP

The butyryl ACP complex may then react with malonyl ACP complex with further elongation of the chain by two carbon atoms to give caproyl ACP. Chain elongation takes place by successive reactions with malonyl coenzyme A until the palmityl ACP complex is produced, when it ceases. Palmitic acid may be liberated by the action of a deacylase. We may write the overall reaction as follows:

$$CH_3.COS.CoA + 7COOH.(CH_2)COS.CoA + \boxed{14 \text{ NADPH} (+H^+)}$$

Acetyl CoA   Malonyl CoA

$$\downarrow$$

$$CH_3.(CH_2)_{14}.COO^- \quad + 7CO_2 + 14NADP^+ + 6H_2O + 8HS.CoA$$

Palmitate   Coenzyme A

Mitochondrial synthesis of fatty acids

This system requires ATP, reduced NAD^+ and reduced $NADP^+$. It involves incorporation of acetyl coenzyme A into medium and long chain fatty acids. It is doubtful whether *de novo* synthesis from acetyl coenzyme A takes place. The pathway may be illustrated as shown:

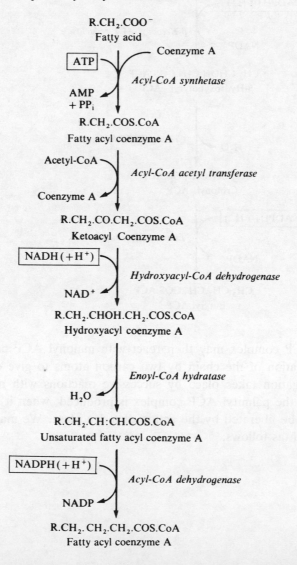

R.CH$_2$.COO$^-$
Fatty acid

Coenzyme A

$\boxed{ATP}$

Acyl-CoA synthetase

AMP
+ PP$_i$

R.CH$_2$.COS.CoA
Fatty acyl coenzyme A

Acetyl-CoA

Acyl-CoA acetyl transferase

Coenzyme A

R.CH$_2$.CO.CH$_2$.COS.CoA
Ketoacyl Coenzyme A

$\boxed{NADH(+H^+)}$

Hydroxyacyl-CoA dehydrogenase

NAD$^+$

R.CH$_2$.CHOH.CH$_2$.COS.CoA
Hydroxyacyl coenzyme A

Enoyl-CoA hydratase

H$_2$O

R.CH$_2$.CH:CH.COS.CoA
Unsaturated fatty acyl coenzyme A

$\boxed{NADPH(+H^+)}$

Acyl-CoA dehydrogenase

NADP

R.CH$_2$.CH$_2$.CH$_2$.COS.CoA
Fatty acyl coenzyme A

The products of this system are staturated acids with 18, 20, 22 and 24 carbon atoms produced, usually from palmitic acid synthesised in the cytoplasmic system. Elongation of both long chain saturated and unsaturated fatty acids may also take place in the endoplasmic reticulum, with malonyl CoA serving as the source of the two carbon fragments.

Unsaturated acids with one double bond may be synthesised in the body from saturated acids of the same chain length. Thus palmitoleic and oleic acids are formed from palmitic and stearic acids respectively:

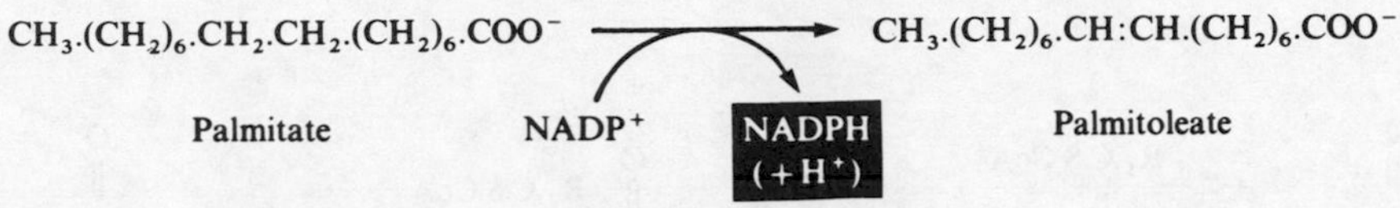

$$CH_3.(CH_2)_6.CH_2.CH_2.(CH_2)_6.COO^- \longrightarrow CH_3.(CH_2)_6.CH:CH.(CH_2)_6.COO^-$$

| Palmitate | NADP$^+$ | NADPH (+H$^+$) | Palmitoleate |

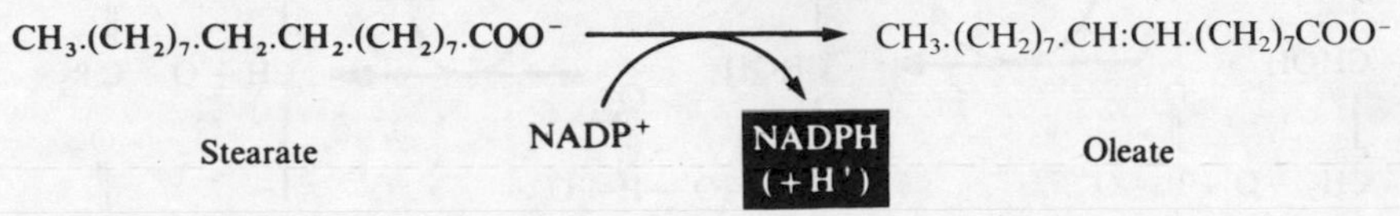

$$CH_3.(CH_2)_7.CH_2.CH_2.(CH_2)_7.COO^- \longrightarrow CH_3.(CH_2)_7.CH:CH.(CH_2)_7COO^-$$

| Stearate | NADP$^+$ | NADPH (+H$^+$) | Oleate |

In mammals, polyenoic acids are formed from palmitoleate, oleate and, where the double bonds lie between the terminal methyl group and the seventh carbon, from linoleate and linolenate. The latter two cannot be synthesised by mammals and are thus dietary essentials. The desaturation reactions involved in polyenoic acid production are NADP$^+$ linked, as shown above, and take place mainly in the liver.

Synthesis *of* L-glycerol 3-phosphate

The usual precursor is dihydroxyacetone phosphate produced by the aldolase reaction of the glycolytic pathway. This is reduced by the NAD-linked glycerol-3-phosphate dehydrogenase:

$$
\begin{array}{ccc}
CH_2OH & & CH_2OH \\
| & \xrightarrow{\text{\textit{Glycerol-3-phosphate dehydrogenase}}} & | \\
C=O & & CHOH \\
| \quad\quad O & & | \quad\quad O \\
| \quad\quad \parallel & & | \quad\quad \parallel \\
CH_2-O-P-O^- & & CH_2-O-P-O^- \\
\quad\quad\quad | & & \quad\quad\quad | \\
\quad\quad\quad O^- & & \quad\quad\quad O^-
\end{array}
$$

NADH (+H$^+$)   NAD$^+$

Dihydroxyacetone phosphate   L-Glycerol-3-phosphate

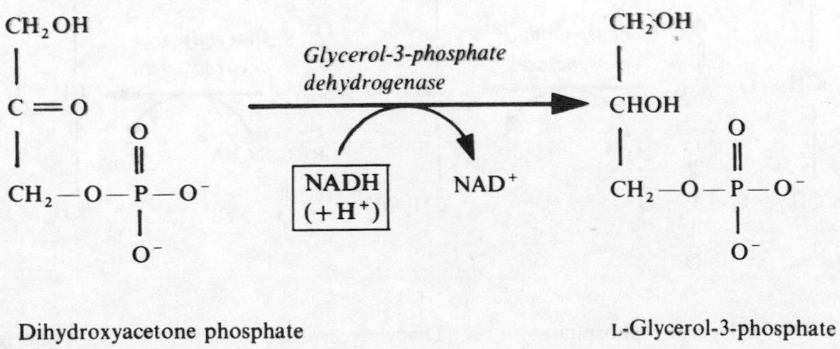

It may also be formed from free glycerol produced by decomposition of triacylglycerols by glycerol kinase, a reaction requiring ATP.

Synthesis of triacylglycerols

The first stage is the acylation, in the presence of glycerolphosphate acyl-transferase, of the free alcohol groups of the glycerol-3-phosphate by two molecules of fatty acyl-CoA to yield a phosphatidic acid.

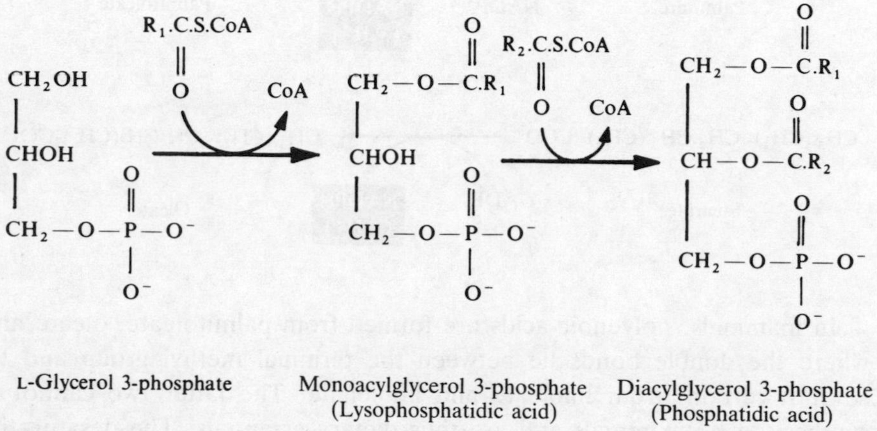

L-Glycerol 3-phosphate   Monoacylglycerol 3-phosphate   Diacylglycerol 3-phosphate
                          (Lysophosphatidic acid)   (Phosphatidic acid)

The reaction occurs preferentially with acids containing 16 and 18 carbon atoms.

The phosphatidic acid is then hydrolysed to give a diacylglycerol which then reacts with a third fatty acyl CoA to give a triacylglycerol:

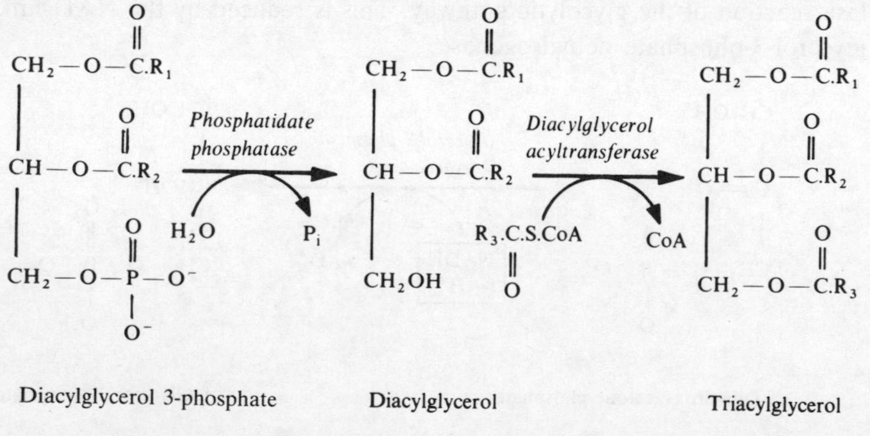

Diacylglycerol 3-phosphate   Diacylglycerol   Triacylglycerol

Direct synthesis of triacylglycerols from monoacylglycerols takes place in the intestinal mucosa of higher animals.

The efficiency of fat synthesis may be calculated from the pathways described. The calculation for the synthesis of tripalmitin by the cytoplasmic system would be:

	Energy expended kJ	*Energy retained kJ*
8 moles acetate	6996.0	
8 moles acetate to acetyl CoA	1363.2	
7 moles acetyl CoA to malonyl CoA	596.4	
7 additions of malonyl CoA	3 578.4	
Energy for 1 mole of palmitate	12 534.0	
Energy for 3 moles of palmitate	37 602.0	
½ mole glucose	1 401.5	
½ mole glucose to dihydroxyacetone phosphate	85.2	
1 mole dihydroxyacetone phosphate to 1 mole L-glycerol 3-phosphate	255.6	
Energy for mole L-glycerol 3-phosphate	1 742.3	
Total energy for 1 mole tripalmitin	39 344.3	
Energy stored in 1 mole tripalmitin		32 037.0

$$\text{Efficiency of synthesis} = \frac{32\ 037.0}{39\ 344.3} = 0.81$$

CARBOHYDRATE SYNTHESIS

The formation of glucose from simpler molecules such as propionic and keto acids has been discussed already. Glucose itself serves as the source material for the synthesis of two other important carbohydrates. These are the chief storage carbohydrate, glycogen, and milk sugar or lactose, the synthesis of which is specific to the mammary gland of the lactating animal.

Glycogen synthesis

Glycogen is a complex polysaccharide made up of condensed glucose residues (Ch. 2), and has the ability to add on further glucose units when these are available in the body. The actual source material for glycogen formation

is uridine diphosphate glucose (UDPG) which is produced from a variety
of sources as shown in Fig. 9.9.

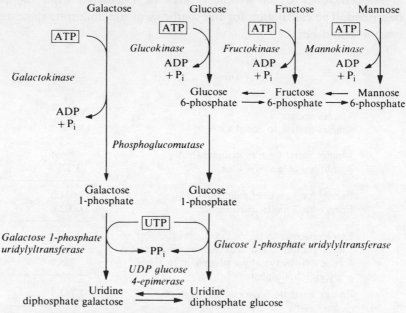

Fig. 9.9.   Formation of uridine diphosphate glucose.

Glycogen is produced by reaction of the uridine diphosphate glucose with
primer molecules, the most active of which is glycogen itself. Molecules with
as few as four glucose residues may serve as primers but reaction rate is
slow. As the complexity of the primer increases so does the reaction rate.
The synthesis involves reaction of uridine diphosphate glucose with the
fourth hydroxyl group of the non-reducing end of the primer chain in the
presence of glycogen synthase:

$$\text{UDP-glucose} + (\text{glucose})_n \xrightarrow{\textit{Glycogen synthetase}} (\text{Glucose})_{n+1} + \text{UDP}$$

The 1,6 linkages responsible for the branching within the glycogen
molecule are formed by transfer of a terminal oligosaccharide fragment of
six or seven glucose residues from the end of the glycogen chain to a 6-
hydroxyl group in a glucose residue in the interior of the chain. This takes
place in the presence of glycogen synthetase (1,4–1,6 transglucosylase).

Lactose synthesis

Lactose, the sugar of milk, is produced in large quantities in the mammary
gland of lactating animals. It is formed by condensation of one glucose and

one galactose molecule. A supply of glucose is readily available but the galactose has to be synthesised, virtually in its entirety from glucose, which involves a configurational change at carbon atom 4. The glucose is first converted to glucose-1-phosphate and then to uridine diphosphate glucose, from which uridine diphosphate galactose is produced by the action of UDP galactose-4-epimerase as shown:

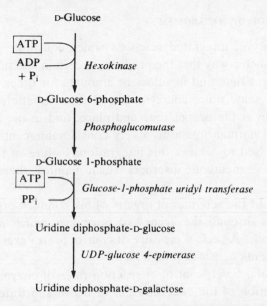

Lactose is then formed by the action of the UDP-D-galactose with D-glucose in the presence of the lactose synthetase system:

$$\text{UDP-galactose} + \text{D-glucose} \xrightarrow{\textit{Lactose synthetase}} \text{UDP} + \text{lactose}$$

The synthetase system is a complex of the enzyme galactosyl transferase with α-lactalbumin. The enzyme is present in the non-lactating mammary gland but is feebly active only, with D-glucose as the galactosyl acceptor. With the onset of lactation α-lactalbumin is produced in the gland and, in its presence, the enzyme becomes highly active in catalysing the transfer.

The energetic efficiency of lactose synthesis may be assessed as follows:

	kJ
2 moles glucose	5606
2 moles glucose to 2 moles glucose-1-phosphate	170.4
1 mole glucose-1-phosphate to galactose-1-phosphate	85.2
Energy required for 1 mole lactose	5861.6
Energy retained in 1 mole lactose	5648.4

$$\text{Energetic efficiency} = \frac{5648.4}{5861.6}$$

$$= \quad 0.96$$

Estimates of the energetic efficiencies of the synthetic processes described, although interesting, should not be given too much weight since their validity depends upon a number of factors including assumptions of complete coupling and ideal conditions, as well as availability of source materials.

CONTROL OF METABOLISM

The multiplicity of integrated reactions which constitute metabolism must take place in such a way that the products of certain reactions become available at the right time and in sufficient amounts for other reactions to take place. At the same time, energy required for synthetic reactions must be made available at the proper time and place, and in the required form. It is clear that in normal organisms a very considerable control of metabolism must be exercised to achieve this integration. Failure of the control mechanisms results in metabolic disorders which manifest themselves in various diseases.

Integration of the activities of organs within complex organisms is largely brought about through the agency of hormones which alter the rates of specific reactions. A certain measure of control is also exerted by circulating levels of nutrients or waste products.

At the cellular level, control of reaction rates of enzyme catalysed reactions is a function of the concentration of active substrate and enzyme. A shortage of substrate may be the result of failure of a previous reaction in the sequence. Alternatively there may be adequate amounts of substrate present in the cell but it may be physically unavailable. This is because within the cell there are a number of membrane bound compartments—e.g. nuclei, mitochondria and lysosomes, which contain various concentrations of substrates and whose membranes show variable permeabilities to the substrates. Enzymes may be rendered physically unavailable for the same reason. Substrates may be chemically unavailable owing to the extreme specificity of many enzymes and the existence of different forms of substrate. Thus oxalacetate may exist in four forms, only one of which is subject to enzyme attack. Enzyme activity is also controlled by a form of chemical compartmentation which may arise in three ways:

(a) the enzyme may exist as an inactive zymogen which must be irreversibly, cleaved before it may perform its catalytic function, e.g. trypsin is produced by cleavage of trypsinogen by enterokinase.

(b) a number of interconvertible enzyme systems are known, one only of the enzymes being active, e.g. glycogen phosphorylase may exist in the inactive non-phosphorylated, or the active phosphorylated form.

(c) protein-protein complexes occur where one, the specifier protein, transforms the activity of the other. One of the best known examples is the role of α-lactalbumin in the lactose synthetase system (page 197).

The balance of inhibitor and activator factors within the cell will affect enzyme activity as will the balance of enzyme synthesis and degradation.

Probably the most common regulatory mechanism operating in metabolism, is feedback inhibition, in which the activity of an enzyme is inhibited by the presence of an end product of the reaction or the pathway. In the synthesis of valine from pyruvic acid, for example, the first step is the formation of acetolactic acid, catalysed by acetolactate synthetase: the activity of this enzyme and the rate of valine formation are reduced by the presence of valine. Another example is the conversion of ornithine to citrulline, catalysed by ornithine carbamoyltransferase (see p. 180). The activity of the transferase is inhibited by arginine. Also at the cellular level, accumulation of an end product may exert some control over the rate of a metabolic pathway by a simple mass action effect. For example, the rate of glucose breakdown via the Embden-Meyerhof pathway is controlled by the reaction

$$\text{1,3-diphosphoglycerate} + \text{ADP} + P_1 \rightarrow \text{3-phosphoglycerate} + \text{ATP}$$

When ATP is being consumed rapidly its breakdown ensures a plentiful supply of ADP and phosphoric acid; the reaction thus proceeds rapidly from left to right. If, on the other hand, ATP is not being used, the supply of ADP and inorganic phosphate is reduced and so is the speed of the reaction.

FURTHER READING

R. M. Denton and C. I. Pagson, 1976. *Metabolic Regulation*. Chapman and Hall, London.
A. L. Lehninger, 1982 *Principles of Biochemistry*, 1st edn. Worth Publ., New York.
M. J. Swenson (ed.), 1977. *Dukes' Physiology of Domestic Animals*, 9th edn. Cornell University Press, Ithaca and London.
L. Stryer, 1981. *Biochemistry*, 2nd edn. W. H. Freeman, San Francisco.
A. White, P. Handler and E. L. Smith, 1978. *Principles of Biochemistry*, 6th edn. McGraw-Hill, New York.

10

Evaluation of foods
(A) Digestibility

This chapter marks a change from qualitative to quantitative nutrition. Those preceding it have shown which substances are required by animals, how these substances are supplied in foods and the manner in which they are utilised. This chapter and those immediately following are concerned with the assessment, first, of the quantities in which nutrients are supplied by foods, and second, of the quantities in which they are required by different classes of farm animals.

The potential value of a food for supplying a particular nutrient can be determined by chemical analysis, but the actual value of the food to the animal can be arrived at only after making allowances for the inevitable losses that occur during digestion, absorption and metabolism. The first tax imposed on a food is that represented by the part of it which is not absorbed and is excreted in the faeces.

The digestibility of a food is most accurately defined as that proportion which is not excreted in the faeces and which is, therefore, assumed to be absorbed by the animal. It is commonly expressed in terms of dry matter and as a coefficient or a percentage. For example, if a cow ate 9 kg of hay containing 8 kg of dry matter and excreted 3 kg of dry matter in its faeces, the digestibility of the hay dry matter would be

$$\frac{8-3}{8} = 0.625 \quad \text{or} \quad \frac{8-3}{8} \times 100 = 62.5 \text{ per cent.}$$

Coefficients could be calculated in the same way for each constituent of the dry matter. Although the proportion of the food not excreted in the faeces is commonly assumed to be equal to that which is absorbed from the digestive tract, there are objections to this assumption, which will be discussed later.

MEASUREMENT OF DIGESTIBILITY

In a digestibility trial, the food under investigation is given to the animal in known amounts and the output of faeces measured. More than one animal is used, firstly because animals, even when of the same species, age and sex, differ slightly in their digestive ability, and secondly because replication allows more opportunity for detecting errors of measurement.

In trials with mammals, male animals are preferred to females because it is easier to collect faeces and urine separately with the male. They should be docile and in good health. Small animals are confined in metabolism cages which separate faeces and urine by an arrangement of sieves, but larger animals such as cattle are fitted with harness and faeces collection bags made of rubber or of a similar impervious material. For females a special device channels faeces into the bag while diverting urine. Similar equipment can be used for sheep.

For poultry, the determination of digestibility is complicated by the fact that faeces and urine are voided from a single orifice, the cloaca. The compounds present in urine are mainly nitrogenous, and faeces and urine can be separated chemically if the nitrogenous compounds of urine can be separated from those of faeces. The separation is based either on the fact that most urine nitrogen is in the form of uric acid or that most faecal nitrogen is present as true protein. It is also possible to alter the fowl's anatomy by surgery so that faeces and urine are voided separately.

The food required for the trial should if possible be thoroughly mixed beforehand, to obtain uniform composition. It is then given to the animals for at least a week before collection of faeces begins, in order to accustom the animals to the diet and to clear from the tract the residues of previous foods. This preliminary period is followed by a period when food intake and faecal output are recorded. This experimental period is usually 5 to 14 days long, with longer periods giving greater accuracy. With simple-stomached animals the faeces resulting from a particular input of food can be identified by adding an indigestible coloured substance such as ferric oxide or carmine to the first and last meals of the experimental period; the beginning and the end of faecal collection are then delayed until the dye appears in and disappears from the excreta. With ruminants this method is not successful because the dyed meal mixes with others in the rumen, and instead an arbitrary time-lag of 24 to 48 hours is normally allowed for the passage of food residues, i.e. the measurement of faecal output begins 1 to 2 days after that of food intake, and continues for the same period after measurement of food intake has ended.

In all digestibility trials, and particularly those with ruminants, it is highly desirable that meals should be given at the same time each day and that the amounts of food eaten should not vary from day to day. When intake is irregular there is the possibility, for example, that if the last meal of the experimental period is unusually large the subsequent increase in faecal output may be delayed until after the end of faecal collection. In this situation the output of faeces resulting from the measured intake of food will be underestimated and digestibility overestimated. The trial is completed by analysing samples of the food used and the faeces collected.

Table 10.1 gives an example of a digestibility trial in which sheep were fed on hay for a preliminary period of ten days and an experimental period of ten days. The results are the means for three animals. The first figure in the last line of the table, the concentration of digestible organic matter in dry matter, is often expressed as a percentage (in this example, 51.5) and termed the 'D' value of the food.

In this example the food in question was roughage and could be given to the animals as the sole item of diet. Concentrated foods, however, may

TABLE 10.1   Results of a digestibility trial in which three sheep were fed on hay

1.   The average quantity of hay dry matter eaten was 1.63 kg per head per day, and the average quantity of dry matter excreted in the faeces was 0.76 kg per head per day. Hay and faeces were analysed, with the following results (g/kg dry matter):

	Organic matter	Crude protein	Ether extract	Crude fibre	N-free extractives
Hay	919	93	15	350	461
Faeces	870	110	15	317	428

2.   From these figures the quantities of the dry matter and its components which were consumed, excreted and, by difference, digested were calculated as follows (kg):

	Dry matter	Organic matter	Crude protein	Ether extract	Crude fibre	N-free extractives
Consumed	1.63	1.50	0.151	0.024	0.57	0.75
Excreted	0.76	0.66	0.084	0.011	0.24	0.33
Digested	0.87	0.84	0.067	0.013	0.33	0.42

3.   The digestibility coefficients were calculated by expressing the weights digested as proportions of the weights consumed:

Dry matter	Organic matter	Crude protein	Ether extract	Crude fibre	N-free extractives
0.534	0.560	0.444	0.541	0.579	0.560

4.   Finally, the composition of the hay was calculated in terms of digestible nutrients, with the following results (g/kg dry matter):

Digestible organic matter	Digestible crude protein	Digestible ether extract	Digestible crude fibre	Digestible N-free extractives
515	41	8	203	258

cause digestive disturbances if given alone to ruminants, and their digestibility is often determined by giving them in combination with a roughage of known digestibility. Thus the hay of the example could have been used in a second trial in which the sheep also received 0.50 kg of oats per day. If the dry matter content of the oats was 900 g/kg daily dry matter intake would increase by 0.45 kg, and if the output of faecal dry matter increased by 0.15 kg the digestibility of the dry matter in oats would be calculated as

$$\frac{0.45 - 0.15}{0.45} = 0.667$$

SPECIAL METHODS FOR MEASURING DIGESTIBILITY

Indicator methods

In some circumstances the lack of suitable equipment or the particular nature of the trial may make it impracticable to measure directly either food intake or faeces output, or both. For instance, when animals are fed as a group it is impossible to measure the intake of each individual. Digestibility can still be measured, however, if the food contains some substance which is known to be completely indigestible. If the concentrations of this indicator substance in the food and in small samples of the faeces of each animal are then determined, the ratio between these concentrations gives an estimate of digestibility. For example, if the concentration of the indicator increased from 10 g/kg in the food dry matter to 20 g/kg in the faeces, this would mean that half of the dry matter had been digested and absorbed. In equation form,

$$\text{Digestibility} = \frac{\text{g indicator/kg faeces} - \text{g indicator/kg food}}{\text{g indicator/kg faeces}}$$

The indicator may be a natural constituent of the food or be a chemical mixed into it. It is difficult to mix chemicals with foods like hay, but an indigestible constituent such as lignin may be used. Other indicators in use today are fractions of the food known as indigestible acid-detergent fibre and acid-insoluble ash (the latter being mainly silica). The indicator most commonly added to foods is chromium in the form of chromic oxide, Cr_2O_3. Chromic oxide is very insoluble and hence indigestible; moreover, chromium is unlikely to be present as a major natural constituent of foods.

Measuring the digestibility of herbage eaten by grazing animals presents a special problem. In theory, natural herbage constituents like lignin can be employed as indicators. In practice, this application of indicator methods is complicated by the difficulty of obtaining samples of the food (i.e. pasture herbage) which are truly representative of that consumed. Animals graze selectively, preferring young plant to old and leaf to stem, and a sample of the sward cut with a mower is therefore likely to contain a higher concentration of fibrous constituents (including lignin) than the herbage harvested by the animal. One way of obtaining representative samples is to use an animal with an oesophageal fistula (an opening from the lumen of the oesophagus to the skin surface). When this is closed by a plug, food passes normally between mouth and stomach; when the plug is temporarily removed, herbage consumed can be collected in a bag hung below the fistula. Samples of grazed herbage obtained in this way can then be analysed, together with samples of faeces, for the indicator.

Laboratory methods of estimating digestibility

Since digestibility trials are laborious to perform, there have been numerous attempts made to determine the digestibility of foods by reproducing in the laboratory the reactions which take place in the alimentary tract of the animal. Digestion in non-ruminants is not easily simulated in its entirety, but the digestibility of food protein may be determined from its susceptibility to attack *in vitro* by pepsin and hydrochloric acid.

The digestibility of foods for ruminants can be measured quite accurately in the laboratory by treating them first with rumen liquor and then with pepsin. During the first stage of this so-called 'two-stage *in vitro*' method a finely ground sample of the food is incubated for 48 hours with buffered rumen liquor in a tube under anaerobic conditions. In the second stage the bacteria are killed by acidifying with hydrochloric acid to $pH2$ and are then digested by incubating them with pepsin for a further 48 hours. The insoluble residue is filtered off, dried and ignited, and its organic matter subtracted from that present in the food to provide an estimate of digestible organic matter. Digestibility determined *in vitro* is generally slightly lower than that determined *in vivo*, and corrective equations are required to relate one measure to the other; an example is illustrated in Fig. 10.1(a).

This technique is now used in the analysis of farm roughages for advisory purposes, and for determining the digestibility of small samples such as those available to the plant breeder. A further application is found in estimating the digestibility of grazed pasture herbage when this is collected from an animal with an oesophageal fistula as described above.

The rumen liquor used in the first stage of this laboratory procedure may vary in its fermentative characteristics according to the diet of the animal from which it is obtained. In an attempt to obtain more repeatable estimates of digestibility, rumen liquor is sometimes replaced by fungal cellulase preparations. Figure 10.1(b) shows how incubation with pepsin followed by

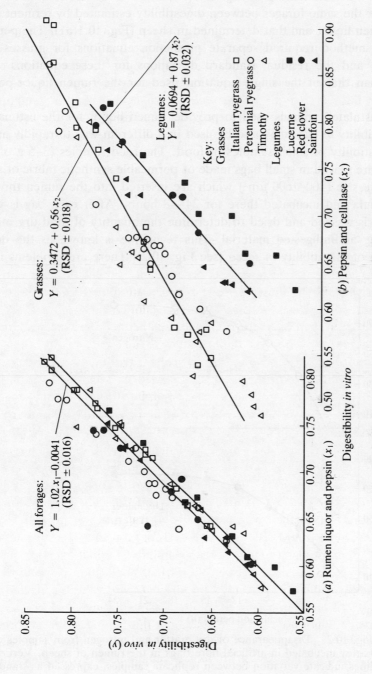

Fig. 10.1 Laboratory methods for estimating the dry matter digestibility of forages. (*a*) Incubation in rumen liquor followed by digestion with pepsin (*b*) Digestion with pepsin followed by digestion with cellulase (From R. A. Terry *et al.*, 1978 *J. Br. Grassld Soc.*, **22**, 13.)

Key:
Grasses
□ Italian ryegrass
○ Perennial ryegrass
△ Timothy
Legumes
■ Lucerne
● Red clover
◀ Sainfoin

All forages:
$Y = 1.02 x_1 - 0.0041$
(RSD ± 0.016)

(*a*) Rumen liquor and pepsin (x_1)

Grasses:
$Y = 0.3472 + 0.56 x_2$
(RSD ± 0.018)

Legumes:
$Y = 0.0694 + 0.87 x_2$
(RSD ± 0.032)

(*b*) Pepsin and cellulase (x_2)

Digestibility *in vitro*

Digestibility *in vivo* (*y*)

incubation with cellulase can be used to estimate the digestibility of forages to sheep. The relationship in Fig. 10.1(b), however, is less close than that found for the same forages between digestibility estimated by fermentation with rumen liquor and that determined in sheep (Fig. 10.1(a)). The pepsin-cellulase method required separate prediction equations for grasses and legumes, and the residual standard deviations for these equations were larger than that of the single equation used for the rumen liquor-pepsin method.

The fistulated animals used to provide rumen liquor for the estimation of digestibility *in vitro* may also be used in a different way to rapidly assess the digestibility of small samples of food. The food samples (3–5 g of dry matter) are placed in small bags made of permeable synthetic fabric of standard pore size (400–1600 μm^2) which are inserted into the rumen through the cannula and incubated there for 24–48 hours. After each bag is withdrawn it is washed and dried to determine the quantity of food dry matter remaining as undigested material. This technique is known as the determination of digestibility *in sacco* (see Fig. 10.2). There are problems in its

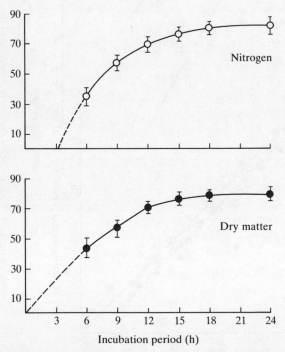

Fig. 10.2. Disappearance of dry matter and nitrogen from samples of barley incubated in artificial fibre bags in the rumen of sheep. Vertical lines indicate variation between replicate samples, expressed as standard deviations. (From A. Z. Mehrez and E. R. Ørskov, 1977, *J. Agric. Sci., Camb.*, **88**, 645.)

use, arising particularly from the need to select an appropriate period of incubation, but it has specialised applications in the assessment of protein breakdown·in the rumen (see Ch. 13) and in the study of rate of digestion (Ch. 8).

VALIDITY OF DIGESTIBILITY COEFFICIENTS

The assumption that the proportion of food digested and absorbed can be determined by subtracting the part excreted in the faeces is open to question on two counts. The first is that in ruminants the methane arising from the fermentation of carbohydrates is lost by eructation, and not absorbed. This loss leads to overestimation of the digestible carbohydrate and digestible energy content of ruminant foods.

More serious errors are introduced by the fact that, as discussed in Chapter 8, not all the faeces are actual undigested food residues. Part of the faecal material is contributed by enzymes and other substances secreted into the gut and not reabsorbed, and by cellular material abraded from the lining of the gut. Thus if a pig, for example, is fed on a nitrogen-free diet it continues to excrete nitrogen in the faeces. Since this nitrogen is derived from the body and not directly from the food, it is known as the metabolic faecal nitrogen; the amounts in which it is excreted are approximately proportional to the animal's dry matter intake. Faeces also contain appreciable quantities of ether-extractable substances and minerals which are of metabolic origin. Some of the ash fraction of faeces is contributed by mineral elements secreted into the gut, because the faeces serve as the route of true excretion of certain minerals, particularly calcium.

The excretion in faeces of substances not arising directly from the food leads to underestimation of the proportion of the food actually absorbed by the animal. The values obtained in digestibility trials are therefore called *apparent digestibility* coefficients to distinguish them from the coefficients of *true digestibility*. The relationship between apparent and true digestibility is illustrated later (p. 210). In practice, true digestibility coefficients are difficult to determine, because the fractions of the faeces attributable to the food and to the animal are in most cases indistinguishable from one another. Apparent coefficients of digestibility of organic constituents of foods are satisfactory for many purposes, and they do represent the net result of the ingestion of food. Apparent coefficients for some mineral elements, however, may be quite meaningless (see p. 214).

DIGESTION AND DIGESTIBILITY IN THE VARIOUS SECTIONS OF THE TRACT

As explained in Chapter 8, nutrients may be absorbed from several parts of the digestive tract. Even in non-ruminants, absorption occurs in two distinctly different parts, the small and large intestines, and in ruminants volatile fatty acids are absorbed from the rumen. A food constituent that is digested (and absorbed) at one site may give rise to nutrients differing quite considerably from those resulting from its digestion at another site, and the nutritive value to the animal of that constituent will therefore depend not only on the extent to which it is digested (i.e. its digestibility) but also on the site of digestion. For example, a carbohydrate such as starch may be fermented in the rumen to volatile fatty acids (and methane) or digested in the small intestine to glucose.

The digestibility of foods in successive sections of the digestive tract is most conveniently measured by exteriorizing the tract at selected points. The creation of an oesophageal fistula has already been described. A fistula may also be formed between the posterior dorsal sac of the rumen and the animal's flank. Such a fistula is normally fitted with a permanent rubber or plastic tube capped with a screw-topped stopper, the whole being known as a cannula. Removal of the stopper allows samples of rumen contents to be taken. Similar, but smaller, cannulae may be inserted in the abomasum or at selected points of the small or large intestines. The intestines may also be fitted with a re-entrant cannula. For this the intestine is severed and both ends are brought close to the skin surface and joined by a tube running outside the animal. With this tube in position the digesta flow normally from the proximal to the distal portions of the intestine. However if the tube is opened, digesta may be collected from the proximal part, measured, sampled, and returned to the distal part. The re-entrant cannula therefore allows the flow of digesta to be measured directly, and hence allows digestibility to be calculated in the same way as if faeces were collected. With simple cannulae, which permit digesta to be sampled but not weighed, digestibility may be estimated by the indicator techniques described on page 203.

An example of the use of cannulated animals to measure digestion in the successive portions of the digestive tract of sheep is shown in Table 10.2.

The sheep were cannulated at the duodenum and terminal ileum, thus allowing digestion to be partitioned among the stomach, the small intestine, and the large intestine. The overall digestibility of the organic matter of the pelleted grass (0.78) was considerably lower than that of the chopped grass (0.83), microbial digestion of the former being partially transferred from the stomach (i.e. the rumen) to the large intestine. These differences were even more apparent for cellulose digestion, to which the small intestine made a negligible contribution.

In Chapter 8 reference was made to the degradation of dietary protein

TABLE 10.2   Digestion of chopped or ground and pelleted dried grass in successive portions of the alimentary tract of sheep
(After D. E. Beever, J. P. Coelho da Silva, J. H. D. Prescott and D. G. Armstrong, 1972. *Br. J. Nutr.* **28**, 347)

Food constituent:	Organic matter		Cellulose	
Form of grass:	Chopped	Pelleted	Chopped	Pelleted
Proportion digested in:				
Stomach	0.52	0.45	0.80	0.56
Small intestine	0.27	0.20	0.02	−0.02
Large intestine	0.04	0.13	0.05	0.23
Overall	0.83	0.78	0.87	0.77

in the rumen. The nitrogen of food protein entering the rumen may leave it in the same form if the protein escapes degradation, or, if the protein is degraded, the nitrogen may leave the rumen either as microbial protein, if it is re-utilised, or as ammonia if it is not. The fate of dietary protein (or nitrogen) in the ruminant may be determined by collecting digesta from successive portions of the digestive tract. Table 10.3 contrasts the digestion

TABLE 10.3   Protein digestion and absorption by sheep given 800 g organic matter per day from one of two types of ryegrass
(from J. C. MacRae and M. J. Ulyatt, 1974. *J. agric. Sci., Camb.* **82**, 309)

		Perennial ryegrass	Short-rotation ryegrass
Total N	(1) In feed	37.8	34.9
(g/day)	(2) At duodenum	27.8	31.7
	(3) At terminal ileum	9.0	9.3
	(4) In faeces	5.8	6.7
Proportion	(5) In stomach	0.26	0.09
of feed	(6) In small intestine	0.50	0.64
N digested	(7) In large intestine	0.08	0.07
	(8) Overall	0.84	0.80
Protein N absorbed (g/day)[†]	(9) In small intestine	15.0	19.1
Amino acid N absorbed (g/day)	(10) In small intestine	14.6	18.3

† protein calculated as 6.25 (non-ammonia nitrogen)

of the nitrogen in two types of ryegrass. Although the two grasses were similar in total nitrogen content (line 1) considerably more nitrogen was 'lost' from the stomach—presumably by absorption of ammonia—with the first grass (line 1–2). Conversely, with this grass less nitrogen was absorbed in the small intestine, expressed either as total nitrogen (line 2–3), protein (line 9) or amino acids (line 10). The difference between total nitrogen and protein nitrogen absorption reflects absorption of ammonia reaching the intestine, and that between protein and amino acid nitrogen would be largely due to the nucleic acid nitrogen of microbial protein. A further loss of nitrogen, again presumably as ammonia, occurred in the large intestine. The net outcome was that although the short-rotation ryegrass contained slightly less nitrogen, of slightly lower overall digestibility, it provided the sheep with about 25 per cent more in terms of absorbed amino acids than did the perennial ryegrass. With the latter, of the total nitrogen apparently absorbed (32 g/day), less than half (14.6 g/day) was in the form of amino acids.

FACTORS AFFECTING DIGESTIBILITY

Food composition

The digestibility of a food is closely related to its chemical composition, and a food like barley, which varies relatively little in composition from one sample to another, will show little variation in digestibility. Other foods, particularly fresh or conserved herbages, are much less constant in composition and therefore vary more in digestibility. The fibre fraction of a food has the greatest influence on its digestibility, and both the amount and chemical composition of the fibre are important.

Modern methods of food analysis attempt to distinguish fractions of cell walls and cell contents. The cell contents of forages are determined by extracting them with neutral detergent solution; the residue from this extraction (neutral detergent fibre) is considered to consist predominantly of cell walls. The cell wall fraction may be further divided into hemicellulose (extractable with acid detergent solution) and cellulose plus lignin (i.e. acid detergent fibre). Cell contents are almost completely digestible, but the digestibility of cell walls depends on the extent to which they are lignified (i.e. on the lignin content of acid detergent fibre).

The apparent digestibility of crude protein is dependent on the concentration of protein (or N) in the food, even though the *true* digestibility may be constant. The reason for this is that the metabolic faecal N represents a constant tax upon the dietary N, and has a greater effect on apparent digestibility with foods lower in N content. This is illustrated in Fig. 10.3.

In pigs, metabolic faecal N is equivalent to about 6 g protein per kg food

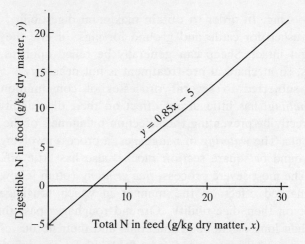

Fig. 10.3. The influence of metabolic faecal nitrogen on nitrogen digestibility in ruminants. From the equation, true digestibility of N is constant, at 0.85 and metabolic faecal N excretion is 5 g/kg dry matter intake. Apparent digestibility of N increases with increasing N content of food; when the latter is 10 g/kg, apparent digestibility of N is 3.5/10 = 0.35, and when N content of food is 30 g/kg, apparent digestibility of N is 20.5/30 = 0.68.

DM, but in ruminants it is much higher, being equivalent to 30 g protein per kg feed DM. The higher value for ruminants is due to the microbial residues from the rumen and hind gut. An important consequence of this high value for ruminants is that foods that are particularly low in protein (such as cereal straws) have negative coefficients for the apparent digestibility of protein, and their consumption reduces the protein reserves of the animal.

Ration composition

The digestibility of a food is influenced not only by its own composition, but also by the composition of other foods consumed with it. This *associative* effect of foods represents a serious objection to the determination of the digestibility of concentrates by difference as described on p. 203. For example, if a roughage (DM digestibility 0.6) and a concentrate (0.8) contribute equal parts of a mixed ration, the digestibility of that ration will not necessarily be 0.7. As discussed in Chapter 8, fermentation of soluble carbohydrate of the concentrate may modify the rumen environment and depress the digestibility of the cellulose and other fibre components of the roughage.

Preparation of foods

The commonest treatments applied to foods are chopping, chaffing, crushing or grinding, and cooking. In order to obtain maximum digestibility cereal grains should be crushed for cattle and ground for pigs, or else they may pass through the gut intact. Sheep can generally be relied upon to chew whole cereal grains, so mechanical pre-treatment is not necessary.

Roughages are subjected to several processes of comminution: the mildest process, *chaffing*, has little direct effect on their digestibility, but may reduce it indirectly by preventing the selection by animals of the more digestible components. The *wafering* of roughages, a process involving their compression into round or square section blocks, also has little effect on their digestibility. The most severe process, *fine grinding* (often followed by *pelleting*), has a marked effect on the manner in which roughages are digested and hence on their digestibility. Ground roughages pass through the rumen faster than long or chopped materials and their fibrous components may be less completely fermented (see Table 10.2). The grinding of roughages therefore reduces the digestibility of their crude fibre by as much as 20 percentage units, and of the dry matter as a whole by 5–15 percentage units. The depression is often greatest for roughages intrinsically low in digestibility, and may be exaggerated by the effect of level of feeding on digestibility since grinding increases the acceptability of roughages to ruminants (see Ch. 16).

Roughages such as the cereal straws, in which the cellulose is mixed or bound with a high proportion of lignin, may be treated chemically to separate the two components. The treatment processes and their effects are described in detail in Chapter 19. The chemicals used are mainly alkalis (sodium and ammonium hydroxides) and they improve the dry matter digestibility of cereal straws quite dramatically, from 0.4 to 0.5–0.7.

Foods are sometimes subjected to *heat treatment* to improve their digestibility. A traditional form of heat treatment is the boiling of potatoes for pigs, but heat may also be applied to other foods as steam or by microwave irradiation (a process known as 'micronisation'). Applied to the cereals, such treatments cause relatively small increases in digestibility, although sorghum appears to be more responsive than other grains. Heat treatments are most effective in improving digestibility when used for the specific purpose of inactivating the digestive enzyme inhibitors that are present in some feeds. The best examples of these inhibitors are found in protein concentrates such as groundnut meal (see Ch. 22). Potatoes, and root crops such as swedes (*Brassica napus*), possess trypsin inhibitors which are inactivated by heat treatment; the benefit of such treatment is not so much the improvement in digestibility as the transfer of protein digestion from fermentation in the caecum to normal enzyme digestion in the small intestine.

Animal factors

Digestibility is more a property of the food than of the consumer, but this is not to say that a food given to different animals will always be digested to the same extent. The most important animal factor is the species of the consumer. Foods low in fibre are equally well digested by ruminants and non-ruminants, but more fibrous foods are better digested by ruminants. Apparent digestibility coefficients for protein are frequently higher for pigs because their excretion of metabolic faecal nitrogen is smaller than that of ruminants. Differences in digestive ability between sheep and cattle are small and of little importance with most diets; however, highly digestible foods—such as cereal grains—tend to be more efficiently digested by sheep, and poorly digested foods—such as low-quality roughages—tend to be better digested by cattle.

Level of feeding

An increase in the quantity of a food eaten by an animal generally causes a faster rate of passage of digesta. The food is then exposed to the action of digestive enzymes for a shorter period, and there may be a reduction in its digestibility. As one would expect, reductions in digestibility due to increased rates of passage are greatest for the slowly digested components of foods, namely the cell-wall components.

Level of feeding is often expressed in multiples of the quantity of food required by the animal for what is called body maintenance (i.e. the quantity which prevents the animal from losing weight but does not allow growth). This level is defined as unity; in ruminants fed to appetite, the level can rise to 2.0–2.5 times the maintenance level in growing or fattening animals, and to 3.0–5.0 times maintenance in lactating animals. Increasing the level of feeding by one unit (e.g. from maintenance to twice that level) reduces the digestibility of roughages such as hay, silage and grazed pasture by only a small proportion (0.01–0.02). With diets for ruminants containing finer food particles (e.g. mixtures of roughages and concentrates) the reduction in digestibility per unit increase in feeding level is greater (0.02–0.03); this would mean, for example, that the digestibility of the dry matter of a typical dairy cow diet might fall from 0.75 at the maintenance level of feeding to 0.70 at 3.0 times the maintenance level. Falls of this magnitude may be due to exaggerated associative effects of foods, and there is some evidence that they can be prevented by raising the protein content of the diet. The greatest reductions in digestibility with increasing feeding level are found with ground and pelleted roughages and also some fibrous by-products (0.05 units per unit increase in level).

In non-ruminants, feeding levels rise to 2.0–3.0 times maintenance in poultry, 3.0–4.0 times maintenance in growing pigs and 4.0–6.0 times maintenance in lactating sows, yet there is little effect of feeding level on digestibility of conventional (i.e. low-fibre) diets.

TOTAL DIGESTIBLE NUTRIENTS SYSTEM FOR EVALUATING FOODS

Tables giving the proximate composition of foods generally include values for digestible composition. It should now be clear that the latter values are average figures, not biological constants, and may be inaccurate when applied to a particular sample of food given to a particular animal. Average digestibility coefficients must therefore be used with caution, especially when the food in question is one which may show considerable variations in composition. Digestibility data are important nevertheless, for the digestible composition of foods is used in several systems of food evaluation as a basis for calculating their energy-supplying power. The simplest of these systems is that in which foods are evaluated in terms of Total Digestible Nutrients (TDN). In this system a value is calculated for each food as follows:

TDN (per 100 kg) = kg dig. crude protein + kg dig. crude fibre +
         kg dig. N-free extractives + 2.25 (kg dig. ether
         extract)

Thus the hay quoted in Table 10.1 would have a TDN value of:

$$4.1 + 20.3 + 25.8 + (2.25 \times 0.8) = 52$$

The ether extract is multiplied by 2.25 because the energy value of fats is approximately two-and-a-quarter times that of carbohydrates. The TDN system has been used extensively in the United States, but not in Britain; it is now being replaced by more complicated systems of food evaluation. Its disadvantage is that it fails to take into account that the efficiency with which the digested energy-yielding nutrients are metabolised and made available to the animal varies from one food to another (see Ch. 12).

AVAILABILITY OF MINERAL ELEMENTS

There are endogenous faecal losses of most minerals, particularly calcium, phosphorus, magnesium and iron, and apparent digestibility coefficients for such elements thus have little significance. The measure of importance is therefore true digestibility, which for minerals is commonly called 'availability'. To measure the availability of a mineral one must generally distinguish between the portion in the faeces which represents unabsorbed material, and the portion which represents material discharged into the gut from the tissues. This distinction may be made by labelling an element within the body, and hence the portion secreted into the gut, with a radio-active isotope.

Mineral elements in digesta exist in three forms, as metallic ions in solution, as constituents of metallo-organic complexes in solution, or as constituents of insoluble substances. Those present in the first form are readily absorbed and those in the third form are not absorbed at all. The metallo-organic complexes, some of which are chelates (see p. 91), may in some cases be absorbed. For some minerals there are opportunities for conversion from one form to another, so broadly speaking the availability of an element will depend on the form in which it occurs in the food and on the extent to which conditions in the gut favour conversion from one form to another. Thus sodium and potassium, which occur in digesta almost entirely as ions, have an availability close to 1.0. At the other extreme, copper occurs almost entirely as soluble or insoluble complexes, and its availability is generally less than 0.1. Phosphorus, to take a further example, is present in many foods as a constituent of phytic acid (see p. 95) and its availability depends on the presence of phytases—of microbial or animal origin—in the digestive tract. A potent factor controlling the interconversion of soluble and insoluble forms of mineral elements is the pH of the digesta. In addition there may be specific agents which bind mineral elements and thus prevent their absorption. For example, calcium may be precipitated by oxalates and copper by sulphides.

The availability of minerals is commonly high in young animals fed on milk and milk products, but declines as the diet changes to solid foods. An additional complication is that the absorption, and hence apparent availability, of some mineral elements is partly determined by the animal's need for them. Iron absorption, discussed in Chapter 8, is the clearest example of this effect, but in ruminants calcium absorption also appears to be determined by the animal's need.

While no attempt is made here or in Chapter 8 to provide a complete list of the factors which influence the availability of minerals, those mentioned serve to illustrate why no attempt is made in tables of food composition to give values for availability comparable to the digestibility coefficients commonly quoted for organic nutrients. Because the availability of the minerals in a particular food depends so much on the other constituents of the diet and on the animal to which it is given, average values for individual foods would be of little significance.

FURTHER READING

Nutrition Society, 1977. Methods for evaluating feeds for large farm animals, *Proc. Nutr. Soc.*, **36**, 169–225.

B. H. Schneider and W. P. Flatt, 1975. *The Evaluation of Feeds through Digestibility Experiments*. University of Georgia Press, Athens.

P. J. Van Soest, 1982. *Nutritional Ecology of the Ruminant*. O & B Books, Corvallis, Oregon.

11

Evaluation of foods
(B) Energy content of foods and the partition of food energy within the animal

The major organic nutrients are required by animals as materials for the construction of body tissues and the synthesis of such products as milk and eggs, and they are needed also as sources of energy for work done by the animal. A unifying feature of these diverse functions is that they all involve a transfer of energy, and this applies both when chemical energy is converted into mechanical or heat energy, as when nutrients are oxidised, and when chemical energy is converted from one form to another, as for example when body fat is synthesised from food carbohydrate. The ability of a food to supply energy is therefore of great importance in determining its nutritive value. The purpose of this chapter and the next is to discuss the fate of food energy in the animal body, the measurement of energy metabolism and the expression of the energy value of foods.

DEMAND FOR ENERGY

An animal deprived of food continues to require energy for those functions of the body immediately necessary for life—for the mechanical work of essential muscular activity, for chemical work such as the movement of dissolved substances against concentration gradients, and for the synthesis of expended body constituents such as enzymes and hormones. In the starving animal the energy required for these purposes is obtained by the catabolism of the body's reserves, first of glycogen, then of fat and protein. In the fed animal the primary demand on the energy of the food is in meeting this requirement for body maintenance and so preventing the catabolism of the animal's tissues.

When the chemical energy of the food is used for muscular and chemical work involved in maintenance, the animal does no work on its surroundings and the energy used is converted into heat. Energy so used is regarded as having been expended, since heat energy is useful to the animal only in maintaining body temperature. In a fasting animal the quantity of heat produced is equal to the energy of the tissue catabolised and when measured under specific conditions is known as the animal's basal metabolism. In chapter 14 it will be shown how estimates of basal metabolism are used in assessing the maintenance energy requirements of animals.

Energy supplied by the food in excess of that needed for maintenance is used for the various forms of production. (More correctly, the nutrients represented by this energy are so used.) A young growing animal will store energy principally in the protein of its new tissues, whereas an adult will store relatively more energy in fat and a lactating animal will transfer food energy into the energy contained in milk constituents. Some other forms of production are the performance of muscular work and the formation of wool and eggs. No function, not even body maintenance, can be said to have absolute priority for food energy. A young animal receiving adequate protein but insufficient energy for maintenance may still store protein while drawing on its reserves of fat. Similarly, some wool growth continues to take place in animals with sub-maintenance intakes of energy, and even in fasted animals.

SUPPLY OF ENERGY

Gross energy of foods

The animal obtains energy from its food. The quantity of chemical energy present in a food is measured by converting it into heat energy, and determining the heat produced. This conversion is carried out by oxidising the food by burning it; the quantity of heat resulting from the complete oxidation of unit weight of a food is known as the gross energy or heat of combustion of that food.

Gross energy is measured in an apparatus known as a bomb calorimeter, which in its simplest form consists of a strong metal chamber (the bomb) resting in an insulated tank of water. The food sample is placed in the bomb, and oxygen admitted under pressure. The temperature of the water is taken, and the sample is then ignited electrically. The heat produced by the oxidation is absorbed by the bomb and the surrounding water, and when equilibrium is reached the temperature of the water is taken again. The quantity of heat produced is then calculated from the rise in temperature and the weights and specific heats of the water and the bomb.

The bomb calorimeter can be used to determine the gross energy content of whole foods or of their constituents, and of animal tissues and excretory

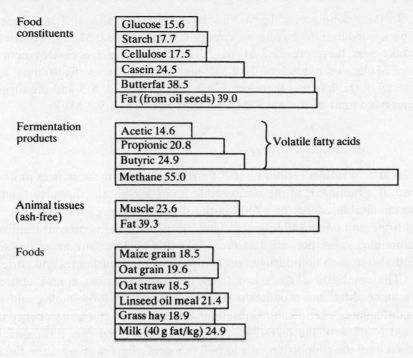

Fig. 11.1. Some typical gross energy values (MJ/kg dry matter).

products. Some typical gross energy values are shown in Fig. 11.1. Fats contain about two and a half times as much energy as carbohydrates, the difference reflecting the larger ratio of carbon plus hydrogen to oxygen in the former (i.e. fats are in a lower state of oxidation and are therefore capable of yielding more energy when oxidised). Proteins also have a higher gross energy value than carbohydrates. In spite of these differences among food constituents, the predominance of the carbohydrates means that the foods of farm animals vary little in gross energy content. Only foods rich in fat, such as linseed oil meal with 90 g ether extract/kg, have high values, and only those rich in ash, which has no calorific value, are much lower than average. Most common foods contain about 18.5 MJ/kg DM.

Digestible energy

The gross energy value of a food is an inaccurate estimate of the energy actually available to the animal, because it fails to take into account the losses of energy occurring during digestion and metabolism. The first source of loss to be considered is that of the energy contained in the faeces. The apparently digestible energy of a food is the gross energy less the energy contained in the faeces which result from any particular input of that food.

In the example of a digestibility trial given in Table 10.1, the sheep ate 1.63 kg of hay DM having an energy content of 18.0 MJ/kg. Total energy intake was therefore 29.3 MJ/day. The 0.76 kg of faeces DM contained 18.7 MJ/kg, or a total of 14.2 MJ/day. The apparent digestibility of the energy of the hay was therefore $(29.3–14.2)/29.3 = 0.515$ and the digestible energy content of the hay DM was $18.0 \times 0.515 = 9.3$ MJ/kg.

Metabolisable energy

The animal suffers further losses of energy-containing substances in its urine and, if a ruminant, in the combustible gases leaving the digestive tract. The metabolisable energy of a food is the digestible energy less the energy lost in urine and combustible gases. The energy of urine is present in nitrogen-containing substances such as urea, hippuric acid, creatinine and allantoin, and also in such non-nitrogenous compounds as glucuronates and citric acid.

The combustible gases lost from the rumen consist almost entirely of methane. Methane production is closely related to food intake, and at the maintenance level of nutrition about 8 per cent of the gross energy of the food (c. 12% of the digestible energy) is lost as methane. At higher levels of feeding the proportion falls to 6–7 per cent of gross energy, the fall being most marked for highly digestible foods.

The metabolisable energy value of a food is determined in a feeding trial similar to a digestibility trial, but in which urine and methane, as well as faeces, are collected. Metabolism cages for sheep and pigs incorporate a device for collecting urine. The urine of cattle is caught in rubber urinals attached below the abdomen for males and over the vulva for females, and is piped by gravity or suction to a collection vessel. When methane production is measured the animal is usually kept in an airtight container known as a respiration chamber. The operation of such chambers is described in more detail later (p. 229).

When no respiration chamber is available, methane production can be estimated as 8 per cent of gross energy intake. In addition, it is possible to estimate the metabolisable energy values of ruminant foods from digestible energy values by multiplying by 0.8. This implies that, on average, about 20 per cent of the energy apparently digested is excreted in the urine and as methane.

For poultry, metabolisable energy is measured more easily than digestible energy, because the faeces and urine are voided together. A rapid, standardised method has been developed for determining the metabolisable energy value of poultry foods. Cockerels are fasted until their digestive tract is empty, then force fed with a single meal of the food under investigation. Excreta are then collected until all residues of the single meal have been voided. At the same time the small quantities of excreta voided by fasted birds are collected, as a measure of endogenous losses. The energy of these endogenous losses is deducted from the energy of the excreta of the fed birds, so the estimate of metabolisable energy obtained is a *true* rather than *apparent* value (see p. 207); it is known as true metabolisable energy (TME), and is not directly comparable with measures of the metabolisable energy of foods obtained by other means.

Factors affecting the metabolisable energy values of foods. Table 11.1 shows metabolisable energy values for a number of foods. It is clear that, of the sources of energy loss so far considered, the faecal loss is by far the most important. Even for a food of high digestibility such as barley twice as much energy is lost in the faeces as in the urine and methane. The main factors affecting the metabolisable energy value of a food are therefore those which influence its digestibility. These have been discussed earlier (Ch. 10), and the emphasis here will be on urine and methane losses.

The metabolisable energy value of a food will obviously vary according to the species of animal to which it is given. In non-ruminants energy losses as methane are negligible, which means that for foods such as concentrates, which are digested to much the same extent by ruminants and non-ruminants, metabolisable energy values will be greater for the non-ruminants.

TABLE 11.1 Metabolisable energy values of some typical foods
(uncorrected values—see text—expressed as MJ per kg food dry matter)

Animal	Food	Gross energy	Energy lost in: Faeces	Urine	Methane	ME
Fowl	Maize	18.4	2.2		–	16.2
	Wheat	18.1	2.8		–	15.3
	Barley	18.2	4.9		–	13.3
Pig	Maize	18.9	1.6	0.4	–	16.9
	Oats	19.4	5.5	0.6	–	13.3
	Barley	17.5	2.8	0.5	–	14.2
	Coconut cake meal	19.0	6.4	2.6	–	10.0
Sheep	Barley	18.5	3.0	0.6	2.0	12.9
	Dried ryegrass (young)	19.5	3.4	1.5	1.6	13.0
	Dried ryegrass (mature)	19.0	7.1	0.6	1.4	9.9
	Grass hay (young)	18.0	5.4	0.9	1.5	10.2
	Grass hay (mature)	17.9	7.6	0.5	1.4	8.4
	Grass silage	19.0	5.0	0.9	1.5	11.6
Cattle	Maize	18.9	2.8	0.8	1.3	14.0
	Barley	18.3	4.1	0.8	1.1	12.3
	Wheat bran	19.0	6.0	1.0	1.4	10.6
	Lucerne hay	18.3	8.2	1.0	1.3	7.8

This is illustrated in Table 11.1, where values for barley given to cattle, sheep, pigs and fowls are compared. Differences between cattle and sheep in urine and methane losses of energy are small and of no significance.

The metabolisable energy value of a food will vary according to whether the amino acids it supplies are retained by the animal for protein synthesis, or are deaminated and their nitrogen excreted in the urine as urea. For this reason, metabolisable energy values are sometimes corrected to zero nitrogen balance by deducting for each 1 g of nitrogen retained either 28 kJ (pigs), 31 kJ (ruminants) or 34 kJ (poultry). The factor most appropriate to each species of animal depends on the extent to which nitrogen is excreted as urea (gross energy 23 kJ/g nitrogen) and as other compounds (e.g. uric acid, 28 kJ/g nitrogen). If an animal is excreting more nitrogen in its urine than it is absorbing from its food (i.e. is in negative nitrogen balance—see Ch. 13), some of the urine nitrogen is not derived from the food, and in this case the metabolisable energy value must be subjected to a positive correction.

The manner in which the food is prepared may in some cases affect its metabolisable energy value. For ruminants the grinding and pelleting of roughages leads to an increase in faecal losses of energy, but this may be partly offset by a reduction in methane production. For poultry the grinding of cereals has no consistent effect on metabolisable energy values.

As discussed earlier (Ch. 10) increases in the level of feeding of ruminants may cause an appreciable reduction in the digestibility of their food, and hence in its metabolisable energy value. Increases in faecal energy loss may be partially offset by reductions in losses of energy in methane and urine. Nevertheless, for finely ground roughages and the mixed roughage and concentrate diets, metabolisable energy value is reduced by increases in level of feeding.

In theory it should be possible to prevent the production of methane in the rumen and thereby avoid losing 8 per cent of gross energy intake in this form. In practice it is possible to suppress methane production by adding antimicrobial drugs to the diet (one effective chemical is chloroform), but the consequences are not consistently favourable. Energy may be diverted to another gaseous by-product, hydrogen (see Fig. 8.3); furthermore, the rumen micro-organisms may adapt to the presence of the drug and revert to the synthesis of methane. The most consistently effective methane suppressant is the coccidiostat, monensin which is widely used as a feed-borne growth stimulant for beef cattle.

Heat increment of foods

The ingestion of food by an animal is followed by losses of energy not only as the chemical energy of its solid, liquid and gaseous excreta but also as heat. Animals are continuously producing heat and losing it to their surroundings, either directly by radiation, conduction and convection, or indirectly by the evaporation of water. If a fasting animal is given food, within a few hours its heat production will increase above the level represented by basal metabolism. This increase is known as the heat increment of the food; it is quite marked in Man after a large meal. The heat increment may be expressed in absolute terms (MJ/kg food DM), or relatively as a proportion of the gross or metabolisable energy. Unless the animal is in a particularly cold environment, this heat energy is of no value to it, and must be considered, like the energy of the excreta, as a tax on the energy of the food.

The main cause of the heat increment is the energetic inefficiency of the reactions by which absorbed nutrients are metabolised. For example, it was shown in Chapter 9 that if glucose is oxidised in the formation of ATP, the efficiency of free energy capture is only about 0.69, 0.31 being lost as heat. Similar inefficiency is apparent in syntheses of body constituents. The linking of one amino acid to another, for example, requires the expenditure of four pyrophosphate 'high energy' bonds, and if the ATP which provides these is obtained through glucose oxidation, about 2.5 MJ of energy will be lost as heat for each kg of protein formed. Protein synthesis, it should be noted, occurs not only in growing animals, but also in those kept at a maintenance level. In the latter, protein synthesis is a part of the process of protein turnover (see p. 185).

A further part of the heat increment is attributable to the processes of digestion. Energy is used for the mastication of food and for its propulsion through the alimentary tract, and since chemical energy used for work performed within the body is converted into heat there will be a consequent increase in the animal's heat production. Another example of work done within the body is in the movement of substances (e.g. Na^+ and K^+ ions) against concentration gradients; this requires the expenditure of high energy phosphate bonds. In ruminant animals particularly, some heat arises from the activity of the micro-organisms of the alimentary tract; this is known as the heat of fermentation. It is estimated to amount to about 5–10 per cent of the gross energy of the food.

New techniques for studying energy metabolism allow the heat production of individual organs, etc. to be measured. It has been estimated, for example, that 30–40 per cent of the heat increment of ruminants is generated by the activities of the gastro-intestinal tract.

Net energy and energy retention

The deduction of the heat increment of a food from its metabolisable energy gives the net energy value of the food. The net energy of a food is that

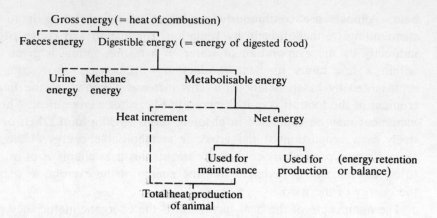

Fig. 11.2. The partition of food energy in the animal.

energy which is available to the animal for useful purposes, i.e. for body maintenance and for the various forms of production.

Net energy used for maintenance is mainly used to perform work within the body, and will leave the animal as heat. That used for growth and fattening and for milk, egg or wool production is either stored in the body or leaves it as chemical energy, and the quantity so used is referred to as the animal's energy retention.

The fate of the gross energy of foods is summarised in Fig. 11.2. It is important to understand that of the heat lost by the animal only a part, the heat increment of the food, is truly waste energy which can be regarded as a direct tax on the food energy. The heat resulting from the energy used for body maintenance is considered to represent energy which has been used by the animal and degraded into a useless form during the process of utilisation.

ANIMAL CALORIMETRY: METHODS FOR MEASURING HEAT PRODUCTION AND ENERGY RETENTION

In order to study the extent to which the metabolisable energy of the food is utilised by the animal, it is necessary to measure either the animal's heat production or else its energy retention. Examination of Fig. 11.2 will make it clear that, if one of these quantities is known, the other can be determined by subtracting the known one from the metabolisable energy. Heat production can be measured directly by physical methods; an animal calorimeter is required and the process is known as direct calorimetry. Alternatively, heat production can be estimated from the respiratory exchange of the animal; for this a respiration chamber is normally used and the approach is one of indirect calorimetry. Respiration chambers can also be used to estimate energy retention rather than heat production, by the procedure known as the carbon and nitrogen balance trial.

Direct calorimetry

Animals do not store heat, except for relatively short periods of time, and when measurements are made over periods of 24 hours or longer it is generally safe to assume that the quantity of heat lost from the animal is equal to the quantity produced. Heat is lost from the body principally by radiation, conduction and convection from the body surface and by evaporation of water from the skin and lungs. The animal calorimeter is basically an airtight and insulated chamber. Evaporative losses of heat are measured by recording the volume of air drawn through the chamber and its moisture content on entry and exit. In most early calorimeters the sensible heat loss (i.e. that lost by radiation, conduction and convection) was taken up in water circulated through coils within the chamber; the quantity of heat removed from the chamber could then be computed from the rate of flow of the water and the difference between its entry and exit temperatures. In a more recent type, the gradient layer calorimeter, the quantity of heat is measured electrically as it passes through the wall of the chamber. This type of calorimeter lends itself well to automation, and both sensible and evaporative losses of heat can be recorded automatically. Most calorimeters incorporate apparatus for measuring respiratory exchange and can therefore be used for indirect calorimetry as well.

The heat increment of the food under investigation is determined as the difference in the heat production of the animal at two levels of intake, as shown in Fig. 11.3. Two levels are needed because a part of the animal's heat production is contributed by its basal metabolism. An increase in food intake causes total heat production to rise, but the basal metabolism is assumed to remain the same. The increase in heat production can thus be attributed to the heat increment of the extra food given.

In the example shown in Fig. 11.3, the food was given at levels supplying 40 and 100 MJ metabolisable energy. The increment of 60 MJ (BD on the figure) was associated with an increase in heat production, CD, of 24 MJ. The heat increment as a fraction was:

$$CD/BD \text{ or } 24/60 = 0.4$$

It is also possible to make the lower level of intake zero, and to estimate the heat increment as the difference in heat production between the basal (or fasting) metabolism and that produced in the fed animal. In the example of Fig. 11.3 this method gives the heat increment as $16/40 = 0.4$.

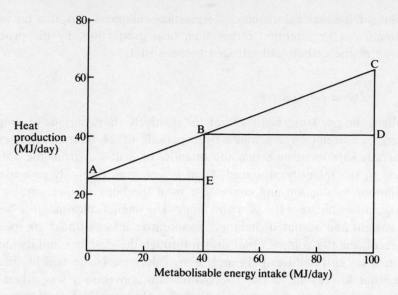

Fig. 11.3. The difference method for estimating the heat increment of foods. A is the basal metabolism and B and C represent heat production at metabolisable energy intakes of 40 and 100 MJ respectively. For the sake of simplicity the relation between heat production and metabolisable energy intake is shown here as being linear, i.e. ABC is a straight line; however, as explained later in the chapter, this is not usually the case.

If a single food is being investigated, it may be given as the sole item of diet at both levels. If the food is one which would not normally be given alone, the lower level may be obtained by giving a basal ration and the higher level by the same basal ration plus some of the food under investigation. For example, the heat increment of barley eaten by sheep could be measured by feeding the sheep first on a basal ration of hay and then on an equal amount of the same hay plus some of the barley.

Animal calorimeters are expensive to build and the earlier types required much labour to operate them. Because of this most animal calorimetry today is carried out by the indirect method described below.

Indirect calorimetry by the measurement of respiratory exchange

The substances which are oxidised in the body, and whose energy is therefore converted into heat, fall mainly into the three nutrient classes of carbohydrates, fats and proteins. The overall reaction for the oxidation of a carbohydrate such as glucose is:

$$C_6H_{12}O_6 + 6O_2 \rightarrow 6CO_2 + 6H_2O + 2.82 \text{ MJ} \qquad [1]$$

and for the oxidation of the typical fat, tripalmitin, is:

$$C_3H_5(OOC.C_{15}H_{31})_3 + 72.5O_2 \rightarrow 51CO_2 + 49H_2O + 32.04 \text{ MJ} \quad [2]$$

In an animal obtaining all its energy by the oxidation of glucose, the utilisation of one litre of oxygen would lead to production of $2820/(6 \times 22.4) = 20.95$ kJ of heat; for mixtures of carbohydrates an average value is 21.12 kJ/l. Such values are known as thermal equivalents of oxygen, and are used in indirect calorimetry to estimate heat production from oxygen consumption. For an animal catabolising mixtures of fats alone, the thermal equivalent of oxygen is 19.61 kJ/l (cf. 19.73 kJ/l calculated from equation [2] above).

Animals do not normally obtain energy exclusively from either carbohydrate or fat. They oxidise a mixture of these (and of protein also), so that in order to apply the appropriate thermal equivalent when converting oxygen consumption to heat production it is necessary to know how much of the oxygen is used for each nutrient. The proportions are calculated from what is known as the respiratory quotient (RQ). This is the ratio between the volume of carbon dioxide produced by the animal and the volume of oxygen used. Since, under the same conditions of temperature and pressure, equal volumes of gases contain equal numbers of molecules, the RQ can be calculated from the molecules of carbon dioxide produced and oxygen used. From equation [1] the RQ for carbohydrate is calculated as $6CO_2/6O_2 = 1$, and from equation [2] that of the fat, tripalmitin, as $51CO_2/72.5O_2 = 0.70$. If the RQ of an animal is known, the proportions of fat and carbohydrate oxidised can then be determined from standard tables. For example, an RQ of 0.9 indicates the oxidation of a mixture of 67.5 per cent carbohydrate and 32.5 per cent fat, and the thermal equivalent of oxygen for such a mixture is 20.60 kJ/l.

The mixture oxidised generally includes protein. The quantity of protein catabolised can be estimated from the output of nitrogen in the urine, 0.16 g of urinary N being excreted for each gram of protein. The heat of combustion of protein (i.e. the heat produced when it is completely oxidised) varies according to the amino acid proportions but averages 22.2 kJ/g. Protein, however, is incompletely oxidised in animals because the body cannot oxidise nitrogen, and the average amount of heat produced by the catabolism of 1 g of protein is 18.0 kJ. For each gram of protein oxidised, 0.77 l of carbon dioxide is produced and 0.96 l of oxygen used, giving an RQ of 0.8.

Heat is produced not only when organic nutrients are oxidised but also when they are used for the synthesis of tissue materials. It has been found, however, that the quantities of heat produced during these syntheses bear the same relation to the respiratory exchange as they do when the nutrients are completely oxidised.

The relationship between respiratory exchange and heat production is disturbed if the oxidation of carbohydrate and fat is incomplete. This situation arises in the metabolic disorder known as ketosis, when fatty acids are not completely oxidised to carbon dioxide and water, and carbon and hydrogen leave the body as ketones or ketone-like substances. Incomplete oxidation occurs also under normal conditions in ruminants, in which an end-product of carbohydrate fermentation in the rumen is methane. In practice heat production calculated from respiratory exchange in ruminants is corrected for this effect by the deduction of 2 kJ for each litre of methane. An alternative means of overcoming difficulties of this kind is to calculate heat production from oxygen consumption alone. If a respiratory quotient of 0.82 and a thermal equivalent of 20.0 are assumed, departures from this RQ of between 0.7 and 1.0 cause a maximum bias of no more than 3.5 per cent in the estimate of heat production. A further simplication is possible in respect of protein metabolism. The thermal equivalent of oxygen used for protein oxidation is 18.8 kJ/l, not very different from the value of 20.0 assumed for carbohydrate and fat oxidation. If only a small proportion of the heat production is caused by protein oxidation it is unnecessary to assess it separately, and so urinary nitrogen output need not be measured.

An example of the calculation of heat production from respiratory exchange is shown in Table 11.2.

TABLE 11.2 Calculation of the heat production of a calf from values for its respiratory exchange and urinary nitrogen excretion
(After K. L. Blaxter, N. McC. Graham and J. A. F. Rook, 1955. *J. agric. Sci., Camb.*, **45**, 10)

Results of the experiment (per 24 hours)		
Oxygen consumed		392.0 l
Carbon dioxide produced		310.7 l
Nitrogen excreted in urine		14.8 g
Heat from protein metabolism		
Protein oxidised	(14.8×6.25)	92.5 g
Heat produced	(92.5×18.0)	1665 kJ
Oxygen used	(92.5×0.96)	88.8 l
Carbon dioxide produced	(92.5×0.77)	71.2 l
Heat from carbohydrate and fat metabolism		
Oxygen used	$(392.0 - 88.8)$	303.2 l
Carbon dioxide produced	$(310.7 - 71.2)$	239.5 l
Non-protein respiratory quotient		0.79
Thermal equivalent of oxygen when RQ = 0.79		20.0 kJ/l
Heat produced	(303.2×20.0)	6064 kJ
Total heat produced	$(1665 + 6064)$	7729 kJ

Apparatus for measuring respiratory exchange

The apparatus most commonly used for farm animals is a respiration chamber, which may be of either the open circuit or closed circuit type. In both types the central feature is an airtight container for the animal which incorporates devices allowing the feeding and watering (and even milking) of the animal, and the collection of its faeces and urine, without the chamber air mixing with the atmosphere. In the open circuit type (Fig. 11.4a), air is drawn through the chamber and then discharged. During its passage it increases in carbon dioxide content and decreases in oxygen, and the amounts of carbon dioxide produced and oxygen used can be calculated by comparing the volume and composition of the air entering and leaving the chamber. In the closed circuit type (Fig. 11.4b), air is circulated from the chamber through carbon dioxide and water absorbers and back to the chamber. Oxygen used by the animal is replaced by a metered supply of the pure gas. The weight of carbon dioxide produced is determined as the increase in weight of the absorbent (generally potassium hydroxide or soda lime). Both oxygen consumption and carbon dioxide production must

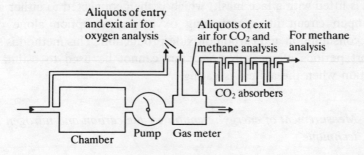

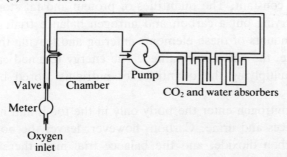

Fig. 11.4. Diagrams of respiration chambers.

be corrected for any differences in the amounts present in the circuit air at the beginning and end of the experiment. Methane is allowed to accumulate in the circuit air, and the amount present is determined at the end of the experiment.

Both types have their disadvantages. The open circuit requires elaborate apparatus for measuring and sampling the air, and very accurate gas analysis. Great accuracy is needed in order to measure the rather small differences in composition between in- and outgoing air. These disadvantages do not obtain with the closed circuit apparatus, where the greatest drawback is the large quantities of absorbents needed. A cow, for example, requires about 100 kg soda lime to absorb the carbon dioxide it produces each day, and 250 kg silica gel to absorb water vapour. Novel cattle chambers built at the Rowett Research Institute, Aberdeen, provide the advantages of both open and closed circuit chambers. For periods of perhaps 30 minutes the chambers operate on a closed circuit without gas absorption, thus allowing an appreciable decrement in oxygen and appreciable increments in carbon dioxide and methane. Then for a brief period (*c.* 3 min) the chambers operate on open circuit, the air drawn from them being metered and sampled for analysis.

Respiratory exchange can be measured without an animal chamber if the subject is fitted with a face mask, which is then connected to either a closed or an open circuit for determining oxygen consumption alone or both oxygen consumption and carbon dioxide production. This method is suitable for short periods of measurement, but cannot be used to estimate heat production when the animal is eating.

Measurement of energy retention by the carbon and nitrogen balance technique

The main forms in which energy is stored by the growing and fattening animal are protein and fat, for the carbohydrate reserves of the body are small and relatively constant. The quantities of protein and fat stored can be estimated by carrying out a carbon and nitrogen balance trial; that is, by measuring the amounts of these elements entering and leaving the body and so, by difference, the amounts retained. The energy retained can then be calculated by multiplying the quantities of nutrients stored by their calorific values.

Both carbon and nitrogen enter the body only in the food, and nitrogen leaves it only in faeces and urine. Carbon, however, leaves the body also in methane and carbon dioxide, and the balance trial must therefore be carried out in a respiration chamber. The procedure for calculating energy retention from carbon and nitrogen balance data is best illustrated by considering an animal in which storage of both fat and protein is taking place. In such an animal intakes of carbon and nitrogen will be greater than the quantities excreted, and the animal is said to be in positive balance with respect to these elements. The quantity of protein stored is calculated by multiplying the nitrogen balance by 1000/160 (= 6.25) for body protein is assumed to contain 160 g N/kg. It also contains 512 g C/kg, and the amount of carbon stored as protein can therefore be computed. The remaining carbon is stored as fat, which contains 746 g C/kg. Fat storage is therefore calculated by dividing the carbon balance, less that stored as protein, by 0.746. The energy present in the protein and fat stored is then calculated by using average calorific values for body tissue. These values vary from one species to another; for cattle and sheep those now recommended are 39.3 MJ/kg for fat and 23.6 MJ/kg for protein. An example of this method of calculating energy retention (and heat production) is shown in Table 11.3.

Other methods for measuring energy retention

Because calorimetric experiments require elaborate apparatus and can be conducted with only small numbers of animals, numerous attempts have been, and are still being, made to measure energy retention in other ways. In many feeding trials the animals' intakes of digestible or metabolisable

TABLE 11.3 Calculation of the energy retention and heat production of a sheep from its carbon and nitrogen balance
(After K. L. Blaxter and N. McC. Graham, 1955. *J. agric. Sci., Camb.*, **46**, 292)

Results of the experiment (per 24 hours)	C (g)	N (g)	Energy (MJ)
Intake	684.5	41.67	28.41
Excretion in faeces	279.3	13.96	11.47
Excretion in urine	33.6	25.41	1.50
Excretion as methane	20.3	—	1.49
Excretion as CO_2	278.0	—	—
Balance	73.3	2.30	
Intake of metabolisable energy	—	—	13.95
Protein and fat storage			
Protein stored		(2.30 × 6.25)	14.4 g
Carbon stored as protein		(14.4 × 0.512)	7.4 g
Carbon stored as fat		(73.3 − 7.4)	65.9 g
Fat stored		(65.9 ÷ 0.746)	88.3 g
Energy retention and heat production			
Energy stored as protein		(14.4 × 23.6)	0.34 MJ
Energy stored as fat		(88.3 × 39.3)	3.47 MJ
Total energy retention		(0.34 + 3.47)	3.81 MJ
Heat production		(13.95 − 3.81)	10.14 MJ

energy can be measured satisfactorily but their energy retention can be estimated only from changes in liveweight. Weight changes, however, provide inaccurate estimates of energy retention, firstly because they may represent no more than changes in the contents of the gut or bladder, and secondly because the energy content of true tissue gain can vary over a wide range according to its proportions of bone, muscle and fat (see Ch. 14). These objections are only partly overcome in experiments where the energy retention is in the form of milk or eggs, whose energy is easily measured, for retention in these products is almost invariably accompanied by retention in other tissues (e.g. milking cows are normally increasing or decreasing in liveweight and in energy content).

Energy retention can, however, be measured in feeding trials if the energy content of the animal is estimated at the beginning and end of the experiment. In the comparative slaughter method this is done by dividing the animals into two groups and slaughtering one (the sample slaughter group) at the beginning of trial. The energy content of the animals slaughtered is determined by bomb calorimetry, the samples used being taken either from the whole, minced body or from the tissues of the body after these have been separated by dissection. A relationship is then obtained between the liveweight of the animals and their energy content, and this is used to predict the initial energy content of animals in the second group. The latter are slaughtered at the end of the trial and treated in the same manner as those in the sample slaughter group, and their energy gains can then be calculated.

The comparative slaughter method requires no elaborate apparatus, but is obviously expensive and laborious when applied to larger animal species. The method becomes less costly if body composition, and hence energy content, can be measured in the living animal, or, failing that, in the whole, undissected carcass. Several chemical methods have been developed for estimating body composition *in vivo*. For most of them the principle employed is that the lean body mass of animals (i.e. empty body weight less weight of fat) is in several respects reasonably constant in composition. For example, in cattle 1 kg lean body mass contains 729 g water, 216 g protein and 53 g ash. This means that if the weight of water in the living animal can be measured, the weights of protein and ash can then be estimated; in addition, if total weight is known, weight of fat can be estimated by subtracting the lean body mass. In practice total body water can be estimated by so-called 'dilution' techniques, in which a known quantity of a marker substance is injected into the animal, allowed to equilibrate with body water and its equilibrium concentration determined. The marker substances most commonly used are water containing the radioactive isotope of hydrogen, tritium, or its heavy isotope, deuterium. One difficulty with these techniques is that the markers mix not only with actual body water, but also with water present in the gut (in ruminants as much as 30% of total body water may be in gut contents). A second chemical method for estimating body composition *in vivo* is based on the constant concentration of potassium in the lean body mass.

The composition of a carcass is often estimated without dissection or chemical analysis from its specific gravity. Fat has an appreciably lower specific gravity than bone and muscle, and the fatter the carcass the lower will be its specific gravity. The specific gravity of a carcass is determined by weighing it in air and in water, but this method has technical difficulties (e.g. air trapped under water) that make it imprecise. Nevertheless, estimates of energy utilisation obtained by comparative slaughter and specific gravity measurements have been used in the USA to establish a complete cattle feeding system (see Ch. 12).

FACTORS AFFECTING THE UTILISATION OF METABOLISABLE ENERGY

The efficiency of utilisation of metabolisable energy in the animal is defined as that proportion of it which is retained, or

$$\frac{\text{change in energy retention}}{\text{change in metabolisable energy intake}}$$

It is therefore the complement of the heat increment when the latter is expressed as a proportion of the metabolisable energy. For example, if food supplying 10 MJ metabolisable energy were given to an animal and energy retention increased by 6 MJ, the efficiency of utilisation would be 0.6 (and the heat increment 0.4 of the metabolisable energy).

The efficiency with which metabolisable energy is utilised depends on the interaction of two principal factors, these being the nature of the chemical compounds in which the metabolisable energy is contained and the purpose for which these compounds are used by the animal.

Utilisation of metabolisable energy for maintenance

For maintenance purposes the animal oxidises the nutrients absorbed from its food principally to provide energy for work. If it is given no food, it obtains this energy mainly by the oxidation of body fat. When food is given, but in quantities insufficient to provide all the energy needed for maintenance, the task of providing ATP is partially transferred from the reserves of body fat to the nutrients absorbed. If the energy contained in these nutrients can be transferred to ATP as efficiently as can that in body fat, no extra heat will be produced by the animal apart from that associated specifically with the consumption, digestion and absorption of food. (Heat of fermentation comes into this category, and also work of digestion, that is, heat arising from the energy used in the mastication of food and its propulsion through the gut, in the absorption of nutrients and in their transport to the tissues.)

The efficiency of free energy capture when body fats are oxidised and ATP is formed can be calculated from the reactions shown in Chapter 9 to be of the order of 0.67. For glucose, to take an example of a nutrient, the efficiency is similar, at about 0.70. One would therefore expect that glucose given to a fasting animal would be utilised without any increase in heat production, or in other words with apparent (calorimetric) efficiency of 1.0. Table 11.4 shows that this is approximately true. In sheep the efficiency is reduced through fermentation losses if the glucose passes into the rumen, but these losses are avoided if it is infused directly into the abomasum.

Table 11.4 shows also that dietary fat is used for maintenance with high

TABLE 11.4   Efficiency of utilisation for maintenance of metabolisable energy in various nutrients and foods

	Ruminant	*Pig, etc.**	*Fowl*
Food constituents			
Glucose	0.94 (1.00)†	0.95	0.89
Starch		0.88	0.97
Olive oil		0.97	
Casein	(0.82)	0.76	0.84
Fermentation products			
Acetic acid	0.59		
Propionic acid	0.86		
Butyric acid	0.76		
Mixture A‡	0.87		
Mixture B	0.86		
Concentrates			
Maize	0.80		
Balanced diets			0.81
Roughages			
Dried ryegrass (young)	0.78		
Dried ryegrass (mature)	0.74		
Meadow hay	0.70		
Lucerne hay	0.82		
Grass silages	0.65–0.71		

* Including dog and rat
† values in parentheses are from administration per duodenum.
‡ Mixture A : acetic 0.25, propionic 0.45, butyric 0.30
‡ Mixture B : acetic 0.75, propionic 0.15, butyric 0.10

energetic efficiency, as one would expect. When protein is used to provide energy for maintenance, however, there is an appreciable heat increment of about 0.2, which is in part attributable to the energy required for urea synthesis (see Ch. 9). In ruminants, energy for maintenance is absorbed largely in the form of volatile fatty acids. Experiments in which the pure acids have been infused singly into the rumens of fasting sheep have shown that there are differences between them in the efficiency with which their energy is utilised (Table 11.4). But when the acids are combined into mixtures representing the extremes likely to be found in the rumen, the efficiency of utilisation is uniform and high. Nevertheless the efficiency is still less than that for glucose, and this discrepancy, together with the energy lost through heat of fermentation in ruminants, leads one to expect that metabolisable energy will be utilised more efficiently for maintenance in those animals in which it is absorbed in the form of glucose than in ruminants.

Very few experiments have been carried out to determine the efficiency with which the metabolisable energy in foods is used for maintenance, and these few have been restricted almost entirely to ruminant animals fed on roughages. A selection of the results is given in Table 11.4.

Most of the metabolisable energy of the foods shown would have been absorbed in the form of volatile fatty acids. Efficiency of utilisation is less, however, than for synthetic mixtures of these acids, since with whole foods heat losses are increased by heat of fermentation and by the energy used for work of digestion. In spite of this, the metabolisable energy in these foods was used with quite high efficiency.

Utilisation of metabolisable energy for productive purposes

Although energy may be stored by animals in a wide variety of products—in body fat, muscle, milk, eggs and wool—the energy of these products is contained mainly in fat and protein (only in milk is much energy stored as carbohydrate). The efficiency with which metabolisable energy is used for productive purposes therefore depends largely on the energetic efficiency of the metabolic pathways involved in the synthesis of fat and protein from absorbed nutrients. These pathways have been outlined in Chapter 9. In general the synthesis of either fat or protein is a more complicated process than its catabolism, in the same way that the construction of a building is more difficult than its demolition. Not only must the building materials be present in the right proportions, but they must arrive on the scene at the right time, and the absence of a particular material may prevent or seriously impair the whole process. Thus it was shown in Chapter 9 that fatty acid synthesis is dependent on a supply of reduced $NADP^+$. Because of the greater complexity of synthetic processes it is more difficult to estimate their theoretical efficiency.

In Chapter 9, the synthesis of a triacylglycerol from acetate and glucose was shown to have a theoretical efficiency of 0.81, a higher value would be expected for fat synthesis from absorbed triacylglycerols. In protein synthesis, the energy cost of linking amino acids together is relatively small and if these are present in the right proportions, the theoretical efficiency of protein synthesis is about 0.88 (Ch. 9). However, if some amino acids have to be synthesized while others undergo deamination, efficiency will be considerably less; as discussed later, the observed efficiency of protein synthesis is generally much lower than the observed efficiency of fat synthesis. The synthesis of lactose from glucose can be achieved with an efficiency of 0.94 (Ch. 9), but in the dairy cow the glucose so utilised will largely be formed from propionic acid, or possibly from amino acids (gluconeogenesis), and the efficiency of lactose synthesis will be lower.

The figures given above are all calculated from the appropriate metabolic pathways, and in relating them to the efficiency with which metabolisable energy will be utilised it is important to remember that they will be reduced by those energy losses mentioned earlier (p. 223) which are directly referable to the consumption, digestion and absorption of food.

Measuring in an animal the energetic efficiency with which a single substance such as protein, or a single product such as milk, is synthesised is complicated by the fact mentioned previously that animals seldom store energy as single substances or even as single products. An exception however is the mature, non-lactating and non-pregnant animal, for which it is considered that almost all energy storage is in the form of body fat. Early studies of energy utilisation were largely confined to mature animals, but more recent experiments have included young, growing animals (i.e. storing much protein, as well as fat), and also animals producing milk, eggs and foetal tissues.

Table 11.5 gives examples of the efficiency with which metabolisable energy in various nutrients and foods respectively has been found to be utilised for fattening in mature animals. In general there is reasonable agreement between the values shown in the tables and the theoretical considerations given previously. In the case of the pure nutrients, non-ruminants utilise fat more efficiently than carbohydrate, and carbohydrate more efficiently than protein. As in Table 11.4 the figures for ruminants are lower than those for non-ruminants; the lower value for groundnut oil in ruminants may be due to its interference with rumen fermentation (p. 154). The efficiency of utilisation of volatile fatty acids, especially acetic, is problematical. Earlier experiments suggested that acetic acid was poorly utilised for fattening by sheep, and that this poor efficiency was conveyed to mixtures of the acids (as shown in Table 11.5). Later work has shown that acetic acid can be more efficiently utilised, especially when it is added to concentrate diets (i.e. diets fermented in the rumen to high-propionic mixtures). Nevertheless, the energy of nutrients is always utilised less efficiently when the nutrients are fermented in the rumen than when they are digested in the intestine. The difference is illustrated for glucose and casein in Table 11.5; it is due to heat of fermentation and possibly to inefficient utilisation of energy in volatile fatty acids. Table 11.5 also shows that, for production as for maintenance, metabolisable energy is utilised more efficiently by the non-ruminant species than by ruminants.

TABLE 11.5 Efficiency of utilisation for fattening of metabolisable energy in various nutrients and foods

	Ruminant	Pig, etc.	Fowl
Food constituents			
Glucose	0.54 (0.72)*	0.74	
Sucrose	0.58	0.75	0.75
Starch	0.64	0.76	0.78
Cellulose	0.61	0.71	
Groundnut oil	0.58	0.86	0.78
Mixed proteins	0.51	0.62	0.55
Casein	0.50 (0.65)		0.60
Fermentation products			
Acetic acid	0.33–0.60	0.60	
Propionic acid	0.56		
Butyric acid	0.62		
Mixture A†	0.58		
Mixture B	0.32		
Lactic acid	0.75		
Ethanol	0.72		
Concentrates			
Barley	0.60	0.77	0.73
Oats	0.61	0.68	0.73
Maize	0.62	0.78	0.74
Groundnut meal	0.54	0.58	0.64
Soya bean meal	0.48	0.57	0.64
Roughages			
Dried ryegrass (young)	0.52		
Dried ryegrass (mature)	0.34		
Meadow hay	0.30		
Lucerne hay	0.52		
Grass silages	0.21–0.61		
Wheat straw	0.24		

* Values in parentheses are from administration per duodenum
† Mixture A : acetic 0.25, propionic 0.45. butyric 0.30
  Mixture B : acetic 0.75, propionic 0.15, butyric 0.10

A comparison of Tables 11.4 and 11.5 reveals two striking differences. In the first place, efficiency values for maintenance are all appreciably higher than the corresponding values for fattening. The main reason for this is that heat increments of nutrients or foods measured below maintenance do not represent the true inefficiency of energy conversion but inefficiency relative to that of the utilisation of body reserves (mainly fat). In a fasting animal, the heat production may be envisaged as consisting of two components, that arising from the transfer of energy from fat to ATP (see Ch. 9) and that arising from the utilisation of ATP to maintain essential body processes. When the animal is given food, nutrients derived from it will replace body fat as generators of ATP, and if they are less efficient in this regard, the first component of heat production will increase. One would expect that the replacement of body fat by dietary fat would have little effect on heat production, and this is confirmed by efficiency values close to unity in Table 11.4 for dietary fats. As food intake increases further, the metabolisable energy in excess of that needed for maintenance has to bear the full cost of nutrient metabolism, hence the invariably lower values for efficiency of utilisation in Table 11.5.

The above explanation is illustrated in Fig. 11.5. As metabolisable energy intake rises from zero to the maintenance level (i.e. distance AC in the figure) heat production rises by distance BC. At maintenance the apparent heat increment is therefore BC/AC, or in other words basal metabolism is the base line for the calculation. However, the increase in metabolisable

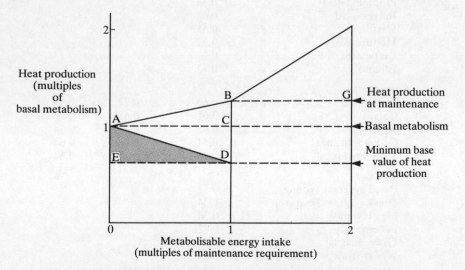

Fig. 11.5.   Relationship between heat production and metabolisable energy intake. Total heat production is represented by line ABF. Other points are explained in the text.

energy intake to the maintenance level is also accompanied by a decline in the heat produced from metabolism of body fat (triangle ADE). The proportion BD/AC therefore provides a more realistic estimate of the efficiency with which the metabolisable energy of the food is used for maintenance. The base line in this case (DE) is known as the *minimum base value of heat production*. Above maintenance the heat increment is given by the expression FG/BG.

The second important difference between efficiency values for maintenance and for fattening lies in the variability encountered in foods for ruminants. In Table 11.4 the variability is quite small, the extreme values being silage, 0.65, and lucerne hay, 0.82, a difference of 24 per cent of the mean. In Table 11.5, however, the range for foods comparable with those of Table 11.4 is from 0.21 for silage to 0.62 for maize, a difference of 99 per cent of the mean. It appears that, while foods generally are used less efficiently for fattening than for maintenance, the ratio between the two efficiency values is not the same for all foods. Thus in ruminants the metabolisable energy of maize is used $100 \times 0.62/0.80$ or 78 per cent as efficiently for fattening as for maintenance, whereas the comparable value for meadow hay is $100 \times 0.30/0.70$ or 43 per cent. In general, as the metabolisable energy concentration of foods declines, the efficiency of utilisation of the metabolisable energy also declines, but this fall in efficiency is relatively greater when energy is used for production than when it is used for maintenance.

The poor utilisation for fattening of the metabolisable energy of low quality foods, such as straw, was originally attributed to the so-called 'work of digestion' required to break them down and expel their indigestible residues. Later, the high proportion of acetic acid produced in the rumen fermentation of such foods was considered to be responsible, but as discussed above, acetic acid is not now thought to be so inefficiently utilised. Site of digestion (see Ch. 8) is another factor to be considered, because fibrous roughages are digested to a very large extent in the rumen rather than in the intestine. It is interesting to note that the efficiency of utilisation of metabolisable energy in low quality roughages is much improved if they are ground and pelleted, and that the improvement is associated with a partial transference of digestion from rumen to intestine (see p. 209). However, the grinding and pelleting of roughages may also reduce the proportion of acetic acid in the volatile fatty acids arising from their rumen fermentation (see p. 149). Thus it seems likely that the poor utilisation of metabolisable energy in low quality feeds is due to several factors.

Alternative models for energy retention and energy intake. In Fig. 11.3 the relationship between heat production and metabolisable energy intake was assumed to be described by a single straight line, which implied that the efficiency of utilisation of metabolisable energy was the same above and below the maintenance level. In Fig. 11.5, a more complex model with two straight lines meeting at maintenance was employed, the implication here being that efficiency of utilisation was less above maintenance than below maintenance. Figure 11.5 can be re-drawn with energy retention on the vertical axis in place of heat production; again there are two straight lines (Fig. 11.6(a)). However, the three points used to construct Fig. 11.6(a) (or 11.5) could also be joined by a single curve, as in Fig. 11.6(b). The implication of the curve is that the efficiency of utilisation of metabolisable energy is greatest at low energy intakes and gradually decreases as energy intake is increased. The curvilinear model seems likely to describe better than the 'double rectilinear' model the balance between anabolic and catabolic processes that is struck at the maintenance point, but the latter model is more commonly employed in practice.

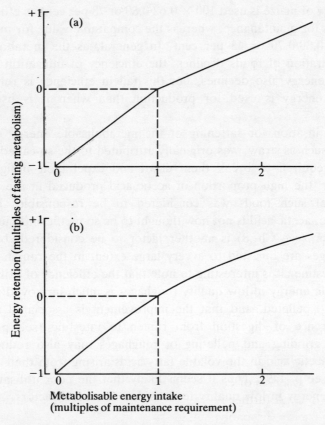

Fig. 11.6. Alternative models for energy retention and energy intake: (*a*) the double rectilinear model (*b*) the curvilinear model.

Utilisation of metabolisable energy for growth. In mature animals, most of the energy retained is stored as fat, but in younger, growing animals a considerable proportion (up to 60%) may be stored as protein.

The equations in Chapter 9 indicate that protein synthesis is energetically more efficient than fat synthesis (0.88 versus 0.81). However, the deposition of these materials in the animal is not just a matter of synthesis but is a resultant of the two processes of synthesis and breakdown. Proteins in body tissues are continuously being broken down and resynthesised and this 'turn-over' of proteins means that the calorimetric efficiency of protein deposition is much lower than the theoretical efficiency of protein synthesis. Calorimetric efficiency of protein synthesis is commonly stated to be about 0.55 in non-ruminants, but lower (0.4) for ruminants. In contrast, the turnover of fat in body tissues is much slower than that of protein, and the calorimetric efficiency of fat deposition is therefore close to the theoretical efficiency of fat synthesis; values of 0.7–0.75 are commonly quoted for all species.

It thus appears that the efficiency of utilisation of metabolisable energy is likely to be lower in young, growing animals (storing energy as protein) than it would be in older fattening stock. In practice the differences are small, except perhaps in the case of ruminants having unusually high rates of protein deposition.

Utilisation of metabolisable energy for milk or egg production. These additional forms of animal production rarely occur alone and are usually accompanied by gains or losses of fat or protein from the body of the lactating cow or laying hen. This means that estimates of the partial efficiency of milk or egg synthesis must usually be made by a mathematical partition of the metabolisable energy utilised. As rather complex diets are required for these syntheses, it is not possible to give efficiency coefficients for single nutrients, as was done for maintenance and fattening in Tables 11.4 and 11.5.

Since the first edition of this book was written, many energy balance trials have been carried out with lactating cows at the U.S. Department of Agriculture laboratory at Beltsville by W. P. Flatt and his colleagues, and in the Netherlands by A. J. H. van Es. Analyses of the results of these trials have shown the average efficiency of utilisation of metabolisable energy to be 0.60, and to vary little with the nature of the diet. These analyses also suggest that body reserves of energy can be transferred to milk energy with an efficiency of 0.80. A further point of interest is that lactating animals that are also storing energy, probably mainly as fat, can achieve these stores with unusually high efficiency of utilisation of metabolisable energy (0.60).

The greater efficiency of the lactating animal, compared with the fattening animal, is probably due in part to simpler forms of energy storage in milk, namely lactose and short-chain fatty acids. Milk protein may be efficiently synthesised because it is rapidly removed from the body, and is not required to take part in the turnover of amino acids experienced by most body proteins.

The laying hen has been estimated to utilise metabolisable energy for egg synthesis with an efficiency of 0.60–0.70; for egg protein synthesis, efficiency is estimated to be 0.45–0.50 and for egg lipids, 0.75–0.80. Body tissue synthesis in laying hens is achieved also with high efficiency (0.75–0.80).

Other factors affecting the utilisation of metabolisable energy

Associative effects. In Chapter 10 it was explained that the digestibility of a food can vary according to the nature of the ration in which it is included. Associative effects of this kind have also been observed in the utilisation of metabolisable energy. In one experiment it was found that the metabolisable energy of maize meal was utilised for fattening with an apparent efficiency varying between 0.58 and 0.74 according to the nature of the basal ration to which it was added. In ruminants, such differences are likely to arise through variations in the effect of the food on the manner in which the whole ration is digested, and hence on the form in which metabolisable energy is absorbed. The implication is that values for the efficiency of utilisation of metabolisable energy for individual foods are of limited significance.

Balance of nutrients. The effect of the nutrient proportions of the food has been partly covered in earlier parts of this section. A fattening animal will tend to use metabolisable energy more efficiently if it is provided as carbohydrate than if it is provided as protein. Similarly, if a growing animal is provided with insufficient protein or with insufficient of a particular amino acid, it will tend to store energy as fat rather than as protein, and the efficiency with which it utilises metabolisable energy will probably be altered.

Deficiencies of minerals and vitamins, too, can interfere with energy utilisation. Thus a deficiency of phosphorus has been shown to reduce efficiency of utilisation by about 10 per cent in cattle. This effect is hardly surprising in view of the vital role of phosphorus in the energy-yielding reactions of intermediary metabolism.

FURTHER READING

K. L. Blaxter, 1967. *The Energy Metabolism of Ruminants*. Hutchinson, London.

A. Ekern and F. Sundstøl, 1982. *Energy Metabolism of Farm Animals*, Proceedings of the 9th European Association for Animal Production Symposium (EAAP Publication No. 29). (See also the proceedings of other symposia in this series.)

J. T. Reid, C. D. White, R. Anrique and A. Fortin, 1980. Nutritional energetics of livestock: some present boundaries of knowledge and future research needs. *Journal of Animal Science*, 51, 1393–1415.

12

Evaluation of foods

(C) Systems for expressing the energy value of foods

For the farmer the essential steps in the scientific rationing of animals are, firstly, the assessment of their nutrient requirements, and secondly, the selection of foods which can supply these requirements. This balancing of demand and supply is made separately for each nutrient, and in many cases the nutrients given first consideration are those supplying energy. There are good reasons why energy should receive priority. In the first place the energy-supplying nutrients are those present in the food in greatest quantity; this means that if a diet has been devised to meet other nutrient requirements first and is then found to be deficient in energy, a major revision of its constituents will probably be needed. In contrast, a deficiency of a mineral or vitamin can often be rectified very simply by adding a small quantity of a concentrated source.

A further feature of the energy-containing nutrients which distinguishes them from others is the manner in which the performance of an animal, when measured as liveweight gain or milk or egg production, responds to changes in the quantities supplied. Whereas an animal with a given level of performance, a steer gaining 1 kg/day, for example, will respond eventually if the allowance of any one nutrient is reduced below the quantity required for this level of performance, increases in single nutrients above the requirement generally have little effect. Increasing the quantity of vitamin A supplied, for example, to twice the requirement is unlikely to affect the steer's liveweight gain (although it may increase its vitamin A reserves). But if energy intake alone is increased the animal will attempt to retain more energy, either partly as protein if nitrogen intake is adequate, or entirely as fat, and its liveweight gain will increase. In effect, energy intake is the pacemaker of production, as animals tend to show a continuous response to changes in the quantities supplied.

If other nutrients are present in amounts only just sufficient for the animal's requirements, the response to an increase in energy intake is likely to be an undesirable one. Increasing the storage of body fat is likely to increase the need for minerals and vitamins associated with the enzyme systems involved in fat synthesis, and so to precipitate a deficiency of those substances. It is therefore important to maintain a correct balance between energy, the pacemaker, and other nutrients in the diet.

CLASSIFICATION OF ENERGY SYSTEMS

Energy systems contain two essential elements, these being measures of the energy values of foods and measures of the energy requirements of animals. The latter are most commonly expressed in the simplest way possible, as the absolute quantities of energy gained or lost by animals. For example, if an animal gains 1 kg of flesh with a heat of combustion of 15 MJ, its energy requirement is stated as 15 MJ/kg of gain. Likewise, if the same animal when fasted loses 30 MJ/day, its maintenance requirement is stated to be 30 MJ/day.

The energy value of foods, however, may be stated as net, metabolisable, digestible (or even gross) energy. Net energy would seem to be the most logical choice, because it could be readily equated with the energy requirements of animals. For example, if a feed containing 5 MJ net energy/kg were added to the maintenance ration of the animal referred to above, the quantity required to promote 1 kg of gain could readily be calculated as $15/5 = 3$ kg feed. But *net energy* systems of the kind exemplified here are not as simple as they appear to be, their chief complication being that any given food has a different net energy value for different animal functions. For example, we saw in the previous chapter that the metabolisable energy of maize (14 MJ/kg DM) is utilised by ruminants with an efficiency of 0.80 for maintenance and 0.62 for growth; its net energy value is therefore $14 \times 0.80 = 11.2$ MJ/kg DM for maintenance and $14 \times 0.62 = 8.7$ MJ/kg DM for growth. In theory, this means that tables of feeding value would need to include two (or more) net energy values for each food; in practice, it has meant that various approximations are used in net energy systems. These are described later.

When compared to the net energy value of a food, its metabolisable energy value comes closer to being a unique measure (although metabolisable energy will still differ according to animal species and to such lesser factors as level of feeding), and metabolisable energy has therefore been used as the primary measure of food value in several energy systems. Its use presents the obvious difficulty that food values and animal requirements cannot be directly equated with one another. To bring them together, an additional element is needed for the system, which may be called an 'interface'. The interface is essentially a method of calculating the equivalence of metabolisable and net energy for any specific rationing situation. Thus, in the example given above, if the energy value of the food was expressed as 10 MJ metabolisable energy per kg the interface element required would be an estimate of k, the efficiency of utilisation of metabolisable energy for growth and fattening; if this were 0.5, the net energy value of the food would be calculated as $10 \times 0.5 = 5$ MJ/kg, as before.

Before some actual energy systems are discussed, two further preliminary points must be made. The first is that the assessment of the energy value of a food requires a laborious and complicated procedure, which cannot be applied, for example, to samples of hay or silage brought by a farmer to an advisory chemist. For this reason an essential feature of most systems is a method for predicting energy value from some more easily measured characteristic of the food, such as its gross or digestible composition. Secondly, it is important to realise that energy systems for ruminants are more complicated than those used for pigs or poultry. The main reasons for the greater complexity of systems for ruminants are the greater variety of foods involved, and the wider range of options for digestion provided by the gut compartments.

ENERGY SYSTEMS FOR RUMINANTS

Early net energy systems

About the year 1900 two net energy systems were developed, by H. P. Armsby at the University of Pennsylvania and by O. Kellner at the Möckern Experiment Station in Germany. Both were based on calorimetric experiments in which cattle were given specific feeds at two or more levels of feeding. The experiments differed in the techniques used, since Armsby employed a direct calorimeter to estimate heat production whereas Kellner used respiration chambers to measure energy retention from carbon and nitrogen balance. They also differed in the feeding levels applied. Armsby generally comparing two levels below maintenance and Kellner, two above maintenance; this meant that Armsby measured net energy values for maintenance whereas Kellner measured net energy values for fattening. Both men generalised their systems by providing methods for predicting the net energy values of foods from their digestible nutrients.

Armsby's *estimated net energy values* were quoted in successive editions of the one-time standard work on the feeding of livestock in the USA, Morrison's *Feeds and Feeding*, but appear not to have been used much in practice. Kellner's system, however, was until recently used extensively in

European countries, including Britain, and still survives in a modified form in East Germany. A full account of the system as it was used in Britain is given in earlier editions of this book, but the essential details are as follows.

Believing that farmers would have difficulty in using energy units, Kellner did not express his net energy values absolutely (as calories or joules) but relative to the net energy value of that common food constituent, starch. Starch was found to have a net energy value for fattening of 2.36 Mcal (9.9 MJ)/kg, whereas barley, for example, had a net energy value of 1.91 Mcal/kg. Kellner calculated the value 1.91/2.36 = 0.81 and called it the *starch equivalent* of barley (i.e. 0.81 kg starch equivalent/kg barley, or 81 kg/100 kg). Weights (kg) of starch equivalent may be converted to net energy for fattening (MJ) by multiplying them by 9.9. Kellner recognised that net energy values differed according to whether a food was used for fattening, maintenance or milk production, but instead of giving each food three values he decided instead to modify his figures for animal requirements, and express them all in terms of net energy for fattening. The basis for this modification, when expressed in modern terminology, was that on *average* (for feeds in general) k_f, k_m and k_l were in ratio 1.0 : 1.30 : 1.25. If an animal required 30 MJ (i.e. 3 kg starch equivalent) for maintenance, this value was divided by 1.30 to express it in terms of net energy for fattening, giving 23 MJ or 2.3 kg starch equivalent, and a similar adjustment was made for energy required for milk production. As shown in Chapter 11, these ratios vary considerably from one food to another; for example, the ratio k_f:k_m is generally wider for poorer quality foods. Kellner's approximation therefore led to errors in his original system, and in the modern East German version of this system, an alternative method is used to correct NE values of foods (see p. 252).

Metabolisable energy systems used in Britain

The energy system now used for ruminants in Britain was devised by K. L. Blaxter, incorporated in the U.K. Agricultural Research Council's 1965 publication *Nutrient Requirements of Farm Livestock: No. 2. Ruminants*, and slightly modified for practical use by the Ministry of Agriculture, Fisheries and Food in their 1975 Technical Bulletin No. 33. A refined form of the original version has now been published in the second (1980) edition of the Agricultural Research Council manual mentioned above, and Technical Bulletin No. 33 was revised in 1984 to become Reference Book No. 433. The main features of the system are as follows.

The energy values of foods are expressed in terms of metabolisable energy, and the metabolisable energy value of a ration is calculated by adding up the contributions of individual feed constituents. The energy requirements of animals are expressed in absolute terms, as net energy. The essential feature of the interface is a series of equations to predict efficiency of utilisation of metabolisable energy for maintenance, growth and lactation (Table 12.1). The predictions are made from the metabolisable energy concentration of the diet, although this is expressed as the fraction ME/GE (sometimes called 'metabolisability') rather than as MJ/kg. Metabolisability may be converted to MJ ME/kg DM by multiplying it by 18.4, the mean concentration of gross energy in food dry matter (although this factor is too high for foods of high ash content, and too low for those of high protein or lipid content). The efficiency values in Table 12.1 illustrate several points made earlier (in this chapter and Ch. 11). Although k_m and k_l vary with metabolisability (q_m), they vary much less than does k_f. To put this another way, for low quality foods ($q_m = 0.4$), k_f is only 50 per cent of k_m, whereas for high quality foods ($q_m = 0.7$) k_f is 74 per cent of k_m.

TABLE 12.1   Efficiency of utilisation of metabolisable energy by ruminants for maintenance, growth and milk production

Metabolisability (q_m)	0.4	0.5	0.6	0.7
Metabolisable energy concentration (MJ/kg DM)	7.4	9.2	11.0	12.9
Maintenance (k_m)	0.643	0.678	0.714	0.750
Growth and fattening (k_f)	0.318	0.396	0.474	0.552
Lactation (k_l)	0.560	0.595	0.630	0.665

Equations : $k_m = 0.35q_m + 0.503$
$k_f = 0.78q_m + 0.006$
$k_l = 0.35q_m + 0.420$

Example (a). Prediction of performance in fattening cattle. The system can now be illustrated by an example. Suppose a 300 kg steer is to be fed on a ration of 4.5 kg hay (containing 4 kg DM) and 2.2 kg maize (2 kg DM). This would supply the following quantity of metabolisable energy:

Food	Dry matter	Metabolisable energy	
	(kg/day)	MJ/kg DM	MJ/day
Hay	4	8	32
Maize	2	14	28
	6		60

The concentration of metabolisable energy in dry matter would be $60/6 = 10$ MJ/kg, and metabolisability would be $10/18.4 = 0.54$. For $q_m = 0.54$, Table 12.1 gives $k_m = 0.692$ and $k_f = 0.427$.

If the daily maintenance requirement (fasting metabolism) of the steer is 23 MJ net energy, its maintenance requirement in terms of metabolisable energy will be $23/0.692 = 33$ MJ. The metabolisable energy *not* required for maintenance will therefore be $60 - 33 = 27$ MJ, and of this amount $27 \times 0.427 = 11.5$ MJ will be retained by the animal as the energy of its liveweight gain. The final step in the calculation is to convert energy retention to liveweight gain. For this example, the energy content of gain is assumed to be 15 MJ/kg, and the liveweight gain of the steer will therefore be $11.5/15 = 0.77$ kg/day.

There are some refinements of the system which have not been used in this example, but are described in the publications listed above. Although the equations given in Table 12.1 are intended to apply to all classes of food it is known that greater accuracy of prediction of the efficiency values can be achieved by using slightly different equations for contrasting food types; for example, ground and pelleted roughages are better utilised for growth (i.e. higher k_f) than are unprocessed roughages having the same metabolisable energy concentration. A further complication arises from the tendency for the metabolisability of a food to be depressed at higher feeding levels (see p. 222), and the system used in the UK to calculate energy requirements for growth in ruminants includes a correction for this.

Example (b). Ration formulation for fattening cattle. The first example was called 'prediction of performance' because the starting point was a specific ration and the value finally predicted was animal performance, expressed as liveweight gain. However, energy systems are frequently operated in reverse, to formulate rations required for specific levels of animal performance. If the metabolisable energy concentration of the ration is known beforehand, as would be the case if it were a single food such as grass silage or a complete diet having fixed proportions of silage and concentrates, then the calculation of the quantity required is quite straightforward. But if the metabolisable energy concentration is not specified there is no unique solution to the problem and the calculations have to be repeated (i.e. 'iterated') until an acceptable solution is found. The alternative is to employ what is known as the variable net energy system, which was devised by F. V. MacHardy, a Canadian working in Edinburgh, and is described in Reference Book No. 433 of the Ministry of Agriculture, Fisheries and Food (see Further Reading).

For an example of ration formulation by this system, let us begin with the endpoint of Example (a). Suppose a steer weighing 300 kg is to be grown at a rate of 0.77 kg/day on a ration of 3 kg hay (i.e less than in Example (a)) plus an unknown quantity of maize. The foods are assumed to have the same energy concentrations as in the previous example, and the problem is to predict the quantity of maize required.

The first step is to convert the energy content of the gain ($0.77 \times 15 = 11.55$ MJ) to a requirement for metabolisable energy, but this cannot be done without an estimate of k_f, which has to be derived from the metabolisable energy concentration of the (as yet undefined) diet. The solution is to work in terms of net energy and to modify the interface between animal requirement and feed supply. The animal requirement for net energy is 23 MJ (fasting metabolism) + 11.55 MJ (energy content of gain) = 34.55 MJ. For the modified interface we need a combined measure of the efficiency of utilisation of metabolisable energy for the combined functions of maintenance and production (k_{mp}). For any given feed this will depend on the proportions of net energy required for maintenance and production respectively, and these proportions are defined in terms of the animal production level (APL):

$$APL = (NE_m + NE_p)/NE_m$$

In our example:

$$APL = (23 + 11.55)/23 = 1.50$$

The basic formula for k_{mp} is:

$$k_{mp} = (NE_m + NE_p)/(NE_m/k_m + NE_p/k_p)$$

When expressed in terms of APL this becomes:

$$k_{mp} = APL \left/ \left(\frac{1}{k_m} + \frac{APL - 1}{k_p} \right) \right.$$

For the hay (8 MJ ME/kg DM or $q_m = 0.43$), k_m and k_p are calculated from the equations of Table 12.1 to be 0.654 and 0.348, hence k_{mp} is 0.506. The net energy value of hay (NE_{mp}) in this rationing situation is therefore $8 \times 0.506 = 4.05$ MJ/kg DM. From similar calculations, maize (14 MJ ME/kg DM; $q_m = 0.76$; $k_m = 0.77$; $k_p = 0.60$) is calculated to have $k_{mp} = 0.70$ and $NE_{mp} = 9.85$ MJ/kg DM.

Thus 3 kg hay (2.7 kg DM) will supply 10.94 MJ NE/day, leaving $34.55 - 10.94 = 23.61$ MJ NE/day to be supplied by maize. The required quantity of maize DM is therefore $23.61/9.85 = 2.40$ kg, or 2.7 kg maize. The ration required is therefore 3 kg hay and 2.7 kg maize per day.

The variable net energy system can be simplified by tabulating values for NE_{mp}. At any given APL, a food of known M/D will have a single value for NE_{mp} (see Appendix Table 8.4).

Example (*c*). *Rationing of dairy cows*. As shown in Table 12.1 efficiency of utilisation of metabolisable energy for milk production (k_l) varies with M/D. This means that for the most accurate rationing of lactating cows, calculations of the kind illustrated for fattening cattle must be undertaken. In practice, some simplification is possible because the metabolisable energy concentration of dairy cow diets does not normally vary over such a wide range as does that of fattening cattle, and it is therefore reasonable to assume constant values for both k_l and k_m. For the system currently used in Britain k_m is assumed to be 0.72 and k_l, 0.62, and ration calculation becomes a matter of simple arithmetic.

As an example, let us consider a 500 kg cow producing 20 kg of milk (containing 40 g fat/kg) per day. The maintenance requirement of the cow is known to be 34 MJ NE, hence 34/0.72 = 47 MJ ME/day. The energy content of milk varies with its fat content (see Ch. 15), and for milk with 40 g fat/kg is 3.13 MJ/kg, hence the requirement in terms of metabolisable energy is 3.13/0.62 = 5.05 MJ/kg. The total requirement of the cow will therefore be 47 + (20 × 5.05) = 148 MJ ME. If the M/D value of its diet was 11 MJ/kg DM, the total quantity of dry matter required would be 148/11 = 13.5 kg/day.

Ration calculations for dairy cows are often complicated by changes in their energy reserves. If a cow is gaining weight and storing energy as fat, her requirements have to be considered in terms of three components, maintenance, milk synthesis and fattening. Conversely, if she is losing weight, then allowance must be made for the energy contributed from her fat reserves to milk synthesis or body maintenance. As for growing ruminants, the system for dairy cattle includes a correction factor for the reduction in metabolisable energy value of foods at higher levels of feeding.

Alternative energy systems for ruminants

Although scientists in different countries agree on the definition of the components of energy systems (e.g. metabolisable energy, fasting metabolism, etc.), they have failed to produce a universally accepted system. In Britain, as explained immediately above, the energy values of foods are expressed in terms of metabolisable energy, the energy requirements of animals are expressed in terms of net energy, and a complex 'interface' is used to bring the two together. This system, although described as a metabolisable energy system, is in effect a net energy system. In other countries, simpler net energy systems (and also 'true' metabolisable energy systems) have been developed for ruminants, and these will now be outlined.

Because of its early origins, in Kellner's system, the East German net energy system is described first. Foods are evaluated in terms of net energy for fattening, which may be calculated from digestible nutrients. For example, for cattle the net energy values of the digestible nutrients are as follows (kcal/g): protein 1.71, fat 7.52, crude fibre or nitrogen-free extractives 2.01. For sheep (and also for pigs and poultry) different factors are used. The net energy content of a food is expressed, not as joules or calories, but as energy feed units (EF), the unit for cattle being equal to 2.5 kcal or 10.5 kJ net energy for fattening. The digestibility coefficient of the energy of the food is tabulated with EF, and when the coefficient for a complete ration (containing one or more foods) is found to be less than 0.67 (i.e. $M/D < 10$ MJ/kg), EF is reduced. For example, when energy digestibility is 0.60, EF is multiplied by 0.93, and when the coefficient is 0.50, EF is multiplied by 0.82. The purpose of this correction is to make allowance for the fact discussed above, that when present in low concentration, metabolisable or digestible energy is utilised (i.e. converted to net energy) with low efficiency. Animal requirements in the East German system are also expressed in terms of EF.

In Western Europe, systems with many features in common have been adopted for the Netherlands, France, Switzerland and West Germany. The Dutch system will be described as an example. The metabolisable energy content of foods is calculated from digestible nutrients and is then converted to a net energy value. For growing animals, the basis for this conversion is that the Animal Production Level is assumed to be constant, at 1.5, and hence that k_{mp}, as explained earlier in this chapter, has a unique value for a food of known metabolisable energy concentration. Each food can therefore be given a single net energy value for maintenance and production (NE_{mp}), but this is converted to a unit value by dividing it by the presumed NE_{mp} of barley (6.9 MJ/kg, or about 8 MJ/kg DM). For lactating cows a corresponding net energy value for maintenance and lactation is calculated by assuming that k_l is 0.60 when $ME/GE = 0.57$ (i.e. when $M/D = 10.5$ MJ/kg) and varies by 0.4 per unit change in ME/GE. For example, if $M/D = 11.5$ MJ/kg and $ME/GE = 0.62$, $k_l = 0.60 + [0.4 \times (0.62 - 0.57)] = 0.62$ and the net energy value of the food for lactation (called NE_l) $= 0.62 \times 11.5 = 7.1$ MJ/kg DM. The calculation is further complicated by reducing the predicted value of NE_l by 2.5 per cent, to allow for the normally high feeding level of dairy cows, and by converting NE_l to a unit value (barley again being assumed to contain 6.9 MJ NE_l/kg). In addition to being used for cows, NE_l values are used in the rationing of younger dairy animals (i.e. heifers being reared as dairy herd replacements).

The simplifying assumptions used in the Dutch scheme to calculate the net energy values of foods are therefore that (a) for growing animals, APL $= 1.5$, and (b) for growing heifers, $k_{mp} = k_l$. A third assumption, which is not stated but is implicit in the system, is that $k_m = k_l$. All three are liable to introduce inaccuracies and are allowed for by adjusting the net energy requirements of animals. For example, with regard to assumption (b) above, it is recognised that in heifers growing slowly (and therefore using most of their energy intake for maintenance) k_l is an underestimate of k_{mp}, whereas in heifers growing rapidly, k_l is an overestimate of k_{mp}. The net energy requirements of slow growing heifers have therefore been reduced, and those of fast growing heifers increased, to cancel out the biases in the estimation of the net energy values of their foods.

The French system is the same in principle as the Dutch system but differs in detail. For example, the metabolisable energy value of foods is calculated by a different procedure, different corrections are used for level of feeding, and the unit value of barley for growing animals (9 MJ NE_{mp}/kg DM) differs from that used in the Dutch system (8 MJ).

Several countries in Scandinavia developed a net energy system expressed in unit values (Scandinavian feeds units, SFU) and related to Kellner's starch equivalent system. This system is still used in Norway and Denmark, and in the latter country the methods used to calculate feed energy values are legally defined. In Sweden, however, the system of choice is one in which both feed values and animal requirements are expressed as metabolisable energy. The feed values are based on direct experimentation, but the origins of the animal requirement data are obscure.

Much the same could be said of the Total Digestible Nutrients (TDN) system which was until recently the preferred system of the United States. During the last 15 years, however, the United States has changed to net energy systems for beef and dairy cattle. For beef cattle, feeds are given two net energy values, for maintenance (NE_m) and gain (NE_{gain}), these being calculated from the metabolisable energy content of each food. Both sets of values are considerably lower than those derived by the UK system, the difference being particularly large for foods low in M/D. Net energy values for lactation are calculated from TDN, digestible or metabolisable energy by an equation similar to that used in the Dutch system. (For example, foods containing 10 and 12 MJ ME/kg DM are calculated to contain 6.0 and 7.1 MJ NE_l in the American system, and 5.8 and 7.2 MJ NE_l in the Dutch system.) Net energy requirements for both maintenance and milk synthesis are expressed as NE_l, as they are in the Dutch and related European systems.

Energy systems for ruminants: retrospect and prospect

The first edition of this book was published at a time (1966) when the older energy systems for ruminants, such as Kellner's starch equivalent system, were being re-evaluated in the light of newly-acquired calorimetric data. In

East Germany, Kellner's system had been re-modelled, and in Britain a new system had recently been introduced by the Agricultural Research Council. The hope at the time was that new information on energy utilisation would bring more accurate systems which might be less complicated and therefore more acceptable for international use. In the event this hope has not been fully realised. Greater accuracy may perhaps have been achieved, but at the expense of greater complexity and an increase in the number of systems in use.

The energy systems for ruminants which have been most improved since 1966 have been those for the lactating cow. As the reader will now appreciate, energy systems can accommodate the dairy cow better than they can

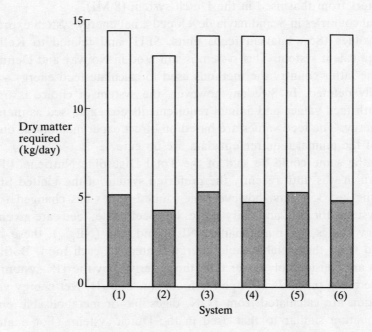

Fig. 12.1.   Comparison of energy systems for predicting the requirement of dairy cows fed on a standard diet containing equal parts of roughages and concentrates. Requirements for maintenance alone are shown by the shaded areas and for maintenance plus the production of 20 kg milk by the complete areas. (After R. Steg and Y. van der Honing, 1979, *Report of the Dutch Institute for Cattle Feeding Research, Hoorn, No. 49*).

Key to systems:
(1) Metabolisable energy (UK)
(2) Scandinavian feed units (Denmark)
(3) Total digestible nutrients (USA)
(4) Energy feed units (East Germany)
(5) Net energy for lactation (USA)
(6) Feed units for lactation (Netherlands)

growing and fattening cattle (or sheep). Efficiency of utilisation of metabolisable energy varies less in the cow than in the steer, and the energy content of the product, milk, is less variable and more predictable than is the case with liveweight gain. The various systems for dairy cows have been compared in Fig. 12.1, by taking a specimen diet and using the procedures of each system to predict first the energy value of the diet, and second, the quantities required for specified levels of production; the conclusion drawn is that systems for dairy cows do not differ greatly.

Figure 12.2 illustrates a different method for comparing energy systems,

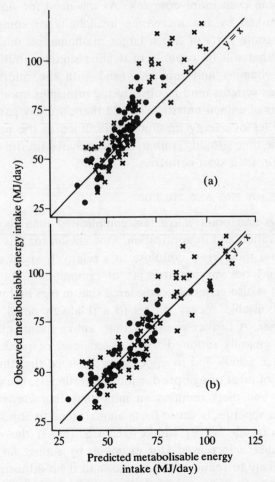

Fig. 12.2. Comparison of metabolisable energy intake observed in beef cattle with intake predicted from animal performance by two systems: (*a*) the metabolisable energy system of the UK Agricultural Research Council (*b*) the net energy system of the US National Research Council. (From J. P. Joyce *et al.*, 1975, *N.Z. J. Agric. Res.*, **18**, 295).

in this instance for beef cattle. A series of input-output data mainly for grass-fed beef cattle provided a basis for comparing actual output with that predicted from input by the use of the two systems. The conclusion drawn was that the US net energy system predicted the performance of the cattle more accurately than did the 1965 version of the UK Agricultural Research Council system.

It must be admitted, however, that comparisons of the kinds illustrated above are seldom sufficiently conclusive for the nutritionists of one country to discard their own system and adopt *in toto* that of another country.

In the future, as energy systems are modified to incorporate new findings they are likely to become even more complex. As the need for simplicity of calculation is diminished by the increasing availability of computers, energy systems may become parts of much larger mathematical models of nutrient requirements, that will be capable of dealing simultaneously with energy, amino acids, vitamins and minerals (and with the interactions between them). Complex systems tend to obscure the principles upon which they are based. Students of animal nutrition should therefore pay particular attention to the principles of energy metabolism outlined in the previous chapter, and at the same time should familiarise themselves with the energy systems currently used in their own countries.

ENERGY SYSTEMS FOR PIGS AND POULTRY

Energy systems for pigs and poultry are less complicated than those for ruminants, both in derivation and in application. One reason for this is that pigs and poultry, because they digest cellulose to a relatively small extent, are limited to a range of foods varying little in, for example, metabolisable energy concentration. It is also generally considered that in pigs and poultry the utilisation of metabolisable energy differs to a relatively small extent from one food to another. A further reason is that, unlike ruminants, pigs and poultry are less commonly rationed for a stated level of production. Poultry, in particular, are usually fed to appetite in the hope that they will achieve maximum rates of meat or egg production. Nevertheless, the energy concentration of foods and diets remains an important characteristic for poultry (and pigs) fed to appetite, because these animals tend to adjust their intakes to provide a constant energy intake (see Ch. 16). If the energy concentration of their diet is increased (for example, by adding fat) non-ruminant animals are likely to reduce their intake, and if no adjustment is made in the concentration of amino acids and other nutrients, they are liable to suffer deficiencies. In the practical rationing of pigs and poultry, one component of energy systems, namely the energy concentration of foods, is quite important, whereas the other component, the energy requirements of the animals, is considered to be much less important.

Pigs

A net energy system for pigs was developed by G. Fingerling, Kellner's successor at the Möckern Experiment Station and this, like Kellner's system for ruminants, has recently been revised by scientists in East Germany. In West Germany also a net energy system for pigs has been introduced recently. In most European countries, however, energy systems for pigs are based on metabolisable energy. The US National Research Council gives values for both digestible and metabolisable energy, but the UK Agricultural Research Council continues to state the energy concentration of foods for pigs in terms of digestible energy only.

Poultry

Net energy values of foods for poultry were determined 40–50 years ago in the United States by G. S. Fraps, who used the method of comparative slaughter to measure energy retention in young chickens. The figures obtained were called *productive energy* values, in order to emphasise that they were net energy values for growth, not maintenance. Fraps also devised methods for predicting the productive energy values of foods from their gross or digestible nutrients.

Fraps's productive energy values have never been widely employed and in most countries metabolisable energy has been used to express the energy value of poultry foods. As for pig foods it has been argued that the metabolisable energy of poultry foods is used with reasonably constant efficiency. The recent expansion of calorimetric studies of poultry has yielded results that do not support this argument; in particular, metabolisable energy provided by dietary fat has been shown to be more efficiently utilised than that supplied by carbohydrate. There have therefore been further proposals for a net energy system for poultry, but these have not been adopted. Metabolisable energy is very easily measured in poultry, since faeces and urine are voided together, and this is undoubtedly a strong factor favouring its retention in energy systems for poultry.

PREDICTING THE ENERGY VALUE OF FOODS

The precision of any energy system depends on the accuracy with which the energy content of individual foods or diets can be estimated. The metabolisable energy content of grass silage, for example, can vary from perhaps 8 to 12 MJ/kg DM according to the type of grass from which it is made and the silage-making practices employed. A farmer wishing to assess the energy value of his silage might use a reference book to identify its type and to provide an appropriate value for MJ ME/kg DM (see Appendix Table 1 for examples). The assessment can be made more precisely, however, if the silage is subjected to chemical analysis. The analytical data can either be used to 'type' the silage more accurately, or better, they can be fitted into equations that have been devised to predict energy value from chemical composition. For example, the following equation allows the metabolisable energy value of silages to be predicted from their content of modified acid detergent fibre (MADF, g/kg DM):

$$ME(MJ/kg\ DM) = 15.33 - 0.0152MADF$$

An alternative to chemical analysis is the assessment of digestibility by fermentation *in vitro* (see p. 204) and the prediction of metabolisable energy value from the digestible organic matter content of the food. For roughages given to ruminants a simple formula that is commonly used is:

$$ME(MJ/kg\ DM) = 0.016DOM$$

where DOM = g digestible organic matter/kg DM.

The most accurate estimate of the metabolisable energy value of a food is obtained by carrying out a metabolism trial with the type of animal for which that food is intended. Although this would not be feasible for the individual silage sample considered above, it is the systematic assessment of many food samples that provides the regression equations used to predict energy value from chemical analysis.

For cereal grains and other concentrated foods, prediction of energy value is easier than for silage because such foods are less variable in chemical composition; appropriate values can therefore be taken from tables without difficulty. Nevertheless, there are occasions when the energy values of concentrates need to be predicted from their chemical composition or other characteristics. For example, many countries are introducing legislation to require compound feed manufacturers to declare the energy value (typically, metabolisable energy value) of their products, and a simple method of checking such values is needed. A United Kingdom committee (see the article by G. Alderman listed in Further Reading) has recently recommended the following equations for predicting the energy content of compound feeds for poultry, pigs and ruminants.

Poultry:

$$Apparent\ ME\ (MJ/kg) = 0.034EE + 0.0165CP$$
$$+ 0.0172STA + 0.0158SUG$$

Pigs:

$$DE\ (MJ/kg) = 17.38 + 0.0105CP + 0.0114EE - 0.0317CF - 0.0402TA$$

Ruminants:

$$\text{ME (MJ/kg DM)} = 11.78 + 0.00654\text{CP} + 0.000665\text{EE}^2 - 0.00414\text{EE} \times \text{CF} - 0.0118\text{TA}$$

where EE = ether extract, CP = crude protein, STA = starch, SUG = sugar, CF = crude fibre, TA = total ash (all expressed as g/kg dry matter).

Note that the three equations differ in principle. The poultry equation assigns logically derived factors to each food constituent; for example, the factor for ether extract indicates that $0.0345/0.0393 = 0.88$ of the gross energy of ether extract in poultry feeds is metabolisable. The pig equation starts with a constant value and adjusts it upwards for the ether extract and protein content of feeds, or downwards for crude fibre and ash. The ruminant equation is similar in that it starts with a constant value and adjusts it for protein and ash, but the plus value of ether extract is reduced by an assumed interaction between ether extract and fibre (as ether extract depresses fibre digestion in the rumen—see p. 154).

FURTHER READING

Agricultural Research Council, 1980. *The Nutrient Requirements of Ruminant Livestock*. Commonwealth Agricultural Bureaux, Farnham Royal.

G. Alderman, 1985. Prediction of the energy value of compound feeds, in *Recent Advances in Animal Nutrition—1985*, W. Haresign and D. J. A. Cole (eds). Butterworths, London.

T. R. Morris and B. M. Freeman, 1974. *Energy Requirements of Poultry*. Poultry Science Symposium no. 9. British Poultry Science Ltd.

Ministry of Agriculture, Fisheries and Food, 1984. *Energy Allowances and Feeding Systems for Ruminants*, Reference Book 433. HMSO, London.

K. Nehring and G. F. W. Haenlein, 1973. Feeding evaluation and ration calculation based on net energy, *J. Anim. Sci.*, **36**, 949.

A. J. H. van Es, 1978. Feed evaluation for ruminants. 1. The systems in use from May 1977 onwards in the Netherlands, *Livest. Prod. Sci.*, **5**, 331.

Agricultural Research Council, 1981. *The Nutrient Requirements of Pigs*. Commonwealth Agricultural Bureaux, Farnham Royal.

National Academy of Sciences—National Research Council, 1978. *Nutrient Requirements of Dairy Cattle*, 5th rev. edn. NRC, Washington. (See also companion publications for other domestic animals.)

Historical references

H. P. Armsby, 1917. *The Nutrition of Farm Animals*. Macmillan, New York.

O. Kellner, 1926. *The Scientific Feeding of Farm Animals*, (translated by W. Goodwin) 2nd edn. Duckworth, London.

13

Evaluation of foods
(D) Protein

Proteins are made up of amino acids, the classification of which into dispensable and indispensable has already been described in Chapter 4, and in Chapter 9 under 'Protein Synthesis'. For food to be used with maximum efficiency, the animal must receive the indispensable amino acids in the correct quantities, and sufficient of the dispensable amino acids to meet the metabolic demands must also be available. Simple-stomached animals such as pigs and poultry obtain these acids from the breakdown of proteins during digestion and absorption, whereas in ruminant animals the position is more complex: considerable degradation and synthesis of protein occur in the rumen, so that the material which finally becomes available for digestion by the animal may differ considerably from that originally present in the food. Different approaches to the evaluation of protein sources are therefore necessary for ruminant and for non-ruminant animals.

At first, foods were evaluated as sources of protein by rather crude and simple methods, which took no account of the species of animal for which the foods were intended. These methods are still in use and will be dealt with first, and then the more refined methods of assessing protein quality will be discussed for simple-stomached and ruminant animals separately.

Crude protein (CP)

Most of the nitrogen required by the animal is used for protein synthesis. Most of the food nitrogen also is present as protein, and it is convenient and almost universal for the nitrogen requirements of animals, and the nitrogen status of foods, to be stated in terms of protein. Chemically, the protein of a food is calculated from its nitrogen content, determined by a modification of the classical Kjeldahl technique; this gives a figure for most forms of nitrogen although nitrites, nitrates and certain cyclic nitrogen compounds require special techniques for their recovery. Two assumptions are made in calculating the protein content from the nitrogen: firstly, that all the nitrogen of the food is present as protein, and secondly, that all food protein contains 160 gN/kg. The nitrogen content of the food is then expressed in terms of crude protein (CP) calculated as follows:

$$CP(g/kg) = gN/kg \times 1000/160$$

or more commonly

$$CP(g/kg) = gN/kg \times 6.25.$$

Both of these assumptions are unsound. Different food proteins have different nitrogen contents, and therefore different factors should be used in the conversion of nitrogen to protein for individual foods. Table 13.1 shows the nitrogen content of a number of common proteins together with the appropriate nitrogen conversion factors. Although fundamentally unsound, the use of an average conversion factor of 6.25 for food proteins is justified in practice since protein requirements of farm animals, expressed in terms of N × 6.25, are requirements for nitrogen and not protein *per se*. The publication of tables of protein contents based on true conversion factors could lead to considerable confusion and inefficiency in feeding. The assumption that the whole of the food nitrogen is present as protein is also false, since many simple nitrogenous compounds such as amides, amino acids, glycosides, alkaloids, ammonium salts and compound lipids may be present (see Ch. 4). Quantitatively however only the amides and amino acids are important, and these are present in large amounts in only a few foods, such as young grass, silage and immature root crops.

About 95 per cent of the nitrogen of most mature seeds is present as true protein, whereas leaves, stems and roots, and storage organs such as pota-

TABLE 13.1 Factors for converting nitrogen to crude protein (After D. B. Jones, 1931. *U.S.D.A. Circ.* No. 183)

Food protein	Nitrogen (g/kg)	Conversion factor
Cottonseed	188.7	5.30
Soya bean	175.1	5.71
Barley	171.5	5.83
Maize	160.0	6.25
Oats	171.5	5.83
Wheat	171.5	5.83
Egg	160.0	6.25
Meat	160.0	6.25
Milk	156.8	6.38

toes and carrots, have 80–90, 60, and 30–40 per cent in this form, respectively. In the diets of pigs and poultry, cereals and oilseed meals predominate, and in these there is little non-protein nitrogen. Hence in practice there is little to be gained from attempting to distinguish between the two types of nitrogen, particularly as a considerable proportion of the non-protein fraction can be utilised for amino acid synthesis by the animal.

Tradition apart, there seems little justification for the continued use of the term *crude protein* in nutrition. The expression of animal requirements and of the status of foods in terms of nitrogen would be more logical and would prevent confusion.

True protein (TP)

When true protein needs to be determined, it can be separated from non-protein nitrogenous compounds by precipitation with cupric hydroxide, or in some plant material by heat coagulation. The protein is then filtered and the residue subjected to a Kjeldahl determination.

Digestible crude protein (DCP)

The crude protein figure provides a measure of the nitrogen present in the food, but gives little indication of its usefulness to the animal. Before the food becomes available to the animal it must undergo digestion, during which it is broken down to simpler substances which are absorbed into the body. The digestible protein in a food may be determined by digestibility trials, as described in Chapter 10. Such trials give figures for 'apparent' and not 'true' digestibility. It is assumed that the difference between the quantities of nitrogen in food and faeces represents the quantity absorbed in utilisable form by the body, and that all the nitrogen which appears in the faeces is of dietary origin. In the light of the fate of ruminal ammonia (Ch. 8), and the presence in faeces of nitrogen of metabolic origin (Ch. 10) these assumptions are obviously untenable. The figures for apparently digestible protein are generally lower than the true values, but since the loss of metabolic faecal nitrogen is inevitable, they are considered to be a more realistic measure of nutritive value. Usually no attempt is made to determine the true digestibility, and the coefficients in everyday use are apparent values.

Protein digestibility is sometimes estimated by incubating the food with acid pepsin for 48 hours at 37 °C. The protein content of the insoluble residue is then estimated by the Kjeldahl method, and the digestible protein calculated by subtracting this value from the protein content of the food. The method is of limited value, since it involves the action of only one enzyme instead of the many of the digestive tract, and the results vary with fineness of grinding, concentration of enzyme and conditions of drying of the food. At best it may be useful for comparative purposes.

MEASURES OF PROTEIN QUALITY FOR MONOGASTRIC ANIMALS

Digestible protein figures are not entirely satisfactory assessments of a protein, because the efficiency with which the absorbed protein is used differs considerably from one source to another. In order to take this into account, methods of evaluating proteins have been devised which are based on the response of experimental animals to the protein under consideration.

Protein efficiency ratio (PER)

The protein efficiency ratio normally uses growth of the rat as a measure of the nutritive value of dietary proteins. It is defined as the weight gain per unit weight of protein eaten, and may be calculated by using the following formula:

$$\text{PER} = \frac{\text{gain in body weight (g)}}{\text{protein consumed (g)}}$$

The value obtained is affected by the age and sex of the rat, length of the assay period and the level of protein. Usually a diet containing 100 g protein/kg and eaten to appetite by male rats is used, and the assay period is four weeks. PER values are frequently stated relative to a standard casein with an assigned PER. A modification of the PER method, where the weight gain of the experimental group is compared with a group on a protein-free diet, gives the net protein retention (NPR) calculated as follows:

$$\text{NPR} = \frac{\text{weight gain of TPG} - \text{weight loss of NPG}}{\text{weight of protein consumed}}$$

Where TPG = group fed on test protein; and NPG = non-protein group. The NPR method is claimed to give more accurate results than the PER method.

Gross protein value (GPV)

The liveweight gains of chicks receiving a basal diet containing 80 g crude protein/kg are compared with those of chicks receiving the basal diet plus 30 g/kg of a test protein, and of others receiving the basal diet plus 30 g/kg of casein. The extra liveweight gain per unit of supplementary test protein, stated as a proportion of the extra liveweight gain per unit of supplementary casein, is the gross protein value of the test protein, i.e.

$$\text{GPV} = \frac{A}{A^\circ}$$

where A is g increased weight gain/g test protein, and A° is g increased weight gain/g casein.

Determinations of the PER, NPR or GPV require a regular supply of standard young animals, which have to be looked after individually during the experimental period. Thus these methods of protein evaluation are expensive and require considerable technical resources. In addition they depend upon measurement of liveweight gains, which may not be related to protein stored. A more accurate evaluation of protein may be obtained by using the results of nitrogen balance experiments. In such experiments the nitrogen consumed in the food is measured as well as that voided in the faeces, urine and any other nitrogen-containing product such as milk, wool or eggs. Where the nitrogen intake is equal to the output, the animal is in nitrogen equilibrium. Where the intake exceeds the outgo, it is in positive nitrogen balance. Where outgo exceeds intake, the animal is in negative balance. Table 13.2 illustrates the calculation of a nitrogen balance for a 42 kg pig given a diet containing soya bean meal as the protein source. The pig was in positive nitrogen balance, storing 10.77 g nitrogen/day. Comparisons of balance trials with comparative slaughter data have shown that balance determinations usually overestimate nitrogen retention, particularly with large animals. Further, retentions are often recorded with little or no change in body weight. These discrepancies are generally considered to be due to errors inherent in the conduct of balance trials.

TABLE 13.2   Nitrogen balance for a Hampshire pig on a soya bean meal diet (From D. G. Armstrong and H. H. Mitchell, 1955. *J. Anim. Sci.*, **14**, 53)

	Daily intake (g)	Daily outgo (g)
Food nitrogen	19.82	—
Faecal nitrogen	—	2.02
Urinary nitrogen	—	7.03
Nitrogen retained by the body	—	10.77
	19.82	19.82

Balance = + 10.77 g/day

Protein replacement value (PRV)

This value measures the extent to which a test protein will give the same balance as an equal amount of a standard protein. Two nitrogen balance determinations are carried out, one for a standard such as egg or milk protein, which is of high quality, and one for the protein under investigation. The PRV is calculated as follows:

$$\text{PRV} = \frac{A - B}{\text{N intake}}$$

where A = N balance for standard protein in mg/basal kJ, B = N balance for protein under investigation in mg/basal kJ. The method can also be used to compare two proteins under similar conditions, when no standard value for replacement is required.

The PRV measures the efficiency of utilisation of the protein given to the animal. Other methods measure the utilisation of digested and absorbed protein.

Biological value (BV)

This is a direct measure of the proportion of the food protein which can be utilised by the animal for synthesising body tissues and compounds, and may be defined as the proportion of the nitrogen absorbed which is retained by the animal. A balance trial is conducted in which nitrogen intake and urinary and faecal excretions of nitrogen are measured, and the results are used to calculate the biological value as follows:

$$\text{BV} = \frac{\text{N intake} - (\text{faecal N} + \text{urinary N})}{\text{N intake} - \text{faecal N}}$$

Part of the nitrogen of the faeces, the *metabolic faecal nitrogen*, is not derived directly from the food. Urinary nitrogen also contains a proportion of nitrogen, known as the *endogenous urinary nitrogen*, which is not directly derived from food. The significance of the endogenous urinary nitrogen is discussed in more detail later (Ch. 14). Briefly, it is nitrogen derived from irreversible reactions involved in the breakdown and replacement of various protein structures and secretions. The existence in both faeces and urine of nitrogen fractions whose excretion is independent of food nitrogen is most convincingly demonstrated by the fact that some nitrogen is excreted when the animal is given a nitrogen-free diet. Such diets, adequate in all other respects, provide a means of measuring the magnitude of endogenous urinary ntirogen and metabolic faecal nitrogen (see Ch. 14). Since both fractions represent nitrogen which has been utilised by the animal, rather than nitrogen which cannot be utilised, it is obvious that their exclusion from the faecal and urinary fractions in the formula given above will yield a more precise estimate of biological value. The revised formula is

$$BV = \frac{N \text{ intake} = (\text{faecal N} - \text{MFN}) - (\text{urinary N} - \text{EUN})}{N \text{ intake} - (\text{faecal N} - \text{MFN})}$$

where MFN = metabolic faecal nitrogen and EUN = endogenous urinary nitrogen.

'Biological values' are normally calculated by this equation, whereas those derived from the former equation are usually termed 'apparent biological values'. In determining biological value, as much as possible of the dietary protein should be provided by the protein under test. Protein intake must be sufficient to allow adequate nitrogen retention, but must not be in excess of that required for maximum retention; if the latter level were exceeded, the general amino acid catabolism resulting would depress the estimate of biological value. For the same reason sufficient non-nitrogenous nutrients must be given to prevent catabolism of protein to provide energy. The diet must also be adequate in other respects. Table 13.3 shows an example of the calculation of a biological value from nitrogen balance data, and in Table 13.4 biological values are given for the proteins of some typical foods.

Such biological values are for the combined functions of maintenance, meaning the replacement of existing proteins, and growth, or the formation of new tissues. A biological value for maintenance alone may be calculated from nitrogen balance data. A linear relationship exists between nitrogen intake and nitrogen balance below equilibrium (Fig. 14.2) which is represented by the following equation:

$$y = bx - a,$$

where y = nitrogen balance,
 x = nitrogen absorbed, $\Big\}$ mg nitrogen/basal kJ;
 a = nitrogen loss at zero intake

b, the nitrogen balance index, represents that fraction of the absorbed nitrogen which is retained in the body, and is equal to the biological value for maintenance.

The usefulness of a protein to an animal will depend upon its digestibility as well as its biological value. The product of these two values is the proportion of the nitrogen intake which is retained, and is termed the *net protein utilisation* (NPU). The product of the NPU and the percentage crude protein is the *net protein value* (NPV) of the food, and is a measure of the protein actually available for metabolism by the animal.

TABLE 13.3   Calculation of biological value of a protein for maintenance and
growth of the rat
(After H. H. Mitchell, 1924. *J. biol. Chem.*, **58**, 873)

Food consumed daily (g)	6.00
Nitrogen in food (per cent.)	1.043
Daily nitrogen intake (mg)	62.6
Total nitrogen excreted daily in urine (mg)	32.8
Endogenous nitrogen excreted daily in urine (mg)	22.0
Total nitrogen excreted daily in faeces (mg)	20.9
Metabolic faecal nitrogen excreted daily (mg)	10.7

$$BV = \frac{62.6 - (20.9 - 10.7) - (32.8 - 22.0)}{62.6 - (20.9 - 10.7)}$$
$$= 0.79$$

TABLE 13.4   Biological values of the protein in various foods
for maintenance and growth for the growing pig
(From D. G. Armstrong and H. H. Mitchell, 1955. *J. Anim. Sci.*, **14**, 53)

Food	BV
Milk	0.95–0.97
Fish meal	0.74–0.89
Soya bean meal	0.63–0.76
Cottonseed meal	0.63
Linseed meal	0.61
Maize	0.49–0.61
Barley	0.57–0.71
Peas	0.62–0.65

The amino acid mixtures absorbed by the animal are required for the synthesis of body proteins. The efficiency with which this synthesis can be effected depends partly on how closely the amino acid proportions of the absorbed mixture resemble those of body proteins, and partly on the extent to which these proportions can be modified. The biological value of a food protein, therefore, depends upon the number and kind of amino acids present in the molecule: the nearer the food protein approaches the body proteins in amino acid make-up, the higher will be the biological value. Animals have little ability to store amino acids in the free state, and if an amino acid is not immediately required for protein synthesis it is readily broken down and either transformed into a dispensable amino acid which is needed by the animal, or used as an energy source. Since indispensable amino acids cannot be effectively synthesised in the animal body, an imbalance of these in the diet leads to a wastage. Food proteins with either a deficiency or an excess of any particular amino acid will tend to have a low biological value.

If we consider two food proteins, one deficient in lysine and rich in methionine and the other deficient in methionine but containing an excess of lysine, then if these proteins are given separately to young pigs, they will both have low biological values because of the imbalance of these two indispensable amino acids. If however the two proteins are given together, then the mixture of indispensable amino acids will be better balanced and the mixture will have a higher biological value than either protein given alone. Such proteins supplement each other. In practice, and for a similar reason, it often happens that a diet containing a large variety of proteins has a higher biological value than a diet containing only a few. This also explains why biological values for individual foods cannot be applied when mixtures of foods are used, since clearly the resultant biological value of a mixture is not simply a mean of the individual components. For the same reason it is impossible to predict the value of a protein, as a supplement to a given diet, from its biological value.

Animal proteins generally have higher biological values than plant proteins, although there are exceptions such as gelatine, which is deficient in several indispensable amino acids.

The amino acid composition of a given food protein will be relatively constant (see Appendix Table 4), but that of the protein to be synthesised will vary considerably with the type of animal and the various functions it has to perform. For the normal growth of rats, pigs and chicks for example, lysine, tryptophan, histidine, methionine, phenylalanine, leucine, isoleucine, threonine, valine and arginine are dietary essentials. Man does not require histidine, and chicks need glycine and proline, as well as those required by the rat, to ensure optimum growth. On the other hand arginine is not a dietary requirement for maintenance of the rat or pig.

The situation is further complicated by the fact that some amino acids can be replaced in part by others; for example methionine can be partly replaced by cystine, and similarly tyrosine can partly replace phenylalanine. In such cases the two amino acids are frequently considered together in assessing the animal's requirements. It is obvious that no single figure for biological value will suffice as a measure of the nutritive value of a food protein for different animals and different functions. The difference in amino acid requirements for young pigs compared with those for laying hens, for example, is shown in the Appendix Tables 13 and 14. The consequent need for multiple figures limits the use of the biological value concept in practice.

The biological measures described, reflect the content of the limiting amino acid in the protein. Changes in the levels of other amino acids will not affect the values until one or other of them becomes limiting. Thus in milk protein with an excess of lysine, change in lysine content will not affect the biological measures until it is reduced to such a level that it itself becomes the limiting acid. The biological measures are thus of limited value in assessing the effects of various processes, such as heat treatment, on nutritive value.

Since biological value is dependent primarily upon indispensable amino acid make-up, it would seem logical to assess the nutritive value of a protein by estimating its indispensable amino acid constitution and then comparing this with the known amino acid requirements of a particular class of animal. Application of modern chromatographic techniques coupled with automated procedures allows relatively quick and convenient resolution of mixtures of amino acids. However, the acid hydrolysis used to produce such mixtures from proteins destroys practically all the tryptophan and a considerable proportion of the cystine and methionine present. Tryptophan has to be released by a separate alkaline hydrolysis while cystine and methionine have to be oxidised to cysteic acid and methionine sulphone to ensure their quantitative recovery. Losses of amino acids and production of artefacts, which are greater with foods of high carbohydrate content, are reduced if the acid hydrolysis is carried out *in vaccuo*. The accurate resolution of the amino acid constitution of proteins does, therefore, still present a considerable technical problem. Evaluations of proteins made by dealing with each amino acid individually would be laborious and are inconvenient, and several attempts have been made to state the results of amino acid analyses in a more useful and convenient form.

Chemical score

In this concept it is considered that the quality of a protein is decided by that constituent indispensable amino acid which is in greatest deficit when compared with a standard. The standard generally used is egg protein, but many workers now use a defined amino acid mixture—the FAO Recommended Reference Amino Acid Pattern. The content of each of the indispensable amino acids of a protein is expressed as a proportion of that in the standard, the lowest proportion being taken as the score. In wheat protein, for example, the indispensable amino acid in greatest deficit is lysine. The contents of lysine in egg and wheat proteins are 72 and 27 g/kg respectively, and the chemical score for wheat protein is therefore $27/72 = 0.37$. Such values correlate well with the biological values for rats and human beings but not for poultry. They are useful for grouping proteins into categories, but suffer a serious disadvantage in that no account is taken of the deficiencies of acids other than the acid in greatest deficit.

The essential amino acid index (EAAI)

Here the amounts of all the ten indispensable amino acids present are considered. It may be defined as the geometric mean of the egg ratios of these acids, and is calculated as

$$\text{EAAI} = \sqrt[n]{\frac{a}{a_e} \times \frac{b}{b_e} \times \frac{c}{c_e} \times \cdots \times \frac{j}{j_e}}$$

where a, b, c, ... j = concentrations (g/kg) of the indispensable amino acids in the food protein, a_e, b_e, c_e ... j_e = concentrations of the same amino acids in egg protein, and n = the number of amino acids entering into the calculation.

The index has the advantage of predicting the effects of supplementation in combinations of proteins. It has the disadvantage that proteins of very different amino acid composition may have the same or a very similar index.

Both the chemical score and the essential amino acid index are based upon gross amino acid composition. A more logical approach would be to use figures for the amino acids available to the animal. Such figures may be obtained in several ways. Determinations of digestibility *in vivo* involve amino acid analyses of food and faeces. The figures so obtained are suspect because the faeces contain varying amounts of amino acids not present in the food, mainly as the result of microbial synthesis in the hind gut. This drawback may be overcome by determining amino acids in the digesta at the terminal ileum instead of the faeces. The resulting ileal digestibility provides the best currently available estimate of the digestibility of an amino acid. Digestibility trials *in vivo* are laborious and time-consuming and require considerable technical resources and skill. Digestibility determinations *in vitro* involve the action of one or at most a few enzymes and are thus not strictly comparable with the action *in vivo*, which involves a series of enzymes—although there is a relationship between the two. Even when an acceptable measure of digestibility is available, it should be borne in mind that it is not synonymous with availability.

Biological assay of available amino acids

The available amino acid content of a food protein may be assayed by measuring the liveweight gain, food conversion efficiency or nitrogen retention of animals given the intact protein as a supplement to a diet deficient only in the acid under investigation. The chick is the usual experimental animal, and the response to the test material is compared with responses obtained with supplements of pure amino acids. The method has been used successfully for lysine, methionine and cystine but, in addition to the usual disadvantages associated with biological methods—time, technical expertise and supply of suitable animals—there is the major problem of constructing diets deficient in specific amino acids but adequate in other respects.

Microbiological methods

Certain micro-organisms have amino acid requirements similar to those of higher animals and have been used for the evaluation of food proteins. The methods are based on measuring the growth of the micro-organisms in culture media which include the protein under test. Best results have been obtained with *Streptococcus zymogenes* and *Tetrahymena pyriformis*. The former is used after an acid or enzymatic predigestion of the food protein and estimates of the availability of lysine and methionine have been shown to agree well with chick assays and measures of NPU. *T. pyriformis* has intrinsic proteolytic activity and is used without predigestion for soluble proteins. An improved method, using predigestion with the enzyme pronase and measuring response in terms of the tetrahymanol content of the culture medium, has been shown to give results for available lysine, methionine and tryptophan, which correlate well with those from biological assays. Tetrahymanol the characteristic pentacyclic terpene synthesised by *T. pyriformis* is determined by gas-liquid chromatography.

Chemical methods

It would be ideal if simple chemical procedures could be used to determine the availability of amino acids, provided the results correlated well with those of accepted biological methods. The most widely used method is that for 'FDNB-reactive lysine' which was originally proposed by Dr K. J. Carpenter, and has since undergone modifications both by Carpenter and other workers. The protein is allowed to react under alkaline conditions with fluoro-2, 4-dinitrobenzene (FDNB). Subsequent acid hydrolysis yields a series of yellow dinitrophenyl (DNP) compounds, including ϵDNP-lysine. The αDNP amino acids are removed by ether extraction, leaving an aqueous layer coloured primarily by ϵDNP-lysine, the concentration of which can be measured. There may, however, be interfering colours present in which case the ϵDNP-lysine is treated with methoxycarbonyl chloride and removed by ether extraction, leaving a blank whose colour may be read and allowed for in calculating the available lysine. In practice the method has been found to correlate well with biological procedures in evaluating proteins as supplements to diets where lysine is limiting, such as those containing high proportions of cereals. The correlation with biological procedures has also been good with diets containing protein of animal origin only although recent data indicate a lack of agreement with the value of heat treated proteins for chicks. This may be due to destruction of sulphur-containing amino acids not reflected in the available lysine value. With vegetable-protein and high-carbohydrate diets the method is not so satisfactory, the results obtained being too low because of the destruction of the coloured lysine derivative during acid hydrolysis. Various modifications of the hydrolysis conditions have been proposed to counter this but none has been completely satisfactory. An alternative method of estimating available lysine has been proposed in which it is measured as the difference between the total lysine obtained after acid hydrolysis and the bound lysine in an acid hydrolysate after pre-treatment with FDNB. Since the method involves column chromatography for the estimation of lysine it does not lend itself to routine application. Much of the attraction of the Carpenter technique is thus lost.

The gross protein value is probably the most commonly used biological method for evaluating proteins. Gross protein values measure the ability of proteins to supplement diets consisting largely of cereals, and they correlate well with FDNB-reactive lysine figures, as shown in Fig. 13.1.

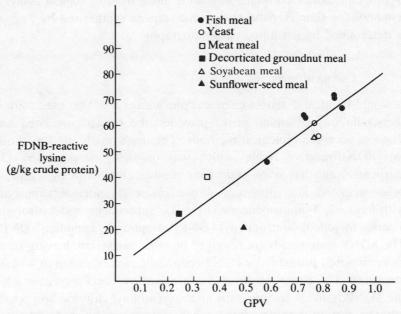

Fig 13.1. Relationship between gross protein value and FDNB—reactive lysine. (After K. J. Carpenter and A. A. Woodham, 1974, *Br. J. Nutr.*, **32**, 647.)

Dye-binding methods have been widely used for estimating protein in foods such as cereals and milk. The methods are rapid and reproducible and attempts have been made to use them for measuring total basic amino acids and reactive lysine. The latter requires blocking of the ε-amino group to prevent its reaction with the dye. The dye Orange G has been used, along with 2,4,6-trinitrobenzene sulphonic acid and propionic anhydride as blocking agents, and has proved effective for estimating the lysine content of cereals but less so for fish and meat meals.

The increased use of synthetic lysine in foods presents a further problem. The amino acid has both amino groups available for reaction with FDNB. The resulting compound is soluble in ether and is therefore not estimated by the Carpenter technique.

Measures of protein quality based on amino acid patterns or on the limiting amino acid do not always show the expected agreement with the results of evaluations with animals. Among the factors which may be responsible for this are amino acid imbalance, amino acid toxicity and amino acid antagonism. Even small increases in the concentration of certain amino acids can sometimes increase the amounts of others needed to maintain growth rate. Certain amino acids may be toxic. The acids most likely to be so are methionine, cystine, tyrosine, tryptophan and histidine. However, toxic levels of acids are higher than those likely to occur in normal diets. An excess of either leucine or lysine may produce deleterious effects owing to an antagonistic reaction. Thus, addition of 20 g/kg of L-leucine to an isoleucine-deficient diet has been shown to increase the isoleucine requirement of rats. Similarly, the arginine requirement of rats may be increased by lysine. Allowances for these effects must be made in the interpretation of such concepts as *chemical score*.

There is considerable evidence that growing animals, such as young rats and chicks, do not fulfil their growth potential if the dietary nitrogen is entirely in the form of indispensable amino acids. Additional nitrogen is required and is best supplied as a mixture of dispensable amino acids although glutamate, alanine and ammonium citrate are effective individually. This is a further factor for consideration, when interpreting protein evaluation tests based on one or more indispensable amino acids.

Measures of food protein used in the practical feeding of pigs and poultry

The difficulties in assessing the value of proteins in the diet will by now be apparent from the variety of methods that have been proposed, all of which have considerable limitations. A crude protein figure is useful, because the digestibility of the proteins in foods commonly given to pigs and poultry is fairly constant. More recently DCP has been used increasingly.

The quality of the proteins in a particular food is indicated by stating the contents of all the indispensable amino acids, or only of those most likely to be deficient. In practice pig and poultry diets are based largely on cereals, and assessment of the protein value of foods for such animals is then a question of measuring their ability to supplement the amino acid deficiencies of the cereals. The main deficiency in such cases is that of lysine and methionine, so that the most useful measures of protein quality are those which reflect the available lysine or methionine content of the food. The determination of available lysine is now accepted as a routine procedure for animal protein foods in many laboratories.

The most recent method of evaluating dietary protein for growing pigs is based on the concept of 'ideal protein'. This is a modification of the chemical score, with the amino acid pattern of lean muscle protein serving as the reference pattern (Table 13.5).

TABLE 13.5   Recommended balance of amino acids (g/kg) in ideal protein (From The nutrient requirements of pigs, *ARC Tech. Rev.*, 1981)

Lysine	70
Methionine + cystine	35*
Threonine	42
Tryptophan	10
Isoleucine	38
Leucine	70
Histidine	23
Phenylalanine + tyrosine	67†
Valine	49
Dispensable amino acids	596

* At least half should be methionine.
† At least half should be phenylalanine.

If the main limiting amino acid in the dietary protein was lysine at 50 g/kg CP then the score would be $50/70 = 0.70$ and the ideal protein content, 700 g/kg CP. A diet with 170 g/kg of this protein would then supply $170 \times 0.7 = 119$ g ideal protein/kg.

Ideal protein is conceived as providing the indispensable amino acids in the proportions required by the pig and as having the correct balance between the indispensable and dispensable amino acids. Thus, once the balance of amino acids has been described in this way, there is no need to tabulate requirements for individual amino acids. The statement of the requirement for ideal protein automatically fixes the requirement for each acid.

The concept of ideal protein makes no allowance for the oversupply of individual amino acids, which could, in certain circumstances, be detrimental. It is usual, therefore, to limit the concentration of any one acid to less than 1.2 of that in the reference protein. It is assumed in calculating the requirement for ideal protein, that the proportions of amino acids made available to the metabolic body pool are the same as those in the dietary protein. This will not be so if the acids are digested, absorbed, or utilised, with differing efficiencies, Requirements are usually stated in terms of apparently digested ideal protein (ADIP) and a digestibility of 0.75 is assumed in transforming the ideal protein supply to an ADIP supply. As more information becomes available on ileal digestibility values these will probably be used instead of the single assumed figure. Currently, protein allowances and the protein values of foods for sows are stated in terms of DCP and individual amino acids. There seems to be no reason why the ideal protein system should not be applied to lactating sows, with the amino acid pattern of sow milk protein being used as the reference. For pregnant sows, the lack of data on amino acid accretion during pregnancy would seem to preclude the adoption of the concept in the near future. For poultry, evaluation of protein sources is based upon their contents of the three major limiting amino acids, lysine, methionine and trytophan. It is generally assumed that diets adequate in these acids will automatically provide sufficient amounts of the others.

MEASURES OF PROTEIN QUALITY FOR RUMINANTS

Proteins in foods for ruminant animals are traditionally evaluated in terms of crude protein or digestible crude protein. Realisation that the crude protein fraction contained variable amounts of non-protein nitrogen led to the use of true protein instead of crude protein, but this was unsatisfactory since no allowance was made for the nutritive value of the non-protein nitrogen fraction. The concept of *protein equivalent* (PE), introduced in 1925 and now no longer used in this context, was an attempt to overcome this difficulty by allowing the non-protein nitrogen fraction half the nutritive value of the true protein.

The term protein equivalent is currently used in connection with foods containing urea. Such foods must by law be sold with a statement of their content of protein equivalent of urea. This means the amount of urea nitrogen multiplied by 6.25.

In the situation where large numbers of foods have to evaluated on a routine basis, determination of DCP by means of digestibility trials (Ch. 10) is impracticable. In such cases the first stage of evaluation is the determination of the crude protein content of the food. For concentrate foods, which have a relatively constant composition, digestibility coefficients available from tables are then used to calculate figures for digestible crude protein. With roughages, a different approach is usually taken owing to their greater variability and the relatively greater importance of metabolic faecal nitrogen in materials of low protein content. In this case regression equations for DCP on CP are used to calculate the former. A typical equation,

$$DCP(g/kg\ DM) = CP(g/kg\ DM) \times 0.9115 - 36.7$$

is widely used for grasses, hays and silages. The use of such equations may lead to the allocation of negative DCP values to certain low protein (> 40g CP/kg DM) roughages such as cereal straws. For many years there has been considerable dissatisfaction with the use of DCP for evaluating food proteins. This has its roots in the extensive degradative and synthetic activities of the rumen micro-organisms described in Chapter 8. Rumen micro-organisms are responsible for providing the major part of the energy requirements of the host animal by transforming dietary carbohydrates to acetate, propionate and butyrate. In order to do so, and to exploit fully the energy potential of the food, they must grow and multiply and this involves large-scale synthesis of microbial protein. The nitrogen for this is obtained, in the form of amino acids and ammonia, by breakdown of the nitrogen fraction of the food. Satisfying the demands of the rumen micro-organisms for readily available nitrogen is a major function of the diet, and to this end a certain proportion of the nitrogen fraction must be readily degradable.

Microbial protein eventually passes to the abomasum and small intestine where it is broken down to amino acids. These are subsequently absorbed and utilised by the host animal for satisfying its nitrogen requirements at tissue level. Much of the dietary protein which escapes breakdown in the rumen (undegradable) suffers the same fate.

The rumen micro-organisms are capable of synthesising all the indispensable as well as the dispensable amino acids; hence the mixture of amino acids eventually entering the animal's blood may bear no relationship to the amino acid make-up of the original diet. Although the spectrum of amino acids is thought to vary according to the nature of the dietary nitrogen, microbial protein is generally considered to have a biological value of about 0.8, whether it is of bacterial or protozoal origin. The digestibility of bacterial protein is lower, 0.74 compared with 0.91 for protozoal protein, and thus has an NPU of about 0.59. The value of a protein can therefore be affected by conditions in the rumen; low pH values, for example, tend to reduce protozoal activity while increasing that of certain bacteria. In many cases conversion of dietary protein of low biological value to microbial protein may be a distinct advantage since it renders the host animal almost independent of a dietary source of indispensable amino acids. For the better quality sources, this will not be so owing to the lowered digestibility of the bacterial protein. Despite the relatively high quality of microbial protein its production, even from dietary protein of low biological value, may be a disadvantage owing to wastage. As described in Chapter 8, not all the dietary nitrogen reaches the small intestine. If deamination is rapid, ammonia is produced in the rumen faster than the micro-organisms can 'trap' it for the synthesis of amino acids, and some is absorbed and then excreted as urea. The efficiency of nitrogen capture depends upon the degradability of the nitrogenous constituents of the diet and also upon the synchronous provision of readily available energy in the form of the carbohydrate of the diet. The speed and extent to which protein is broken down depends upon such factors as the surface area available for microbial attack, the physical consistency and chemical nature of the protein, and the protective action of other constituents. Susceptibility to breakdown is thus a characteristic of the protein itself and should be capable of being quantified. Claims have been made that solubility of the protein is correlated with ease of breakdown but these do not survive critical examination. Thus casein, which is readily degraded in the rumen, is not readily soluble while albumin, which is resistant to breakdown, is readily soluble. It has been suggested that a major factor in deciding susceptibility to breakdown is the amino acid sequence within the protein. If this is so then the nature of the microbially-produced rumen peptidases is of considerable importance and it would seem doubtful whether any simple laboratory test for susceptibility to breakdown is possible.

It would appear that the value of a protein to a ruminant depends as much upon overall dietary considerations as upon the nature of the protein itself. The value will thus differ for any given dietary situation, and any attempt to allocate a single accurate value to any protein source is foredoomed to failure. Measurement of ruminal ammonia concentration reflects the balance achieved between protein breakdown and synthesis under particular dietary circumstances. Providing conditions of determination were standardised a ruminal ammonia figure could be useful for predicting the value of proteins in particular dietary combinations without the considerable commitment involved in balance trials.

The concept of digestible protein fails to cope adequately with the complexities of the nitrogen economy of the ruminant animal. Thus it regards microbial faecal nitrogen as having been of no value to the animal, which is not so. Microbial nitrogen represents nitrogen which has been utilised to satisfy the demands of the microbial population which form part of the legitimate requirement of the animal for nitrogen. Further it does not distinguish between absorption of amino acid-N and the disappearance of non-useful forms of nitrogen such as ammonia from the gut, does not relate nitrogen requirements to energy intake, ignores the need for a certain proportion of the nitrogen to be degradable and the importance of a dietary energy supply for the rumen micro-organisms. Several alternative systems have been proposed. Chief of these are the 'Metabolisable Protein' system used in the USA and that proposed by the Agricultural Research Council (ARC) for the UK and based on 'Rumen degradable and undegradable dietary protein'. Metabolisable protein is that part of the dietary protein which is absorbed by the host animal and is available for use at tissue level. It consists partly of dietary true protein which has escaped degradation in the rumen but which has been broken down to amino acids which are subsequently absorbed from the small intestine. Microbial protein, synthesised in the rumen, similarly contributes to metabolisable protein. This is illustrated schematically in Fig. 13.2.

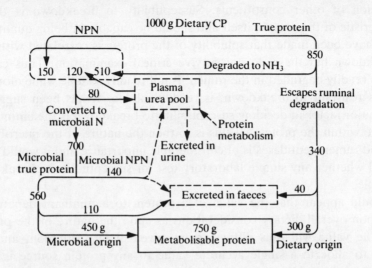

Fig. 13.2.   Calculation of 'Metabolisable Protein' of diet. (From L. D. Satter and R. E. Roffler, 1977, in *Protein Metabolism and Nutrition*, EAAP Pub. No. 22).

In this instance 1000 g of dietary protein yields 750 g of metabolisable protein, but this depends upon the validity of certain assumptions particularly the proportion of dietary crude protein present in non-protein form, the degradability of dietary true protein, and the efficiency of synthesis of microbial protein, which is determined by the supply of energy readily available to the rumen micro-organisms.

In the system of protein allowances, proposed for the UK (ARC, 1984), foods are evaluated in terms of rumen degradable protein (RDP) i.e. that available to the micro-organisms, and undegradable dietary protein (UDP) which escapes degradation in the rumen but which undergoes digestion and absorption in the lower gut, and utilisation at tissue level. In calculating allowances, assumptions have to be made concerning microbial nitrogen requirements, efficiency of NPN capture by rumen micro-organisms, digestibility of protein in the small intestine and utilisation of absorbed nitrogen at tissue level. These are dealt with in Chapter 14.

Central to both the American and UK systems is the concept of degradability, or the extent to which dietary protein is degraded in the rumen, since this affects the value of the protein as a source of nitrogen to the rumen micro-organisms, and of amino acids for absorption from the small intestine.

The proportion of protein escaping breakdown in the rumen may be estimated *in vivo* by measuring dietary nitrogen intake, and the non-ammonia nitrogen and microbial nitrogen passing the duodenum. Degradability of nitrogen is then expressed as:

$$\text{Degradability} = 1 - \frac{\text{Non-ammonia duodenal N} - \text{microbial N}}{\text{Dietary N intake}}$$

The method requires accurate measurement of duodenal flow and microbial nitrogen. The former, which requires the use of a dual phase marker system, has a large coefficient of variation (between animals) and many published values must be suspect owing to the small number of animals used in their determination. Microbial nitrogen in duodenal nitrogen is usually identified by means of marker substances such as diaminopimelic acid (DAPA), aminoethylphosphoric acid (AEPA), ribonucleic acid and ^{35}S, ^{32}P and ^{15}N labelled amino acids. The concentration of marker in the micro-organisms is measured in a sample of rumen fluid. Different markers may give results which vary widely, sometimes by as much as 100 per cent and frequently by 20–30 per cent. The assumption that the micro-organisms isolated from rumen fluid are representative of those in the duodenum is of doubtful validity since the latter include organisms normally adherent to food particles or the rumen epithelium.

The formula for calculating degradability given above ignores the fact that duodenal nitrogen contains a significant fraction which is of endogenous origin. It would be more accurate if degradability was calculated as follows:

$$\text{Degradability} = 1 - \frac{\text{Non-ammonia duodenal N} - (\text{Microbial N} + \text{endogenous N})}{\text{Dietary N intake}}$$

The endogenous fraction constitutes about 50 to 200 g/kg of duodenal nitrogen but is difficult to quantify. A value of 150 g/kg is frequently assumed. Measurements of degradability are thus subject to possible error owing to uncertainties in measuring duodenal flow, microbial nitrogen and the endogeneous nitrogen and, in addition, are affected by dietary considerations such as level of feeding and the size and frequency of meals. It has been calculated that estimates of degradability may vary over a range of 0.3 to 0.35 owing to errors of determination alone. Despite its inadequacies this technique remains the only method, currently available, for providing an absolute measure of protein degradability and a standard against which other methods have to be assessed.

A method of estimating protein degradation in the rumen by incubation of the food in synthetic fibre bags suspended in the rumen is often used. The degradability figure is calculated as the difference between the nitrogen initially present in the bag and that present after incubation, stated as a proportion of the initial nitrogen:

$$\text{Degradability} = \frac{\text{Initial nitrogen} - \text{nitrogen after incubation}}{\text{Initial nitrogen}}$$

The technique is subject to several inherent sources of error which must be controlled if reproducible results are to be obtained. Chief of these are sample size, bag size and porosity of the bag material. A basic assumption of this method is that disappearance of nitrogen from the bag (reflecting virtually solubility in rumen fluid) is synonymous with degradability. While small amounts of food protein which are solubilised may leave the rumen without being degraded, and in individual cases solubility and breakdown are not highly correlated, most of the available evidence would confirm the basic assumption. However, attempts to correlate the disappearance of nitrogen from bags with degradability data *in vivo* have not been encouraging. Another factor which affects the extent of degradation is the time of incubation and there has been considerable uncertainty concerning the appropriate time at which measurement of degradability should be made.

When protein disappearance (p) is regressed on time, p increases but at a reducing rate. The relationship may be described by an equation of the following form:

$$p = a + b\,(1 - e^{-ct})$$

in which a, b and c are constants which may be fitted by an iterative least-squares procedure. The relationshihp is illustrated in Fig. 13.3.

In Fig. 13.3, a is the intercept on the y-axis, b is the difference between a and the asymptote and c is the rate of disappearance of the protein.

Protein degradation will depend upon the rate of passage of the protein through the rumen, and the effective degradability taking this into account may be defined as:

$$P = a + (bc/(c + k))(1 - e^{-(c+k)t})$$

in which k is the rate of passage from the rumen to the abomasum. As the time of incubation increases, the fraction of the protein remaining in the rumen falls to zero as does the rate of degradation, and P may then be defined as follows:

$$P = a + bc/(c + k)$$

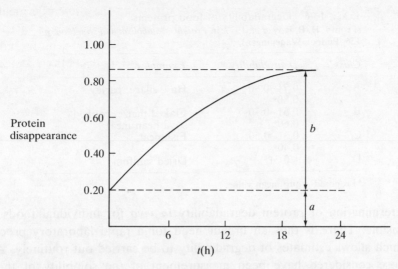

Fig. 13.3. Relationship of protein disappearance with time of incubation.

Retention times may be determind by treatment of the protein with dichromate. The treatment renders the protein completely indigestible, there is no loss of chromium from the protein subsequent to treatment, and particle size distribution is not affected. The rate of dilution of chromium in samples of rumen contents taken over a period of time can therefore provide estimates of the rate of passage of the protein from the rumen.

For the purposes of feeding in practice, it has been suggested that the following rate constants may be appropriate for different classes of animal:

	k
Cattle and sheep on ground diets with no roughage	0.0002
Low-yielding dairy cows and sheep and cattle on mixed diets	0.0005
High-yielding dairy cows	0.0008

Thus, for a sample of food protein with $a = 0.20$, $b = 0.80$ and $c = 0.00082$, the appropriate degradability for cattle on mixed diets would be

$$0.2 + \left(\frac{0.80 \times 0.00082}{0.00082 + 0.0005}\right) = 0.697$$

For a high-yielding dairy cow it would be 0.605.

Published values for degradable protein are extremely variable and in the proposed UK system no attempt is made to allocate values to individual foods. Instead they are classified into four broad classes as shown in Table 13.6.

In view of the quoted ranges, use of the suggested mean values must be attended by considerable error. It will be equally obvious that routine

TABLE 13.6   Degradability of food proteins.
(From J. H. B. Roy *et al*, 1977, in *Protein Metabolism and Nutrition*, p. 126. Pudoc., Wageningen).

Class	Degradability	Examples
A	0.71–0.90 0.80*	Hay, silage, barley
B	0.51–0.70 0.60*	Flaked maize, cooked soya bean meal
C	0.31–0.50 0.40*	Fishmeal
D	< 0.31	Dried sainfoin

*Proposed group mean value.

determination of protein degradability *in vivo* for individual foods is not feasible. There is thus an urgent need for a rapid laboratory procedure, which allows estimates of degradability to be carried out routinely. Among those considered have been measurement of the solubility of the food protein in rumen liquor, sodium chloride solutions, artificial saliva, enzyme solutions and buffer solutions. None of these have, as yet, provided a satisfactory technique and, despite the efforts being made, it may be that no simple test can be devised which will enable quantitative prediction of the fate of a protein in such a complex system as the rumen. Table 13.7 shows some examples of laboratory techniques for determining rumen degradability of food proteins, which have been investigated:

TABLE 13.7   Laboratory methods for determining protein degradability
(From R. J. Crawford *et al*., 1978. *J. Anim. Sci.* **46**, 1768.

Laboratory Method	In vivo comparison	Correlation coefficient (r)	Population
1. Solubility in 10 % Wise Burroughs mineral buffer	2 hour degradation in 'Dacron' bags suspended in the rumen of the steer	0.66	15 miscellaneous concentrates, 7 hays and 6 silages
2. Solubility in 0.15 M sodium chloride		0.47	
3. Solubility in autoclaved rumen liquor		0.54	

For silages alone, r was 0.94 with the buffer solution, which would appear to be promising, but it must be remembered that the comparison is with a technique concerning which many authorities would have considerable reservations. Furthermore the buffer solution is highly complex, contains micro-amounts of trace minerals and has a high level of background nitrogen, which makes it difficult to prepare and use.

Solubilisation of the protein by purified proteolytic enzymes from fungi, and bacteria such as *Streptomyces griseus* and *Bacteroides amylophilus*, shows some promise as a method for estimating degradability, but further work needs to be done on the methodology.

Attempts have been made to describe the degradable nitrogen of foods more accurately by laboratory methods. In one such model, quickly degradable nitrogen (QDN) is defined as that which is soluble in cold water under specified conditions, and the efficiency of microbial capture of this nitrogen is assumed to be 0.8. Slowly degradable nitrogen (SDN) is calculated as follows:

$$SDN = 0.85 (N - (QDN - ADIN)) - 3.1$$

in which ADIN is acid-detergent insoluble nitrogen. The effective degradable nitrogen is then 0.8 QDN + SDN and degradability is (0.8 QDN + SDN)/N intake.

Both the American and UK systems provide a logical scientific basis for formulating protein allowances for ruminants, and for evaluating foods as protein sources. As at present constituted, they make certain questionable assumptions, and suffer the major disadvantage that no satisfactory method exists for routinely determining protein degradability. The general approach of these systems is highly commendable and their introduction in practice, after validation under field conditions, should be encouraged.

FURTHER READING

A. E. Bender, R. Kihlberg, B. Lofquist and L. Munck (eds), 1970. *Evaluation of Novel Protein Products*. Pergamon Press, Oxford.

Agricultural Research Council, 1980.*The Nutrient Requirement of Ruminant Livestock*. Commonwealth Agricultural Bureaux, Farnham Royal.

Agricultural Research Council, 1984. *The Nutrient Requirements of Ruminant Livestock*, Supplement No. 1. Commonwealth Agricultural Bureaux, Farnham Royal.

Evaluation of Protein Quality, 1963. National Academy of Sciences, National Research Council, Washington, D.C.

E. L. Miller, I. H. Pike and A. J. H. Van E S (eds), *Protein Contribution of Feedstuffs for Ruminants: application to feed formulation*. Butterworth Scientific, London.

H. H. Mitchell, 1962. *Comparative Nutrition of Man and Domestic Animals*, Vol. 1. Academic Press, New York and London.

Proc. 2nd Int. Symposium on Protein Metabolism and Nutrition 1977. EAAP Publication No. 22.

Proc. 3rd. Int. Symposium on Protein Metabolism and Nutrition, 1980. EAAP Publication No. 27.

Proc. 4th Int. Symposium on Protein Metabolism and Nutrition, 1983. EAAP Publication No. 31.

14

Feeding standards for maintenance and growth

Statements of the amounts of nutrients required by animals are described by the general term, *feeding standards*. Two other terms used in the same context are *nutrient requirement* and *allowance*. Neither is strictly defined, but a rough distinction between them is that, if the requirement is a statement of what animals on average require for a particular function, the allowance is greater than this amount by a safety margin designed principally to allow for variations in requirement between individual animals.

Feeding standards may be expressed in quantities of nutrients or in dietary proportions. Thus the phosphorus requirement of a 50 kg pig might be stated as 11 g P/day or as 5 g P/kg of the diet. The former method of expression is used mainly for animals given exact quantities of foods, the latter for animals fed to appetite. Various units are used for feeding standards. For example, the energy requirements of ruminants may be stated in terms of net energy, metabolisable energy or TDN, and their protein requirements in terms of crude protein, digestible crude protein or a combination of rumen degradable and undegradable protein. It is obviously desirable that the units used in the standards should be the same as those used in the evaluation of foods. Standards may be given separately for each function of the animal or as overall figures for the combined functions. The requirements of dairy cows, for example, are often given separately for maintenance and for milk production, but those for growing chickens are for maintenance and growth combined. In some cases the requirements for single functions are not known; this is true particularly of vitamin and trace element requirements.

As mentioned above, the translation of a requirement into an allowance which is to be used in feeding practice is accompanied by the addition of a safety factor. The justification for such safety factors is illustrated by the following example. Suppose that in cattle of 500 kg liveweight the requirement for energy for maintenance is found to range in individuals from 30 to 36 MJ net energy per day, with a mean value of 33 MJ. Although some of the variation may be caused by inaccuracies in the methods of measurement used, much of it will undoubtedly reflect real differences between animals. This being so, the adoption of the mean estimate of requirement, 33 MJ, as the allowance to be used in practice, will result in some cattle being over- and others under-fed. Underfeeding is regarded as the greater evil, and so a safety factor may be added to the requirement when calculating the allowance to be recommended. This safety factor will be designed to ensure that no animals, or only those with an exceptionally high requirement, will be underfed. It may be an arbitrary addition or, better, one based mathematically on the expected variation between animals; the larger this variation, the greater should be the safety factor. Safety factors have been criticised on the grounds that over-feeding, say, 90 per cent of the population to ensure that the remaining 10 per cent are not grossly underfed is a wasteful procedure.

Variations between animals, and also between samples of a food, must always be borne in mind when applying feeding standards. Such variations mean that the application of standards to individual animals and individual samples of foods must inevitably be attended by inaccuracies. For this reason feeding standards should be considered as guides to feeding practice, not as inflexible rules; they do not replace the art of the stockman in the finer adjustment of food intake to animal performance. Feeding standards, however, are not restricted in application to feeding individual animals; they can be used on a larger, farm scale to calculate, for example, the total winter feed required by a dairy herd, and even on a national scale to assist in planning food imports.

The object in this chapter and the next will be to discuss the scientific basis for feeding standards and to describe briefly how they are determined. Some tables of feeding standards are included in the Appendix. For many years the standards used in the United Kingdom were those published in the Ministry of Agriculture's bulletins *Rations for Livestock* and *Poultry Nutrition*. Between 1960 and 1965 the UK Agricultural Research Council undertook a comprehensive review of standards, and published its recommendations in three comprehensive reviews, for poultry, ruminants and pigs, respectively. Since 1970, all three review publications have been revised, the new edition for poultry appearing in 1975, that for ruminants in 1980 and that for pigs in 1981 (see Further Reading). In some cases it has been found necessary to simplify or modify these standards for practical application (for example, by adding safety margins, as discussed above). For pig and poultry standards, such changes have been minimal. For ruminants, standards for minerals and vitamins have been generally accepted for use in practice, although doubts have been expressed about the adequacy of the Agricultural Research Council's latest (1980) proposals for calcium and phosphorus requirements. The 1965 energy standards for ruminants were brought into use in a slightly modified form in Ministry of Agriculture Technical Bulletin No. 33 (now called Reference Book No. 433), and this publication is currently being updated to take account of the Agricultural Research Council's 1980 review. The protein standards for ruminants proposed in 1965 were found to be difficult to apply in practice and were never adopted. In 1980 they were recalculated in terms of rumen degradable and undegradable protein (see Ch. 13), and are now being adapted for practical application.

Many other countries have manuals of feeding standards. The standards used in the United States are summarised in a series of booklets published by the National Research Council under the general title *Nutrient Requirements of Domestic Animals*.

Feeding Standards for Maintenance

An animal is in a state of maintenance when its body composition remains constant, when it does not give rise to any product such as milk and does not perform any work on its surroundings. As farm animals are rarely kept in this non-productive state, it might seem to be only of academic interest to determine nutrient requirements for maintenance; but the total requirements of several classes of animals, notably dairy cows, are arrived at factorially by summing requirements calculated separately for maintenance and for production. Consequently a knowledge of the maintenance needs of animals is of practical as well as theoretical significance. The relative importance of requirements for maintenance is illustrated in Table 14.1, which shows the proportion of their total energy requirements used for this purpose by various classes of animal.

Animals deprived of food are forced to draw upon their body reserves to meet their nutrient requirements for maintenance. We have seen already that a fasted animal must oxidise reserves of nutrients to provide the energy needed for such essential processes as respiration and the circulation of the blood. Since the energy so utilised leaves the body, as heat, the animal is then in a state of negative energy balance. The same holds true for other nutrients: an animal fed on a protein-free diet continues to lose nitrogen in its faeces and urine, and is therefore in negative nitrogen balance. The purpose of a maintenance ration is to prevent this drain on the body tissues, and the maintenance requirement of a nutrient can therefore be defined as the quantity which must be supplied in the diet so that the animal experi-

TABLE 14.1 Approximate proportions of the total energy requirements of animals which are contributed by their requirements for maintenance

	Requirement (MJ net energy) for:		Maintenance as a percentage of total
	Maintenance	Production	
Daily values			
Dairy cow weighing 500 kg and producing 20 kg milk	32	63	34
Steer weighing 300 kg and gaining 1 kg	23	16	59
Pig weighing 50 kg and gaining 0.75 kg	7	10	41
Broiler chicken weighing 1 kg and gaining 35 g	0.50	0.32	61
Annual values			
Dairy cow weighing 500 kg, producing a calf of 35 kg and 5 000 kg milk	12 200	16 000	43
Sow weighing 200 kg, producing 16 piglets, each 1.5 kg at birth, and 750 kg milk	7 100	4 600	61
Hen weighing 2 kg, producing 250 eggs	190	95	67

ences neither net gain nor net loss of that nutrient. The requirement for maintenance is thus the minimum quantity promoting zero balance. (The qualification 'minimum' is necessary, because if the animal is unable to store the nutrient in question, increasing the quantity supplied above that required for maintenance will still result in zero balance.)

Energy requirements for maintenance

Basal and fasting metabolism. It was explained earlier (p. 218) that energy expended for the maintenance of an animal is converted into heat and leaves the body in this form. The quantity of heat arising in this way is known as the animal's *basal metabolism*, and its measurement provides a direct estimate of the quantity of net energy which the animal must obtain from its food in order to meet the demand for maintenance. The measurement of basal metabolism is complicated by the fact that the heat produced by the animal does not come only from this source, but comes also from the digestion and metabolism of food constituents (the heat increment of feeding), and from the voluntary muscular activity of the animal. Heat production may be further increased if the animal is kept in a cold environment (see p. 293).

When basal metabolism is measured, the complicating effect of the heat increment of feeding is removed by depriving the animal of food. The period of fast required for the digestion and metabolism of previous meals to be completed varies considerably with species. In Man an overnight fast is sufficient, but in ruminant animals digestion, absorption and metabolism continue for several days after feeding ceases, and a fast of at least four days would be needed. The same period is recommended for the pig, and two days for the fowl. There are a number of criteria for establishing whether the animal has reached the postabsorptive state. If heat production can be measured continuously, the most satisfactory indication is the decline in heat production to a steady, constant level. A second indication is given by the respiratory quotient (p. 227). In fasting, the oxidation mixture gradually changes from absorbed fat, carbohydrate and protein to body fat and some body protein. This replacement in the mixture of carbohydrate by fat is accompanied by a decline in the non-protein respiratory quotient, and when the theoretical value for fat (0.7) is reached it may be assumed that energy is being obtained only from body reserves. In ruminants an additional indication that the postabsorptive state has been reached is a decline in methane production (and therefore digestive activity) to a very low level.

The contribution of voluntary muscular activity to heat production can be reduced to a low level when basal metabolism is measured in human subjects, but in farm animals the cooperation needed to obtain a state of complete relaxation can rarely be achieved. Fasting may limit activity, but even the small activity represented by standing as opposed to lying is sufficient to increase heat production. Consequently the term *fasting metabolism* is to be preferred to basal metabolism in studies with farm animals, since strict basal conditions are unlikely to be observed.

A term used in conjunction with fasting metabolism is *fasting catabolism*. This includes the relatively small quantities of energy lost by fasting animals in their urine.

Some typical values for fasting metabolism are given in Table 14.2. As one would expect, the values are greater for large than for small animals, but column 2 shows that per unit of liveweight, fasting metabolism is greater in small animals. At an early stage in the study of basal metabolism it was recognised that fasting heat production is more nearly proportional to the surface area of animals than to their weight, and it became customary to compare values for animals of different sizes by expressing them in relation to surface area (column 3 of Table 14.2). The surface area of animals is obviously difficult to measure, and methods were therefore devised for predicting it from their body weight. The basis for such methods is that, in bodies of the same shape and of equal density, surface area is proportional to the two-thirds power of weight ($W^{0.67}$). The logical development of this approach was to omit the calculation of surface area and to express fasting metabolism in relation to $W^{0.67}$. When the relation between fasting metabolism and body weight was examined further, however, it was found that the closest relationship was between metabolism and $W^{0.73}$ not $W^{0.67}$. The function $W^{0.73}$ was used as a reference base for fasting metabolism of farm animals until 1964, when it was decided to round off the exponent to 0.75 (see column 4 of Table 14.2).

There has been considerable discussion whether surface area or $W^{0.75}$ (often called *metabolic liveweight*) is the better base. This will not be repeated here, but is contained in the books listed at the end of the chapter. Mathematically there is nothing to choose between the two bases, for their relationships with fasting metabolism are equally close.

The fasting metabolism of adult animals ranging in size from mice to elephants was found by S. Brody to have an average value of 70 kcal/kg $W^{0.73}$ per day; the approximate equivalent is 0.27 MJ/kg $W^{0.75}$ per day. There are, however, considerable variations from species to species, as shown in Table 14.2. For example, cattle tend to have a fasting metabolism about 15 per cent higher than the interspecies mean, and sheep a fasting metabolism 15 per cent lower. There are also variations within species, notably those due to age and sex. Fasting metabolism per unit of metabolic liveweight is higher in young animals than in old, being for example 0.39 MJ/kg $W^{0.75}$ per day in a young calf, but only 0.32 MJ/kg $W^{0.75}$ in a mature cow; it is also 15 per cent higher in male cattle than in females or castrated males.

Estimating maintenance energy requirements from measurements other than those of fasting metabolism. The quantity of energy required for maintenance is, by definition, that which promotes energy equilibrium (zero

TABLE 14.2 Some typical values for the fasting metabolism of adult animals of various species

Animal	Live-weight (kg)	Fasting metabolism (MJ/day)			
		Per animal (1)	Per kg liveweight (W) (2)	Per sq. metre surface area (3)	Per kg $W^{0.75}$ (4)
Cow	500	34.1	0.068	7.0	0.32
Pig	70	7.5	0.107	5.1	0.31
Man	70	7.1	0.101	3.9	0.29
Sheep	50	4.3	0.086	3.6	0.23
Fowl	2	0.60	0.300	—	0.36
Rat	0.3	0.12	0.400	3.6	0.30

energy balance). This quantity can be estimated directly in fed, as opposed to fasted, animals if the energy content of their food is known and their energy balance can be measured. In theory the quantities of food given could be adjusted until the animals were in exact energy equilibrium, but in practice it is easier to allow them to make small gains or losses and then to use a model of the kind depicted in Fig. 11.6(a) to estimate the energy intake required for equilibrium. For example, suppose a 300 kg steer was given 3.3 kg DM/day in the form of a food with M/D = 11 MJ/kg DM and with k_f = 0.5. If the steer retained 2 MJ/day, its maintenance requirement for metabolisable energy would be calculated as:

$$(3.3 \times 11) - (2/0.5) = 32.3 \text{ MJ ME/day.}$$

This approach can also be followed in feeding trials in which animals are not kept in calorimeters. The animals are given known quantities of food energy, and their liveweights and liveweight gains or losses are measured. The partition of energy intake between that used for maintenance and that used for liveweight gain can be made in two ways. The more simple method involves the use of known feeding standards for liveweight gain. The alternative is to analyse the figures for energy intake (I), liveweight (W) and liveweight gain (G) by solving equations of the form

$$I = a \, W^{0.75} + bG.$$

The coefficients a and b then provide estimates of the quantities of food energy used for maintenance and for each unit of liveweight gain respectively. This form of analysis can be extended to animals with more than one type of production, such as dairy cows, by adding extra terms to the right-hand side of the equation.

The main objection to determining energy requirements for maintenance (and also for production) in this way is that liveweight changes may fail to give a correct measure of energy balance. It is possible, however, to put the method on a sounder 'energy' basis by using the comparative slaughter technique to estimate changes in the energy content of the animals.

Fasting metabolism as a basis for estimating maintenance requirements. The 'feeding trial' method of estimating maintenance requirements has the advantage of being applied to animals kept under normal farm conditions, rather than under the somewhat unnatural conditions represented by a fast in a calorimeter. It is often difficult to translate values for fasting metabolism into practical maintenance requirements. One factor to be taken into account is that animals on the farm commonly use more energy for voluntary muscular activity. Another factor is that productive livestock must operate with more intense metabolism than fasted animals and thereby incur a higher maintenance cost. Thirdly, animals on the farm experience greater extremes of climate and may need to use energy specifically to maintain their normal body temperature. The first two factors are discussed below, and the effects of climate on energy requirements for maintenance are discussed on p. 293.

Energy used for voluntary muscular activity is regarded as part of the maintenance requirement and should therefore be taken into account when values for fasting metabolism are translated into allowances for maintenance. The activity of animals kept tied up in farm buildings is probably little greater than that of an animal kept in a calorimeter. However, the farm animal might spend more of its time in a standing position, and it has been found in cattle that heat production during standing is about 12 per cent greater than during lying. In grazing animals the maintenance requirement will be increased by the energy costs of locomotion and the muscular activity involved in harvesting grass. The energy cost of walking on the level is quite small. For example, sheep require 2.6 kJ per kg body weight per km travelled, and so a 50 kg sheep, walking perhaps 2 km/day, would require 260 kJ for this purpose (equivalent to 6% of its fasting metabolism). A summation of the energy costs of the grazing animal's activities suggests that they would be equivalent to 25–50 per cent of its fasting metabolism.

Table 14.3 shows the maintenance requirements of sheep estimated indoors (by two methods) and out-of-doors. The two estimates made with fed sheep exceed that obtained with fasted animals by about the proportions suggested above. However, it is not unusual for such 'practical' estimates of maintenance requirements, obtained by the regression method, to exceed fasting estimates by a much greater proportion. The difference is often particularly large for grazing animals, but this may be due to the difficulties of measuring their food intake accurately.

Animals on a high plane of nutrition tend to have a higher maintenance requirement for energy than those kept on a lower plane. This can be shown by subjecting animals previously on high or low planes to a sudden fast; the

TABLE 14.3 Maintenance requirements for energy of a 50 kg sheep
(After J. P. Langlands *et al.*, 1963. *Anim. Prod.*, 5, 1 and 11)

Basis for estimate	*Maintenance requirement (MJ net energy per day)*
Sheep fasted in calorimeter	4.4
Sheep fed indoors and requirements partitioned by equations of the type shown on page 290	4.8
Sheep grazing out-of-doors and requirements partitioned by equations	5.8

former will have a fasting heat production perhaps 20 per cent greater than that of the latter. The phenomenon can also be demonstrated by adjusting an animal's food intake to keep its liveweight constant; as time passes and it becomes adapted to a low plane of nutrition, less and less food is needed to maintain its weight. High-plane animals tend to have larger internal organs (gut and liver) and as these are metabolically active their greater size raises the animal's maintenance requirement. Another factor is that low-plane animals tend to be physically less active (i.e. they conserve their energy). Maintenance requirements based on fasting metabolism are therefore likely to be too low for productive animals. But this effect is also likely to lead to an underestimation of the efficiency of utilisation of metabolisable energy above maintenance (and give rise to a curvilinear relationship between energy intake and energy retention: see Fig. 11.6). Opinions vary on the extent to which these errors cancel one another, but in a new Australian system for expressing the energy requirements of ruminants, special provision is made for the higher maintenance energy requirements of productive animals.

Present standards. The maintenance requirements of cattle were estimated by the Agricultural Research Council (in its 1980 publication) from data for fasting metabolism (F, MJ/day). For steers and heifers the equation used was:

$$F = 0.53(W/1.08)^{0.67}$$

Liveweight (W) is reduced to an estimated fasted weight by dividing by 1.08, and metabolic weight is calculated with the power 0.67 rather than the more usual 0.75. An allowance for the minimal activity expected of housed animals is calculated as $0.0043W$. Thus the net energy requirement for maintenance of a 500 kg cow would be calculated as:

$$0.53 (500/1.08)^{0.67} + 0.0043 \times 500 = 34.5 \text{ MJ/day}$$

If the cow's diet contained 11 MJ ME/kg DM, k_m (Table 12.1) would be 0.714, and the cow's requirement for metabolisable energy would be:

$$34.5/0.714 = 48.3 \text{ MJ ME/day}$$

Maintenance requirements of bulls are considered to be 15 per cent higher than those of steers and heifers of the same weight. For sheep the equation for predicting the maintenance requirement (as net energy) for ewes and wethers is:

$$0.226(W/1.08)^{0.75} + 0.0106W$$

From this a 60 kg ewe is calculated to require 5.2 MJ NE/day for maintenance, or if $q_m = 0.714$, its requirement could alternatively be expressed as 7.3 MJ ME/day. As with cattle, intact males (rams) are considered to require 15 per cent more energy for maintenance.

The energy requirements of pigs and poultry are normally stated for maintenance and growth combined, although some theoretical standards for maintenance have been calculated by the UK Agricultural Research Council. For example, the maintenance requirements of pigs as MJ net energy per day may be calculated as $0.719W^{0.63}$, which is equivalent to about $0.458W^{0.75}$. For laying hens the maintenance requirement is on average about 0.490 MJ ME kg $W^{0.75}$/day.

The influence of climate on requirements of energy for maintenance

Mammals and birds are *homeotherms*, which means that they attempt to keep their body temperature constant. Animals produce heat continuously and, if they are to maintain a constant body temperature, must lose it to their surroundings. The routes by which they may lose heat are by radiation, conduction and convection from their body surface, and by the evaporation of water from both their body surface and their lungs. The rate at which heat is lost depends in the first instance on the difference in temperature between the animal and its surroundings; for farm animals the rectal temperature, which is slightly lower than the deep body temperature, lies in the range 36–43 °C. The rate of heat loss is influenced also by characteristics of the animal, such as the insulation provided by its tissues and coat, and by such characteristics of the environment as air velocity, relative humidity and solar radiation. In effect, the rate of heat loss is determined by a complex interaction of factors contributed by the animal and its environment, but it is only the effect of air temperature which has been reasonably well investigated. For housed animals, of course, air temperature is the most significant feature of the environment.

The animal must achieve homeothermy in the face of potential variations in both heat production and rates of heat loss, and it has only partial control over these variations. In general, the animal's primary response to a change in its environment is to maintain the rate of heat loss at its former level by selective adjustment of the temperature gradients in and around its body; only if these measures are insufficient will it alter its heat production. Let us consider the case of a pig kept under basal conditions, i.e. fasting, resting and at a 'comfortable' air temperature of 25 °C (Fig. 14.1, line A). If the air temperature is then gradually reduced, the pig is liable to lose heat more rapidly and hence to suffer a fall in deep body temperature. Its primary defence against this is principally to restrict the flow of blood from the interior of the body to the skin, thus reducing skin temperature and restoring to its former level the temperature gradient from skin to air. As the fall in air temperature continues, however, a stage is reached when the pig can maintain its body temperature only by increasing its heat production. This increase might be achieved through the muscular activity of shivering. The temperature below which heat production is increased is known as the *critical temperature*. In Fig. 14.1 this is 20 °C. Heat produced specifically to maintain body temperature represents an extra drain on the fasting animal's energy reserves and must be regarded as a part of the requirement of energy for maintenance. Thus we should expect that at temperatures below 20 °C a pig would need a higher energy intake to promote energy equilibrium than it would at higher temperatures. In a pig given food (Fig. 14.1, line B), however, the situation is rather different. At the 'comfortable' temperature of 25 °C, heat production is greater in the fed than in the fasting pig because of the heat increment of the food. This means that the air temperature can fall to considerably below 20 °C before the fed pig needs to shiver and produce more heat. This lower temperature, 5 °C in the example, is sometimes known as the *effective critical temperature*. While it is obviously of greater practical significance than the critical temperature recorded for the fasted animal, even the effective critical temperature has limitations as a guide to whether or not animals are likely to have their energy requirements increased by their environment. The effective critical temperature will vary with the quantity of food consumed

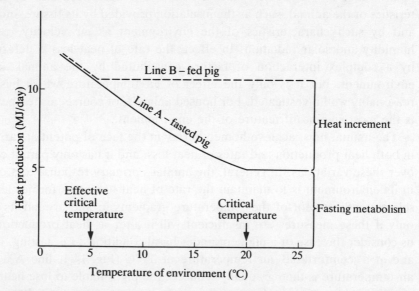

Fig. 14.1.   Effect of environmental temperature on the heat production of the pig.

and with the efficiency with which it is metabolised. In the pig it will also vary from time to time during the day, for the contribution to total heat production of the heat increment of the food will be greater shortly after a meal than immediately before it. (In ruminants there will be less diurnal variation because absorption and metabolism are more continuous than in simple-stomached animals.) A further factor to be considered is that animals may regulate their heat loss by huddling together; at temperatures below the critical the total heat loss of a group of animals is less if they lie together than if they keep apart.

When one turns to consider the effect of the environment on the energy transactions of animals kept out of doors, where there are more variables, the situation becomes decidedly more complex. The complexities are well illustrated by the experiments of K. L. Blaxter, who has shown that the effective critical temperature of Cheviot sheep fed at a maintenance level of nutrition varies between −3 and 24 °C according to the length of their fleece (10–100 mm) and the wind speed to which they are exposed (0–24 km/h). For sheep with a 55-/mm fleece, the effective critical temperature was estimated to increase from 1 °C when there was no air movement to 16 °C at a wind speed of 24 km/h. In sheep kept on the uplands of Britain there is little doubt that the requirement of energy for maintenance is frequently increased by the environment, but it is difficult to assess the size of the increase.

The farm animals that are most likely to suffer cold stress are the newborn lamb, calf and pig. As they are small, their surface area is large in relation to their bodyweight (and heat production). Their tissue insulation may be low because of a lack of subcutaneous fat, and in the lamb the insulation of the coat may be reduced because it is wet. If the newborn animal fails to suck milk from its mother, its heat increment of feeding will be low. The lamb and calf have a special tissue for generating heat which is known as brown adipose tissue. Deposits of 'brown fat' are located at strategic points of the young animal's body, such as the shoulders and abdomen. Fat globules are stored in actively metabolising cells, and the tissue is well supplied with blood. When the fat is metabolised, the oxidation is 'uncoupled', energy being released as heat instead of being captured as ATP. The heat so generated is carried to other parts of the body by the blood. The protective action of brown adipose tissue in young animals is limited by their small fat reserves, and it is essential that young lambs and calves should receive food (i.e. milk) as soon after birth as possible.

In a cold environment the animal's problem is one of conserving heat, but in a hot environment the problem becomes that of disposing of the heat produced. As air temperature increases, less heat can be lost by the sensible routes of radiation, conduction and convection, and more has to be dissipated by the evaporative routes of the respiratory tract and the skin. Pigs and poultry are poorly equipped for evaporative losses, and in a hot environment must often resort to reducing heat production by diminishing food intake. Ruminants can increase evaporative losses by both panting and sweating, but the sheep—on account of its fleece—relies more on the respiratory route.

Protein requirements for maintenance

An animal placed on a nitrogen-free, but otherwise adequate, ration continues to loss nitrogen in its faeces and urine. The nitrogen of the faeces, as described earlier (Ch. 10), arises from the enzyme and cell residues of the digestive tract, and from microbial residues. If the animal continues to eat, it must continue to lose nitrogen in this fashion.

It is less obvious, perhaps, why an animal on a nitrogen-free diet should lose nitrogen in its urine. In part this excretion represents nitrogen which has been incorporated into materials subsequently expended, and which cannot be recovered by the body for re-use. Thus the creatine of muscles is eventually converted into creatinine, which is excreted in the urine. But the greater part of the nitrogen in the urine of animals not receiving food nitrogen is (in mammals) in the form of urea, the typical by-product of amino acid catabolism. The significance of this urea nitrogen was not fully appreciated until experiments with isotopically labelled amino acids had shown that the proteins of body tissues are not static but are constantly being broken down and replaced. The turnover rate of body proteins varies considerably from one tissue to another; proteins are replaced at intervals of hours or days in the intestine and liver, whereas in bone and nerve the interval may be one of months or even years. The amino acids released when body proteins are broken down form a pool from which the replacement proteins are synthesised; a particular amino acid molecule may therefore be present one day in a protein of the liver, for example, and the next day in muscle protein. In effect the body proteins exchange amino acids among themselves. Like the synthesis of body proteins from absorbed amino acids (Ch. 13), however, this domestic traffic in amino acids is not completely efficient. Acids liberated from one protein may fail to be incorporated in another and are catabolised, their amino groups thus yielding the urea which is excreted in the urine.

When an animal is first placed on a nitrogen-free diet, the quantity of nitrogen in its urine may fall progressively for several days before stabilising at a lower level, and when nitrogen is re-introduced into the diet there is a similar lag in the re-establishment of equilibrium. This suggests that the animal possesses a protein reserve which can be drawn upon in times of scarcity of dietary nitrogen and restored in times of plenty. In times of scarcity, the tissues most readily depleted of protein are those in which the proteins are most labile, such as the liver. Depletion of liver nitrogen is accompanied by some reduction in enzyme activity, and the reserve protein is therefore envisaged as a 'working reserve' which consists of the cytoplasmic proteins themselves.

Once the reserve protein has been depleted, the urinary nitrogen excretion of an animal deprived of food nitrogen reaches a minimal and approximately constant level. (This level will be maintained only if energy intake is adequate, for if tissue proteins are catabolised specifically to provide energy, urinary nitrogen excretion will rise again.) The nitrogen excreted at this minimal level is known as the *endogenous urinary nitrogen*, and represents the smallest loss of body nitrogen commensurate with the continuing existence of the animal. The endogenous urinary nitrogen excretion can therefore be used to estimate the nitrogen (or protein) required by the animal for maintenance. It is analogous to the basal metabolism, and in fact there is a relationship between the two. The proportionality commonly quoted is 2 mg endogenous urinary nitrogen per kcal basal metabolism (about 500 mg/MJ). For adult ruminants, however, the ratio is appreciably lower, at 300–400 mg endogenous urinary nitrogen per MJ of fasting metabolism. The reason for this is that ruminants on diets devoid of, or deficient in, nitrogen are capable of recycling urea to the rumen or large intestine, and nitrogen that would in a non-ruminant be excreted in the urine is therefore excreted as nitrogen of microbial residues. The *total* endogenous nitrogen excretion (i.e. urinary plus faecal nitrogen) of ruminants is calculated as 350 mg N/kg $W^{0.75}$ per day and is equivalent to 1000–1500 mg/MJ fasting metabolism; it is therefore two to three times as great as in non-ruminants.

When nitrogen is re-introduced into the diet, the quantity of nitrogen excreted in the urine increases through the wastage of amino acids incurred in utilising the food protein. Urinary nitrogen in excess of the endogenous portion is known as the *exogenous urinary nitrogen*. This name implies that such nitrogen is of food origin as opposed to body origin; but, with the exception of the creatinine fraction of the endogenous portion, it is doubtful whether such a strict division of urinary nitrogen is justified. It is more realistic to regard the so-called exogenous fraction as the extension of an existing loss of nitrogen rather than as an additional source of loss, for it probably reflects mainly an increase in turnover and wastage of amino acids.

The quantity of nitrogen or of protein required for maintenance is that which will balance the metabolic faecal and endogenous urinary losses of nitrogen (and also the small dermal losses of nitrogen occurring in scurf, hair and sweat). The two most commonly employed methods for estimating this quantity are analogous to those used for determining the quantity of energy needed for maintenance. The first, analogous to the determination of the fasting catabolism, involves measuring the animal's losses of nitrogen when it is fed on a nitrogen-free diet, and calculating the quantity of food nitrogen required to balance these losses. In the second method, the quantity of food nitrogen required is determined directly by finding the minimum intake which produces nitrogen equilibrium. This method is similar to that used for estimating maintenance energy requirements (p. 289).

Estimating protein requirements for maintenance from total endogenous nitrogen and other obligatory losses of nitrogen (the factorial method). The starting point in the factorial calculation is the total endogenous nitrogen excretion. For example, for a cow of 600 kg, this loss might be 42 g/day. Loss of N in hair and scurf would add 2 g/day, giving a total of 44 g/day, which is equivalent to $6.25 \times 44 = 275$ g protein. The efficiency with which this might be replaced by amino acids absorbed from the small intestine is expressed by the biological value (see Ch. 13). For cattle the biological value of protein digested and absorbed is frequently assumed to be relatively constant, at 0.8. The cow's requirement for protein digested and absorbed from the intestine would therefore be $275/0.8 = 344$ g/day. This protein could be supplied as either microbial protein or undegraded food protein. If we assumed that it would all be supplied as microbial protein then the quantity required could be calculated by assuming the true digestibility of microbial protein to be 0.85 and its amino acid content to be 0.8 of its total protein content (i.e. 0.2 nucleic acid). This quantity would be $344/(0.85 \times 0.8) = 506$ g.

The quantity of microbial protein produced in the rumen depends on the quantity of organic matter fermented, which in turn depends on the quantity of organic matter consumed or, alternatively, on metabolisable energy intake. A suggested relationship is that 8.3 g microbial protein is produced for each 1 MJ of metabolisable energy consumed. Thus if the cow's metabolisable energy intake was equal to its maintenance requirement for metabolisable energy (61 MJ), the quantity of microbial protein supplied to the intestine of the cow would be $8.3 \times 61 = 506$ g. This is exactly equal to the cow's protein requirement calculated above, and one may therefore conclude that if the protein requirements of the rumen microbes (i.e. 506 g/day) were satisfied, then the host animal's requirements for maintenance would also be satisfied. In this situation it would be desirable to supply all the dietary protein in a highly degradable form (i.e. degradability = 1.0). If degradability were lower, 0.7 for example, the quantity of protein in the diet would have to be increased to $506/0.7 = 723$ g/day, to ensure an adequate supply of protein for the rumen microbes.

The final step in the calculation might be to estimate the minimum crude protein content of the cow's diet. If the diet contained 11 MJ metabolisable energy per kg DM, dry matter intake would be $61/11 = 5.54$ kg/day. If the degradability of the crude protein was 1.0 (which is unlikely), the minimum crude protein concentration required would be $506/5.54 = 91$ g/kg DM; if degradability was 0.7, the required concentration would be 130 g/kg.

It should be noted that undegraded protein reaching the intestine from food with protein degradability < 1.0 simply increases the surplus of protein available over protein required at this point. We shall see from a later example that whereas this undegraded protein may be of no value to the ruminant with a relatively low protein requirement (i.e. for maintenance only), it becomes an essential contribution to the protein supply of the animal with a high requirement.

Estimating protein requirements from nitrogen balance trials. If animals are fed on a number of rations which supply equal amounts of dry matter and energy, but varying amounts of protein, their nitrogen balances can be expected to follow a pattern of the type shown in Fig. 14.2. As nitrogen intake increases from a low level, there is a gradual reduction in the negative balance until the point of equilibrium is reached. The extent to which further increments of food nitrogen promote nitrogen storage will depend on the age of the animals and the supply of other nutrients. A stage is eventually reached, however, when further increments of protein fail to promote further nitrogen retention, and the curve then becomes horizontal. Mature animals may store very little nitrogen, in which case the curve will become horizontal at a nitrogen balance close to zero (the lowest horizontal line in Fig. 14.2).

When maintenance protein requirements are determined in nitrogen balance trials, sufficient negative and positive balances are obtained, by using rations supplying different quantities of protein, to enable lines like those of Fig. 14.2 to be constructed. Care is needed in interpreting positive balances, since these may be represented by points on the horizontal part of the curve. A further difficulty to be borne in mind is that the protein intake required to promote equilibrium will depend to some extent on the previous nutrition of the animals. If the animals are well supplied with reserve protein, a higher intake will be needed to maintain equilibrium than if their reserves are depleted.

Present standards. Protein standards used in the UK for ruminants are currently being revised. New standards will be based on the concepts of rumen degradability of food protein and of subsequent digestion, absorption and metabolism of microbial and undegraded food protein that were

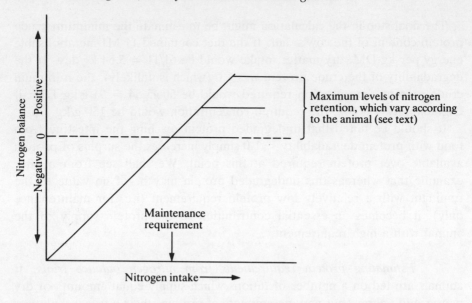

Fig. 14.2.   Stylized representation of the assessment of protein requirements for maintenance by means of nitrogen balance trials.

discussed in Chapter 13. A system based on these concepts was introduced by the Agricultural Research Council in 1980, and some of the factors used in the system were revised in 1984. Some interim standards based on the Agricultural Research Council system are included in the Appendix tables.

Standards for maintenance of ruminants are based on total endogenous losses of nitrogen, as illustrated in the example of pp. 298–9. As this example shows, the quantity of food protein required by the ruminant for the maintenance of its own tissues is likely to be about the quantity required by its rumen micro-organisms. The latter quantity is related to the metabolisable energy intake of the animal. This means that at or about the maintenance level of feeding, requirements for protein expressed as dietary concentrations depend on only two factors, the metabolisable energy concentration of the diet and the degradability of dietary protein (Table 14.4). The quantity of protein required in degraded form in the rumen for each 1 MJ of the diet is 8.3 g, so if the diet contains 9.2 MJ ME/kg DM (first line of Table 14.4), the concentration of rumen degradable protein will be 8.3 × 9.2 = 76 g/kg DM. If the dietary protein is completely degradable the concentration of protein required in the diet will also be 76 g/kg DM, but if the degradability of dietary protein is only 0.6, for example, the concentration of protein required rises to 127 g/kg DM.

For pigs and poultry, protein standards are expressed for maintenance and growth combined, and a discussion of them is delayed until later in this chapter.

TABLE 14.4 Dietary crude protein concentrations meeting the requirements of rumen micro-organisms and approximating to the maintenance requirement of the host animal (g/kg DM)
(Agricultural Research Council London, 1984, *The Nutrient Requirements of Ruminant Livestock, Supplement No. 1.* Commonwealth Agricultural Bureaux, Farnham Royal).

Metabolisability (q)	*ME (MJ/kg DM)*	*Degradability of dietary protein*			
		0.6	*0.7*	*0.8*	*1.0*
0.5	9.2	127	109	95	76
0.6	11.0	152	130	114	91
0.7	12.9	178	153	134	107

FEEDING STANDARDS FOR GROWTH AND WOOL PRODUCTION

Growth and development

The crudest but most common measure of growth in farm animals is change in liveweight. It is crude because changes in liveweight include changes in the weights of intestinal contents, which in ruminants may often account for 20 per cent of liveweight gain. The typical pattern of liveweight growth is shown in Fig. 14.3. During the foetal period and from birth to about puberty the rate of growth accelerates; after puberty it decelerates and reaches a very low value as the mature weight is approached. As animals grow they do not simply increase in weight and size—they also show what is termed development. By this we mean that the various parts of the body grow at different rates, so that its proportions change as the animal matures. The relationship of the weight of a part of the body to the weight of the whole can be described by an equation of the form:

$$y = bx^a$$
$$\text{or } \log y = \log b + a \log x$$

(where y = the weight of the part, x = the weight of the whole, a is known as the growth coefficient, and b is a constant). This equation was first proposed by J. S. Huxley in 1932 and is known as the allometric equation. The 'part' referred to may be a region of the body (such as the head or a commercial joint of meat), an individual organ or muscle, a dissectible tissue (bone, muscle or fat) or a chemical component (water, protein, ether extract or ash). Similarly, the 'whole' may be the whole body of the animal, its empty body, its carcass or possibly a fraction of the body such as the fat-

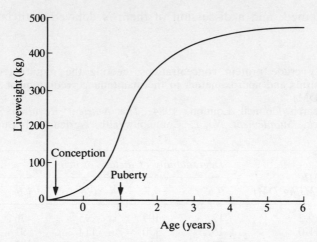

Fig. 14.3.   The typical sigmoid growth curve as it would appear for the dairy cow.

free body mass. In this chapter, which is concerned with nutrient requirements, the animal is considered mainly in terms of the chemical composition
(and energy value) of its whole or empty body. The nutritionist, however,
must concern himself also with the dissectible or anatomical composition of
the carcass, since these bear more relation to the meat value of animals.

The growth coefficient, a, is a measure of the rate of growth of the part
relative to the rate of growth of the whole. If it has a value greater than
unity, the part is growing relatively faster than the whole, and the part's
contribution to the whole is increasing; the part in question is thus described
as a late-maturing part of the body. Conversely, if its growth coefficient is
less than unity, a part is said to be early-maturing.

Figure 14.4 shows a logarithmic plot of the weight of water, protein and
fat, and the quantity of energy, in the bodies of sheep against empty body
weight. The value on each line is the growth coefficient for the component
represented. Protein and water have coefficients less than unity and are thus
early maturing components, whereas fat has a coefficient considerably
greater than unity and is thus late maturing. The energy content of the body
is largely determined by its fat content, so energy too has a high coefficient.

In Fig. 14.5 the weights of the dissected components of pig carcasses—
bone, muscle and fat—have been plotted against empty body weight. As
in Fig. 14.4, logarithmic scales have been used, and the growth coefficients
are shown. Bone appears as the earliest maturing tissue, followed by muscle
and then fat.

Developmental changes in body proportions are important in nutrition
because they influence the animal's nutrient requirements. For example,

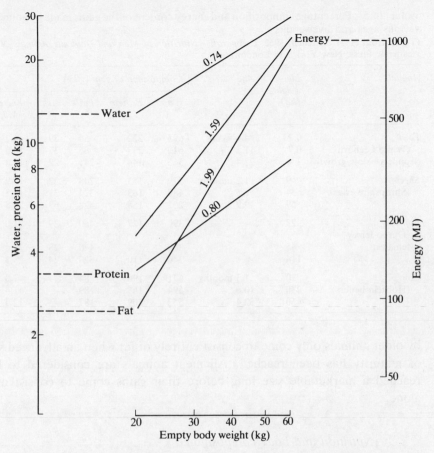

Fig. 14.4. Growth coefficients for water, protein, fat (ether extract) and energy in the whole empty body of sheep. (From J. T. Reid *et al.*, 1968. In *Body composition in Animals and Man*, Publ. Nat. Acad. Sci. No. 1598).

from the data used for Fig. 14.4 one may calculate that the empty body weight gains of sheep growing from 20 to 30 kg contain 133 g protein, 313 g fat and 16.3 MJ/kg. In contrast, the gains of a sheep growing from 50 to 60 kg are calculated to contain 113 g protein, 682 g fat and 26.0 MJ/kg.

The composition of gains made by several species is shown in Table 14.5. As an animal ages the general tendency is for there to be a progressive decline in the water content of its gains, a slight fall in protein content and a marked rise in fat content. These changes lead to a rise in the energy content of each unit of gain. The term 'fattening' is used in agricultural practice to describe the later stages of growth, but it is misleading to think of muscle and bone growth occurring as a separate and earlier phase. The gains of young animals almost invariably contain some fat, and those made

TABLE 14.5   Percentage composition and energy content of the gains made by animals of various ages and liveweights
(Taken from H. H. Mitchell, 1962. *Comparative Nutrition of Man and Domestic Animals*, Vol. 1. Academic Press, New York and London)

Animal	Live-weight (kg)	Age	Composition of gain (g/kg)				
			Water	Protein	Fat	Ash	Energy (MJ/kg)
Fowl	0.23	4.4 weeks	695	222	56	39	6.2
(White Leghorn	0.7	11.5 "	619	233	86	37	10.0
pullets—slow growth)	1.4	22.4 "	565	144	251	22	12.8
Sheep	9	1.2 months	579	153	248	22	13.9
(Shropshire ewes)	34	6.5 "	480	163	324	31	16.5
	59	19.9 "	251	158	528	63	20.8
Pig	23	—	390	127	460	29	21.0
(Duroc-Jersey							
females)	45	—	380	124	470	28	21.4
	114	—	340	110	520	24	23.3
Cow	70	1.3 months	671	190	84	—	7.8
(Holstein heifers)	230	10.6 "	594	165	189	—	11.4
	450	32.4 "	552	209	187	—	12.3

by older animals only come to consist entirely of fat when an advanced stage of maturity has been reached. All meat animals are considered to have reached a marketable size long before their gains come to consist of fat alone.

Nutrition and body composition

If Huxley's allometric principle applied exactly to the individual members of a species (the pig, for example) the farmer would be able to produce pigs having a specified carcass composition simply by growing them to a specified weight. A rapid survey of pigs or other farm animals (or of our own species) soon convinces us, however, that body composition is not exclusively determined by body weight, and hence that quality control in meat production is not as simple as it might be. Breeding and sex are recognised as factors influencing body composition but nutrition is particularly important in this regard.

The growth curve of an animal is obviously dependent on its level of feeding. If the level is high, growth is rapid and the animal reaches a specified weight at an early age. A reduction in the level of feeding will cause the curve to flatten and perhaps even to reverse its slope (i.e. the animal loses weight). Thus the shape of the curve may depart so widely from the idealised form shown in Fig. 14.3, that one is caused to wonder whether the composition of an animal depends solely on its weight or whether the route by which that weight has been reached is also important. Qualitative variations in nutrition give cause for similar speculation; it seems unlikely that an animal grown to a specified weight on a diet grossly deficient in protein will have the same composition as one which has received adequate protein.

Experiments have demonstrated the *general* validity of the allometric principle but have at the same time shown that specific departures from it may be large enough to be important in agriculture. An extreme example is provided by pigs that were first given such limited quantities of food that they weighed only 5–6 kg at one year old and were then fed liberally, and grew rapidly. When they reached their mature weight, these animals were remarkably similar in composition to animals whose growth had not been interrupted. However, they were not exactly the same, being smaller skeletally and fatter than normal pigs.

A further example of the general applicability of the allometric principle is provided by Fig. 14.5. But this figure also provides an example of specific

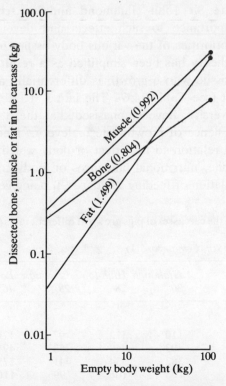

Fig. 14.5. Growth coefficients for bone, muscle and fat in the carcasses of pigs. (Calculated from the data of C. P. McMeekan and R. W. Pomeroy by N. M. Tulloh, 1964. In *Carcass, Composition and Appraisal of Meat Animals*. Commonwealth Scientific and Industrial Research Organisation, Australia.)

departures from the principle. Twenty of the pigs represented came from an experiment in which animals were grown to 90 kg liveweight at low or high rates of gain. Some of the pigs were changed from one rate to the other half-way through the experiment (high-low and low-high) and others were kept on the same rate throughout (high-high and low-low). Table 14.6 shows the composition of their carcasses.

The greatest differences between the pigs were in their dissectible fat content. The pigs which grew fast throughout (high-high) had a fat content 108g/kg higher than those which grew slowly throughout (low-low), and there was a difference of the same order between the low-high and high-low groups. But the lower part of Table 14.6 shows that if the effects of differences in fat content are removed, the groups differed to only a small extent in their proportions of other tissues. In effect, the pigs that grew faster were small but fat at 90 kg, whereas those that grew more slowly were larger-framed but thinner.

The significance of the effects of nutrition on body composition remains a matter for debate. The late Sir John Hammond and his school at Cambridge attached great importance to such effects and developed a complex theory regarding the priorities of the various body tissues for food nutrients. More recently this theory has been simplified as a result of the demonstration that the main tissue whose growth is differentially affected by variations in nutrition is fat (see Table 14.6). The late J. T. Reid, and his colleagues at Cornell University, have emphasised that the effect on body composition of age (and hence of growth rate or level of feeding) is very small when considered in relation to the effect of body weight. Thus the debate continues. Meanwhile, nutritional influences on body composition are recognised in the calculation of feeding standards, it being generally

TABLE 14.6   Composition (g/kg) of the carcasses of pigs grown at different rates (see text) and killed at 90 kg
(After C. P. McMeekan, 1940. *J. agric. Sci., Camb.*, **30**, 511)

Growth rate: Age at slaughter (weeks):	High-high 20	High-low 28	Low-high 28	Low-low 46
Composition of whole carcass				
Bone	110	112	97	124
Muscle	403	449	363	491
Fat	383	334	441	275
Skin, etc.	105	106	99	110
Composition of fat-free carcass				
Bone	178	168	174	171
Muscle	653	674	649	677
Skin, etc.	170	160	177	152

assumed that the gains made by faster-growing animals contain greater concentrations of fat and energy, and lesser concentrations of water, protein and ash.

Energy requirements for growth

Ruminants. The UK Agricultural Research Council has collated data for large numbers of cattle and sheep that were slaughtered at various ages and weights and subjected to physical dissection and chemical analysis. For both species, the main determinant of the energy content of liveweight gain was found to be liveweight. For cattle, additional influences of rate of gain, mature size and sex were recognised and incorporated in feeding standards, some of which are illustrated in Fig. 14.6.

Figure 14.6 shows first the major effect of liveweight on energy content of gain. For example, in steers of a medium-sized breed growing rapidly (the middle line of Fig. 14.6a), energy content of gain rises from 9 MJ/kg at 100 kg liveweight to 19 MJ/kg at 400 kg liveweight. However, steers of the same type growing slowly (middle line of Fig. 14.6b) make gains containing less fat and hence less energy; for example, at 400 kg liveweight their gains contain 17 MJ/kg. The influence of breed is a reflection of mature size. For example (Fig. 14.6a), a steer of a small breed such as an Aberdeen

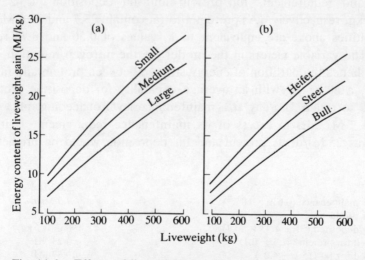

Fig. 14.6. Effects of liveweight, mature size, rate of gain and sex on the energy contents of gains made by cattle. (Agricultural Research Council, 1980. *The Nutrient Requirements of Ruminant Livestock.* Commonwealth Agricultural Bureaux.) (*a*) Steers of small, medium and large breeds growing at 1.0 kg/day. (*b*) Heifer, steer and bull of a medium-sized breed, all growing at 0.5 kg/day.

Angus is approaching its mature weight at 400 kg liveweight, and its gains contain a high concentration of fat and hence of energy (21.5 MJ/kg if it is gaining 1.0 kg/day). A steer of a large breed, such as a Charolais, is far from mature at 400 kg and its gains at that weight therefore contain less fat and less energy (15.9 MJ/kg if it is gaining 1.0 kg/day). The differences between the sexes (Fig. 14.6b) are that, for animals of the same weight and breed and growing at the same rate, the gains of heifers have a higher concentration of energy than those of steers, and the gains of steers have a higher concentration than those of bulls. Thus at any given weight, animals having gains of the highest energy concentration will be fast-growing heifers of small breeds, while those with gains of the lowest energy concentration will be slow-growing bulls of large breeds.

The data for sheep showed comparable effects of liveweight and sex on the energy content of gains, but no clear effect of rate of gain. Breed differences were small, except that Merinos tended to have gains containing higher concentrations of energy than those of other breeds.

Pigs. In the UK, the Agricultural Research Council has devised energy feeding standards for growing pigs by using a factorial model. The model is designed principally for performance prediction (i.e. predicting growth rate for a given energy intake). The fixed factors in the model are an assumed maintenance requirement of 0.639 MJ ME/kg $W^{0.67}$/day (see p. 288) and requirements for protein and fat deposition of 42.3 and 53.5 MJ/kg, respectively. (As protein and fat contain 23.7 and 39.6 MJ GE, the quantities above are equivalent to k_f values of 0.56 and 0.74 respectively.) The variable factors in the model are the nitrogen retention of the pig, and hence the partition of energy storage between protein and fat. For example, a 60 kg pig with an average capability for depositing protein is reckoned to retain 31 g/day at a maintenance level of feeding, plus 4.43 g for each 1 MJ ME in excess of its maintenance requirement. If its ME intake was 25 MJ/d, its protein and fat deposition would be predicted as follows:

ME intake	25.0 MJ/d
ME for maintenance ($0.639 \times 60^{0.67}$)	9.9 MJ/d
ME for production	15.1 MJ/d
Protein deposited ($31 + 4.43 \times 15.1$)	98 g (0.1 kg)
ME used for protein (42.3×0.1)	4.23 MJ
ME used for fat ($15.1 - 4.23$)	10.87 MJ
Fat deposited ($10.87/53.5$)	0.203 kg

The weight of lean tissue deposited can be estimated from its known protein content (213 g/kg; see p. 232) to be $98/0.213 = 460$ g (0.46 kg). Total empty body gain is therefore $0.46 + 0.21 = 0.67$ kg/day. The energy content of gain is calculated to be $(0.1 \times 23.7) + (0.20 \times 39.6) = 10.3$ MJ, or 15.4 MJ/kg.

For practical use, information of the kind provided by this model may be used to devise feeding scales for pigs that allow rapid growth without excessive fat deposition. The fastest growth is achieved by allowing the pigs to eat to appetite, and some types of pig can eat to appetite from birth until slaughter at 'bacon weight' (90 kg) without laying down too much fat. With many pigs, however, food intake must be restricted during growth from when they reach 45 kg, or even 25 kg, if a sufficiently lean carcass is to be obtained. Feeding scales have therefore been devised which stipulate the quantities of food or digestible energy to be given at different liveweights. A typical scale for a diet containing 12.5 MJ digestible energy per kg would rise gradually from 1.2 kg at a liveweight of 20 kg to 2.2 kg at 50 kg liveweight and 2.4 kg at 90 kg. Such restriction of intake slows down the rate of growth, but it reduces the fat content of the carcass and makes the carcass more acceptable to the bacon curer.

Poultry. With the possible exception of birds reared for breeding (see Ch. 15), growing poultry are normally fed to appetite, and feeding standards for them are therefore expressed, not as amounts of nutrients, but as the nutrient proportions of the diet (Appendix Table 14).

As explained in Chapter 16, the quantities of food eaten by poultry are inversely related to the concentration of energy in their diets. This means that if the energy concentration of a diet is increased without change in the concentration of, for example, protein, and birds begin to eat less of the diet, then although their energy intake may remain approximately at the former level, their intake of protein will fall. The birds may then be deficient in protein. To generalise: a nutrient concentration which is adequate for a diet of low energy content may be inadequate for one richer in energy. It follows that feeding standards stated as nutrient concentrations are satisfactory only when applied to diets with a particular energy concentration. The standards of Appendix Table 14 for chicks up to six weeks of age apply to diets containing 11.5 MJ metabolisable energy per kg, and need to be adjusted for diets containing more or less energy. Some adjustments are discussed later in this chapter (p. 312).

Protein requirements for growth

Ruminants. In feeding standards for growing animals, protein requirements for growth are usually incorporated into a single value for maintenance and growth combined. Older standards, expressed in terms of digestible crude protein, were often based on the results of feeding trials in which animals were fed on rations supplying differing amounts of protein and their responses measured in terms either of liveweight gain or of nitrogen retention. In such trials, rations supplying equally liberal quantities of energy, minerals and vitamins, and varying but more restricted quantities of protein, are compared; the minimum protein level giving maximum growth or nitrogen retention is taken as the estimate of the requirement. An example of such a nitrogen balance trial is shown in Fig. 14.7. Calves were fed on rations supplying from 93 to 230 g digestible crude protein per day, and maximum nitrogen retention was achieved with a minimum intake of about 190 g digestible crude protein per day.

In the factorial system of estimating protein requirements described earlier (p. 298), the protein content of liveweight gain is added to the estimates of endogenous losses. For example, suppose a 20-kg lamb growing at 0.2 kg/day is estimated to have endogenous losses of nitrogen equivalent to 21 g protein day, and suppose its liveweight gains are found from serial slaughter experiments to contain 170 g protein/kg. Its net requirement for protein will therefore be:

$$21 + (0.2 \times 170) = 55 \text{ g}$$

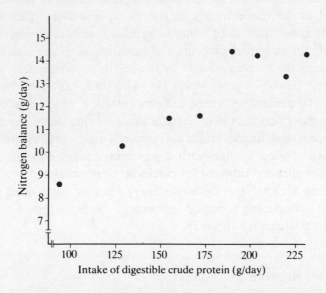

Fig. 14.7.   Determining the protein requirement of calves of 50 kg liveweight and gaining about 0.35 kg per day from their nitrogen balance. (Plotted from the results of F. G. Whitelaw, T. R. Preston and R. D. Ndumbe, 1961. *Anim. Prod.*, 3, 121.)

The factors used in the previous calculation (biological value, 0.80; digestibility of protein 0.85) are then used to calculate the quantity of protein
required to enter the small intestine:

$$55/(0.80 \times 0.85) = 81 \text{ g}$$

If the animal's requirement for metabolisable energy is found from
appropriate tables to be 8.4 MJ, the quantity of microbial protein supplied
to the small intestine will be:

$$8.4 \times 8.3 = 70 \text{ g}$$

and of this $0.8 \times 70 = 56$ g will be true protein (and the remainder, nucleic
acids). Thus, $81 - 56 = 25$ g of the protein entering the small intestine must
be supplied as undegraded food protein. The animal's minimum requirement for protein will therefore be $25 + 70 = 95$ g and this will be sufficient
only if degradability is $70/95 = 0.74$.

If the actual degradability (dg) is less than the optimum degradability
($_d$g; in this case 0.74), then the quantity of dietary protein will need to be
multiplied by $_d$g/dg. For example, if the actual degradability was 0.10 below
the optimum (i.e. 0.64), the dietary requirement for protein would be raised
from 95 to 110 g/day. In practice, however, it might be more appropriate
to raise the degradability of dietary protein to the optimum, for example
by adding urea, which has a degradability of 1.0.

If the actual degradability is greater than the optimum, then the rumen
micro-organisms will be supplied with an excess of protein that will be lost
from the rumen as ammonia. Furthermore, the remaining undegradable
protein will be insufficient to supplement the supply to the small intestine
of microbial protein. This problem can be overcome either by raising the
allowance of dietary protein, or by reducing protein degradability. If the
former method is adopted, the requirement can be recalculated by multiplying by $(1 - {_d}g)/(1 - dg)$. For example, if the actual degradability was
0.10 above the optimum (i.e. 0.84), the requirement would rise from 95 to
154 g/day.

The previous example of the factorial calculation of protein requirement
(p. 298) was for maintenance only, and the animal's requirement was equal
to that of its rumen micro-organisms. In contrast, a fast-growing lamb has
a net requirement for protein that is high in relation to its energy requirement. Consequently, protein synthesised in the rumen is not sufficient to
meet the animal's requirement, and there is therefore a substantial
additional requirement for undegradable food protein. For the animal at
maintenance, optimal degradability of dietary protein was 1.0, whereas for
the growing lamb optimum degradability was 0.74. A further point to note
is that when actual degradability is higher than optimal degradability,

protein intake has to be much greater to supply sufficient undegradable protein.

Pigs and poultry. In addition to a general need for protein, non-ruminant animals have specific dietary requirements for ten or so indispensable amino acids. During the past 30 years many experiments have been carried out to determine quantitative requirements for indispensable amino acids, and feeding standards expressed in terms of total protein are normally supplemented by (or even completely replaced by) standards for all or some of these amino acids.

Requirements may also be stated in terms of 'ideal protein' (i.e. protein containing indispensable amino acids in exactly the proportions required by the animal; see Table 13.5).

Standards expressed as total protein alone may nevertheless be justified for animals whose diets are compounded from a small number of feeds of known amino acid composition. For example, in the USA growing pigs are fed largely on combinations of maize and soya bean meal, and in Britain pig diets are based on barley, soya bean meal and fish meal. Requirements of pigs for total protein are determined in feeding trials in which growth rate is the main criterion of adequacy. But a protein level high enough to give maximum growth may not promote optimum carcass composition, so additional measurements of carcass quality, and sometimes of nitrogen balance, are commonly included in the feeding trials. For both pigs and poultry, total protein requirements are stated as a concentration of protein in the diet. The proportion required falls as the animals become older and store relatively less protein and more fat in each unit of gain.

It must be emphasised that standards expressed only as total protein are of limited value in the feeding of pigs and poultry unless there is some stipulation as to the protein quality of the diet. In pig diets, protein quality is often limited by deficiencies of one or two of the indispensable amino acids, and if standards for total protein are accompanied by standards for these acids, they become much more meaningful. In addition, qualifying a standard for total protein by stating that certain proportions of it must be in the form of these critical amino acids may allow the standard to be set at a lower level than formerly.

The requirements of pigs and poultry for indispensable amino acids. The requirement for an indispensable amino acid is assessed by giving diets containing different levels of the acid in question but equal levels of the remaining acids and measuring growth or nitrogen retention. Diets differing in their content of just one amino acid may be prepared from foods naturally deficient in it, to which are added graded amounts of the pure acid. Figure 14.8 shows the outcome of an experiment with chicks in which a diet low in lysine was supplemented in this way so as to give diets ranging in lysine content from 7 to 14 g/kg. The lysine requirement of the chick was concluded from this experiment to be 11 g/kg of the diet. In other experiments it has been found more convenient to use synthetic diets in which much or all of the nitrogen is in the form of the pure amino acids.

Standards for the indispensable amino acid requirements of chicks, turkey poults and young pigs have now been devised, and some of these are given in Appendix Tables 13 and 14. At present they must be regarded only as approximate standards, for there are considerable complications to defining requirements for amino acids. Thus requirements are influenced by interactions among the indispensable amino acids themselves, between indispensable and dispensable acids and between amino acids and other nutrients. For chicks, the requirement for glycine is increased by low concentrations in the diet of methionine, arginine or B-complex vitamins. Interactions may be due to the conversion of one amino acid into another. If cystine, or its metabolically active form cysteine, is deficient in the diet it is synthesised by the animal from methionine. The requirement for methionine is therefore partially dependent on the cystine (or cysteine) content of the diet, and the two acids are usually considered together (i.e. the requirement is stated for methionine plus cystine). It should be noted, however, that the acids are not mutually interconvertible; methionine is not synthesised from cystine and therefore a part of the total requirement must always be met by methionine. Phenylalanine and the dispensable acid tyro-

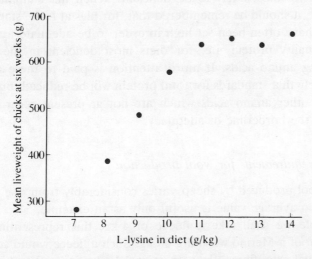

Fig. 14.8. Growth of chicks given diets containing different levels of lysine. (Plotted from the data of H. M. Edwards, L. C. Norris and G. F. Heuser, 1956. *Poultry Sci.*, **35**, 385.)

sine, have a similar relationship. In the chick, glycine and serine are interconvertible.

Further complications are introduced by relationships between amino acid requirements and the total protein content of the diet. If the latter is altered to compensate for a change in energy content, then the amino acid requirements will also change. For this reason, amino acid requirements are sometimes expressed as g amino acid/MJ metabolisable energy.

The application in feeding practice of standards for as many as ten or eleven amino acids is likely to be rather laborious. How necessary is it, then, to consider all of them when devising diets for pigs and poultry? In theory there is almost unlimited scope for adjusting the proportions of dietary constituents (including synthetic acids) until the indispensable amino acid contents of the diet exactly equal those demanded by the standards. In practice it is usually found that the amino acid mixtures provided by the diet are out of proportion, and hence inefficiently utilised, because one or two amino acids are very markedly deficient. In consequence, the degree of success achieved in applying the standards depends particularly on whether or not the requirements of these 'limiting' amino acids are met. By comparing requirements for amino acids with the amino acid contents of typical diets, it can be shown that for pigs the acid likely to be most deficient is lysine. For chicks the 'first-limiting' amino acid is most commonly methionine, although lysine and perhaps arginine may also be deficient.

In practice therefore it may be sufficient to ensure that diets for pigs and poultry contain, first, adequate total protein, and second, adequate contents of those amino acids most likely to be deficient. When this simplification is made, however, it should be remembered that, for pigs at least, standards for total protein have often been set high in order to be adequate for diets containing low-quality protein, i.e. for diets most deficient in the more frequently limiting amino acids. If more attention is paid to these amino acids it seems likely that standards for total protein will be reduced, bringing into prominence other amino acids which are not at present regarded as likely to be near the borderline of adequacy.

Nutrient requirements for wool production

The weight of wool produced by sheep varies considerably from one breed to another, and an average value is useful only as an example.

For an example we shall take a fleece of 4 kg, this representing the annual production of a Merino weighing 50 kg. Such a fleece would contain about 3 kg of actual wool fibres, the remaining 1 kg being wool wax, suint, dirt and water. Wool wax is produced by the sebaceous glands, and consists mainly of esters of cholesterol and other alcohols with the acids normally found in glycerides and other aliphatic acids. Suint, the secretion of the sudoriferous glands, is a mixture of inorganic salts, potassium soaps and potassium salts of lower fatty acids.

The wool fibres consist almost entirely of the protein, wool keratin. To grow in one year a fleece containing 3 kg protein the sheep would need to deposit a daily average (i.e. ignoring seasonal fluctuations in wool growth) of about 8 g protein or 1.3 g nitrogen. If this latter figure is compared with the 6.6 g nitrogen which a sheep of 50 kg might lose daily as endogenous nitrogen, it will be seen that in proportion to its requirement for maintenance the sheep's nitrogen requirement for wool growth is small. These figures however do not tell the whole story, since the efficiency with which absorbed amino acids are used for wool synthesis is likely to be much less than that with which they are used for maintenance. Keratin is characterised by its high content of the sulphur-containing amino acid, cystine, which although not an indispensable amino acid is synthesised from the indispensable acid, methionine. The efficiency with which food protein can be converted into wool is therefore likely to depend on their respective proportions of cystine and methionine. Keratin contains 100–120 g/kg of these acids, compared with the 20–30 g/kg found in plant proteins and in the microbial proteins synthesised in the rumen, and so the biological value of food protein for wool growth is likely to be not greater than 0.3.

With regard to energy requirements for wool growth, a sheep producing a fleece of 4 kg would retain about 0.23 MJ/day. The efficiency with which metabolisable energy is transferred to wool is not known precisely but is estimated to be about 0.18. Thus the sheep would require 0.23/0.18 = 1.3 MJ ME/day for wool growth and this may be compared with its maintenance requirement of approximately 6 MJ ME/day.

While nutrient requirements for wool production are, even for protein, quantitatively small, it must not be supposed that maximum wool growth will take place at a level of nutrition only slightly above the maintenance level. Wool growth reflects the general level of nutrition of the sheep. At sub-maintenance levels, when the sheep is losing weight, its wool continues to grow, although slowly. As the plane of nutrition improves and the sheep gains in weight, so wool growth too increases. There appears to be a maximum rate of growth for wool, varying from sheep to sheep within a range as great as 5 to 40 g day.

The dependence of wool growth rate on the plane of nutrition (i.e. energy intake) of the sheep is due in part to the association between energy intake and the synthesis of microbial protein. The real determinant of wool growth rate is likely to be the quantity of protein digested and absorbed in the small intestine of the sheep, and it has been shown, for example, that a Merino must absorb 120–150 g protein day to achieve its maximum rate of wool growth. In a ewe with a metabolisable energy intake of 12 MJ/day (i.e. twice its maintenance requirement), only 101 g microbial protein would be synthesised per day, and only $101 \times 0.8 \times 0.85 = 69$ g would be absorbed as amino acids. To achieve maximum growth of wool the sheep is therefore dependent on a good source of undegraded food protein. In practice this is likely to be supplied by the consumption of large quantities of protein in pasture herbage. This may be relatively highly degradable but can still supply much undegradable protein. For example, a ewe might consume 250 g protein/day, of which 0.3 (75 g) would be undegraded and $0.85 \times 75 = 64$ g would be absorbed in the small intestine. Nevertheless, wool growth in sheep is considerably increased by protein supplements protected from rumen degradation, such as formalin-treated casein. As would be anticipated, the most effective of such supplements are those rich in the sulphur-containing amino acids.

Wool quality is influenced by the nutrition of the sheep. High levels of nutrition increase the diameter of the fibres, and it is significant that the finer wools come from the nutritionally less favourable areas of land. Periods of starvation may cause an abrupt reduction in wool growth; this leaves a weak point in each fibre and is responsible for the fault in fleeces with the self-explanatory name of 'break'. An early sign of copper deficiency in sheep is a loss of 'crimp' or waviness in wool; this is accompanied by a general deterioration in quality, the wool losing its elasticity and its affinity for dyes.

MINERAL AND VITAMIN REQUIREMENTS FOR MAINTENANCE AND GROWTH

This section is concerned with the general principles governing the determination of feeding standards for minerals and vitamins. No attempt is made to discuss individual nutrients, since to do so would result in duplicating the material of Chapters 5 and 6.

Minerals

Animals deprived of a dietary supply of mineral elements continue to excrete these nutrients. Those elements which occur in the body mainly as constituents of organic compounds, such as the iron of haemoglobin and the iodine of thyroxine, are released from these compounds when they are expended or 'worn out'. To a variable but often large extent the elements so liberated are re-utilised, but re-utilisation is never complete and a proportion of each mineral will be lost from the body in the faeces and urine and through the skin. Of those elements which occur in inorganic forms, such as calcium, sodium, potassium and magnesium, there are losses in the urine such as those arising from the maintenance of the acid-base balance of the animal, and losses in the faeces occurring through secretions into the gut which are not reabsorbed. Because they suffer all these endogenous losses, animals require minerals for maintenance.

Endogenous losses of minerals are often small, however, in relation to the mineral content of the body. A pig of 30 kg, whose body contains about 230 g calcium, suffers endogenous losses amounting to 0.9 g of the element per day, and thus needs to replace daily about 0.4 per cent of its body calcium. The same pig contains about 40 g sodium and needs to replace 0.036 g or 0.09 per cent each day. In contrast, the pig would need to replace about 0.7 per cent of its body nitrogen each day.

The approaches used in assessing requirements for minerals are the same as those used in determining standards for energy and protein. A 'theoretical' approach is provided by the factorial method, and 'practical' estimates of requirement by balance and growth trials. Since the mineral contents of foods are expressed as the total or gross amounts present, requirements are stated in the same terms. The standards must therefore make allowance for the differences in mineral availability that occur between different species and age classes of animal (see Ch. 10).

Factorial estimates of mineral requirements. The *net requirement* of a mineral element for maintenance plus growth is calculated as the sum of the endogenous losses and the quantity retained. To determine the *dietary requirement*, the net requirement is divided by an average value for availability (expressed as a decimal). For example, a heifer of 300 kg liveweight gaining 0.5 kg/day might have endogenous losses of calcium of 5 g/day and retain 6 g day: its net requirement would therefore be 11 g calcium day. For an animal of this weight an average value for the availability of calcium would be about 0.65 and the animal's dietary requirement for calcium would therefore be $11/0.65 = 17$ g/day.

The difficulties involved in the factorial approach to mineral requirements are the same as those associated with factorial estimates of protein requirement. While the mineral composition of liveweight gains may readily (if laboriously) be determined by carcass analysis, the assessment of endogenous losses, and hence also of availability, is more difficult. Diets for ruminants which are completely free of an element are particularly difficult to prepare. Perhaps because of these difficulties of technique, theoretical estimates of mineral requirements often do not agree with practical estimates.

Growth and balance trials. When mineral requirements are assessed by comparing the effects on the animal of diets supplying different quantities of an element, the great problem is to establish a satisfactory criterion of adequacy. That intake which is sufficient to prevent clinical signs of deficiency may be insufficient to support maximum growth. Of the bone-forming elements, the intake which gives a maximum rate of liveweight gain may yet be inadequate if judged by the strength of bone produced. The position is further complicated by the mineral reserves of the animal. If these are large at the beginning of an experiment of short duration, they may be sufficient to allow normal health and growth even if the diet is inadequate, and it is therefore desirable that mineral balance should be determined directly or assessed indirectly by analysis of selected tissues. Even balance trials however may be difficult to interpret, since if the element is one for which the animal has great storage capacity, a dietary allowance which promotes less than maximum retention may still be quite adequate. In long-term experiments, then, such as those continuing for one or more annual cycles of the dairy cow, health and productivity alone may provide reliable indications of minimum mineral requirements. For growing animals on the other hand trials must usually be of shorter duration, and measurements of liveweight gain should be supplemented by measurements of mineral retention.

Present standards. The requirements for minerals given in the Appendix Tables are based partly on factorial calculations and partly on the results of feeding trials. For all species the elements which have been the subject of most investigation have been calcium and phosphorus, these being the elements most likely to be deficient in the diet. *Net* requirements for calcium and phosphorus, relative to requirements for other nutrients, tend to decline as the animal ages and bone growth slows down, but *dietary* requirements decline less with age because the availability of these elements is reduced as the animal matures. It should be noted that within small ranges in weight mineral requirements are considered to be proportional to liveweight, not to metabolic liveweight.

Vitamins

There are no endogenous losses on which to base factorial estimates of vitamin requirements, and standards must therefore be derived from the results of feeding trials. As in the assessment of mineral requirements, it may be difficult in these trials to select the criteria by which the allowances compared are judged adequate or inadequate. The main criteria are again growth rate and freedom from signs of deficiency, deficiency being detected either by visual examination of the animals or by physiological tests such as determination of the vitamin levels in the blood. Vitamin storage also may be assessed, either from actual analyses of tissues or from such indirect evidence of tissue saturation as is provided by excretion of the vitamin in the urine. The difficulties involved in assessing requirements are illustrated in Table 14.7, which shows that the apparent requirement can vary widely according to the criterion of adequacy preferred.

In practice, vitamin allowances must be at least high enough to prevent signs of deficiency and not to restrict the rate of growth. Higher allowances, which promote storage or higher circulatory levels of the vitamin, are justifiable only if they can be shown in the long term to influence the health and productivity of animals in a way which does not become apparent in the short-term experiments by which requirements must often be assessed. Some storage can be justified, since in most animals there may be fluctuations in both the need for the vitamins and the supply of them, and allowances are usually set at levels permitting the maintenance of stores.

Requirements for the fat-soluble vitamins, in older animals at least, are considered to be proportional to body weight. Those for the B group however, which are more intimately concerned with metabolism, vary with food intake in general, or in some cases with intake of specific nutrients. Thus the requirement for thiamin which is particularly concerned with carbohydrate metabolism, varies according to the relative importance of carbohydrate and fat in the diet. For similar reasons riboflavin requirements are increased by a high protein intake. Requirements vary also according to the extent to which B vitamins are synthesised in the alimentary tract. In herbivores sufficient synthesis occurs to make the animal independent of dietary supplies. In pigs and poultry considerable synthesis takes place in the lower gut, but the vitamins produced may fail to be absorbed: the contribution of the intestinal synthesis may then depend on whether the animals are free to practise coprophagy (the eating of faeces). A final point of importance concerns the availability of vitamins. Requirements are often determined from diets containing synthetic sources of the vitamins, whose availability may well be higher than that of the vitamins of natural foods. Although little is known about vitamin availability, a well documented

TABLE 14.7   The vitamin A requirement of calves
(From the data of J. M. Lewis and L. T. Wilson, 1945. *J. Nutr.*, **30**, 467)

Minimum requirement for:	*Vitamin A (i.u./kg liveweight/day)*
Prevention of night blindness	32
Optimal growth	64
Limited storage of vitamin A	250
Maximal blood level of vitamin A	500

instance of non- availability is provided by the nicotinic acid of cereals, some of which is in a bound form not available to pigs.

FURTHER READING

J. L. Black and P. J Reis (eds), 1979. *Physiological and Environmental Limitations to Wool Growth*. University of New England, Armidale, NSW, Australia.

K. N. Boorman and B. J. Wilson (eds), 1977. *Growth and Poultry Meat Production*. Longman, Edinburgh.

S. Brody, 1945. *Bioenergetics and Growth*. Reinhold, New York.

G. A. Lodge and G. E. Lamming, 1968. *Growth and Development of Mammals*. Butterworths, London.

H. H. Mitchell, 1962, 1964. *Comparative Nutrition of Man and the Domestic Animals* 2 vols. Academic Press, New York and London.

J. T. Reid (ed.), 1968. *Body Composition in Animals and Man*. Publ. National Research Council, Washington, No. 1598.

See also the publications listed in the Appendix.

15

Feeding standards for reproduction and lactation

The reproductive cycle may be considered to consist of three phases: the first phase, which is important in both sexes, comprises the production of ova and spermatozoa, and is dealt with in the first part of this chapter. Nutrient requirements for these processes in mammals are small compared with the requirements for egg production in birds, and the feeding of laying poultry is therefore separately treated in the second part of this chapter. Nutrient requirements for the second phase of the cycle, pregnancy, are discussed in the third part of the chapter, while the final section deals with the third phase, lactation.

In the female mammal, the quantities of nutrients required in excess of those needed for maintenance are small for the first phase, moderate for the second and large for the third. Consequently, nutrient requirements fluctuate considerably during the reproductive cycle, especially when there is an interval between weaning and the next conception. Under natural conditions, such fluctuations in nutrient demand are partly matched with fluctuations in the food supply, but both in the wild and in intensive animal husbandry the mother often has to act as a buffer by depleting her body reserves in times of high demand and poor supply and restoring them when conditions are more favourable. However, in intensive husbandry, opportunities for the repletion of reserves are limited because one reproductive cycle follows closely—or even overlaps—the previous cycle. Thus dairy cows may conceive when producing 30 kg of milk per day, and pigs (and now some sheep) are expected to complete two reproductive cycles a year. In these conditions feeding standards must be applied more exactly and less reliance placed on body reserves.

REPRODUCTION

Two general points regarding nutrition and reproduction may be made at this stage. The first is that many of the effects of nutrition on reproductive performance are transmitted via the endocrine system, particularly via the hormones produced in the anterior pituitary. In a few instances the effect of a dietary deficiency can be related directly to a reduced output of a particular hormone, and will be corrected as successfully by hormone therapy as by improving the diet. In others the involvement of the endocrine system is less easily demonstrated; the effect on reproduction may be a consequence of a change in the rate at which a hormone is destroyed rather than the rate at which it is secreted, or of an altered sensitivity to the hormone in the organ responding to it.

The second general point is that in farm animals the effects of nutrition on reproduction are known only very imperfectly. The research needed is difficult to carry out because the response to a diet is often slow to develop, and experiments must therefore be of long duration, perhaps several generations. Many animals may be needed if, for example, the fertility of a number of bulls has to be tested by mating each to 20–30 cows.

EFFECTS OF NUTRITION ON THE INITIATION AND MAINTENANCE OF REPRODUCTIVE ABILITY

Puberty in cattle is markedly influenced by the level of nutrition at which animals have been reared. In general terms, the faster an animal grows, the earlier will it reach sexual maturity. In cattle, puberty occurs at a particular liveweight or body size rather than at a fixed age. This is illustrated in Table

TABLE 15.1   Age and size at puberty of Holstein cattle reared on different planes of nutrition

Sex	Plane of nutrition (per cent. of accepted standard for TDN)	At puberty		
		Age (weeks)	Weight (kg)	Height at withers (cm)
Females*	High (129)	37	270	108
	Medium (93)	49	271	113
	Low (61)	72	241	113
Males†	High (150)	37	292	116
	Medium (100)	43	262	116
	Low (66)	51	236	114

*From A. M. Sorenson *et al.*, 1959. *Bull. Cornell Univ. agric. Exp. Stn*, No. 936.
†From R. W. Bratton *et al.*, 1959. *Bull. Cornell Univ. agric. Exp. Stn*, No. 940.

15.1, which shows the effects of three planes of nutrition on the initiation of reproductive ability in dairy cattle. Although in both sexes there were considerable differences in age at puberty between the three treatments, differences in liveweight and in body size (as reflected in the measurement of height at withers) were much smaller. The attainment of puberty in sheep is complicated by their seasonal breeding pattern. Spring-born ewe lambs that are well nourished will reach puberty in the early autumn of the same year. Moderately fed lambs will also reach puberty in the same year, but later in the breeding season and at a lower liveweight. Poorly fed lambs will fail to come into oestrus until the following breeding season (i.e. at 18 months of age).

In pigs, on the other hand, high planes of nutrition do not advance puberty to any marked extent. The primary determinants of puberty in gilts, are age (170–220 days), breed (crossbred gilts reach puberty about 20 days before purebreds), and the age at which gilt meets boar (a sudden meeting after about 165 days of age may induce first oestrus).

In practice, the factor which decides when an animal is to be first used for breeding is body size, and at puberty animals are usually considered to be too small for breeding. Thus although heifers of the larger dairy breeds may be capable of conceiving at 7 months of age, they are not normally mated until they are at least 15 months old. The tendency today is for cattle, sheep and pigs of both sexes to be mated when relatively young, which means that in the female the nutrient demands of pregnancy are added to those of growth. Inadequate nutrition during pregnancy is liable to retard foetal growth and to delay the attainment of mature size by the mother. Incomplete skeletal development is particularly dangerous because it may lead to difficulties at parturition.

Rapid growth and the earlier attainment of a size appropriate to breeding has the economic advantage of reducing the non-productive part of the animal's life. With meat-producing animals a further advantage is that a high plane of nutrition in early life allows the selection for breeding purposes of the individuals which respond to liberal feeding most favourably in terms of growth, and which may therefore be expected to produce fast-growing offspring. But there are also some disadvantages of rapid growth in breeding stock, especially if there is excessive fat deposition. In dairy cattle, fatness in early life may prejudice the development of milk-secreting tissue, and there is also some evidence that rapid early growth reduces the useful life of cows. Over fat gilts do not mate as readily as normal animals, and during pregnancy may suffer more embryonic mortality. The rearing of breeding stock is a matter requiring more long-term research; at present the best recommendation is that such animals should be fed at a plane of nutrition which allows rapid increase in size without excessive fat deposition.

Feeding of male animals

In mammals, the spermatozoa and ova and the secretions associated with them represent only very small quantities of matter. The average ejaculate of the bull, for example, contains 0.5 g of dry matter. It therefore seems reasonable to suppose that nutrient requirements for the production of spermatozoa and ova are likely to be inappreciable compared with the requirements for maintenance and for processes such as growth and lactation.

If this were so, one would expect that adult male animals kept only for semen production would require no more than a maintenance ration appropriate to their species and size. There is insufficient experimental evidence on which to base feeding standards for breeding males, but in practice such animals are given food well in excess of that required for maintenance in females of the same weight. There is no reliable evidence that high planes of nutrition are beneficial for male fertility, though it is recognised that underfeeding has deleterious effects (see below). The liberal feeding of males probably reflects the natural desire of farmers not to risk underfeeding and so jeopardise the reproductive performance of the whole herd or flock. Males, however, do have a higher fasting metabolism and therefore a higher energy requirement for maintenance than do females and castrates.

Flushing

A low plane of nutrition may reduce the secretion of gonadotrophic hormones and hence affect fertility. In sheep, there is often a period of several months between weaning and remating during which the ewes may be on a low plane of nutrition. It has been found that ewes which have been better fed during this period, and for this or other reasons are in better condition at the start of the breeding season, are more likely to have multiple ovulations and hence bear twins or triplets. This effect is utilised in the practice known as *flushing*, in which ewes are transferred from a maintenance level of feeding to a higher level for 4–6 weeks, beginning 2–3 weeks before the start of the breeding season. Flushing often increases the lambing percentage (lambs born per 100 ewes) by 10–20 per cent. The present view of the practice is that its effect is associated more with *improved* than with *improving* body condition, and could just as well be obtained by preventing the depletion of reserves. However, in ewes initially in poor condition the dynamic effect may be as important as the static.

Flushing is also used to increase litter size in gilts, the improved level of feeding being imposed for about 10 days before first mating. In cows, only a single ovulation is normally required, and therefore flushing is not needed. It is recommended, however, that both dairy and beef cows should be fed to gain weight from about day 70 of lactation so that reconception is achieved without difficulty.

Effects of prolonged under- or overfeeding of breeding animals

Animals given a sub-maintenance ration eventually show some reduction in fertility. In males this may be brought about by a decreased output of spermatozoa or by a smaller output of the accessory secretions. In females continued underfeeding leads to a cessation of ovarian function; the farm animals most likely to suffer in this way are heifers kept on inadequate rations during the winter feeding period. It should be stressed, however, that underfeeding has to be severe and prolonged to exert its full effect. Thus a bull kept on a starvation ration and losing weight at the rate of 0.9 kg/day was at the end of 14 weeks still capable of producing semen containing normal spermatozoa, but the production of fructose and citric acid in the accessory secretions was much reduced by this treatment.

The evidence for a causative association between *over*feeding and impaired reproductive ability is less convincing. Very fat animals frequently are sterile, but the two conditions, fatness and sterility, may both be effects of, for example, an endocrine disturbance, rather than one the cause of the other. Fatness and sterility occur together most commonly in sows, and also occur together frequently in show animals. Over-fat sows may continue to produce ova while failing to show signs of oestrus; it has been suggested that the oestrogens intended to be responsible for the latter are absorbed in the fat depots.

Effects of specific nutrient deficiencies on the production of ova and spermatozoa

So far this chapter has been concerned with the effects upon reproduction of variations in quantity of food, effects which in many cases are responses to variations in the supply of energy. Attention will be given in this section to the effects on reproduction of deficiencies, or excesses, of specific nutrients.

Little is known about the influence on reproduction in farm animals of a deficiency of protein. One reason for this is that protein deficiency, because it depresses appetite, is frequently complicated by deficiencies of other nutrients. Protein has been studied more extensively with laboratory animals, where its deficiency eventually leads to reproductive failure. In general the effects of protein deficiency on reproduction appear to be much more severe in growing than in mature animals.

When deficiencies of minerals or vitamins occur in breeding animals, the general signs of deficiency described earlier (see Chs. 5 and 6) usually appear before reproductive ability is seriously affected. In other words, reproductive function is more resistant to these deficiencies than are other bodily activities. The effect of vitamin A deficiency illustrates this point, for although such a deficiency ultimately causes complete failure of reproduction, animals blinded by the deficiency may still be capable either of producing semen or of conceiving. Prolonged deficiency leads eventually in males to degeneration of the testes and in females to keratinisation of the vagina.

Deficiency of vitamin E has a profound effect on reproduction in rats, but the evidence suggests that deficiency of the vitamin does not play any appreciable role as a cause of infertility in cattle and sheep. In pigs, however, it has been reported that vitamin E deficient diets may reduce reproductive performance. There is also evidence from experiments with mature fowls that a prolonged vitamin E deficiency results in sterility in the male and reproductive failure in the female, and in the male sterility may become permanent through the degenerative changes in the testes.

Of the mineral elements, both calcium and phosphorus are important in reproduction, although of the two it is phosphorus whose deficiency is more commonly associated with reproductive failure. (In females this failure may sometimes, but not always, be due to oestrus not taking place.) Phosphorus deficiency arises most commonly in ruminants grazing on herbage deficient in the element, and in such circumstances the failure of reproduction occurs in conjunction with the general signs of phosphorus deficiency described earlier. When low-phosphorus diets have been used in experimental studies, however, there have been cases where reproduction was impaired in animals that were normal in other respects. In sows, manganese deficiency has been shown to interfere with reproduction. A complex interaction between manganese, calcium and phosphorus has therefore been suggested as influencing reproduction in cattle. In male animals, zinc deficiency may impair the production of spermatozoa.

EGG PRODUCTION

Rearing of hens

Birds intended for egg production are commonly fed to appetite during the rearing period, but in recent years the possibilities of restricting food intake have been investigated. Restriction during the rearing period (8–21 weeks of age), to 70–80 per cent of what would be consumed voluntarily, appears to delay the onset of egg production and to retard growth; the practice is also associated with a higher mortality during rearing. However, if such 'restricted' birds are subsequently, during the laying period, fed to appetite they appear to compensate for many of the earlier disadvantages. Once they begin to lay they produce eggs at a slightly faster rate than birds previously fed to appetite, and so by the end of the first laying cycle (i.e. at equal age) will have laid an equal number of eggs. The 'restricted' birds also make up their liveweight deficit. In general, mortality is lower among layers which have been reared on a restricted regime than among birds fed on a normal diet, presumably because more of the weaker birds do not survive during rearing.

Birds reared on a restricted intake naturally yield a saving in food, but they may almost nullify this by eating more food at the beginning of the laying period; the total saving to the time of 50 per cent production (i.e. an egg every two days) might be 5 per cent. One definite advantage of food restriction is a decrease in number of small eggs laid at the start of the production cycle.

Nutrient requirements of laying hens

Good flocks of layers produce an average of about 250 eggs per bird per year (i.e. 70% production). Their eggs weigh on average 57 g and have the chemical composition shown in Table 15.2, and an energy value of about 375 kJ; this information can be used as the basis for a factorial calculation of the nutrient requirements of layers. At one time, laying hens were rationed according to a system in which they were given a certain amount of food per day for maintenance and a certain amount for the estimated egg production, but today they are almost invariably fed to appetite. Feeding standards for layers, as for other classes of poultry, are therefore expressed in terms of nutrient proportions rather than quantities. The requirements of layers are shown in Appendix Table 14.

Energy. A hen weighing 2 kg has a fasting metabolism of about 0.36 MJ/kg $W^{0.75}$ per day, or 0.60 MJ/day, and utilises metabolisable energy for maintenance and production with a combined efficiency of about 0.8. Its requirement for metabolisable energy for maintenance would therefore be $0.60/0.8 = 0.75$ MJ/day and for 70 per cent egg production. $0.375 \times 0.7/0.8 = 0.33$ MJ/day (total 1.08 MJ/day). Maintenance requirements increase as temperature falls; for example, 2 kg birds adapted to 25 °C would require an extra 0.018 MJ/day for each 1 °C fall in temperature below 25 °C. An alternative method of estimating energy requirements is to fit regression equations to data for metabolisable energy intake and the weight, weight change and egg production of hens. From the best known of these equations, that of T. C. Byerly, the maintenance requirement of a 2 kg hen

TABLE 15.2 . Average composition of the hen's egg

	Per kg whole egg	Per egg of 57 g	Proportion of nutrient in edible part of egg
Gross constituents (g)			
Water	668	38.1	1.00
Protein	118	6.7	0.97
Lipid	100	5.7	0.99
Carbohydrate	8	0.5	1.00
Ash	107	6.1	0.04
Amino acids (g)			
Arginine	7.2	0.41	0.97
Histidine	2.6	0.15	
Isoleucine	6.4	0.36	(assumed) for all
Leucine	10.1	0.57	amino acids)
Lysine	7.9	0.45	
Methionine	4.0	0.23	
Phenylalanine	6.0	0.34	
Threonine	5.5	0.31	
Tryptophan	2.2	0.13	
Valine	7.6	0.44	
Major minerals (g)			
Calcium	37.3	2.13	0.01
Phosphorus	2.3	0.13	0.85
Sodium	1.2	0.066	1.00
Potassium	1.3	0.075	1.00
Magnesium	0.8	0.046	0.58
Trace elements (mg)			
Copper	5.0	0.3	1.00
Iodine	0.3	0.02	
Iron	33	1.9	(traces of minor
Manganese	0.3	0.02	elements in shell)
Zinc	16	1.0	
Selenium	5.0	0.3	

is estimated to be 0.97 MJ metabolisable energy per day, the requirement for 70 per cent egg production, 0.55 MJ/day and that for a gain of 1 g/day, 0.014 MJ. Thus a hen with 70 per cent egg production would consume 1.52 MJ ME/day. For the laying hen, as for other classes of livestock, energy requirements measured under practical conditions are higher than those estimated factorially.

In commercial practice, laying hens are usually fed to appetite; their metabolisable energy intake is commensurate with T. C. Byerly's and other similar equations, (i.e. 1.5 MJ/day) and they gain 1–2 g/day. Like younger birds, the hen eats less of diets high in energy concentration and more of those low in. energy concentration, with the result that an increase or decrease of 1 per cent in energy concentration causes a corresponding increase or decrease in energy intake of only 0.5 per cent or less. The usual concentration is 10–12 MJ metabolisable energy per kg, (i.e. 11.5–13.5 MJ/kg DM) and for Appendix Table 14 a concentration of 11.1 MJ/kg (12.8 MJ/kg DM) has been assumed. Concentrations below 10 MJ/kg are likely to depress energy intake sufficiently to reduce egg production, and concentrations greater than 12 MJ/kg usually increase body-weight gain rather than the number of eggs laid, (although egg weight may be greater). The food intake of hens is also affected by the environmental temperature; it falls by 1–2 per cent for each 1° rise in temperature in the range 10–30°.

Protein. Laying hens receiving a diet containing 11.1 MJ metabolisable energy per kg and consuming 110 g food/day require a total protein concentration of about 160 g/kg. The amino acid requirements of layers have not been as accurately defined as have those of chicks, because it is difficult to maintain a satisfactory level of egg production in hens given their protein as mixtures of pure amino acids. The requirements given in Appendix Table 14 are intended to apply to a bird of 1.8 kg at its maximum rate of lay (i.e. 50 g of egg/day). Requirements for methionine (often the first limiting amino acid for layers) and for lysine, isoleucine and tryptophan have been reasonably well established, and for these and other indispensable amino acids there are equations available for predicting requirements factorially. For example, the lysine requirement is predicted by the equation:

$$L = 9.5\ E + 60W$$

where L = available lysine (mg/day)
E = egg production (g/day)
W = bodyweight (kg)

As eggs contain 7.9 g lysine per kg, the factor 9.5 implies that lysine is used for egg production with an efficiency of 7.9/9.5 or 83 per cent. Glycine (or its substitute, serine) is apparently non essential for layers. The dietary concentrations of all amino acids will need to be increased if food intake is less than 110 g/day, and may be reduced if food intake is greater than this amount.

Mineral elements. The laying hen's requirement for calcium is 2–3 times greater than that of the non-layer, because of the large quantities of this element in the eggshell. The minimum requirement for maximum egg production is about 3 g/day, but maximum eggshell thickness is not achieved until calcium intake is increased to 3.8 g/day. The whole quantity of calcium required is commonly included in the mash (meal) or pellets, but if the hen is given a separate source of calcium, as grit, it is capable of adjusting its intake to its requirements. Phosphorus requirements are difficult to define because of uncertainties regarding the availability of phytate phosphorus; it is therefore common for the requirement to be stated as the proportion of inorganic phosphorus to be added to the diet. Other elements which are likely to be deficient in normal diets are sodium, chlorine, iron, iodine, manganese and zinc.

Common salt is generally added to the diet of laying hens and is beneficial in counteracting cannibalism and feather pecking. The requirement of poultry for sodium is met by the provision of 3.8 g NaCl/kg of diet. In excessive amounts salt is definitely harmful, although adult birds can withstand 200 g/kg in the diet if adequate drinking water is available. The iron content of the egg is relatively high (see Table 15.2), and consequently the requirement of the laying hen is large compared with the requirement for maintenance. Excessive iron in the diet is, however, harmful and may give rise to rickets by rendering the phosphorus of the diet unavailable. Iodine and manganese are particularly important for breeding hens, since a deficiency of either leads to a reduction in the hatchability of eggs, and may also reduce the viability of the chicks after hatching. The requirements for manganese are influenced by breed differences as well as by the levels of calcium and phosphorus in the diet; this trace element is more likely to be deficient in diets predominantly rich in maize than in those based on wheat or oats. Zinc deficiency in the diet of laying hens adversely affects egg production and hatchability, and results in the production of weak chicks with a high mortality rate. In the past it is possible that the use of galvanised feeding and drinking troughs was an important source of this element.

Vitamins. An important feature of the vitamin requirements of laying hens is that the minimum amounts required to ensure maximum egg production may be insufficient to provide for the normal growth of the chick, both before and after hatching. Requirements for some vitamins are not yet known, but it appears that for most B vitamins the quantities needed for maximum hatchability are appreciably greater than those for egg production alone. For vitamins A and D this is not so.

The value of β-carotene as a source of vitamin A for poultry depends upon a number of factors, and it has been suggested that in practice this provitamin should be considered as having, on a weight basis, only 33 per cent of the value of vitamin A.

Regarding vitamin D, it should be remembered that D_3 (cholecalciferol) is about 10 times as potent for poultry as D_2 (ergocalciferol).

REQUIREMENTS FOR PREGNANCY

Growth of the foetus

The growth of the foetus is accompanied by the formation of the membranes associated with it, and also by considerable enlargement of the uterus. The quantities of nutrients deposited daily in the uterus and its contents may be

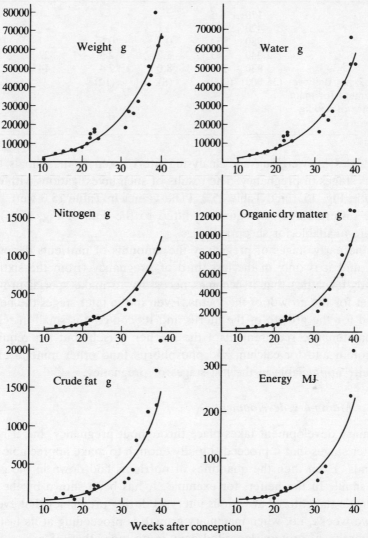

Fig. 15.1. Weight and composition of the pregnant bovine uterus related to time after conception. (From J. Moustgaard, 1959. Nutrition and reproduction in domestic animals, in *Reproduction in Domestic Animals* (ed. H. H. Cole and P. T. Cupps), Vol. II. Academic Press, New York and London.)

TABLE 15.3   Deposition of various nutrients and energy in the uterus and mammary
gland of the cow at different stages of pregnancy
(From values and equations given by J. Moustgaard, 1959, in *Reproduction in Domestic Animals*
H. H. Cole and P. T. Cupps (eds), 2 vols. Academic Press, New York and London)

| Days after conception | Deposited in uterus (per day) | | | | Deposited in mammary gland |
	Energy (kJ)	Protein (g)	Calcium (g)	Phosphorus (g)	Protein (g/day)
100	170	5	—	—	—
150	420	14	0.1	—	—
200	980	34	0.7	0.6	7
250	2 340	83	3.2	2.7	22
280	3 930	144	8.0	7.4	44
(Approx. net daily requirement for maintenance of 450 kg cow)	(35 000)	(200)	(8)	(12)	

determined by weighing and analysing uteri taken from animals killed at
various stages of pregnancy. The results of such investigations with cows are
shown in Fig. 15.1 and Table 15.3. (The values in Table 15.3 were obtained
by the differentiation of equations fitted to the data of Fig. 15.1.) Similar
data are available for sheep and pigs.

In the early stages of pregnancy the amounts of nutrients deposited are
small, and it is only in the last third of pregnancy (from the sixth month
onwards in cattle) that it becomes necessary to make special provision in
the diet for the growth of the foetus. Even in the later stages the net energy
needed for the growth of the uterus and its contents is small in relation to
the maintenance requirement of the mother herself, but net requirements
for protein and for calcium and phosphorus (and other mineral elements)
are quite appreciable in the last stages of pregnancy.

Mammary development

Mammary development takes place throughout pregnancy, but it is only in
the later stages that it proceeds rapidly enough to make appreciable nutrient
demands. Even then the quantities of nutrients laid down in the gland are
quite small. In the heifer, for example, it has been shown by the analysis
of animals slaughtered at various intervals during pregnancy that even in the
last two weeks, i.e. when mammary growth is proceeding at its fastest rate,
the quantity of protein deposited daily is no more than 45 g.

Energy metabolism during pregnancy

If a pregnant animal is given a constant daily allowance of food, its heat production will rise towards the end of gestation. The increase is due mainly to the additional energy required by the foetus for both maintenance and growth. It has been found that metabolisable energy taken in by the mother in addition to her own maintenance requirement is utilised by the foetus with comparatively low efficiency. For each additional 1 MJ, only about 0.13 MJ is retained in the foetus, but the apparent efficiency coefficient of 0.13 is not directly comparable with values for k_f, etc. discussed earlier, because it includes, in effect, the basal heat production of the foetus as well as the heat arising during synthesis of foetal tissue. Nevertheless, the demands of the foetus for maintenance and growth lead to a considerable increase in the energy requirements of the mother. For example, for a cow weighing 500 kg at the beginning of gestation, and requiring 48 MJ metabolisable energy per day for maintenance, the total energy requirement will increase to 90 MJ/day by the end of gestation. Thus the requirement for metabolisable energy in pregnancy is increased by far more than might be deduced from the storage of energy in the uterus and its contents (Table 15.3).

Extra-uterine growth during pregnancy

The liveweight gains made by pregnant animals are often considerably greater than can be accounted for by the products of conception alone. For example, a litter of 10 piglets and its associated membranes may weigh 18 kg at birth, but sows frequently gain over 50 kg during gestation. The difference represents the growth of the mother herself, and sows may in their own tissues deposit 3–4 times as much protein and 5 times as much calcium as is deposited in the products of conception. This *pregnancy anabolism*, as it is sometimes called, is obviously necessary in immature animals which are still growing, but it occurs also in older animals. Frequently much of the weight gained during pregnancy is lost in the ensuing lactation.

Pregnancy anabolism is often encouraged in pigs on the grounds that it increases the birthweight of the young, and that the reserves accumulated allow the sow to milk better and hence promote faster growth in the piglets. Table 15.4 illustrates these effects. It should be noted, however, that greater maternal growth in pregnancy did not increase litter size, nor did the greater birthweight of the piglets improve their viability. Furthermore, the increases in the weight of the piglets at weaning were small in relation to the extra food required by the sows. In general, a weight gain averaging about 15 kg over each of the first three reproductive cycles is sufficient to permit the sow to grow in size without depleting fat reserves, and is also likely to result in the most economic reproductive performance.

In dairy cows a high plane of nutrition in the dry period preceding

TABLE 15.4   Energy intake during pregnancy and weight changes of sows, and their association with reproductive performance
(After F. W. H. Elsley *et al.*, 1969. *Anim. Prod.*, **11**, 225)

	Digestible energy intake in pregnancy (MJ/day)		
	22	33	44
Liveweight change of sows (kg)			
From mating to just after parturition	+12	+32	+53
From parturition to weaning	+1	−13	−25
Piglet performance			
Number born	11.0	11.1	11.0
Number weaned	8.9	8.8	8.2
Birthweight (kg)	1.23	1.36	1.44
Weaning weight (kg)	15.9	16.5	17.2

parturition is claimed to promote the growth of mammary tissue and to increase body reserves, and thus to raise milk yield in the ensuing lactation. This is the theory behind the practice of 'steaming up' cows by giving them increasing quantities of concentrates before calving. However, if the cow is not unduly thin at the end of the previous lactation there seems to be little effect on milk production of increasing her liveweight gain in late pregnancy beyond that representing the foetus and its associated structures; in a 500 kg cow this gain amounts to about 0.8 kg/day over the last six weeks of pregnancy.

Consequences of malnutrition in pregnancy

Malnutrition—meaning both inadequate and excessive intakes of nutrients—may affect pregnancy in several ways. The fertilised egg may die at an early stage (i.e. embryo loss), or later in pregnancy the foetus may develop incorrectly and die; it may then be resorbed *in utero*, expelled before full-term (abortion) or carried to full-term (stillbirth). Less severe malnutrition may reduce the birthweight of the young, and the viability of small offspring may be diminished by their lack of strength or by their inadequate reserves (e.g. of fat). In some circumstances, it is the mother, not the foetus, that suffers from malnutrition. The foetus has a high priority for nutrients and if the mother has a low intake, her reserves will be used to meet the needs of the foetus. This priority is seen most strikingly in the case of iron, for the foetus can be adequately supplied with iron when the mother herself is anaemic. The protection thus afforded the foetus is not absolute, however, and in severe and prolonged deficiencies both foetus and mother will suffer. The degree of protection varies also from one nutrient to another, for although ewes as a result of an insufficient supply of energy may lose 15 kg of body substance during pregnancy and still give birth to normal lambs, an avitaminosis A which is without apparent effect on the ewe herself can lead to serious abnormalities in the young. The effects of underfeeding in pregnancy will also depend on the reserves of the mother, and particularly on the stage of pregnancy at which it occurs. In general, deficiencies are more serious the later they occur in pregnancy, but this rule is not invariable: vitamin A deficiency in early pregnancy, by interfering with the initial development of certain organs, can lead to abnormalities and even death in the young.

Effects on the young Deficiencies of individual nutrients in pregnancy must be severe to cause the death of foetuses; protein and vitamin A are the nutrients most likely to be implicated, although deaths through iodine, calcium, riboflavin and pantothenic acid deficiencies have also been observed. Congenital deformities of nutritional origin often arise from vitamin A deficiency, which causes eye and bone malformations in particular. Iodine deficiency causes goitre in the unborn, and in pigs has been observed to result in a complete lack of hair in the young. Hairlessness can also be caused by an inadequate supply of riboflavin during pregnancy. Copper deficiency in the pregnant ewe leads to the condition of swayback in the lamb, as described earlier (p. 104).

In the early stages of pregnancy, when the nutritional demands of the embryo are still insignificant, the energy intake of the mother may influence embryo survival. There is evidence that in sheep and pigs both very low and very high intakes of energy at this stage may be damaging, especially in females in poor condition at mating. The probable cause is a disturbance of the delicate hormone balance required at this time for implantation of the embryo. In mid-pregnancy, the nutrient requirements of the foetus are still low, but the placenta must grow at this time; if the growth of the placenta is restricted by under-nutrition it will be unable to nourish the foetus adequately in the final stage of pregnancy, and birth weight will be reduced. For sheep and pigs a common recommendation is that feeding for the first two-thirds of pregnancy should be at about the maintenance level. In the last one-third of pregnancy the requirements of the foetus(es) increase rapidly. Variations in birthweight of nutritional origin are usually a reflection of the energy intake of the mother during the later stages of pregnancy. We have seen already (Table 15.4) that in sows, variation in pregnancy anabolism is associated with differences in the weight of the young at birth. Ewes are frequently underfed in late pregnancy and thus lose weight: Table 15.5 illustrates the association between their weight losses and the birthweights of their lambs.

Young animals should be born with reserves of mineral elements, particularly iron and copper, and of vitamins A, D and E, because the milk, which may be the sole item of diet for a time after birth, is frequently poorly supplied with these nutrients. With regard to iron, it appears that if the mother is herself adequately supplied and is not anaemic, the administration of extra iron, whether in her food or by injection, will have no influence on the iron reserves of the newborn. If however, the mother is anaemic these reserves—though not haemoglobin—may be reduced. The copper and fat-soluble vitamin reserves of the newborn are more susceptible to improvement through the nutrition of the mother.

Effects on the mother. The high priority of the foetus for nutrients may mean that the mother is the more severely affected by dietary deficiencies. The ability of the foetus to make the mother anaemic has already been mentioned; this situation is unusual in farm animals because their diets are normally well supplied with iron.

The foetus has a high requirement for carbohydrate, and by virtue of its priority is able to maintain the sugar concentration of its own blood at a level higher than that of the mother. If the glucose supply of the mother is insufficient her blood glucose may fall considerably, to levels at which nerve tissues (which rely on carbohydrate for energy) are affected. This occurs in sheep in the condition known as *pregnancy toxaemia*, which is prevalent in ewes in the last month of pregnancy. Affected animals become dull and lethargic, lose their appetite and show nervous signs such as trembling and holding the head at an unusual angle; in animals showing these signs the mortality rate may be as high as 90 per cent. The disease occurs most frequently in ewes with more than one foetus—whence its alternative name of 'twin lamb disease'—and is most prevalent in times of food

TABLE 15.5   The effects of energy intake during the last six weeks of pregnancy on the liveweight gains of ewes and on the birthweights of their twin lambs (From J. C. Gill and W. Thomson, 1954. *J. agric. Sci., Camb.*, **45**, 229)

Group	Energy intake (MJ ME/day)	Liveweight change of ewes (kg)*	Birthweight of lambs (kg)
1	9.4	− 14.5	4.3
2	12.4	− 12.7	4.8
3	13.9	− 11.4	5.0
4	18.6	− 5.4	5.2

* From 6 weeks before to immediately after parturition.

shortage and when the ewes are subjected to stress in the form of inclement weather or transportation. Loss of appetite is especially common among fat ewes. Blood samples from affected animals usually show, in addition to hypoglycaemia, a marked rise in ketone content and an increase in plasma free fatty acids. In the later stages of the disease the animal may suffer metabolic acidosis and renal failure.

There does not appear to be one single cause of pregnancy toxaemia. The main predisposing factors are undoubtedly the high requirement of the foetus for glucose and possibly a fall in the carbohydrate supply of the mother, which may arise through food shortage or through a decline in appetite in late pregnancy. One biochemical explanation for the disease hinges on the fact that the tricarboxylic acid cycle cannot function correctly without an adequate supply of oxalacetate, which is derived from glucose or such glucogenic substances as propionate, glycerol and certain amino acids. If the oxalacetate supply is curtailed, acetyl CoA, which is derived from fats or from acetate arising through rumen fermentation, is unable to enter the cycle, and so follows an alternative pathway of metabolism which culminates in the formation of acetoacetate, β-hydroxybutyrate and acetone. In pregnancy toxaemia the balance between metabolites needing to enter the cycle is upset by a reduction in glucose availability and an increase in acetyl CoA production, the latter being caused by the animal having to metabolise its reserves of body fat. The clinical signs can thus be attributed both to hypoglycaemia and to the acidosis resulting from hyperketonaemia. An additional factor is that increased production of cortisol by the adrenal cortex in response to stress may reduce the utilisation of glucose; this possibility is supported by the fact that hyperketonaemia may continue after the blood glucose level has been restored to normal.

The disease has been treated by the injection of glucose, by feeding with substances likely to increase blood glucose levels, or by hormone therapy. Only moderate success has been achieved, however, and there is no doubt that the control of pregnancy toxaemia lies in the hands of the shepherd rather than the veterinary surgeon. The condition can be prevented by ensuring an adequate food supply in late pregnancy and by using foods which supply glucose or its precursors rather than acetate, i.e. concentrates rather than roughages.

LACTATION

We are here concerned with the nutrient requirements for milk production, which involves a conversion of nutrients into milk on a large scale and is a considerable biochemical and physiological achievement. A high-yielding dairy cow, for example, may in a single lactation produce 3 to 4 times as

TABLE 15.6   The composition of milks of farm animals (g/kg)

	Fat	Solids-not-fat	Crude protein	Lactose	Calcium	Phos-phorus	Magnesium
Cow	37	90	34	48	1.2	0.9	0.12
Goat	45	87	33	41	1.3	1.1	0.20
Ewe	74	119	55	48	1.6	1.3	0.17
Sow	85	120	58	48	2.5	1.7	0.20

much dry matter in the form of milk as is present in her own body. The raw materials from which the milk constitutents are derived, and the energy for the synthesis of certain of these in the mammary gland, are supplied by the food. The actual requirement for food depends upon the amount and composition of the milk being produced.

Qualitatively the milk of all species is similar in composition, although the detailed constitution of the various fractions such as protein and fat varies from species to species. Table 15.6 shows the typical composition of milks of farm animals.

The major constitutent of milk is water. Dissolved in the water are a wide range of inorganic elements, soluble nitrogenous substances such as amino acids, creatine and urea, the water-soluble protein albumin, together with lactose, enzymes, water-soluble vitamins of the B complex and vitamin C. In colloidal suspension in this solution are inorganic substances, mostly compounds of calcium and phosphorus, and the protein casein, while dispersed throughout this aqueous phase is a suspension of minute milk fat globules. The fat phase contains the true milk triacylglycerols (c. 980 g/kg) in addition to certain fat-associated substances such as phospholipids, cholesterol, the fat-soluble vitamins, pigments, traces of protein and heavy metals. The fat phase is usually referred to simply as 'fat', and the remaining constituents, other than water, are classed as 'solids-not-fat' or 'SNF'.

SOURCES OF THE MILK CONSTITUENTS

All or most of the major milk constituents are synthesised in the mammary gland from various precursors which are selectively absorbed from the blood. The gland also exerts this selective filtering action on certain proteins, minerals and vitamins, which are not elaborated by it but are simply transferred directly from the blood to the milk.

Milk proteins

About 95 per cent of the nitrogen in milk is present as protein, the remainder being present in substances such as urea, creatine and ammonia, which filter from the blood into the milk. In this respect milk functions as an alternative excretory outlet to urine. The protein fraction is dominated by the caseins. In cow's milk there are five of these, α_{s1}-, α_{s2}-, β-, κ- and γ, which together contain about 0.78 of the total milk nitrogen, and the protein in next greatest amount is β-lactoglobulin. The remainder of the fraction is made up of small amounts of α-lactalbumin, bovine serum albumin and the immune globulins, pseudo-globulin and euglobulin, all of which are absorbed directly from the blood.

Amino acids are absorbed by the mammary gland in quantities sufficient to account for the protein synthesised within it. Considerable interconversion of amino acids occurs before synthesis and certain amino acids are important as sources of others. Thus ornithine, which does not appear in milk protein, is absorbed and retained in large quantities by the mammary gland and has been shown to be a precursor of proline, glutamate and aspartate. Synthesis of the carbohydrate moieties of the proteins takes place in the mammary gland, as does phosphorylation of serine and threonine before their incorporation into the caseins.

Lactose

With the exception of traces of glucose, neutral and acid oligosaccharides and galactose, lactose is the only carbohydrate in milk. Chemically a molecule of lactose is produced by the union of one glucose and one galactose residue. The mammary gland contains an α-lactalbumin dependent enzyme system capable of changing glucose to galactose, which may then unite with glucose to give lactose (see Ch. 9). Lactose may also be synthesised from small carbon fragments in the form of acetate, but this is a minor source of the sugar.

Milk fat

Milk fat is a mixture of triacylglycerols containing saturated acids with four to twenty carbon atoms, and a range of unsaturated acids of which the chief is oleic acid; in addition there are small amounts of the more unsaturated linoleic and linolenic acids. The amounts of the fatty acids present vary with the species of the animal, ruminants having a higher proportion of low molecular weight acids in their milk fat than simple-stomached animals, as shown in Table 15.7. In non-ruminants, like the sow, milk fatty acids up to C_{18} are synthesised from blood glucose and acetate, and are also obtained directly from plasma lipids. In ruminants all acids up to C_{10}, are synthesised entirely within the mammary gland from blood acetate and β-hydroxybutyrate. The saturated C_{18} acids are derived from the plasma lipids and the

TABLE 15.7   Fatty acid composition (molar proportions) of milk lipids
(After R. Bickerstaff, 1970. Uptake and metabolism of fat in the lactating mammary
gland, in *Lactation* Ian J. Falconer (ed). Butterworth, London)

Fatty Acid	Cow	Goat	Sow	Mare
Saturated				
Butyric	0.031	0.013	—	0.004
Hexanoic	0.019	0.028	—	0.009
Octanoic	0.008	0.083	—	0.026
Decanoic	0.020	0.129	0.002	0.055
Dodecanoic	0.039	0.036	0.003	0.056
Tetradecanoic	0.106	0.102	0.033	0.070
Hexadecanoic	0.281	0.245	0.303	0.161
Octadecanoic	0.085	0.098	0.040	0.029
Unsaturated				
Hexadecenoic	—	0.009	0.099	0.075
Octadecenoic	0.364	0.233	0.353	0.187
Octadecadienoic	0.037	0.018	0.130	0.076
Octadecatrienoic	—	—	0.025	0.161
Others	—	0.008	—	0.081

intermediate acids partly from both sources. About half the total acids of milk fat arise from each source. The monoene acids are produced in part from analagous saturated acids or from dietary fat, but the di- and triene acids are obtained only from the diet since they cannot be synthesised in the body. Ultimately all the acids of milk fat originate in the products of digestion, but not all do so directly. Some come from endogenous acetate and fatty acids after storage in the body. Milk fat yield is influenced by the balance of fat synthesis and mobilisation. This is under hormonal control but depends upon the balance of glucogenic substances in the products of digestion. Thus a high proportion of propionate, glucose and amino acids stimulates fat deposition in adipose tissue and a reduced supply of fat precursors to the mammary gland.

Glucose does not serve as a fatty acid precursor in ruminants, but is essential for fatty acid synthesis as the source of reduced NADP. It is also a primary source of the glycerol residues of the triacylglycerols in both ruminants and non-ruminants, the remainder coming from plasma acylglycerols and glycerol.

Minerals

The inorganic elements of milk may be divided conveniently into two groups. The first comprises the major elements calcium, phosphorus, sodium, potassium, magnesium and chlorine. The second group, the trace constituents, contains some twenty-five elements whose presence in milk has been well authenticated; these include metals such as aluminium and tin, the metalloids boron, arsenic and silicon, and the halogens fluorine, bromine and iodine. Such substances are present in very small amounts, and their presence in milk is coincidental with their presence in blood; nevertheless they may have an important bearing on the nutritive value of the milk and on the health and well-being of the sucking animal. The inorganic constituents of milk are absorbed directly from the blood by the mammary gland, which shows considerable selectivity; the gland is able to block the entry of some elements such as selenium and fluorine while allowing the passage of others such as zinc and molybdenum. This selectivity may be a considerable disadvantage when it acts against elements whose presence at increased levels in the milk may be desirable. Copper and iron, for example, are both elements important in haemoglobin formation and therefore for the nutrition of the young animal. Yet, despite the fact that the levels of iron and copper in milk are never adequate, they cannot be raised by giving increased amounts to the lactating animal, even when blood levels of these elements are so raised. The iron content of colostrum, the milk produced in the immediate *post partum* period, may be up to fifteen times that of normal milk, but during this time transfer of substances between blood and milk is abnormal.

Vitamins

Vitamins are not synthesised in the mammary gland and those present in milk are absorbed from the blood. Milk has considerable vitamin A potency owing to the presence of both vitamin A and β-carotene. The amounts of vitamins C and D present are very small, while vitamins E and K occur only as traces. There is a large range of B-vitamins in milk, including thiamin, riboflavin, nicotinic acid, B_6, pantothenic acid, biotin, folacin, choline, B_{12} and inositol.

It will be clear from the foregoing that the mammary gland must be provided with a wide range of materials if it is to perform its function of producing milk. Indispensable amino acids must be available, and either a supply of dispensable amino acids or the raw materials for their synthesis must be provided, so that synthesis of specific milk proteins may take place. In addition non-specific milk proteins must be supplied as such. Glucose and acetate are required for lactose and fat synthesis, and minerals and vitamins must be provided in quantities that allow the maintenance of normal levels of these milk constituents. The substances themselves, or the raw materials from which they are produced, have to be supplied either from the food or from the products of microbial activity in the alimentary canal.

NUTRIENT REQUIREMENTS OF THE LACTATING DAIRY COW

The nutrient requirements of the dairy cow for milk production depend upon the amount of milk being produced and upon its composition.

The yield of milk is decided primarily by the breed of the cow. Generally speaking, the order of yield for the main British dairy breeds is: Holstein, Friesian, Ayrshire, Guernsey and Jersey (Table 15.8), but there are considerable intra-breed variations with strain and individuality. Thus certain strains and individuals of a low-yielding breed may often outyield others of a higher-yielding breed. Old cows tend to have higher yields than younger animals, but the main short-term factor affecting milk yield is the stage of lactation. Yield generally increases from parturition to about 35 days and then falls regularly at about 2.5 per cent per week to the end of the lactation. In individual cases, yield frequently reaches a peak earlier in lactation and the fall thereafter is much sharper.

As a result of these factors the yield of milk may vary over a very wide range. Fortunately such variations present little difficulty in assessing the nutrient requirements of the cow, since yield is easily and conveniently measured.

When estimates of yield are necessary for the long term planning of the feeding of the lactating cow, several useful generalisations may be made which allow prediction of yield at a given stage of lactation. Thus peak yield may be calculated as one two-hundredth of the expected lactation yield, or as 1.1 times the yield recorded two weeks *post partum* e.g. a cow yielding 23 kg at this time could be expected to have a peak yield of 25 kg. The assumption of a weekly rate of decline, from peak yield, of 2.5 per cent per week is useful in predicting milk yield, and also in monitoring deviations from normality during the progress of lactation. Such estimates are relatively imprecise and attempts to increase accuracy have resulted in highly sophisticated mathematical descriptions of the changes in yield with lactation. P.D.P. Wood, for example, has suggested that the yield of milk, on

TABLE 15.8   Milk yields of the main British breeds of dairy cows
(From *Dairy Facts & Figures*, 1985)

Breed	Average lactation yield (kg)
Holstein	6252
Friesian	5593
Ayrshire	5043
Guernsey	4022
Jersey	3849

any day *post partum*, may be calculated using equations of the following type:

$$y(n) = an^be^{-cn}$$

when n = week of lactation and a, b and c are constants which depend upon length of lactation, lactation yield, week of lactation in which peak yield is attained and the rate of decline of yield following upon attainment of peak yield. In the equation, a is a scaling factor for yield and b and c define the slopes of the curve before and after peak yield, respectively. A typical curve for a cow yielding 7000 kg milk in 44 weeks is shown in Fig. 15.2. In practice such yield prediction curves have to be modified to take account of the month of calving and the effect of season of the year.

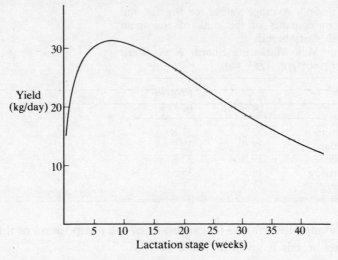

Fig. 15.2   Yield prediction curve for a cow with a lactation yield of 7000 kg milk.

The composition of milk varies with a number of non-nutritional factors. Milking technique may have a profound effect on fat content and thus on total solids content, since incomplete milking may leave a considerable volume of fat-rich residual milk in the udder. Unequal intervals between milkings may reduce yield and fat content when a single interval exceeds sixteen hours, especially with high-yielding cows. Again, diseases, particularly mastitis, may reduce the yield and compositional quality of milk. Lactose and potassium contents are lowered and those of sodium and chloride raised. Changes in fat content are erratic and crude protein shows little change. The net result, depending upon the severity of the infection is a reduction in solids-not-fat and total solids content. In a properly managed herd none of these factors should be of any importance. Certain variations in composition have to be accepted, however, since they are inevitable for a given herd. The factors responsible for these variations are breed, strain, individuality, age of cow and stage of lactation.

Effect of breed, strain within the breed and individuality on milk composition

There is a definite breed order in relation to milk quality which is the reverse of that for milk yield. From Table 15.9 it can be seen that the Jersey produces the highest quality milk, while the high yielding Holstein gives the poorest quality product.

TABLE 15.9 Average values for the fat and protein contents of the milk of the main British dairy breeds
(From Milk Marketing Board, *Rep. Breed and Prod. Org.* 1984–85).

Breed	Fat (g/kg)	Protein (g/kg)
Holstein	37.5	31.8
Friesian	38.0	32.2
Ayrshire	39.0	33.3
Guernsey	46.3	35.5
Jersey	51.8	37.9

A more detailed though less recent report of the composition of the milks of the various breeds is given in Table 15.10.

It is interesting to note the difference in the constitution of the solids-not-fat fraction of the different breeds; the milk from the high-yielding Friesian,

TABLE 15.10 Average values for the detailed composition of the milk of four main British dairy breeds.
(From J. A. F. Rook, 1961. *Dairy Sci. Abstr.*, 23, 251)

Constituent (g/kg)	Ayrshire	Friesian	Guernsey	Shorthorn
Fat	36.9	34.6	44.9	35.3
SNF	88.2	86.1	90.8	87.4
Protein	33.8	32.8	35.7	33.2
Lactose	45.7	44.6	46.2	45.1
Ash	7	7.5	7.7	7.6
Calcium	1.16	1.13	1.30	1.21
Phosphorus	0.93	0.90	1.02	0.96

for example, has a proportionately higher lactose and lower protein content than the milk from the lower yielding Guernsey. Strain and individuality of the cows have an important effect on milk composition and many Friesian cows may average more than 40 g fat/kg and 89 g SNF/kg over a lactation, whereas some Channel Island cows may not match these figures. Thus, the ranges within the four breeds quoted in Table 15.10 were as given in Table 15.11.

TABLE 15.11 Within-breed variation in the composition of cow's milk*. (g/kg).
(From J. A. F. Rook, 1961. *Dairy Sci. Abstr.*, **23**, 251)

	Ayrshire	Friesian	Guernsey	Shorthorn
Fat	35.7–38.7	33.2–37.2	43.1–49.0	33.7–38.1
SNF	86.5–89.4	84.0–87.5	88.2–93.0	85.7–89.0
Lactose	43.7–46.8	43.0–46.0	45.7–47.3	43.8–45.9
Protein	33.0–34.7	32.0–34.4	33.9–37.3	31.6–34.2

* Annual averages for individual herds

Effect of age on milk composition

As the age of the cow increases, so the quality of the milk produced becomes poorer. This is shown for Ayrshire cows in Table 15.12. The regression of solids-not-fat content on age is linear, and the decrease occurs almost equally in lactose and protein. Fat content on the other hand is relatively constant for the first four lactations, and then decreases gradually with age. Studies on commercial herds indicate that, over the first five lactations, there is a linear decline in fat and solids-not-fat contents of about 2 and

TABLE 15.12 Effect of age of cow on the composition of milk (g/kg)
(After R. Waite *et al.*, 1956. *J. Dairy Res.*, **23**, 65)

Lactation	Fat	Solids-not-fat	Crude protein	Lactose
1	41.1	90.1	33.6	47.2
2	40.6	89.2	33.5	46.2
3	40.3	88.2	32.8	45.9
4	40.2	88.4	33.0	45.7
5	39.0	87.2	32.6	45.3
6	39.1	87.4	33.0	44.8
7	39.4	86.7	32.5	44.8
8	38.2	86.5	32.3	44.4
9	40.3	87.0	32.7	44.8
10	38.3	86.6	32.5	44.6
11	37.7	86.1	31.6	44.6

4 g/kg respectively. The age frequency distribution of a herd may profoundly affect the average composition of the mixed-herd milk.

Effect of stage of lactation on milk composition

Advancing lactation has a marked effect on the composition of milk, which is of poorest quality during that period when yield is at its highest. Both fat and solids-not-fat contents are low at this time, and then improve gradually until the last three months of the lactation, when the improvement is more rapid. The changes are shown for Ayrshire cows in Fig. 15.3. Solids-not-fat content fell during the first seven weeks of lactation, being the resultant of the fall in crude protein content from day 15 to day 45 (2.8 g/kg) and the rise in lactose content (0.4 g/kg) which took place during this period. Subsequently the rise in protein content outweighed the fall in lactose content and the solids-not-fat content rose to the end of the lactation. Fat content fell

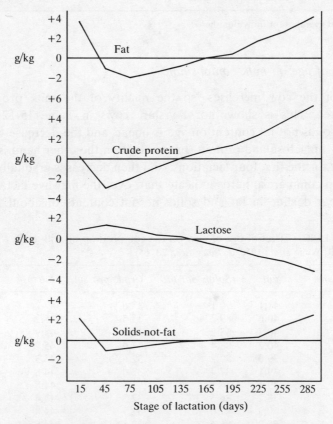

Fig. 15.3. Effect of stage of lactation on the composition of the milk of the dairy cow. (After R. Waite *et al.*, 1956. *J. Dairy Res.*, **23**, 65.)

sharply in early lactation, when yield was rising rapidly, and then continued to fall more slowly until day 75 of lactation. Thereafter fat content rose slowly until day 195, after which the rise was much faster.

It will be obvious that assessment of milk composition is a more difficult task than measurement of milk production, since there are here five main variants to be considered. Modern analytical methods allow of routine milk analysis on a large scale and values for fat, lactose and protein contents of herd bulk milks are now readily available. When analytical results are not available, assumptions are often made concerning the quantitative relationships between constituents, which allow composition to be predicted from the content of a single easily determined constituent, usually fat.

Energy requirements

Energy standards may be derived factorially. This involves a calculation of the gross energy value (EV_l) of the milk being produced, which may be used along with the yield to estimate the net energy requirement for milk production. Determination of the gross energy of milk involves either bomb calorimetry or a detailed chemical analysis; the amounts of fat, carbohydrate and protein are then multiplied by their energy values and the products summed, as illustrated in Table 15.13.

TABLE 15.13   Calculation of the gross energy value of milk

	Component (g/kg)	Gross energy (MJ/kg)	Gross energy (MJ/kg milk)
Fat	40	38.12	1.52
Protein	34	24.52	0.83
Carbohydrate	47	16.54	0.78
Milk			3.13

In its publication, *The Nutrient Requirements of Ruminant Livestock*, 1980, the Agricultural Research Council proposed the following equation for calculating the energy value of milk:

$$EV_l(MJ/kg) = 1.509 + 0.0406F$$

in which F = fat content of milk (g/kg). More accurate assessments may be obtained by including the solids-not-fat content in the prediction equation:

$$EV_l(MJ/kg) = 0.0386F + 0.0206SNF - 0.2353$$

When compositional data are not available, the following energy values based on average fat and solids-not-fat contents may be used (Table 15.14).

The next step in the factorial estimate is the calculation of the amount

TABLE 15.14   Energy values of the milks of the major British breeds of dairy cows

Breed	Energy value (MJ/kg)
Holstein	3.01
Friesian	3.04
Ayrshire	3.09
Guernsey	3.40
Jersey	3.64
SCM*	3.14

* Milk with 40 g fat/kg and 89 g SNF/kg.

of food energy required to provide the estimated net requirement. For this the efficiency of utilisation of food energy for milk production must be known. From the calorimetric work of Forbes, Fries and Kellner, an efficiency of utilisation of metabolisable energy for milk production (k_l) of about 0.70 is indicated. There is evidence that this applies to diets which promote a normal rumen fermentation; the proportion of acetic acid in the rumen volatile fatty acids is then in the range 0.50–0.60. However when the proportion of acetic acid is outside this range, the efficiency of utilisation of metabolisable energy for milk production is less than 0.70 (Fig. 15.4). It would seem that when the proportion of acetic acid is below 0.50 the cow is unable to synthesise sufficient of the lower and medium-chain fatty acids which form a large part of milk fat; when there is more than 0.65 of acetic acid, efficiency is low as in other forms of production.

Estimates of k_l vary widely from 0.51 to 0.81 but the majority cluster around about 0.60 to 0.65. There is considerable evidence that much of the variation is due to differences in energy concentration of the diet. Van Es has suggested that efficiency of utilisation of metabolisable energy for milk production is related to the metabolisability (q_m) of the diet, defined as ME (MJ/kg DM) at the maintenance level, as a proportion of the GE of the diet (MJ/kg DM). His implied relationships for Dutch (a) and American (b) data are:

$$\text{(a)} \quad k_l = 0.385 + 0.38 \, q_m$$
$$\text{(b)} \quad k_l = 0.466 + 0.28 \, q_m$$

where k_l = efficiency of utilisation of ME for milk production at zero weight change. More recently (ARC Tech. Rev. No. 2, 1980) it has been suggested that k_l is best calculated as $0.35 \, q_m + 0.42$. It is common, in deriving energy allowances, to assume that the gross energy value of the dry matter of all foods is constant at 18.4 MJ/kg. The above relationship may then be transformed to give:

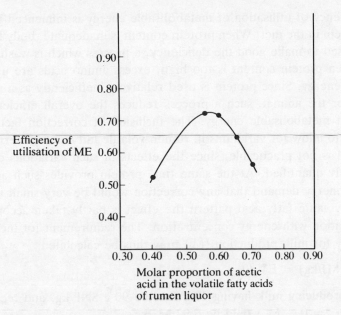

Fig. 15.4. Energetic efficiency of lactation. (After K. L. Blaxter, 1967. *The Energy Metabolism of Ruminants*, p. 259. Hutchinson, London.)

$$k_l = 0.019 \, M/D + 0.42$$

and k_l may be calculated from the energy concentration of the diet. For certain foods such as silage and high fat dairy compounds, which frequently comprise a major part of the diets of lactating cows, the assumption is not valid and the equation based on metabolisability should be used. In this case estimates of the gross energy of foods are needed. Routine determination is not feasible, but values calculated from the proximate composition would be an acceptable alternative:

$$GE \, (MJ/KG) = 0.0226CP + 0.0407EE + 0.0192CF + 0.0177NFE$$

in which CP, EE, CF and NFE are stated as g/kg. When neither calorimetric nor compositional data are available silages should be allocated a gross energy of 19.2 MJ/kg DM, high oil compound foods 19.4 MJ/kg DM and all other foods a value of 18.4 MJ/kg DM. Efficiency of utilisation of ME for milk production will thus vary from 0.61 to 0.67 for diets of q_m 0.55–0.70. The range of energy concentrations encountered with diets for milk production is narrow and it is widely held that a single factor could be adopted without causing significant error. Current UK standards are based on a figure of 0.62.

The efficiency of utilisation of metabolisable energy is influenced by the level of protein in the diet. When protein content is inadequate, body tissues are catabolised to make good the deficiency, a process which is wasteful of energy. When protein content is too high, excess amino acids are used as a source of energy. Since protein is used relatively inefficiently as a source of energy for the animal, such a process reduces the overall efficiency of utilisation of metabolisable energy. The inclusion of correction factors in calculations to allow for variations in rumen volatile fatty acid patterns and protein level is not practicable, since the effects of such variations cannot be adequately quantified. At the same time protein provides such a small part of the energy demand that any correction would be very small and in the case of volatile fatty acid pattern the effect is partly taken account of by its correlation with energy concentration. The requirement for metabolisable energy for milk production (M_l) may thus be calculated:

$$M_l \; (MJ/kg) \; = \; EV_l/(0.35q_m + 0.42)$$

For a cow producing milk having 40 g fat and 90 g SNF/kg, and receiving a diet with $q_m = 0.6$, M_l would be 5.02 MJ/kg.

Lactating cows are usually either gaining or losing body weight. A cow losing weight would be making reserves of energy available to maintain her level of milk production. If on the other hand she were gaining weight, some of the production ration would be diverted from milk production for this purpose. Values quoted in the literature, for the energy value of live-weight gain vary from 20 to 30 MJ/kg and it would appear that 26 would be an acceptable figure. There is general agreement that the efficiency with which metabolisable energy is used for tissue deposition (k_g) in the lactating cow is higher than in the non-lactating animal. Values for k_g of up to 0.84 have been quoted but the majority of published work suggests that k_g is very similar to, but slightly lower than k_1. For this reason the suggestion that $k_g = 0.95 \, k_1$ seems appropriate. Thus, in a cow gaining weight, each kg means that $26/0.95 \, k_1$ MJ dietary metabolisable energy is unavailable for milk production or that this amount of dietary metabolisable energy must be supplied in addition to that required for maintenance and milk production. In net energy terms, each kg of weight gain may be regarded as adding $26/0.95 = 27.36$ MJ to the lactation demand. Published estimates of the efficiency of utilisation, for milk production, of the energy of mobilised body tissue, are high and do not vary greatly. In the light of these results, the figure of 0.84 used in the previous edition has been retained. This means that for each kg of body tissue mobilised, $26 \times 0.84 = 21.84$ MJ of energy are secreted as milk.

As well as the production of milk, the diet of the lactating cow must provide the energy for maintenance. This may be calculated as:

$$E_m(MJ/d) = 0.53(W/1.08)^{0.67} + 0.0043W$$

The coefficient 0.0043, the activity increment, applies to cows living indoors under normal loose housing conditions. It is based on certain assumptions of time spent standing, number of positional changes and distance walked, and is valid only in situations in which these assumptions hold true. Available evidence would indicate that under such conditions a cow would spend about 14 hours standing, would stand up and lie down nine times and would walk about 500 m, during a normal day's activity. The coefficient may then be calculated to be:

Activity	Energy cost	Energy expended (MJ/kg per day)
Standing (14 hours)	10 kJ/kg per day	0.0058
Positional changes (9)	0.26 kJ/kg	0.0023
Walking (0.5 km)	2 J/kg per metre	0.0010
		0.0091

Efficiency of utilisation of dietary metabolisable energy for maintenance (k_m) may be calculated as:

$$k_m = 0.35q_m + 0.503$$

and the requirement for metabolisable energy for maintenance as

$$M_m(MJ/d) = \frac{0.53(W/1.08)^{0.67} + 0.0091W}{0.35q_m + 0.503}$$

In calculating the energy requirements of the dairy cow, cognisance must be taken of the decline in the metabolisable energy content of the diet with increasing level of energy intake. In order to do this the calculated requirement has to be increased accordingly. The procedure, which involves the use of a correction factor is best illustrated by an example.

Example

Calculation of the metabolisable energy allowance for a 600 kg cow producing 30 kg/day of milk containing 40 g fat/kg, and losing 0.4 kg live-weight/day on a diet with $q_m = 0.6$.

$E_m = 0.53(600/1.08)^{0.67} + 0.0091 \times 600$	= 42.04 MJ/day
$k_m = 0.35 \times 0.6 + 0.503$	= 0.713
$E_l = 30(1.509 + 0.0406 \times 40)$	= 93.99 MJ/day
$k_l = 0.35 \times 0.6 + 0.42$	= 0.630
$E_g = -0.4 \times 26$	= −10.4 MJ/day
Net energy spared by weight loss (10.4×0.84)	= 8.74 MJ/day
$M_m = 42.04/0.713$	= 58.96 MJ/day
$M_l = (93.99 - 8.74)/0.630$	= 135.32 MJ/day

Correction factor for level of feeding $(1 + 0.018(M_p/M_m))$ = 1.041
$M_{mp} = (135.32 + 58.96) \times 1.041$ = 202.25 MJ/day
Safety margin = 5%
ME allowance (202.25×1.05) = 212 MJ/day

In experiments in which responses to the additions of energy to the diet have been measured in terms of milk yield, it has been found that part only of the theoretically expected increase in yield has been obtained. The discrepancy is the result of two factors:

(1) Additions of concentrate foods to the diet bring about concomitant decreases in the roughage component so that the increase in energy intake is less than is added in the supplement. The supplementation rate, defined as change in forage intake per unit change in supplement intake, is greater for high-quality forages and at high intakes of supplement. In certain cases it may approach unity and the increase in the intake of dietary energy resulting from supplementation may be very small.

(2) Energy consumed by the lactating animal, over and above that required for maintenance, is partitioned between milk production and body gain. Response to supplements of energy added to the diet is negatively curvilinear in the case of milk yield and positively curvilinear in the case of liveweight gain (Fig. 15.5).

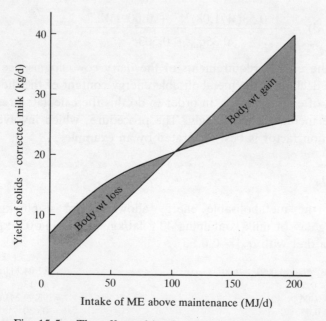

Fig. 15.5. The effect of intake of metabolisable energy (ME) on milk output and bodyweight change. (From W. Broster and C. Thomas, 1981. *Recent Advances in Animal Nutrition*, W. Haresign (ed.), pp. 49–69. Butterworths, London.)

Thus an increase in energy intake will result in increased milk yield, together with a reduction in liveweight loss or an increase in liveweight gain. When these changes are considered along with the true increments of dietary energy, then responses approach the theoretical.

The ability of cows to divert part of their production ration for the growth of their own tissues, or alternatively to supplement the energy available for milk production by the breakdown of these tissues, varies considerably from one individual to another; cows of high-yielding capacity use a higher proportion of a production ration for milk than those of lower potential. Within individuals the tendency is that, as the production ration is increased, the proportion of the energy of the increment which is used for milk production decreases; in other words, the input of energy required per kg increases as the milk yield of the cow increases.

Response, in terms of milk yield, to the addition of concentrate foods to a fixed diet is directly proportional to potential or current milk yield. This is well illustrated by the work of Blaxter who fed energy in excess of the level dictated by the Woodman standards,* to cows of different yield groups. He showed that the response ranged from 0.016 kg milk/MJ ME at a yield of 10 kg/day, to 0.172 kg at 25 kg. A typical response to an increment of dietary energy under such conditions would be 0.14 kg milk/MJ change in metabolisable energy intake (MEI), with responses of 0.003 and 0.01 kg/MJ MEI for yields of fat and solids-not-fat respectively.

Under conditions of *ad libitum* feeding, response does not depend upon potential or current yield and a typical value would be 0.7 kg milk/kg concentrate dry matter. If the true increase in dietary metabolisable energy is measured there is a trend for high-yielding groups of cows to show greater responses.

The major determinant of total lactation yield is peak yield, whether this is achieved as a result of cow potential or feeding practice. Further, responses to increments of energy decline as lactation progresses and, with low and medium planes of nutrition, elevation of energy intake in early lactation results in a 55 per cent residual effect in later lactation. Theoretically therefore, allocation of high levels of concentrate supplements to early lactation, in order to ensure maximum peak yield, should result in higher lactation yields. Experimental evidence does not entirely support this expectation since enhanced persistency may counterbalance a low peak yield. It would appear that when the plane of nutrition is low in relation to potential yield, the allocation of concentrates to early lactation improves lactation yield. When the plane of nutrition is generous in relation to potential, then

---

* H. E. Woodman, 1957. *Rations for Livestock*, MAFF, Bull, No. 48. HMSO, London.

total input of concentrate foods over the lactation is more important than pattern of allocation.

For animals of high potential, it is very difficult to maintain generous levels of feeding in early lactation, when intakes of dry matter are low. Such animals must be provided with high-energy concentrate foods and the finest quality roughages if generous levels of feeding are to be achieved. A major problem at this time is to ensure that the ration does not cause rumen disorders and result in loss of appetite and the production of low fat milk. The proportion of roughage should therefore not be allowed to fall below 0.35 of the diet. Weight lost at this time needs to be replaced before the next lactation and this is usually achieved during the dry period. In the light of recent evidence on the high efficiency of utilisation of energy for body gain in the lactating cow, it may be that the most effective method of replacing lost tissue is by feeding in excess of requirement during late lactation. Accumulation of reserves for use during periods of high yield is also justifiable in view of the high efficiency of the use of such reserves for milk production.

Protein requirements

Factorial procedure. The protein requirements for milk production are currently stated in terms of rumen degradable protein (RDP) and undegradable dietary protein (UDP), and may be estimated by methods analagous to those used to derive energy requirements. The requirement for rumen degradable protein in the lactating cow depends upon the size of the animal, the yield of milk and liveweight change. These factors decide the energy requirement and thus the microbial demand for protein. This may be calculated as 8.34MEI g/day for mixed diets, 8.67MEI g/day for silage based diets and 7.84MEI for diets consisting solely of grass silage. As well as the microbial requirement there is a demand for protein within the animal, at tissue level. This is made up of the following:

(a) A maintenance component which may be calculated as 2.19 g/kg $W^{0.75}$

(b) A milk component calculated as milk protein (g/kg) $\times 0.95$. The protein content of milk is now readily available to most producers in the UK as part of the quality payment schemes administered by the various milk marketing boards. When such figures are not available, protein content may be calculated from the fat content using regression equations such as that of Gaines and Overman:

Protein (g/kg) = 21.7 + 0.31F

F = fat content of milk (g/kg). Alternatively, the protein contents of the milks of the main British breeds of dairy cows given in Table 15.9 may be used. The factor of 0.95 is used because the non-protein nitrogen fraction of the milk is regarded as excretory material which has been used already by the body, and has therefore formed part of a previously satisfied demand.

(c) A dermal component resulting from the loss of hair and scurf, and which may be calculated as $0.1125 \text{ g/kg W}^{0.75}$.

(d) A component reflecting liveweight change. Body tissue is assumed to contain 150 g protein/kg, and each kilogram of gain to increase tissue protein demand by this amount. It is further assumed that tissue protein is used for milk production with an efficiency of 0.75, and that for each kilogram of weight lost, tissue protein demand is reduced by $150 \times 0.75 = 112 \text{ g}$ protein.

As previously discussed, part or all of the tissue protein demand (TP) may be satisfied by the contribution of microbial protein from the rumen. Using the factors of 0.8, 0.8 and 0.85 for the proportion of true protein in microbial protein, biological value and true digestibility of microbial protein respectively, the contribution of microbial protein may be calculated as:

$$RDP \times 0.8 \times 0.8 \times 0.85$$

or $8.34ME \times 0.8 \times 0.8 \times 0.85$

The difference between the tissue demand and the microbial protein supply has to be made good by dietary protein which has escaped breakdown in the rumen. Using the usual factors for biological value and true digestibility the requirement for undegradable dietary protein may be calculated as:

$$UDP \text{ (g/day)} = \frac{TP - (8.34ME \times 0.8 \times 0.8 \times 0.85)}{0.8 \times 0.85}$$

which simplifies to:

$$UDP \text{ (g/day)} = 1.47TP - 6.67ME$$

The protein requirements of a 600 kg cow producing 30 kg of milk containing 32 g CP/kg and losing 0.4 kg W/d may be calculated as follows:

ME requirement (MJ/day)	=	202.3
Requirement for RDP (g/day) = 8.34×202.3	=	1687.2
Protein for maintenance (g/day) = $2.19 \times 600^{0.75}$	=	265.5
Dermal loss (g/day) = $0.1125 \times 600^{0.75}$	=	13.6
Milk production (g/day) = $32 \times 0.95 \times 30$	=	912.0
Tissue loss (g/day) = 0.4×112	=	44.8

$$\text{TP (g/day)} = 265.5 + 13.6 + 912.0 - 44.8 \qquad\qquad = 1146.3$$
$$\text{UDP requirement (g/day)} = 1.47\text{TP} - 6.67\text{ME} \qquad = 335.7$$
$$\text{RDP allowance (g/day)} = 1687.2 \times 1.05 \qquad\qquad = 1772$$
$$\text{UDP allowance (g/day)} = 335.7 \times 1.05 \qquad\qquad = 352$$

The minimum requirement for crude protein is then 1745 + 374 = 2119 g. If the blend of degradable and undegradable protein is not ideal, the requirement may be considerably increased. The attainment of such an ideal blend may allow the use of cheaper sources of protein, or justify the use of expensive protein or the industrial processing needed to modify degradability of unsuitable dietary protein.

Feeding trials. In these, diets are used which are accepted as satisfactory in all respects other than protein, and the minimum intake of protein adequate for maximum production is determined. Such experiments have to be of a long-term nature, since even on deficient diets production may be maintained owing to the cow's ability to utilise her body tissues which contain about 150 g CP/kg. This will result in a negative nitrogen balance, and studies of nitrogen balance are often carried out to supplement the main feeding trial. In treating the results of feeding trials, an allowance is made for protein required for maintenance, and the residue equated to milk production. Estimates of digestible protein requirement based on such trials have varied from 1.75 times that present in milk to as little as 1.25 times in the more recent work. These low levels only apply where the content of crude protein in the diet is of the order of 160 g/kg; where the content is reduced to about 120 g/kg crude protein, the requirement for milk production rises.

Mineral requirements

The net daily maintenance requirement of calcium for the dairy cow is generally accepted to be about 16 mg/kgW. The phosphorus requirement is more problematical with estimates varying from levels of 14 to 28 mg/kg W for animals of about 500 kg liveweight. Evidence from feeding trials supports a value of about 14.5 mg/kgW. In addition to the requirements for maintenance, calcium and phosphorus must be provided for milk production. The concentrations of the two elements in the milks of the main breeds of dairy cows are given in Table 15.10 and these represent the net requirements for milk production. In order to calculate the dietary requirements it is necessary to know the availability of the dietary supplies. Current values for dairy cows are 0.45 for calcium and 0.55 for phosphorus although recent evidence would indicate a much higher availability of 0.68 for calcium and a similar but slightly higher figure of 0.58 for phosphorus. The results of feeding trials suggest that allowances of calcium and phosphorus, considerably lower than indicated by factorial calculations, can be fed for long periods with no ill effects. Thus 25 to 28 g of calcium and 25 g of phosphorus per day have proved adequate for cows producing 4540 kg of milk per annum over four lactations, which implies a dietary requirement of 1.10 to 1.32 g calcium and 1.10 g phosphorus per kg of milk. The requirements given in Appendix Table 7 have been derived by factorial calculation and are probably slightly higher than the minimum requirement, but are considered necessary to ensure a normal life span and satisfactory reproduction.

Balance experiments have shown that even very liberal allowances of calcium and phosphorus are frequently inadequate to meet the needs of the cow for these elements during the early part of the lactation, whereas in the later stages and in the dry period storage of calcium and phosphorus takes place. Figure 15.6, for example, shows the cumulative weekly calcium and phosphorus balances throughout a forty-seven-week lactation for a mature Ayrshire cow producing 5000 kg of milk. Despite the negative balances which occurred over considerable periods early in lactation, there was a net positive balance over the lacation and dry period as a whole. It has therefore become normal practice to consider the complete lactation in assessing calcium and phosphorus requirements; early negative balances are regarded as normal, since no ill effects are evident as long as subsequent replenish-

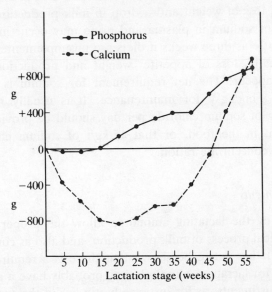

Fig. 15.6. Cumulative balances of calcium and phosphorus during lactation (47 weeks) and dry period. (From H. B. Ellenberger, J. A. Newlander and C. H. Jones, 1931. *Bull. Vt. Agric. Exp. Stn.*, No. 331.)

ment of body reserves takes place, and daily requirements for calcium and phosphorus are recommended on the basis of total production over the lactation. It has been suggested that 45 g of calcium and 60 g of phosphorus per day are adequate for a cow yielding 4500 kg of milk over the lactation; other workers consider that 39 g of calcium and 33 g phosphorus are sufficient for similar yields. Nevertheless, though the lactation approach is satisfactory in many cases, considerable trouble may arise if the allowances used are too low. Where shortage is serious, progressive weakening and breaking of the bones may result, and in less severe cases a premature drying off, which reduces yield and shortens the productive life of the cow. There seems to be little reason why calcium and phosphorus requirements should not be based on daily or weekly yield measurements. In diets which are deficient in phosphorus the ratio of calcium to phosphorus can be important. In practical diets, evidence for the importance of the ratio is conflicting and, in the absence of definitive evidence, the best approach would be to keep the Ca : P ratio between 1 : 1 and 2 : 1. In calculating magnesium allowances, a net requirement for maintenance of 3 mg/kgW may be assumed along with a concentration of 0.125 g/kg in the milk. Availability of dietary magnesium is very low at about 0.17.

Lactating cows are usually given a sodium chloride supplement. This is done by adding the salt to the food or by allowing continuous access to salt licks. The primary need is for sodium rather than chloride, which is more plentiful in normal diets. A deficiency manifests itself in a loss of appetite, rough coat, listlessness, loss of weight and a drop in milk production. Salt hunger and low levels of sodium in plasma and urine may occur in high-yielding cows after as little as three weeks if diets are unsupplemented with sodium chloride. Normally, loss of appetite, weight and production may take about a year to appear. The net requirement for sodium is about 0.60 g/kg of milk and 8 mg/kgW for maintenance. It is usually recommended either that 28 g of sodium chloride per day should be provided in addition to that present in the food, or that 15 kg/t of sodium chloride should be added to the concentrate ration.

Vitamin requirements

Vitamins are required by the lactating animal to allow the proper functioning of the physiological process of milk production, and also as constituents of the milk itself. It has yet to be shown that there is a requirement for vitamins specifically for lactation, although they probably have a role in the synthesis of milk constituents, as for instance biotin has in the synthesis of the fat. Most of the evidence points to the conclusion that, provided levels of vitamins in the diet are sufficient for maintenance, normal growth and reproduction, then no further allowance for lactation need be made. But normal levels of vitamins in the milk must be maintained, and sufficient amounts must be given to allow for this. The B-vitamins are an exception, since an adequate supply becomes available as a result of microbial synthesis in the rumen. Maintenance of normal vitamin levels in milk is particularly important where milk is the sole source of vitamins for the young animal, as for example with the young pig.

Winter milk has a vitamin A potency of about 2000 i.u./kg. Apart from the almost colourless vitamin A, milk contains variable amounts of the precursor β-carotene. This is a red pigment, yellow in dilute solution, as in milk, to which it imparts a rich creamy colour. The vitamin A potency of milk varies widely, being particularly sensitive to changes in dietary levels even though only about 3 per cent of the intake finds its way into the milk. Thus green foods are excellent sources of the provitamin, as is shown by the deep yellow colour and the high potency of the milk produced by grazing cows. Feeding with vitamin A concentrates in excess of levels adequate for reproduction may increase the potency by up to twenty times, but has no effect on the yield or the gross composition of milk. Considerable storage of vitamin A may occur in the body, and these reserves may be tapped to maintain levels in the milk. Since the newborn animal normally has small vitamin A reserves it is almost entirely dependent upon milk for its supply, and it is essential to feed the mother during pregnancy and lactation so as to maintain levels in the milk. No problem arises with cattle and sheep which are given early access to green food but great care is required where this is not so, as for example in winter calving suckler herds. Young pigs, too, are entirely dependent upon milk for a considerable time and the vitamin A levels of sow diets are of great importance. The daily requirement of the lactating cow is 99 i.u./kgW.

There is some evidence that there may be a requirement for β-carotene itself which is quite distinct from its function as the provitamin.

When lactating dairy cows are kept on diets deficient in vitamin D, and irradiation is prevented, deficiency symptoms appear, showing that the vitamin is essential for normal health. There is no evidence however to show a requirement greater than that which supports maintenance and reproduction. The vitamin D potency of milk is largely influenced by the extent of exposure to sunlight. Large intakes are necessary for small increases in the level in milk. Vitamin D has little effect on the negative balances of calcium and phosphorus during early lactation, but very heavy feeding (with 30 000 000 i.u./day) for three to seven days *pre partum* and one day *post partum* has been claimed to control milk fever (see p. 93). The daily requirement of the lactating cow is 10 i.u./kgW.

Dietary intakes of the B-vitamins are of no significance in ruminant animals because of ruminal synthesis. It is probable, however, that a physiological requirement exists for many of them as part of the complicated enzyme systems involved in milk synthesis, and for maintaining normal levels in the milk.

Effect of limitation of food intake on milk production

There is a great deal of evidence to show that reduction of food intake has a profound effect upon both the yield and the composition of milk.

When cows are kept without food, the yield drops to very low levels, of about half a kilogram per milking, within three days. At the same time the solids-not-fat and fat contents rise to about twice their previous levels, the increases being due to a concentration effect resulting from the reduced yield. Less severe limitations reduce yields to a lesser extent; the solids-not-fat content is again lowered but the effect on fat content is variable.

Limitation of the energy part of the diet has a greater effect than does limitation of the protein, on the solids-not-fat content of milk, although it is the protein fraction which is reduced in both cases. Lactose content shows little change, as would be expected of the major determinant of the osmotic pressure of milk. The lowered protein content is probably due mainly to increased gluconeogenesis from amino acids, owing to a reduced propionate supply on the low energy diets. As a result the supply of amino acids to the mammary gland is reduced and so is protein synthesis. It is also conceivable that the low energy supply limits microbial protein synthesis in the rumen and thus the amino acids available to the mammary gland. Throughout the winter feeding period in the United Kingdom there is a decline in milk yield and solids-not-fat content in most herds, the rate of decline being most marked in the later part of the period. The traditional pattern is for both yield and solids-not-fat content to increase when the cows are allowed access to spring pasture. It has been shown experimentally, that where levels of winter feeding are high such increases do not take place and indeed the opposite effect may be produced. It would appear, therefore, that winter feeding of dairy cows is frequently inadequate.

The change to pasture feeding in spring is frequently accompanied by a fall in the fat content of the milk. Spring pastures have a low content of crude fibre and a high content of soluble carbohydrates; other diets having similar characteristics also bring about a decline in milk fat. These diets may have a low ratio of roughage to concentrates or the roughage that is present may be too finely divided (e.g. ground roughage). The effect is well shown in Table 15.15.

The drop in fat content becomes obvious as the proportion of roughage

TABLE 15.15 Comparison of fat contents of milks produced on various diets with those of milk produced on a diet containing 5.4 kg of hay + concentrates (After C. C. Balch *et al.*, 1954. *J. Dairy Res.*, **21**, 172)

Diet	Change in milk fat (g/kg)
3.6 kg hay + concentrates	− 11.6
3.6 kg ground hay + concentrates	− 17.2

in the diet falls below 400 g/kg, and below 100 g/kg herd mean fat content may be below 20 g/kg.

Such lowering of the fat content is usually associated with changes in the fatty acid constitution, with a decrease in the saturated acids and an increase in the unsaturated, particularly the $C_{18:1}$ acids. The changes in fat content and composition are associated with changes in rumen fermentation patterns. Low-fibre diets fail to stimulate salivary secretion and hence diminish the buffering power of the rumen liquor. Such diets are often fermented rapidly giving rise to pronounced peaks of acid production with very low ruminal pH values. As a result the activity of cellulolytic fibre digesting micro-organisms is inhibited and that of various starch utilisers encouraged. These changes are reflected in changes in the balance of volatile fatty acids in the rumen. On high fibre diets the molar proportions of the volatile fatty acids in the rumen contents would be about 0.70 acetic acid, 0.18 propionic acid and 0.12 of butyric acid, plus a number of higher acids present only in small amounts. If the fibre content of the diet is reduced and the proportion of concentrates increased the proportion of acetic acid falls and in extreme cases may be less than 0.4. This fall is usually accompanied by a decrease in butyric acid and an increase in propionic acid which may form more than 0.45 of the total acids present: the concentration of valeric acid may also increase. Diets containing high proportions of carbohydrate which has been treated to increase its availability, as with the starch in flaked maize, are particularly effective in decreasing the ratio of acetate to propionate. This is well shown in Table 15.16.

TABLE 15.16 Relation of dietary cooked starch to milk fat content and to rumen volatile fatty acids (After W. L. Ensor *et al.*, 1959. *J. Dairy Sci.*, **42**, 189)

Diet	Percentage change in fat content	Molar proportions of total rumen volatile fatty acids	
		Acetic acid	Propionic acid
12.4–14.5 kg ground pelleted hay	0	0.68	0.20
12.4 kg ground pelleted hay + 1.8 kg ground maize	− 13	0.62	0.25
12.4 kg ground pelleted hay + 1.8 kg cooked maize	− 53	0.54	0.31

Experiments in which volatile fatty acids have been infused into the rumen of lactating cows have shown that acetate and butyrate increase milk fat content while propionate reduces it.

It has been suggested that if the ratio of acetic to propionic acid in the rumen contents falls below 3 : 1 then low milk fat contents will result. Another suggestion is, that it is the balance of glucogenic to non-glucogenic acids in the rumen contents which is the major determinant of milk fat content. The non-glucogenic ratio (NGR) is defined as:

$$NGR = (A + 2B + V)/P + V$$

in which A, P, B and V are molar proportions of rumen acetate, propionate, butyrate and valerate respectively. There is danger of production of low-fat milk if the ratio falls below 3. On balance it would appear that propionic is the major acid affecting milk fat content with acetic and butyric acids having a lesser effect in the opposite direction.

There is a tendency for dietary fat to be regarded simply as a source of energy. However when rats, young pigs and young ruminants are given diets adequate in energy but are deprived of fat, they fail to grow, develop dermatitis and eventually die. The necessity for dietary fat arises from the fact that it supplies the animal with the essential fatty acids (p. 30). Few studies of the needs of adult farm animals for essential fatty acids have been made, except for hens, probably because natural foods contain adequate amounts and their provision is not a problem in practice. It has been shown, however, that when fat is replaced by an isocaloric amount of starch in the diet of the lactating cow, then milk yields may be lowered. There is also some evidence that diets having 50–70 g/kg of ether extract produce more milk than diets containing less than 40 g/kg. The feeding of diets low in fat does not always depress milk fat content but may instead cause production of fat with a high proportion of acids which are synthesised in the mammary gland, particularly palmitic acid. There is evidence that some fats or oils have specific effects on the yield and composition of the milk. Thus, it has been demonstrated that coconut oil and red palm oil increase the fat content and yield of fat when added to low-fat diets, while groundnut oil increases the yield of milk and of fat without affecting fat content. Care must be taken that certain highly unsaturated fats are not given to excess, or milk fat content may be reduced: cod liver oil at levels of about 200 g/day can reduce fat content by as much as 25 per cent and herring oil has a similar effect. Levels of up to 120 g/kg of fat in the diet have been given to dairy cows without ill effects, but here the fat was highly saturated and was of mammalian origin. The level of fat which may be safely included in the diet of the lactating cow is important in view of its high energy value which makes it convenient for inclusion in high-energy dairy cakes.

The nature of dietary fat can have a profound effect on the composition of milk fat. Diets rich in acids up to palmitic generally increase the proportions of these acids in milk fat, at the expense of the C_{18} acids. Dietary fats rich in saturated and unsaturated acids result in increased yields of oleic and stearic acids with associated decreases in shorter chain acids particularly palmitic. The secretion of linoleic and linolenic acids is not affected, mainly owing to the extensive hydrogenation which occurs in the rumen. There are some indications that the polyunsaturated C_{18} acids can affect the acetate : propionate ratio in the rumen. Soya bean oil, for example, is rich in linoleic acid and can markedly reduce the ratio. Diets which reduce milk fat content have been shown to increase the activity, in adipose tissue, of enzymes involved in fatty acid and triacylglycerol synthesis. At the same time a less pronounced reduction in the activity of such enzymes occurs in mammary tissue. As a result the amount of acetate available for milk fat synthesis is reduced owing to its use in fatty acid synthesis in the adipose tissue. Stimulation of triacylglycerol synthesis would have a similar effect on the level of plasma-free fatty acids. Liver synthesis of low density lipoproteins and the supply of these to the mammary gland for fat synthesis would thus be reduced. Finally, the concurrent changes in the mammary gland would cause the reduced supply of precursors to be used less efficiently for milk fat synthesis. Intravenous infusions of glucose reduce plasma glyceride concentrations and it may be that the effect of low-fat producing diets is due to the increased glucogenic nature of the acid mixture absorbed from the rumen on such diets.

Changes in dietary carbohydrate which reduce milk fat contents, tend to increase milk protein content, if dietary protein supply is adequate. The effect may require about two to three weeks to manifest itself and be of the order of 8 g protein/kg milk. It is probable that the increased propionic acid production, with such diets has a sparing action on certain glucogenic amino acids such as glutamate, and more of these are then available to the mammary gland for protein synthesis. The increased intake of energy *per se*, which usually occurs on such diet would have the same effect.

Reductions in dietary protein level may decrease milk yield and almost invariably non-protein nitrogen content but protein *per se* is little affected until intake of protein falls below 60 per cent of requirement. This is probably the result of an insufficient supply of certain indispensable amino acids. The first limiting amino acid under such circumstances is likely to be methionine, followed by threonine and tryptophan.

NUTRIENT REQUIREMENTS OF THE LACTATING DAIRY GOAT

In addition to the dairy cow, the goat also is used for the commercial production of milk for human consumption. Yield varies with breed (see

TABLE 15.17 Yield and composition of milk of various breeds of goats (after F. Knowles and J. E. Watkin, 1938. *J. Dairy Res.*, **9**, 153)

Breed	Fat (g/kg)	Crude protein (g/kg)	Calcium (g/kg)	Phosphorus (g/kg)	Lactation yield (kg)
Anglo-Nubian	56	38.5	1.56	1.39	840
British Saanen	41	31.0	1.26	1.04	1325
British Alpine	43	32.7	1.37	1.18	1135
British Toggenburg	45	34.1	1.44	1.26	1077

Table 15.17) and with stage of lactation, the peak yield occurring at about four weeks.

A lactation normally lasts for about eight to nine months, during which time up to 1350 kg of milk may be produced. Breed and stage of lactation also affect the compostion of the milk. Total solids content falls to a minimum value at about four months, rises for the succeeding three months, and then remains constant until the close of the lactation. The nutrient requirements of the lactating goat may be derived factorially from estimates of the requirements for maintenance, milk production and liveweight change.

Energy

The net requirement for energy for maintenance under indoor conditions may be calculated as $0.272 \text{ MJ/kgW}^{0.75}$. This should be increased by about 25 per cent for grazing animals. The energy value (EV_1) of goat milk is given by:

$$EV_1 \text{ (MJ/kg)} = 0.04 \text{ F} + 1.66$$

in which F = fat content (g/kg).

When no data are available a value of 3.46 MJ/kg may be adopted. In the absence of definitive information it is suggested that values of 0.7 and 0.62 should be adopted for k_m and k_l. Metabolisable energy requirements for maintenance are then 0.39 and $0.49 \text{ MJ/kgW}^{0.75}$ for animals kept indoors and outdoors respectively. For milk production a requirement of $3.46/0.62 = 5.6 \text{ MJ/kg}$ may be assumed. By analogy with the dairy cow each kilogram of mobilised body tissue may be assumed to contribute the equivalent of 35 MJ of dietary metabolisable energy to the input of the goat.

Protein

The requirement (g/day) for rumen degradable protein may be calculated as 8.34MEI, and that for undegradable dietary protein as 1.47TP-6.67MEI. Tissue protein requirement is made up of that required for maintenance (2.19 g/kgW$^{0.75}$), plus that for milk production (33×0.95 g/kg), and that required for liveweight gain (150 g/kg) or made available by tissue mobilisation (112 g/kg).

Calcium, phosphorus and magnesium

Endogeneous losses (net requirements for maintenance) are 20 mg calcium, 30 mg phosphorus and 3.5 mg magnesium. The net requirements for milk production are 1.3 g calcium, 1.1 g phosphorus and 0.26 g magnesium/kg of milk (Table 15.6). There is a dearth of information on the availability of dietary minerals in the goat, but it is suggested that those for the sheep, 0.51, 0.58 and 0.17 for calcium, phosphorus and magnesium respectively, should be adopted.

Calculation of the requirements of a 50 kg goat producing 5 kg of milk with 40 g fat/kg, and losing 50 gW/day.

$$
\begin{aligned}
M_m \text{ (MJ/day)} &= 0.39 \times 50^{0.75} &&= 7.3 \\
M_l \text{ (MJ/day)} &= 5 \times 5.6 &&= 28.0 \\
M_g \text{ (MJ/day)} &= 0.05 \times 35 &&= -1.8 \\
M_m + M_p \text{ (MJ/day)} &&&= 33.5 \\
\text{Feeding level correction} &= (1 + 0.018 M_p/M_m) &&= 1.064 \\
M_{mp} = \text{(MJ/day)} &= 33.5 \times 1.064 &&= 35.7 \\
\text{RDP (g/day)} &= 8.34 \times 35.7 &&= 298 \\
\text{TP (g/day)} &&&= 192 \\
\text{UDP (g/day)} &= 1.47 \times 192 - 6.67 \times 35.7 &&= 44.1 \\
\text{Ca (g/day)} &= (50 \times 0.02 + 5 \times 1.3)/0.51 &&= 14.7 \\
\text{P (g/day)} &= (50 \times 0.03 + 5 \times 1.1)/0.58 &&= 12.1 \\
\text{Mg (g/day)} &= (50 \times 0.0035 + 5 \times 0.20)/0.17 &&= 6.9
\end{aligned}
$$

NUTRIENT REQUIREMENTS OF THE LACTATING EWE

The lactation of the ewe usually lasts from twelve to twenty weeks, although individuals show very considerable variations. Stage of lactation has a pronounced effect on milk yield, which is at a maximum at the second to third week and then falls steadily, as shown for Suffolk ewes in Fig. 15.7.

It has been calculated that about 38 per cent of the total yield is obtained in the first four weeks of lactation, 30 per cent in the succeeding four weeks, 21 per cent in the next four weeks and 11 per cent in the final four weeks. Comparison of the milk yields of different breeds is difficult, since the data have been obtained under widely differing climatic conditions, levels of feeding and sampling techniques. They indicate, however, that breed differences do exist (see Table 15.18), and that within-breed differences are frequently large.

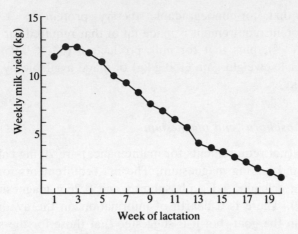

Fig. 15.7.   Effect of stage of lactation on milk yield of the ewe. (From L. R. Wallace, 1948. *J. Agric. Sci., Camb*, **38**, 93.)

TABLE 15.18   Lactation yields of different breeds of sheep
(Based on a 12-week lactation)

Breed	Yield (kg)	
	Twins	Singles
Romney Marsh	148	115
Cheviot	—	91
Border Leicester × Cheviot	211	124
Suffolk	145	94
Hampshire Down	79	75
Scottish Blackface	142	102
Finnish Landrace × Scottish Blackface	206	133

Animals suckling more than one lamb produce more milk than those suckling single lambs. The higher yield is probably due to higher frequency of suckling and greater emptying of the udder, indicating that a single lamb is incapable of removing sufficient milk from the udder to allow the full milking potential of the ewe to be fulfilled.

Data on the composition of the milk of the ewe are relatively few. Such factors as sampling techniques, stage of lactation and milking intervals all affect composition and figures are not strictly comparable and show considerable variation. Thus, published figures for breed average fat and protein contents vary from 50 to 100 g/kg and from 40 to 70 g/kg respectively. Breed differences in composition are illustrated by the figures given in Table 15.19.

TABLE 15.19 Effect of breed on the composition of ewe's milk (g/kg)
(After R. E. Neidig and E. J. Iddings, 1919. *J. agric. Res.*, **17**, 19)

Breed	Fat	Solids-not-fat
Hampshire	71	97.5
Cotswold	77	95.2
Shropshire	81	96.7
Southdown	75	103.6

Stage of lactation, also, affects the composition of ewe's milk, as shown in Fig. 15.8. The changes are similar to those in the dairy cow if allowance is made for the different length of lactation.

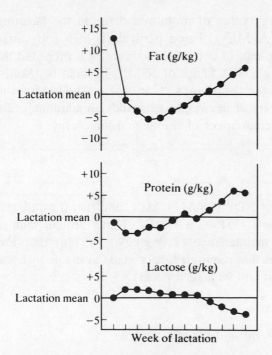

Fig. 15.8. Effect of stage of lactation on milk composition of Finnish Landrace × Blackface ewes. (From J. N. Peart *et al.*, 1972. *J. Agric. Sci., Camb.*, **79**, 303).

The nutrient requirements of the lactating ewe may be derived factorially from estimates of the requirements for maintenance, milk production and liveweight change.

Energy

The net requirement for maintenance (E_m) of housed ewes may be calculated as follows:

$$E_m = 0.226 \, (W/1.08)^{0.75} + 0.0106W$$

in which W = liveweight (kg).

For ewes kept outdoors the activity increment will be greater. For grazing sheep under lowland conditions a value of 0.0225 instead of 0.0106 would appear to be a reasonable approximation and for hill sheep 0.0337.

The energy value of ewe's milk is given by the following equation:

$$EV_l(MJ/kg) = 0.0328F + 0.0025D + 2.20$$

in which F = fat content (g/kg), and D = day of lactation. Alternatively, a value of 4.6 MJ/kg may be assumed when no information on composition is available.

Estimates of the energy value of mobilised tissue in the lactating ewe have varied from 17 to 68 MJ/kg, being particularly high and variable in early lactation. In the absence of definitive evidence, it is proposed that, by analogy with the lactating cow, a figure of 26 MJ/kg should be taken. Each kilogram of mobilised tissue contributes $26 \times 0.84 = 21.84$ MJ net energy as milk, and each kilogram of liveweight gain adds an additional (26/0.95) = 27.36 MJ to the net requirement of the animal for energy.

Protein

The requirement (g/d) for RDP is $8.34 \times MEI$, and that for undegradable dietary protein (g/day) is $1.47TP - 6.67MEI$. Tissue protein requirement is the sum of that for maintenance ($2.19 \, g/kgW^{0.75}$), plus that for milk production, (53 g/kg) plus that required for, or made available by liveweight change (130 g/kg for gain and 98 g/kg for loss).

Calcium, phosphorus and magnesium

Endogenous losses (net requirements for maintenance) for calcium, phosphorus and magnesium are 16, 30 and 3 mg/kg respectively. Net requirements for milk production are 1.6 g calcium, 1.3 g phosphorus and 0.17 g magnesium (Table 15.6). The availability of the minerals may be assumed to be 0.51 for calcium 0.58 for phosphorus and 0.17 for magnesium.

For the purposes of calculating nutrient allowances the milk yields given in Table 15.20 may be adopted.

TABLE 15.20 Suggested milk yields for calculating nutrient allowances for lactating ewes

Type of ewe	Number of lambs	Milk yield			
		12-weeks (kg)	Days 1–28 (kg/d)	Days 29–56 (kg/d)	Days 57–85 (kg/d)
Hill	One	86	1.21	1.09	0.75
	Two	130	1.90	1.63	1.11
Lowland	One	140	2.00	1.80	1.20
	Two	190	2.90	2.31	1.56

Example

Calculation of the nutrient requirements of a 75 kg lowland ewe in the fourth week of lactation, suckling two lambs, receiving a diet of $q_m = 0.60$, and losing 100 g W/day.

E_m (MJ/kg) = $0.226 \times (75/1.08)^{0.75} + 0.0106 \times 75$	6.23
$k_m = 0.35 \times 0.6 + 0.503$	0.713
M_m (MJ/day)	8.74
E_l (MJ/day) = 2.31×4.7	10.86
$k_l = 0.35 \times 0.6 + 0.42$	0.63
M_l (MJ/day)	17.23
M_g (MJ/day) = $(0.1 \times 21.84/0.63)$	−3.47
Level of feeding correction $(1 + 0.018 (M_p/M_m)$	1.0283
M_{mp} (MJ/day) = $(M_m + M_l + M_g) \times 1.0283$	23.13
RDP (g/day) = 8.34×23.13	193.0
UDP (g/day) = $1.47 \times 199 - 6.67 \times 23.13$	138.3
Ca (g/day) = $(75 \times 0.016 + 2.31 \times 1.6)/0.51$	9.6
P (g/day) = $(75 \times 0.03 + 2.31 \times 1.3)/0.58$	9.1
Mg (g/day) = $(75 \times 0.003 + 2.31 \times 0.17)/0.17$	3.6

Ewes which have been severely undernourished during pregnancy show a more rapid decline in milk production during subsequent lactation than do adequately nourished animals. This accords with independent observations of reduced metabolic capacity in ewes severely undernourished during pregnancy. Where restriction is less severe and ewes come to parturition in lean condition they have been shown to milk as well when adequately nourished in the subsequent lactation as do ewes with better condition scores at lambing. Ewes will not maintain high levels of milk production at the expense of body reserves, and even relatively small restrictions of intake depress milk production. Severe restriction of nutrient intake (to provide maintenance only) may reduce milk production by as much as 50 per cent in two to three days. If restriction of intake is continued beyond the time when peak yield is normally achieved then recovery of yield may not be accomplished even if subsequent intake is raised.

NUTRIENT REQUIREMENTS OF THE LACTATING SOW

In most breeding units, lactation length does not exceed six weeks, and many litters are weaned at three weeks of age. Maximum yield of milk occurs at about four weeks and the production falls gradually afterwards, as shown in Table 15.21.

TABLE 15.21 Variation in yield and composition of sow's milk with stage of lactation (After F. W. H. Elsley, 1970. Nutrition and Lactation in the Sow. In *Lactation* Ian R. Falconer (ed.), p.398. Butterworth, London)

Week	1	2	3	4	5	6	7	8
Daily yield (kg)	5.10	6.51	7.12	7.18	6.95	6.59	5.70	4.89
Fat (g/kg)	82.6	83.2	88.4	85.8	83.3	75.2	73.6	73.1
SNF(g/kg)	115.2	113.2	111.8	114.1	117.3	120.5	126.1	129.9
Protein (g/kg)	57.6	54.0	53.1	55.0	59.2	62.3	68.3	73.4
Lactose (g/kg)	49.9	51.5	50.8	50.8	49.0	48.6	47.5	45.6
Ash (g/kg)	7.7	7.7	7.9	8.3	9.1	9.6	10.3	10.9

Fat content rises to the third week and then falls to the end of lactation. Solids-not-fat content is at a minimum at the third week and then rises to the end of lactation, mainly owing to a rise in protein content.

Milk yield varies also with breed, age and litter size. It increases with the number of piglets suckled although yield per piglet decreases as shown in Table 15.22.

TABLE 15.22 Effect of litter size on milk yield in the sow (From F. W. H. Elsley, 1970. Nutrition and Lactation in the Sow, in *Lactation* Ian R. Falconer (ed.), p. 396. Butterworth, London)

Number of pigs	4	5	6	7	8	9	10	11	12
Daily milk yield (kg)									
per litter	4.0	4.8	5.2	5.8	6.6	7.0	7.6	8.2	8.6
per pig	1.0	1.0	0.9	0.9	0.9	0.8	0.8	0.7	0.7

Energy requirements

The net requirement of the lactating sow for dietary energy is the sum of that expended in maintenance, plus the gross energy of the milk less the contribution made by mobilised body tissue. The maintenance requirement may be calculated from the fasting metabolism (0.279 MJ/kgW$^{0.75}$ per day) plus an additional 25 per cent for activity. The gross energy of the milk of the sow may be assumed to be 5.2 MJ/kg, and that of body tissue, 39.4×0.87 MJ/kg (based on the assumptions that the tissue contains 0.87 of fat with a gross energy of 39.4 MJ/kg).

Efficiency of utilisation of dietary digestible energy for maintenance is 0.75, for lactation 0.65 and for the conversion of the energy of tissue to milk energy 0.85.

The dietary requirement for digestible energy for maintenance is then $(1.25 \times 0.279)/0.75 = 0.465$ MJ/kgW$^{0.75}$ for milk production, $(5.2/0.65) = 8.0$ MJ/kg, and each kilogram of weight lost represents a contribution to satisfying energy requirement equivalent to $(39.4 \times 0.87 \times 0.85)/0.65 = 44.83$ MJ/kg. Thus a 200 kg sow with a litter of eight piglets (milk yield = 6.6 kg/day) and losing 0.2 kgW/day would have a digestible energy requirement of:

$$(200^{0.75} \times 0.465) + (6.6 \times 8) - (0.2 \times 44.83) = 68.6 \ (\text{MJ/day})$$

The validity of the results of such factorial calculations depends in turn upon the validity of the assumptions upon which the calculations are based. The critical assumptions in this case are the yield of milk, and that the maintenance requirement of the lactating sow is equal to that of a non-lactating sow. There is evidence that the maintenance requirement is in fact considerably greater, and it has been suggested that, instead of treating maintenance and lactation separately, a gross efficiency of energy conversion into milk of about 0.45 should be used in calculating energy requirements. This would give an estimate of $34.3/0.45 = 76.2$ MJ of digestible energy per day. A similar approach is implicit in the derivation of equations for the regression of milk yield upon energy intake, such as that given below:

Milk yield (MJ/d) $= 0.0015 + 0.47\text{FI}^*$

where FI is the intake of digestibe energy in MJ.

The levels of intake indicated by these calculations are higher than have been used in the recent past. It is now considered, however, that the greatest efficiency of energy utilisation is achieved by giving adequate amounts of energy during lactation after restricted intake during pregnancy. A true gain in sow body weight of 12–15 kg over a complete reproductive cycle (about 30 kg during pregnancy) appears to give optimal reproductive performance and provides body reserves for lactation. When levels of food intake during lactation are designed to meet requirement, sows lose less weight and produce more milk than those animals on a lower plane. On the other hand, piglets reared by sows receiving the higher intakes have failed to show significant advantage. This is partly because the lower yielding sows produced richer milk, and partly because the increased yields were obtained

---

* After F. W. H. Elsley, 1970. *Lactation*, (Ian R. Falconer (ed.). Butterworths, London.

after the third week when creep feed was being eaten and the piglets on the low-yielding sows ate more of this feed than the others. Consumption of creep feed is, however, variable and cannot always be relied upon to make good the deficiencies in the milk yield of the sow; for safety it would be preferable to feed the piglets via the milk rather than by the more efficient, direct creep feeding. There is evidence that low levels of energy during lactation have a cumulative effect and that considerable reduction of milk yield and depletion of subcutaneous fat may occur over three lactations. Various techniques, including, frequent feeding, wet feeding, pelleting, high-energy diets and avoidance of over-fatness at farrowing are used to ensure satisfactory energy intakes in lactation.

Protein requirements

The lactating sow is a highly efficient converter of protein into milk. Most of the available evidence indicates an apparent digestibility of protein in excess of 0.80 and an overall efficiency of protein utilisation for body gain and milk production of 0.7. Estimates of the gross efficiency of conversion of dietary protein to milk protein vary from 0.30 to 0.45.

The net protein requirement for a lactating animal is made up of that required for maintenance plus that secreted in the milk. Estimates of protein requirements for lactation may be based on a figure for gross efficiency of protein utilisation, usually 0.33. A sow producing 6.6 kg of milk/day would secrete 376 g (6.6×57) of protein/day in her milk; the crude protein required in the daily ration would then be 1140 g ($376/0.33$). A meal given at 6 kg/day would thus need to have a crude protein content of 190 g/kg. In recent experiments, raising the protein content of the sow's diet from 140 to 160 or 190 g/kg has not increased milk yield.

Current standards in the UK are derived factorially. Maintenance requirement is assumed to be 0.45 g/kg W and milk to contain 57 g protein/kg. Coupled with an efficiency of utilisation of digested protein of 0.7 and a digestibility of protein of 0.8, this gives a crude protein requirement (g/day) of:

$$0.45 \text{ kg W} + 57Y/0.7 \times 0.8$$

in which Y = milk yield (kg/day). For the 200 kg sow nursing 8 piglets (see above) the requirement for crude protein is ($200 \times 0.45 + 6.6 \times 57$)/$0.7 \times 0.8 = 833$ g/day.

Dietary protein is the sole source of indispensable amino acids for the pig. The indispensable amino acid composition of sow's milk protein is shown in Table 15.23 along with requirements for individual amino acids calculated from them. The estimates are based on an assumed digestibility

TABLE 15.23   Amino acids requirements of sows for milk production

Amino acid	g/kg of milk*	Requirement (g/kg of milk)†
Arginine	3.3	6.11
Histidine	1.2	2.89
Isoleucine	2.4	4.50
Leucine	4.6	8.56
Lysine	4.2	8.93
Methionine	0.8	2.01
Phenylalanine	2.0	3.75
Threonine	2.0	3.75
Valine	2.9	6.25
Tryptophan	0.6	1.08

* Calculated from the data of S. E. Beacom and J. P. Bowland, 1951. *J. Nutr.*, **45**, 419.

† J. T. Reid, 1961. In *Milk: the Mammary Gland and its Secretion* S. K. Kon and A. T. Cowie (eds), vol. 2, p. 70. Academic Press, New York and London.

of 0.80 and an assumed biological value of 0.70. Evidence is accumulating that the high protein levels usually recommended for the diets of lactating sows may be required only because of inadequate protein quality. It has been further demonstrated that the biological value of barley can be raised from about 0.56 to 0.72 by supplementation with 20 g L-lysine per sow per day. This is close to the figure of 0.73 for the biological value of barley and fish meal diets. In the light of this information it has been suggested that diets containing as little as 120 g/kg crude protein may be adequate for milk production, provided that lysine levels are adequate and intake is not less than 5 kg/day for sows suckling eight piglets. Such a low level of dietary crude protein would indicate a gross efficiency of conversion of dietary to milk protein of 0.63, which would be highly optimistic.

Mineral requirements

There is no evidence to suggest that any minerals other than calcium and phosphorus have to be provided in the diet of lactating sows at levels above those necessary for maintenance and normal reproduction. Balance experiments indicate that the gross efficiency of utilisation of calcium and phosphorus for lactation are about 0.47 and 0.5 respectively. Table 15.6 shows that the milk of the sow contains 2.5 g/kg of calcium and 1.7 g/kg of phosphorus. A sow producing 6.6 kg of milk/day would thus be secreting 16.5 g of calcium and 11.2 g of phosphorus. Obligatory losses may be assumed to be 3.2 g calcium/100 kg W and 2 g phosphorus/100 kg W per day. A 200 kg sow nursing eight piglets (see above) would have requirements of (3.2 × 2 + 6.6 × 2.5)/0.47 = 48.7 g Ca/day and (2 × 2 + 6.6 × 1.7)/0.5 = 30.4 g P/day. In a meal given at 6 kg/day this would require 8.1 g Ca/kg and 5.1 g P/kg.

Vitamin requirements

Little information is available concerning the vitamin requirements for lactation of the sow. Those given in Appendix Table 13 are the same as for the pregnant sow, with the exception of vitamins A and D. The assumption is made that levels which allow normal reproduction and maintenance are adequate for lactation.

FURTHER READING

Agricultural Research Council, London, 1980. *The Nutrient Requirements of Ruminant Livestock*, Commonwealth Agricultural Bureau, Farnham Royal.

Agricultural Research Council, London, 1984. *The Nutrient Requirements of Ruminant Livestock*. Supplement. No. 1. Commonwealth Agricultural Bureaux, Farnham Royal.

Agricultural Research Council, London, 1981. *The Nutrient Requirements of Pigs*. Commonwealth Agricultural Bureaux, Farnham Royal.

K. L. Blaxter, 1967. *The Energy Metabolism of Ruminants*. Hutchinson, London.

W. H. Broster and H. Swan (eds), 1979. *Feeding Strategy for the High Yielding Dairy Cow*. Granada Publishing, London.

H. H. Cole and P. T. Cupps (eds), 1977. *Reproduction in Domestic Animals*. Academic Press, New York and London.

D. L. Duncan and G. A. Lodge, 1961. Diet in relation to reproduction and viability of the young. Part III. Pigs. *Tech. Commun. Commonw. Bur. Anim. Nutr.*, No. 21. Aberdeen.

I. R. Falconer (ed.) 1970 *Lactation*. Butterworths, London.

S. K. Kon and A. T. Cowie (eds). 1961. *Milk: the Mammary Gland and its Secretion*. Academic Press, New York and London.

H. H. Mitchell, 1962. *Comparative Nutrition of Man and the Domestic Animals*. Academic Press, New York and London.

J. T. Morgan and D. Lewis (eds), 1962. Nutrition of Pigs and Poultry. Butterworths, London.

J. A. F. Rook and P. C. Thomas (eds), 1983. *Nutritional Physiology of Farm Animals*, Longman, London.

Scientific Principles of Feeding Farm Livestock, 1958. Farmer and Stockbreeder Publications Ltd., London.

H. Swan and W. H. Broster (eds), 1976. *Principles of Cattle Production*. Butterworths, London.

W. Thomson and F. C. Aitken, 1960. Diet in relation to reproduction and viability of the young. Part. II. sheep: world survey of reproduction and review of feeding experiments. *Tech. Commun. Commonw. Bur. Anim. Nutr.*, No. 20. Aberdeen.

16

Voluntary intake of food

In previous chapters attention has been concentrated on the energy and nutrient requirements of farm animals for maintenance and various productive processes. An additional important factor which must be considered is the quantity of food that an animal can consume in a given period of time. The more food an animal consumes each day, the greater will be the opportunity for increasing its daily production. An increase in production, which is obtained by higher food intakes, is usually associated with an increase in overall efficiency of the production process, since maintenance costs are decreased proportionately as productivity rises. There are, however, certain exceptions to the generalisation; for example with some breeds of bacon pigs, excessive intakes of food lead to very fat carcasses which are unacceptable to the consumer and therefore economically undesirable.

Feeding is a complex activity which includes such actions as the search for food, recognition of food and movement towards it, sensory appraisal of food, the initiation of eating and ingestion. In the alimentary tract the food is digested and the nutrients are then absorbed and metabolised. All these movements and processes can influence food intake on a short term basis. In addition it is necessary to consider why, in most mature animals, body weight is maintained more or less constant over long periods of time, even if food is available *ad libitum*. Thus the concept of short- and long-term control of food intake must be considered, the former being concerned with the initiation and cessation of individual meals and the latter with the maintenance of a long-term energy balance. Although many of these control systems are thought to be similar in all classes of farm animals there are important differences among species that depend mainly on the structure and function of their digestive tracts.

FOOD INTAKE IN MONOGASTRIC ANIMALS

Control centres in the central nervous system

Feeding in mammals and birds is controlled by centres in the hypothalamus, situated beneath the cerebrum in the brain. It was originally proposed that there were two centres of activity. The first of these was the feeding centre (lateral hypothalamus), which caused the animal to eat food unless inhibited by the second, the satiety centre (ventromedial hypothalamus), which received signals from the body as a result of consumption of food. Quite simply it was considered that the animal would continue to eat unless the satiety centre received signals which inhibited the activity of the feeding centre. There is little doubt that this is an over simplification and, although the hypothalamus does play an important role in intake regulation, it is now believed that other areas of the central nervous sytem are also involved.

Short-term regulation

Chemostatic theory. The absorption of nutrients from the digestive tract, and the presence of nutrients in the circulating blood, constitute a set of primary signals which may in turn influence the satiety centre of the hypothalamus. A number of blood constituents have been suggested as possible signals including glucose, free fatty acids, peptides, amino acids, vitamins and minerals. Of these, glucose (*glucostatic theory*) has received the most attention. It has long been known that a small dose of insulin, which lowers the concentration of blood glucose, also causes an animal to feel hungry. It is also known that the blood glucose level rises after eating and then falls slowly. It has been suggested that the hypothalamus may contain 'glucoreceptors' that are sensitive to blood glucose, and that the rise in the latter after a meal triggers the glucoreceptors to 'switch off' the animal's desire to eat. This belief is now thought to be too simplistic, and a more recent theory proposes that the difference between the levels of glucose in mammalian arterial and venous blood is the effective signal rather than the blood glucose level itself.

Another metabolite which is thought by many to play a role in influencing intake is the peptide, cholecystokinin, also classed as a hormone. Cholecystokinin is present in the brain and is also released into the gastrointestinal tract when digestive products, such as amino acids and fatty acids, reach the duodenum (see p. 135). Systemic administration of cholecystokinin has been shown to decrease food intake in a number of mammalian species.

The short-term control of intake in the fowl does not seem to be influenced to the same extent by blood glucose or other nutrient levels and it appears that signals are received directly from the crop as is explained later.

Thermostatic theory. This theory proposes that animals eat to keep warm and stop eating to prevent hyperthermia. Heat is produced during the digestion and metabolism of food and it is considered that this heat increment could provide one of the signals used in the short-term regulation of food intake. It has been established that there are thermoreceptors, sensitive to changes in heat, present in the anterior hypothalamus and also peripherally in the skin. Some support for this thermostatic theory is obtained from observations, with a number of species, that food intake increases in cold and decreases in hot environments.

Long-term regulation

The long-term preservation of a relatively constant body weight combined with an animal's desire to return to that body weight if it is altered by starvation or forced feeding, implies that some agent associated with energy storage acts as a signal for the long-term regulation of food intake. One suggestion is that this might be fat deposition. Studies with poultry tend to support this *lipostatic theory* of regulation. Cockerels forced to eat twice their normal intake of food deposited fat in the abdomen and liver. When force-feeding was stopped the birds fasted for 6–10 days, and when voluntary feeding recommenced food intake was low. It was evident from these studies that the birds lost weight when force-feeding ceased, and tissue fat concentration decreased to levels approaching normal after 23 days. The exact mechanism by which the hypothalamus receives the lipostatic signal is not known although a natural steroid may be involved.

In pigs, it would appear that any feedback mechanism from body fat to the controlling centres of feeding is not as sensitive as that in poultry and other animals. This insensitivity may have arisen through early genetic selection for rapid weight gain, when excessive carcass fat was not considered, as it is today, an undesirable characteristic. The natural propensity of the modern pig to fatten is usually counteracted in practice by restricted feeding and also by selection of pigs with a smaller appetite.

Sensory appraisal

The senses of sight, smell, touch and taste, play an important role in stimulating appetite in man, and in influencing the quantity of food ingested at any one meal. It is a common assumption that animals share the same attitudes to food as ourselves but it is now generally accepted that the senses play a less important role in food intake in farm animals than they do in man.

The term palatability is used to describe the degree of readiness with which a particular food is selected and eaten but palatability and food intake are not synonymous. Palatability involves only the senses of smell, touch and taste. Most domestic animals exhibit sniffing behaviour, but the extent to which the sense of smell is necessary in order to locate and select foods is difficult to measure. A variety of aromatic substances, such as dill, aniseed, coriander and fenugreek, are frequently added to animal foods. The inference is that the odour from these spices makes the food more attractive and hence increases intake. Although transitory increases in food intake may occur, there is no evidence to indicate that the effects of these additives are long lasting in terms of overall increased food intake.

Similarly, with the sense of taste, most animals show preferences for certain foods when presented with a choice. Typical is the preference of young pigs for sucrose solutions rather than water. The fowl is indifferent to solutions of the common sugars but finds xylose objectionable, and will not ingest salt solutions in concentrations beyond the capacity of its excretory system. Every species studied has shown considerable individual variability, e.g. in a litter of pigs tested with saccharin solutions of different concentrations, some animals preferred high levels of the sweetener whereas others rejected them.

Physiological factors

The classical experiments of Adolph in 1947 demonstrated that, when the diets of rats were diluted with inert materials to produce a wide range of energy concentration, the animals were able to adjust the amount of food eaten so that their energy intake remained constant. This concept that 'animals eat for calories' has also been shown to apply to poultry and other non-ruminant farm animals. The manner in which chicks respond to diets of differing energy content is illustrated in Table 16.1 in which a normal diet containing 8.95 MJ productive energy (or about 13.2 MJ metabolisible energy) per kg was 'diluted' with increasing proportions of a low-energy constituent, oat-hulls. The most diluted diet had an energy concentration which was only half that of the original and much lower than the range normally experienced by chicks. The chicks responded by eating up to 25 per cent more food, but even so energy intake declined by up to 29 per cent. If the energy content of a diet is increased by the addition of a concentrated source of energy such as fat, chicks respond in the opposite way. They eat less, but the reduction in intake may be insufficient to prevent a rise in

TABLE 16.1   The effects of reducing the energy content of the diet on the food and energy intakes of chicks and on their growth
(After F. W. Hill and L. M. Dansky, 1954. *Poultry Sci.*, **33**, 112)

	Diet No.				
	1	2	3	4	5
Energy content of diet					
Productive energy (MJ/kg)	8.95	7.91	6.82	5.73	4.64
Metabolisable energy (MJ/kg)	13.18	11.59	10.21	8.91	7.45
Metabolisable energy					
(% of diet No. 1)	100	88	78	68	57
Performance of chicks to 11 weeks of age					
(% of result for diet No. 1)					
Total food intake	100	101	113	117	125
Total metabolisable energy intake	100	90	88	80	71
Liveweight gain	100	99	102	98	98
Fat content of carcass at 11 weeks of age					
(% of dry matter)					
(Male chicks only)	26.8	23.2	21.1	18.1	16.1

energy intake. Where extensive diet dilution is carried out, by using low digestibility materials, the ability to adjust the intake may be overcome because gastrointestinal capacity becomes a limiting factor. The crop appears to be concerned with intake in the fowl, since cropectomized birds eat less than normal when feeding time is restricted. Inflation or introduction of inert materials into the crop is known to cause a decrease in food intake. In mammals, distension and tension receptors have been identified in the oesophagus, stomach, duodenum, and small intestine. Distension in these areas of the tract increases the activity in the vagus nerve and in the satiety centre of the hypothalamus.

The general relationship between food intake and energy requirement suggests that, as with energy, intake should vary not directly with liveweight but with metabolic liveweight ($W^{0.75}$). This relationship is generally held to exist although variations can occur depending upon the physiological state of the animal. For example, lactation is usually associated with a marked increase in food intake, and in the rat at the peak of lactation, food intake may be nearly three times that of a non-lactating animal. In sows, the smaller the amount of food given during pregnancy, the greater is the amount consumed during lactation. Reports of changes in food intake during pregnancy are conflicting. There are several reports of increases in intake occurring with the onset of pregnancy in rats, but other reports suggest little or no change.

It would seem reasonable to assume that intake increases with exercise and studies with rats have shown that there is a linear relationship between food intake and the duration of exercise. There is, however, little information about farm animals on this subject.

Nutritional deficiencies

Utilisation by the tissues of the absorbed products of digestion depends upon the efficient functioning of the many enzymes and coenzymes of the various metabolic pathways and dietary deficiencies of indispensable amino acids, vitamins and minerals are likely to affect the intake of food. In poultry, severe deficiencies of amino acids reduce food intake whereas moderate deficiencies, insufficient to affect growth markedly, increase intake. When hens are given a diet containing high concentrations of calcium (30 g/kg), intake is about 25 per cent greater on egg-forming than on non-egg forming days. This large variation does not occur when low-calcium diets are given with calcium being provided separately as calcareous grit. It would appear that laying hens 'eat for calcium'. The effect of specific deficiencies of trace elements, especially cobalt, copper, zinc and manganese, and also vitamins such as retinol, cholecalciferol, thiamin and B_{12} on appetite has already been dealt with in Chapters 5 and 6.

Choice feeding

Animals have precise nutritional requirements, but under natural conditions are faced with a wide variety of foods to choose from, some of which are nutritionally inadequate.

The domestic rat and mouse are known to regulate their intakes of foods to satisfy, as far as the properties of the foods allow, their requirement for energy, protein and certain other nutrients. In studies with farm animals attention has concentrated on poultry, and it has been demonstrated that the domestic fowl has specific appetites for calcium, phosphorus, zinc, thiamin and various amino acids.

The commercial application of the ability of poultry to select foods for their nutritional content uses a whole cereal grain (e.g. whole wheat) and a balancer food which contains relatively high levels of amino acids, vitamins and minerals. The birds are thus allowed to balance the energy:protein ratio of their overall diet. The balancer food is formulated so that equal proportions of the two foods are expected to be eaten, and because milling, mixing and pelleting costs are avoided for the whole cereal, total feeding costs can be reduced. The choice feeding system has been used successfully with large flocks of growing turkeys and growing replacement laying stock. A similar ability to select a balanced diet from foods has been demonstrated in broiler chickens and adult laying hens, but some of these studies have given variable results. The theory that poultry have a control system that allows them to choose suitable amounts of different foods to satisfy their nutritional requirements is regarded as being too simplistic and other factors such as the physical form of ingredients, composition of the food, trough position and previous experience are also likely to be involved.

FOOD INTAKE IN RUMINANTS

Chemostatic, thermostatic and lipostatic regulation

The amount of glucose absorbed from the digestive tract of the ruminant is relatively small and blood glucose levels show little relation to feeding behaviour. It would therefore seem unlikely that a glucostatic mechanism of intake control could apply to ruminants. A more plausible chemostatic mechanism might involve the three major fermentation volatile fatty acids produced in the rumen—acetic, propionic and butyric. Intraruminal injections of acetate and propionate have been shown to depress intake of concentrate diets and it is suggested that receptors for acetate and propionate regulation of feed intake occur on the luminal side of the reticulo-rumen. Butyric acid seems to be less effective than acetate or propionate in reducing intake and, as butyrate is normally metabolised to acetoacetate and β-hydroxybutyrate by the rumen epithelium, it seems unlikely that butyrate *per se* is an important factor in intake regulation.

Ruminants respond to environmental temperature in the same way as monogastric animals, in that prolonged exposure to heat lowers food intake and continued exposure to cold increases it. The shearing of sheep also increases intake. Although this relationship with intake and environmental temperature has been established, it is more difficult to see how short-term relationships between intake and heat increment of individual meals can apply in the case of ruminants, since the peak of heat production frequently occurs after food intake has ceased.

There is sufficient evidence to indicate that fatness reduces intake in cattle. This can be considered as an aspect of energy balance in that the thin animal has a 'requirement' for nutrients for fat synthesis which is reduced in or absent from the fat animal. Another suggestion is that in the very fat animal, the deposition of fat in the abdominal cavity may reduce the space into which the rumen can expand during feeding.

Sensory appraisal

The senses do not appear to have much influence on the overall control of voluntary intake of food by ruminants, but are important in their grazing and eating behaviour. Cattle and sheep prefer to graze young, green herbage to dry, old material, and they prefer leaf to stem. The sense of sight appears to be relatively unimportant in grazing, as animals will graze in the dark and will eat readily even in complete darkness. The senses of smell and possibly taste appear to feature in the behaviour of the grazing animal since contamination of an area of ground by faeces of its own species renders the herbage growing on that area unacceptable. Palatability is probably not an important factor in determining intakes of better quality roughages such as hay or dried grass but it may limit the intake of poor quality foods, like cereal straws.

Physical factors

Size of reticulo-rumen. With diets containing a high proportion of roughage, voluntary intake is limited by the capacity of the reticulo-rumen and by the rate of disappearance of digesta from this organ.

Removing swallowed hay from the reticulo-rumen of cows through a fistula has been shown to increase voluntary hay intake. Conversely, the addition of digesta, in the form of recently ingested hay, to the rumen of cows during a meal caused an immediate decrease in hay intake. Additions of inert material such as sawdust or finely milled polyvinylchloride also decreased hay intake, a finding which lends support to the idea that the physical capacity of the reticulo-rumen is an important factor controlling intake in ruminants. Addition of water to rumen contents during eating does not affect the food intake of adult cattle and sheep, presumably because the water rapidly leaves the rumen, but the same quantity of water held in the rumen in balloons does depress intake. Thus in ruminants it would seem that there is a certain threshold of rumen distension which the animal will not exceed even though its energy requirements have not been satisfied. Stretch and tension receptors present in the reticulo-rumen probably function as a limiting factor when bulky diets are consumed, but are of less importance with concentrate type diets for which chemostatic sensory signals are likely to play a major role in intake regulation.

Digestibility and rate of disappearance of food from the gut. The rate of disappearance of digesta from the reticulo-rumen depends primarily on its rate of digestion and this in turn depends upon the chemical and physical composition of the food eaten. Fibrous foods of low digestibility are broken down slowly because, in the first instance, the rate at which physical comminution can take place is low. Apart from delaying the access of enzymes to the food constituents, slower physical breakdown leads to a more lengthy retention of food in the rumen, for only particles of small size are permitted to pass down the tract. Digestion in the rumen is retarded by the larger quantities of cellulose in fibrous foods, since cellulose is digested relatively slowly. There is, therefore, a relationship between digestibility and rate of digestion which leads in turn to a relationship between digestibility and food consumption.

The nature of the latter relationship is made clear by considering the case of an animal being allowed to eat roughage to appetite at discrete meals. The more digestible the food is, and the faster it is removed from the rumen, the greater will be the space cleared in the rumen in the interval between meals, and the more the animal will be able to eat. (This effect of digestibility on food intake should not be confused with the effect of intake on digestibility mentioned in Chapter 10, whereby reducing the amount of food given causes a small increase in its digestibility.)

An example of the relationship between digestibility and food consumption is shown in Fig. 16.1, which relates to sheep given various roughages to appetite, as the sole item of their diet. Digestibility is expressed as the coefficient for food energy (see Ch. 11), but the values approximate to those for dry matter. Sheep given oat hay with a digestibility of 0.4, ate less than half as much dry matter as when they were given dried grass, with a digestibility of 0.74. The difference in *digestible* dry matter intake is, of course, much greater, that of the dried grass being four times as much.

This relationship between digestibility and food consumption in ruminants is a general one and may be modified by the influence on intake of properties of foods other than digestibility. If roughages are finely ground,

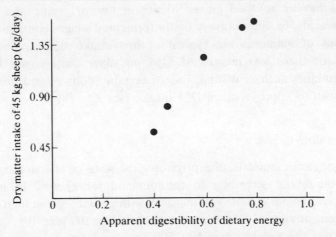

Fig. 16.1. Food consumption and disgestibility in sheep fed on roughages. (After K. L. Blaxter, F. W. Wainman and R. S. Wilson, 1961, *Anim. Prod.*, **3**, 51).

for example, their digestibility is reduced but their consumption is increased, both effects being attributable to the faster rate at which the fine particles leave the rumen and pass on down the gut. When a diet of roughage is supplemented with a concentrate, the increase in total food intake, which generally takes place, is often greater than can be accounted for by the higher digestibility of the supplemented diet. This effect is particularly striking when a protein concentrate, or a source of non-protein nitrogen such as urea, is added to a low protein roughage such as oat or barley straw. The crude protein content of these straws is too low (usually less than 40 g/kg.DM) to maintain an adequate level of microbial activity in the rumen. Supplementation of the roughage with a suitable nitrogen source rectifies the deficiency, resulting in an increased rate of fermentation and a faster flow of digesta from the rumen. In a study with non-lactating dairy cows, the consumption of oat straw was increased by 42 per cent when urea was infused into the rumen at a rate of 150 g/day. With highly digestible roughages, differences in digestibility may have a relatively smaller effect on intake than less digestible materials. Indeed it is possible that, when high levels of digestibility are attained, food consumption in ruminants is no longer limited by the rate at which digesta can be removed from the tract, but is controlled in the manner of non-ruminants.

Silage diets. There is evidence that on some roughage diets ruminants do not eat to a constant rumen-fill, and the general relationship between intake and digestibility is not valid. This is true for certain categories of silage where intakes of DM are lower than those obtained with animals, especially sheep, consuming herbage of similar digestibility and preserved as hay or as dried grass. Silages of low pH value and rich in fermentation acids, or, alternatively badly fermented silages containing high concentrations of ammonia are typical of low intake silages. The exact explanation for these low intakes of DM on silage diets is not known. Ensiling techniques such as wilting, use of certain additives and fine chopping are known to improve silage DM intake (see Ch. 18).

Physiological state

As with monogastric animals, the physiological state of the ruminant will influence food intake according to the demand for energy. In growing animals, the abdominal volume will increase with growth. When high quality diets are given, intakes of DM in cattle of about 90–100 g/kg $W^{0.75}$/day are maintained over the weight range 100–500 kg.

Growing animals offered high-quality foods *ad libitum* will, after a period of underfeeding, gain weight faster than similar animals which have not been so restricted. This compensatory growth has been attributed to both an improved efficiency in the utilisation of food and to an increased DM intake. Although liveweight gain is reduced during restricted feeding, the growth and development of the digestive tract is not, and this is related more closely to age than to liveweight. If at a similar age, animals of the same breed and sex, but differing in liveweight owing to nutritional restriction, consume the same amount of food, the lighter animals will increase in weight faster than the heavier animals because they have a lower maintenance requirement and consequently a greater proportion of energy is available for liveweight gain.

In pregnant animals, two opposing effects influence food intake. The increased need for nutrients for foetal development causes intake to rise. In the later stages of pregnancy the effective volume of the abdominal cavity is reduced as the foetus increases in size, and so is the space available for expansion of the rumen during feeding. As a result intake will be depressed, especially if the diet is predominantly a roughage one.

The increased intake in ruminants with the onset of lactation is well known. This increase is mainly physiological in origin although there may also be a physical effect resulting from the reduction in fat deposits in the abdominal cavity. There is a noticeable lag in the response of food intake to the increased energy demand of lactation. In early lactation, the dairy cow loses weight which is replaced at a later phase of lactation when milk

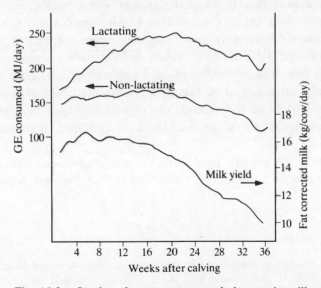

Fig. 16.2. Intake of gross energy and changes in milk production in lactating and non-lactating cows. (After J. B. Hutton, 1963 *Proc. N.Z. Soc. Anim. Prod.*, **23**, 39.)

yields are falling while intakes of DM remain high. These changes are illustrated in Fig. 16.2 for lactating and non-lactating identical twin, Jersey cross bred cows fed exclusively on fresh pasture herbage over a 36 week period. The intakes of gross energy by the lactating cows were about 50 per cent higher than those of the non-lactating animals.

Food intake in practice

A knowledge of the dry matter which an animal will consume, is an essential prerequisite for the accurate formulation of diets to satisfy its nutrient requirements. Many methods of predicting dry matter intake (DMI) have been used. In the simplest case intakes have been related to the weight in kg (W) and sometimes the physiological state of the animal e.g. lactating dairy cows in early lactation may be assumed to eat less (about 28 g DM/kg W) than those later in lactation (about 32 g DM/kg W), while for growing animals the figure may be about 22 g DM/kg W.

Attempts have been made to relate intake to performance e.g. to yield in kg (Y) in lactating dairy cows by means of equations such as;

$$\text{DMI (kg/day)} = 0.025\ W + 0.1\ Y$$

Such approaches are not entirely satisfactory as they ignore certain food characteristics which affect intake, and also interactions between the various foods making up the diet.

Attempts to quantify the effects of the factors which significantly affect the intake of a particular animal can result in highly complex equations and different equations are frequently used depending upon the type of animal and the basic food of the diet. Even when dealing with a single type of animal, the dairy cow, and rations based on a single roughage, grass silage, a whole range of factors has to be taken into account. Chief of these are the liveweight and yield of the animal, the digestibility and the type and extent of fermentation of the silage, and the kind and level of supplementary foods offered.

Such equations are usually produced by multiregression analysis of recorded data and need considerable validation before they are applied in practice.

FURTHER READING

C. A. Baile, 1975. Control of feed intake in ruminants, in *Digestion and Metabolism in the Ruminant*, I. W. McDonald and A. C. I. Warner (eds). Univ. of New England Publishing Unit, Armidale.

J. A. Bines, 1979. Voluntary food intake, in *Feeding Strategy for the High Yielding Dairy Cow*, W. H. Broster and H. Swan (eds). Granada Publishing, London.

K. N. Boorman and B. M. Freeman (eds), 1979. *Food Intake Regulation in Poultry*. British Poultry Science, Edinburgh.

R. C. Campling and I. J. Lean, 1983. Food characteristics that limit voluntary intake, in *Nutritional Physiology of Farm Animals*, J. A. F. Rook and P. C. Thomas (eds). Longman, London.

J. P. Dulphy and C. Demarquilly, 1983. Voluntary feed consumption as an attribute of feeds, in *Feed Information and Animal Production*, G. E. Robards and R. G. Packham (eds). Commonwealth Agricultural Bureaux, Farnham Royal, Slough.

J. M. Forbes, 1983. Physiology of regulation of food intake, in *Nutritional Physiology of Farm Animals*, J. A. F. Rook and P. C. Thomas (eds). Longman, London.

J. M. Forbes, 1986. *The Voluntary Food Intake of Farm Animals*. Butterworths, London.

A. H. Sykes, 1983. Food intake and its control, in *Physiology and Biochemistry of the Domestic Fowl*, B. M. Freeman (ed.). Academic Press, London.

17

Grass and forage crops

The natural food of herbivorous domestic animals is pasture herbage, and for a large part of the year this food forms all or most of the diet. Grasslands may be divided into two main groups,. *natural grassland*, which includes rough and hill grazing, and *cultivated grassland*, which may be further subdivided into permanent and temporary pastures. The latter form part of a rotation of crops, whereas permanent pasture is intended to remain as grass indefinitely. Natural grasslands normally include a large number of species of grasses, legumes and herbs, whereas cultivated grasslands may consist of single species or mixtures of relatively small numbers of species.

Chemical composition of pasture grass

The composition of the dry matter of pasture is very variable; for example, the crude protein content may range from as little as 30 g/kg in very mature herbage to over 300 g/kg in young, heavily-fertilised grass. The crude fibre content is, broadly, related inversely to the crude protein content, and may range from 200 to as much as 400 g/kg in very mature samples.

The moisture content of pasture is of particular importance when a crop is being harvested for conservation; it is high in the early stages of growth, usually between 750 and 850 g/kg, and falls as the plants mature to about 650 g/kg. In addition to stage of growth, weather conditions greatly influence the moisture content.

The water-soluble carbohydrates of grasses include fructans and the sugars glucose, fructose, sucrose, raffinose and stachyose (see Table 17.1). Their total concentration is very variable, ranging from as little as 40 g/kg

DM in some cocksfoot varieties to over 300 g/kg DM in certain varieties of Italian ryegrass. The soluble carbohydrate content is higher in the stems than in the leaves of grasses, concentrations sometimes being as much as three- or fourfold. The highest concentration usually occurs just prior to flowering.

The cellulose content is generally within the range of 200–300 g/kg DM and that of hemicelluloses may vary from 100 to 300 g/kg DM. Both these polysaccharide components increase with maturity; so also does the lignin, which influences adversely the digestibility of the useful nutrients, except for the soluble carbohydrates, which are completely digestible.

Proteins are the main nitrogenous compounds in herbage. Although the total protein content decreases with maturity the relative proportions of amino acids do not alter greatly. Similarly, the amino acid composition of proteins varies little among grass species. This is not surprising as up to half of the cellular protein in grasses is in the form of a single enzyme, ribulose-1,5-biphosphate carboxylase, which plays an important role in the photosynthetic fixation of carbon dioxide. Grass proteins are particularly rich in the amino acid, arginine, and also contain appreciable amounts of glutamic acid and lysine. They have high biological values for growth compared with seed proteins. Methionine is the first and isoleucine is the second limiting amino acid for growth in herbage proteins.

The non-protein nitrogenous fraction of herbage varies with the physiological state of the plant. Generally, the more favourable the growth conditions, the higher is the non-protein N content as well as the total nitrogen value, and as the plants mature the contents of both decrease. The main components of the non-protein N fraction are amino acids, and amides

TABLE 17.1 Composition of the dry matter of a sample of Italian ryegrass cut at a young leafy stage (g/kg).

Proximate composition		Carbohydrates		Nitrogenous components		Other constituents	
Crude protein	190	Glucose	16	Total N	30	Lignin	52
Ether extract	45	Fructose	13	Protein N	27		
Crude fibre	208	Sucrose	45	Non-protein N	3		
Nitrogen-free		*Oligosaccharides	19				
extractives	449	Fructans	70				
Ash	108	Galactan	9				
		Araban	29				
		Xylan	63				
		Cellulose	202				

* Excluding sucrose

such as glutamine and asparagine, which are concerned in protein synthesis; nitrates may also be present, and considerable attention has recently been given to the presence of these in pasture herbage because of their toxic effects on farm animals. Nitrate *per se* is relatively non-toxic to animals. The toxic effect on ruminants is caused by the reduction of nitrate to nitrite in the rumen. Nitrite, but not nitrate, oxidises the ferrous iron of haemoglobin to the ferric state, producing a brown pigment, methaemoglobin, which is incapable of transporting oxygen to the body tissues. Toxic symptoms include trembling, staggering, rapid respiration and death. It has been reported that toxic symptoms may occur in animals grazing herbage containing more than 0.7 g nitrate-N/kg DM, although the lethal concentration is much higher than this. Some authorities have quoted a lethal figure for nitrate-N of 2.2. g/kg DM, whereas others have suggested a value far in excess of this. A sudden intake of nitrate may be particularly dangerous; experimentally this may be brought about by drenching, but may occur in practice when herbage that is normally non-toxic is eaten unusually quickly. Nitrate is sometimes less toxic if the diet also contains soluble carbohydrates. The nitrate content of grasses varies with species, variety and manuring, although the amount present is generally directly related to the crude protein content.

The lipid content of grasses, as determined in the ether extract fraction, is comparatively low and rarely exceeds 60 g/kg DM. The components of this fraction include triacylglycerols, glycolipids, waxes, phospholipids and sterols. Triacylglycerols occur only in small amounts and the major components are galactolipids, which constitute about 60 per cent of the total lipid content. Linolenic acid is the main fatty acid, comprising between 60 and 75 per cent of the total fatty acids present, with linoleic and palmitic acids the next most abundant.

The mineral content of pasture is very variable, depending upon the species, stage of growth, soil type, cultivation conditions and fertiliser application; an indication of the normal range in content of some essential elements is given in Table 17.2.

Green herbage is an exceptionally rich source of carotene, the precursor of vitamin A, and quantities as high as 550 mg/kg may be present in the dry matter of young green crops. Herbage of this type supplies about 100 times the requirement of a grazing cow when eaten in normal quantities.

It has generally been considered that growing plants do not contain vitamin D, although precursors are usually present. Recent studies suggest, however, that vitamin D may be present in herbage but in relatively small amounts. The increased vitamin D content of mature herbage compared with young material may be due in part to the presence of dead leaves in which vitamin D_2 has been produced from irradiated ergosterol.

TABLE 17.2 Ranges of essential mineral contents of temperate pasture grasses

Element	Low	Normal	High
g/kg DM			
Potassium	<12	15–30	> 35
Calcium	< 2.0	2.5–5.0	> 6.0
Phosphorus	< 2.0	2.0 –3.5	> 4.0
Sulphur	< 2.0	2.0–3.5	> 4.0
Magnesium	< 1.0	1.2–2.0	> 2.5
mg/kg DM			
Iron	<45	50–150	>200
Manganese	<30	40–200	>250
Zinc	<10	15–50	> 75
Copper	< 3.0	4.0–8.0	> 10
Molybdenum	<0.40	0.5–3.0	> 5.0
Cobalt	<0.06	0.08–0.25	> 0.30
Selenium	<0.02	0.03–0.20	> 0.25

Most green forage crops are good sources of vitamin E and of many of the B-vitamins, especially riboflavin.

Factors influencing the nutritive value of pasture

Stage of growth. Stage of growth is the most important factor influencing the composition and nutritive value of pasture herbage. As plants grow there is a greater need for structural tissues, and therefore the structural carbohydrates (cellulose and hemicelluloses) and lignin increase. This is reflected in the crude fibre content, which may increase from below 200 g/kg DM in the young plant to as much as 400 g/kg DM in the mature crop. As the plant ages the concentration of protein decreases; there is therefore a reciprocal relationship between crude protein and crude fibre contents in a given species, although this relationship can be upset by the application of nitrogenous fertilisers.

The variations in chemical composition of timothy at different stages of growth are shown in Table 17.3. In addition to the changes in crude protein and carbohydrates, changes also occur in the mineral or ash constituents. The total ash content decreases as the plant matures. This is reflected in the calcium content, which follows a similar pattern to that of the total ash in grasses. The magnesium content is generally high in the early spring, but falls off sharply; during the summer it rises, reaching high values in the autumn.

The digestibility of the organic matter is one of the main factors determining the nutritive value of forage, and this may be as high as 0.85 in

TABLE 17.3   Chemical composition of established timothy at different stages of growth. (After R. Waite and K. N. S. Sastry, 1949. *Emp. J. exp. Agric.* **17**, 179)

		Composition of dry matter						Moisture (g/kg)
		Crude protein (g/kg)	True protein (g/kg)	Crude fibre (g/kg)	Ether extract (g/kg)	Ash (g/kg)	Carotene (mg/kg)	
Date								
May	20	184	165	203	35.4	80.0	274	772
	26	188	157	232	35.6	85.0	273	816
June	2	142	128	263	37.4	71.5	156	790
	10	106	88	287	29.1	77.2	128	806
	16	92	87	308	30.2	68.6	108	761
	23	71	58	326	21.8	65.0	74	722
	30	63	55	304	21.2	59.8	72	655
July	7	68	44	312	17.9	61.2	66	652
	14	55	38	321	16.4	56.2	88	650

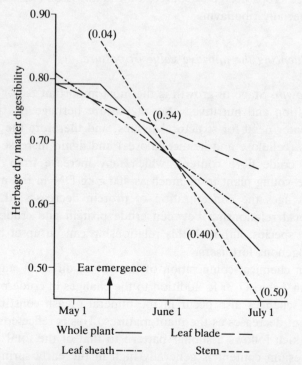

Fig. 17.1.   The digestibility *in vitro* of the dry matter in the whole plant, and in the leaf blade, leaf sheaf, and stem fractions of S.37 cocksfoot during first growth in the spring. Figures in parentheses are the proportions of stem in the whole plant. (After W. F. Raymond, 1969. *Adv. Agron.*, **21**, 21.)

young spring pasture grass and as low as 0.50 in winter foggage. Although there is a relationship between stage of growth and digestibility in that digestibility decreases as plants mature, the relationship is complicated by there being a spring period of up to a month during which the herbage digestibility remains fairly constant. This period has been described as the 'plateau'. The end of this period is associated in some plant species with ear emergence, after which digestibility of organic matter may decrease abruptly. In grasses grown in the UK, this rate of decrease is about 0.005 per day.

Differences in digestibility of grasses are influenced by leaf/stem ratios. The use of *in vitro* fermentation techniques has enabled the digestibility of different fractions of plants to be determined. In very young grass the stem is more digestible than the leaf, but whereas with advancing maturity the digestibility of the leaf fraction decreases very slowly, that of the stem fraction falls rapidly. As plants mature, the stem comprises an increasing proportion of the total herbage and hence has a much greater influence on the digestibility of the whole plant than the leaf. These changes are illustrated in Fig. 17.1.

The decrease in digestibility with stage of growth is also reflected in the net energy values, as shown in Table 17.4 for four cuts of S.23 ryegrass. The

TABLE 17.4   Composition (g/kg DM) and energy values (MJ/kg DM) of four cuts of S 23 ryegrass.
(After D. G. Armstrong, 1960. *Proc. 8th int. Grassld Congr.* p. 485)

	Cut 1 Young leafy	Cut 2 Late leafy	Cut 3 Ear emergence	Cut 4 Full seed
Proximate composition				
Ash	81	85	78	57
Crude protein	186	153	138	97
Ether extract	38	31	30	25
Crude fibre	212	248	258	312
Nitrogen-free extractives	483	483	496	509
Other constituents				
Crude lignin	36	46	55	75
Cellulose	253	284	299	356
Soluble carbohydrates	125	115	115	101
*Energy values**				
Metabolisable energy	13.1	12.2	11.6	8.9
Net energy (for maintenance)	10.3	9.3	8.8	7.3
Net energy (for liveweight gain)	6.9	6.9	5.6	3.8

* Determined with mature sheep

low net energy value of mature herbage is not only due to a low organic matter digestibility, but is also associated with a high concentration of cellulose. This polysaccharide encourages, in the rumen, high levels of acetic acid, which can have a low efficiency of utilisation for the production of depot fat (see Ch. 11).

Species and climate. The Gramineae form a very large family which has been subdivided into 28 tribes of which the six largest contain most pasture grasses of economic importance. All six of these tribes have a wide distribution but their importance in any particular region is determined largely by the temperature and to a lesser extent by the rainfall. The distribution of these tribes, and some agriculturally important members of them, are shown in Table 17.5.

In temperate areas having a reasonably uniform distribution of rainfall, grasses grow and mature relatively slowly and can thus be utilised at an early stage of growth when their nutritive value is high. In warmer climates, however, grasses mature more rapidly, their protein and phosphorus contents falling to very low levels, and their fibre content rising (see Table 17.6). In the wet tropics the herbage available is commonly fibrous but lush (i.e. high in water content); in drier areas the mature herbage becomes desiccated and is grazed as 'standing hay'. In both cases digestibility is low, typical values for tropical herbage being 0.1–0.15 units lower than for temperate herbage. The differences in composition between temperate and tropical grasses are not only a result of climate. Temperate species of grasses belong to the C_3 category of plants in which the three-carbon compound,

TABLE 17.5   The main tribes of the Gramineae, their areas of major importance, and some examples of their grassland species

Tribe	Major areas	Examples
Agrosteae	All temperate	*Agrostis* spp.—bent grasses
		Phleum pratense—timothy
Aveneae	Cold and temperate	*Holcus lanatus*—Yorkshire fog
		Danthonia pilosa—tussock grass
Festuceae	Temperate, particularly USA	*Lolium* spp.—ryegrasses
		Festuca spp.—fescues
		Bromus spp.—bromes
Eragrosteae	Tropical and warm temperate	*Eragrostis curvula*—weeping love grass
Andropogoneae	Tropics, particularly S.E. Asia	*Andropogon gayanus*—gamba
		Hyparrhenia rufa—Jaragua
Paniceae	Tropics and sub-tropics	*Digitaria decumbens*—pangola grass
		Panicum maximum—Guinea grass
		Paspalum dilatatum—dallis grass
		Pennisetum purpureum—elephant grass

TABLE 17.6 Composition (g/kg DM) and ME values (MJ/kg DM) of three cuts of Jaragua grass (*Hyparrhenia* rufa) grown in Brazil.
(From B.˙Göhl, 1981. *Tropical Feeds*. FAO, Rome)

| Stage of growth | Composition | | | | ME |
	Crude protein	Crude fibre	Ether extract	Ash	
Vegetative	92	289	26	149	8.4
Full bloom	35	314	19	136	7.0
Milk stage	28	337	15	115	6.5

phosphoglycerate is an important intermediate in the photosynthetic fixation of carbon dioxide. Most tropical grasses have a C_4 pathway of photosynthesis in which carbon dioxide is first fixed in a reaction involving the four-carbon compound oxalacetate. The low protein contents often found in tropical grasses are an inherent characteristic of C_4 plant metabolism, which is associated with survival under conditions of low soil fertility.

In temperate grasses fructans are the main storage carbohydrates while in tropical species these are replaced by starch.

Another factor of nutritional importance is that the mesophyll cells in the leaves of tropical grasses are more densely packed than those in temperate grasses and intercellular air spaces represent only 3–12 per cent of leaf volume compared with 10–35 per cent in temperate species. This may partly explain why tropical grasses have a higher tensile strength than temperate ones, a feature which results in both a slower mechanical and slower microbial degradation in the rumen. A consequence of this is relatively low voluntary dry matter intakes by ruminants consuming these plants.

The selection of pasture species and varieties is based on such agronomic characters as persistency and productivity, but nutritive value is also taken into account. Varieties within a species generally differ to only a small degree in nutritive value, if the comparison is made at the same stage of growth, but differences between comparable species may be larger. A classical example for the temperate grasses is the difference between British varieties of perennial ryegrass (*Lolium perenne*) and of cocksfoot or orchard grass (*Dactylis glomerata*). At the same stage of growth the cocksfoot variety, S.37, has a lower concentration of soluble carbohydrates and is 0.05–0.06 units lower in dry matter digestibility than the ryegrasses, S.23 and S.24.

In Britain perennial ryegrass is the most important species of sown pastures, but Italian ryegrass (*Lolium multiflorum*), timothy (*Phleum pratense*), cocksfoot and the fescues (*Festuca* spp.) are also common. In older pastures these are accompanied by weed species, particularly meadow grasses (*Poa* spp.), Yorkshire fog (*Holcus lanatus*) and the bents (*Agrostis* spp.). In the uplands, however, some of these weed species, particularly the bents, are valued constituents of the sward.

Plants of the family Leguminosae, especially *Trifolium* spp., make an important contribution to pastures. The nutritional characteristics of this family will be described in a later section of this chapter.

Soils and fertiliser treatment. The type of soil may influence the composition of the pasture, especially its mineral content. Plants normally react to a mineral deficiency in the soil either by limiting their growth or by reducing the concentration of the element in their tissues, or more usually by both.

The acidity of the soil is an important factor which can influence, in particular, the uptake of many trace elements by plants. Both manganese and cobalt are poorly absorbed by plants from calcareous soils, whereas low molybdenum levels of herbage are usually associated with acid soils. Teart (see p. 105), associated with high herbage molybdenum levels, generally occurs on pasture grown on soils derived from Lower Lias clay or limestone.

Liberal dressings of fertilisers can markedly affect the mineral content of plants, and it is also known that the application of nitrogenous fertilisers can increase the crude protein of pasture herbage and influence the amide and nitrate content. Application of nitrogenous fertilisers also depresses the fructan content of temperate grasses. Fertilisers may also affect, indirectly, the nutritive value of a sward by altering the botanical composition. For example legumes do not thrive on a lime-deficient soil, while heavy dressings of nitrogen encourage growth of grasses and at the same time depress clover growth.

Other factors affecting the nutritive value of pasture. Such factors as climate, season and rate of stocking may influence the nutritive value of pasture. The concentration of sugars and fructans, for example, can be influenced markedly by the amount of sunshine received by the plant. Generally on a dull cloudy day the soluble carbohydrate content of grass will be lower than on a fine sunny day. Rainfall can affect the mineral composition of pasture herbage. Calcium, for example tends to accumulate in plants during periods of drought but to be present in smaller concentration when the soil moisture is high; on the other hand phosphorus appears to be present in higher concentrations when the rainfall is high.

Both under-grazing and over-grazing can influence the composition of the pasture. With under-grazing the plants become mature and of low digestibility, resulting in reduced intake of dry matter. Over-stocking will tend to eliminate or weaken many of the best grasses by depriving them of the opportunity for building up and storing reserve nutrients in their roots.

The net energy value of autumn grass is often lower than that of spring grass, even when the two are cut at the same stage of growth and are equal in digestibility or metabolisable energy concentration. For example the dry matter of spring grass containing 11 MJ ME/kg DM was found to supply 5.2 MJ net energy/kg for fattening, whereas later growths of the same metabolisable energy content supplied only 4.3 MJ net energy/kg. This difference cannot at present be attributed to differences in the chemical composition of early- and late-season herbages, although the latter are often lower in soluble carbohydrate content and higher in protein content.

OTHER FORAGE CROPS

Legumes

The family Leguminosae, which contains about 18 000 species are valued for their ability to grow in a symbiotic relationship with nitrogen-fixing bacteria and for their drought resistance. The commonest legumes found in pastures are the clovers (*Trifolium* spp.), the main representatives being red clover (*T. pratense*) and white clover (*T. repens*) in the cooler and wetter regions such as Europe and New Zeland, and subterranean clover (*T. subterraneum*) in drier areas such as southern Australia.

Nutritionally, the clovers are superior to grasses in protein and mineral content (particularly calcium, phosphorus, magnesium, copper and cobalt), and their nutritive value falls less with age. Studies with white clover have shown that the rate of reduction in particle size and the rate of movement of particulate matter from the rumen is more rapid than with grass. Sheep and cattle offered white clover as fresh forage consumed 20 per cent more dry matter compared with grass of the same metabolisable energy content. Similar high voluntary intakes of dry matter have been obtained using red clover and other legumes.

The sugars present in clovers are similar to those found in grasses, the main sugar being sucrose. Fructans are generally absent, but starch is present and concentrations of this polysaccharide as high as 50 g/kg have been reported in the dried leaves of red clover.

Many tropical pastures are deficient in indigenous legumes, but determined efforts are being made to introduce them. In Australia, for example, the South American Centro (*Centrosema pubescens*) has been introduced into pastures of the wet tropical type, and Siratro (*Macroptilium atropurpureum*) has been bred from Mexican cultivars for use in drier areas.

Lucerne or alfalfa (*Medicago sativa*) also occurs in pastures, but like many other legumes is more commonly grown on its own. It is found in warm temperate areas and in many tropical and subtropical countries. The protein content is comparatively high and declines only slowly with maturity (Table 17.7). Under conditions in the UK, lucerne tends to be high in crude fibre, particularly the stem, and at the flowering stage may be as high as 500 g/kg DM. Lucerne varieties are distinguished by the time of flowering, and under UK conditions early flowering types are recommended. These varieties usually flower in the second week of June, but to obtain a cut with an acceptable digestibility, the crop should be first harvested at the early bud stage (end of May), when the expected digestible organic matter (DOM) content would be 620–640 g/kg DM, and subsequently cut at six- to eight-week intervals to give DOM values of 560–600 g/kg DM.

In Britain, the small area of lucerne grown is harvested mainly for artificial drying or made into silage (see Ch. 18), but in other parts of the world, notably the USA (where it is known as alfalfa), the crop is also used for grazing.

Berseem or Egyptian clover (*Trifolium alexandrinum*) is an important legume grown in the Mediterranean area and India. It is valued for its rapid growth in the cooler winter season in the subtropics and for its good recovery after cutting or grazing. It has a nutritive value very similar to that of lucerne.

Sainfoin (*Onobrychis viciifolia*) is a legume of less economic importance than lucerne, and in the UK is confined to a few main areas in the south. In common with most green forages the leaf is richer than the stem in crude protein, ether extract and minerals, especially calcium. Changes which occur in the composition of the plant are mainly due to variation in stem composition and leaf-stem ratio. The crude protein content in the dry matter may vary from 240 g/kg at the early flowering stage to 140 g/kg at full flower. Corresponding crude fibre values at similar stages of growth may be 140 and 270 g/kg DM.

Peas (*Pisum sativum*), beans (*Vicia faba*) and vetches (*V. sativa*) are

TABLE 17.7 Composition of the dry matter of lucerne.
(From MAFF 1975. *Energy Allowances and Feeding Systems for Ruminants, Tech. Bull. 33*, p. 70. HMSO, London)

		Pre-bud	In-bud	Early flower
Crude fibre	(g/kg)	220	282	300
Ash	(g/kg)	120	82	100
Crude protein	(g/kg)	253	205	171
Digestible crude protein	(g/kg)	213	164	130
Digestible organic matter	(g/kg)	670	620	540
Metabolisable energy	(MJ/kg)	10.2	9.4	8.2

sometimes grown as green fodder crops, and if cut at the early flowering stage are similar in nutritive value to other legumes.

A trouble which is frequently encountered in cattle and sheep grazing on legume-dominant pastures is *bloat*, characterised by an accumulation of gas in the rumen. The primary cause of bloat is the retention of the fermentation gases in a stable foam, preventing their elimination by eructation (see p. 149).

Leucaena (*Leucaena leucocephala*), also known as ipil-ipil, is a valuable tropical legume which is cultivated extensively in many parts of the world, notably in SE Asia, Latin America and the West Indies. Unlike the other legumes described in this chapter, leucaena is a shrub or tree. The leaves, young stems, flowers and pods are all excellent sources of protein and minerals. Leaf material of leucaena compares favourably with lucerne in terms of crude protein (250–340 g/kg DM), minerals and is also an excellent source of β-carotene. However, leucaena as the sole source of feed for ruminants can induce weight loss, thyroid dysfunction and alopecia (hair loss) because of the presence of certain toxic components. One of these is the amino acid, mimosine, which occurs in all parts of the plant. If ruminants are gradually introduced to leucaena, rumen micro-oganisms capable of breaking down the mimosine increase so that the toxic action of the amino acid is reduced.

Cereals

Cereals are sometimes grown as green forage crops, either alone or mixed with legumes. Like the grain, the forage is rich in carbohydrate and low in protein, its nutritive value depending mainly on the stage of growth when harvested (see Table 17.8). The crude protein content at the cereal grazing stage is generally within the range of 80–120 g/kg DM. At the time of ear formation the concentration of crude fibre falls as a result of the great increase in soluble carbohydrates.

Sugarcane

Sugarcane (*Saccharum officinarum*) is a tropical or subtropical perennial grass which grows to a height of 4.5 to 6.0 m or more. The crop is processed for its sugar which leaves two by-products, molasses and a fibrous residue termed bagasse. Sugarcane molasses is a high-energy, low-protein food similar in composition to the molasses obtained as a by-product from sugar beet (see p. 433). Bagasse is a high-fibre, low-protein product of very low digestibility which is sometimes mixed with the cane molasses for cattle feeding. The unmolassed bagasse has a digestibility of about 0.28, but this

TABLE 17.8   Composition of whole barley at different stages of growth

| Stage | Composition of dry matter (g/kg) | | | | | | Moisture (g/kg) | Digestibility (in vitro) of organic matter |
	Crude protein	Crude fibre	Ash	Calcium	Phosphorus	Water soluble carbohydrates		
Ear emergence	103	313	75	5.1	2.3	169	809	0.619
Full flower	87	321	65	5.5	2.0	180	800	0.542
Watery kernel	72	286	56	5.2	1.8	249	742	0.578
Milky kernel	67	253	47	4.1	1.7	318	707	0.608
Early dough	56	204	38	3.5	1.6	241	647	0.626
Late dough	60	195	35	2.3	1.7	147	614	0.614
Mature	66	254	35	1.9	2.1	46	575	0.549

* *In vitro values*

can be dramatically increased to about 0.55 by short-term (5 to 15 minutes) treatment with wet steam at 200 °C. Steam treated bagasse, supplemented with urea, has been shown to be suitable as a maintenance feed for beef cows.

In some countries, the whole sugarcane crop is used as a forage for ruminants. The whole crop on a dry matter basis has an ME value of about 9 MJ/kg and a low crude protein content of about 40 g/kg.

Brassicas

The genus *Brassica* comprises some forty species, of which the following are of agricultural importance: kales, cabbages, rapes, turnips and swedes. Some of the brassicas are grown primarily as root crops, and these will be discussed in Chapter 20.

Kales (*Brassica oleracea*). The kales include a very wide variety of plant types which range from short leafy plants, 30 cm high, to types 2 m tall with stems strong enough to be used in building. The commonest short type is thousandhead kale (var. *fruticosa*) and the commonest tall type, marrowstem kale (var. *acephala*)—known in Australasia as 'chou moellier'. They are grown in temperate parts of the world to provide green fodder during winter, but in drier areas they may also be used to supplement summer grazing.

Kales are low in dry matter content (*c*. 140 g/kg) which is rich in protein (*c*. 150 g/kg), water-soluble carbohydrates (*c*. 200–250 g/kg), and calcium (10–20 g/kg), and their digestibility is generally high. The woody stems of marrow-stem kale are lower in digestibility than the rest of the plant and may be rejected by animals.

Rapes. The rapes grown in Britain are usually swede-rapes (*B. napus*) although turnip-rapes (*B. campestris*) also occur. Their nutritive value is similar to that of the kales.

Cabbages (*B. oleracea*, var. *capitata*). These are grown for both human and animal consumption and range in type from 'open-leaved' to 'drumhead'. All have a low proportion of stem and hence are less fibrous then either kales or rapes.

Toxicity of brassica forage crops. All brassicas, whether grown as forage, root or oilseed crops, contain goitrogenic substances (see p. 108).

In the forage crops these are mainly of the thiocyanate type which interferes with the uptake of iodine by the thyroid gland, and whose effects can be overcome by increasing the iodine content of the diet. All animals grazing on forage brassicas may develop goitre to some extent, but the most serious effects are found in lambs born to ewes which have grazed on brassicas during pregnancy; these lambs may be born dead or deformed. It has been suggested (but not adequately confirmed) that cows grazing on kales may secrete sufficient goitrogen in their milk to cause goitre in children drinking it.

Forage brassicas may also cause a haemolytic anaemia in ruminants, in extreme cases of which the haemoglobin content of the blood falls to only a third of its normal value and the red cells are destroyed so rapidly that haemoglobin appears in the urine (haemoglobinuria). The condition is due to the presence in brassicas of the unusual amino acid, S-methylcysteine sulphoxide, which in the rumen is reduced to dimethyl disulphide:

$$2 CH_3 - S.CH_2CHNH_2COOH + H_2 \rightarrow CH_3.S.S.CH_3 + 2CH_3CO.COOH + 2NH_3$$

S-methylcysteine sulphoxide   Dimethyl disulphide   Pyruvic acid

The dimethyl disulphide is known to damage the red cells. Green brassicas contain 10–20 g S-methylcysteine sulphoxide/kg DM. The condition is best avoided by ensuring that kale or rape contributes no more than one-third of the animal's total dry matter intake.

Green tops

Mangel, fodder beet, sugar beet, turnip and swede tops may all be used for feeding farm animals. Care is required in feeding with mangel, fodder and sugar beet tops, since they contain a toxic ingredient which may lead to extensive scouring and distress and in extreme cases death. The risk appears to be reduced by allowing the leaves to wilt. The toxicity has been attributed to oxalic acid and its salts, which are supposed to be reduced or removed by wilting. A recent study casts some doubt on this theory, since the oxalate content of the leaves is practically unaffected by wilting. It is possible that the toxic substances are not oxalates but other factors which are destroyed during wilting.

Swede and turnip tops are safe to feed, and may have a crude protein content in the dry matter as high as 200 g/kg, the digestibility of the organic matter being about 0.70; like kale, rape and cabbage, they may cause haemolytic anaemia in ruminants.

Sugar beet tops generally contain the upper part of the root as well as the green leaves and are more digestible, about 0.77. All these green tops are excellent sources of carotene.

Oestrogenic substances in forage crops

A large number of species of plants are known to contain compounds which have oestrogenic activity. Pasture plants containing these phytoestrogens are mainly of the species, *Trifolium subterraneum* (subterraneum clover), *T. pratense* (red clover), *Medicago sativa* (lucerne) and *M. truncatula* (barrel medic). The oestrogens in *Trifolium* sp. are mainly isoflavones, whereas those in *Medicago* sp. are usually coumestans.

Naturally occurring isoflavones and coumestans have relatively weak oestrogenic activity, but this activity can be increased as a result of metabolism in the rumen. Some plants, e.g. *T. repens* (white clover) are normally non-oestrogenic but when infected with fungi can produce high concentrations of coumestan.

The consumption of oestrogenic pasture plants by sheep leads to severe infertility and post-natal death in lambs. The infertility can persist for long periods after the ewes have been taken off the oestrogenic pastures. The main cause of the infertility is a failure of fertilisation associated with poor sperm penetration to the oviduct. A 'temporary infertility' may occur in ewes grazing oestrogenic pastures at the time of mating. Fertility is restored when sheep are moved to other pastures.

Cattle grazing oestrogenic pastures do not appear to suffer the severe infertility problems that affect sheep.

FURTHER READING

G. W. Butler and R. W. Bailey (eds), 1973. *Chemistry and Biochemistry of Herbage*, Vols 1–3. Academic Press, London.
D. C. Church, 1984. *Livestock Feeds and Feeding*. O and A Books, Corvallis, Oregon.
B. Gohl, 1981. *Tropical Feeds*. FAO, Rome.
J. B. Hacker, (ed), 1982. *Nutritional Limits to Animal Production from Pastures*. Commonwealth Agricultural Bureaux, Farnham Royal.
W. Holmes, (ed), 1980. *Grass, its Production and Utilisation*. Blackwell Scientific Publications, Oxford.
P. J. Skerman, 1977. *Tropical Forage Legumes*. FAO, Rome.
C. R. W. Spedding and E. C. Diekmahns (eds), 1972. *Grasses and Legumes in British Agriculture*. Commonwealth Agricultural Bureaux, Farnham Royal.
P. C. Whiteman, 1980. *Tropical Pasture Science*, Oxford University Press.

18

Silage

Silage is the material produced by the controlled fermentation of a crop of high moisture content. Ensilage is the name given to the process, and the container, if used, is called the silo. The fermentation is controlled either by encouraging the growth of lactic acid bacteria which are present on the fresh herbage, or by restricting fermentation by pre-wilting the crop or by using chemical additives. In all cases, in order to obtain successful preservation, it is necessary to achieve and maintain anaerobic conditions. In practice this is done by chopping the crop during harvesting, by rapid filling of the silo and by adequate consolidation and sealing. Almost any crop can be preserved as silage, although the commonest crops used are grasses, legumes, whole cereals (especially maize), and fruit residues.

The nutritional value of the silage produced depends firstly upon the species and stage of growth of the harvested crop, factors which have already been discussed in the previous chapter, and secondly upon the changes resulting from the activities of the plant and microbial enzymes during the storage period.

SILAGE MICROBIOLOGY

Aerobic fungi and bacteria are the dominant micro-organisms on fresh herbage, but as anaerobic conditions develop in the silo they are replaced by bacteria able to grow in the absence of oxygen, such as species of *Escherichia*, *Klebsiella*, *Bacillus*, *Clostridium*, *Streptococcus*, *Leuconostoc*, *Lactobacillus* and *Pediococcus*. Yeasts are also present and, being facultative anaerobes (i.e. able to grow both aerobically and anaerobically), can survive and proliferate in silage.

The lactic acid bacteria, which are also facultative anaerobes, are normally present on growing crops in small numbers but usually multiply rapidly after harvesting, particularly if the crop is chopped or lacerated. They can be divided into two categories, the homofermentative and the hetero-fermentative lactic acid bacteria (Table 18.1). When the crop is ensiled, the lactic acid bacteria continue to increase, fermenting the water soluble carbo-hydrates (see p. 388) in the crop to organic acids, mainly lactic, which reduce the *p*H value. At a certain critical *p*H level, which varies with moist-ure content, the acids inhibit the growth of other bacteria and at about *p*H 3.8 to 4.0 microbial activity virtually ceases and the material remains stable for as long as anaerobic conditions are maintained.

If a stable *p*H has not been achieved then the saccharolytic clostridia, which are present on the original crop as spores, will multiply; they ferment lactic acid and residual water soluble carbohydrates to butyric acid causing the *p*H to rise. The less acid-tolerant proteolytic clostridia then usually become active, leading to a further increase in *p*H caused by the production of ammonia. Clostridia are particularly sensitive to water availability and they require very wet conditions for active growth. With very wet crops (i.e. those with a DM concentration of about 150 g/kg), even the achievement of a *p*H value as low as 4 may not inhibit their activity.

Another group of organisms, which are normally found on harvested crops and which can multiply in the silo, are the coliform bacteria. They are members of the family Enterobacteriaceae and include *Escherichia* and *Klebsiella* spp. These organisms have been referred to as the 'acetic acid bacteria' since acetic acid is a major product when they ferment sugars. The optimum *p*H for the growth of the coliform bacteria is about 7.0, and they are usually only active in the early stages of fermentation when the *p*H is favourable for their growth.

TABLE 18.1 Some species of lactic acid bacteria commonly found on fresh herbage and in silage

Homofermentative	*Heterofermentative*
Lactobacillus coryneformis	*Lactobacillus brevis*
Lactobacillus plantarum	*Lactobacillus buchneri*
Pediococcus acidilactici	*Lactobacillus fermentum*
Pediococcus pentosaceus	*Lactobacillus viridescens*
Streptococcus faecalis	*Leuconostoc citrovorum*
Streptococcus faecium	*Leuconostoc dextranicum*
Streptococcus lactis	*Leuconostoc mesenteroides*

BIOCHEMICAL CHANGES DURING ENSILAGE

Carbohydrates

Plant respiratory enzyme activity will continue in the ensiled herbage as long as conditions are aerobic and the *p*H is not drastically altered. The water soluble carbohydrates in the crop will be oxidised to carbon dioxide and water, with the production of heat sufficient to cause a considerable rise in temperature of the mass. If the herbage has not been well consolidated during and after filling, air may permeate into the mass and the temperature will continue to rise. If this is not checked then an overheated silage may result.

Under more normal conditions, when anaerobiosis is rapidly achieved, the lactic acid bacteria ferment the water soluble carbohydrates, available after hydrolysis mainly as glucose and fructose, to lactic acid and other products. The homofermentative lactic acid bacteria are more efficient at producing lactic acid from hexose sugars than are the heterofermentative organisms (Table 18.2). Some hydrolysis of hemicelluloses also occurs, liberating pentoses, which may be fermented to lactic acid.

Proteins

In the growing crop, about 75 to 90 per cent of the total N is in the form of protein N. Immediately after harvesting, plant proteases hydrolyse proteins to amino acids and within 12 to 24 h, 20 to 25 per cent of the total N is converted into non-protein N. Although most of the latter is in the form of amino acids some further breakdown of these, particularly glutamic and aspartic acids, by plant decarboxylases usually occurs. Lactic acid bacteria have only a limited ability to attack amino acids and current evidence suggests that only serine and arginine are significantly metabolised, to acetoin and ornithine respectively. If, however, clostridia become dominant, then extensive changes to the amino acids can occur. These changes are brought about by three different types of reactions—deamination, decarboxylation and coupled oxidation/reduction (Stickland) reactions, which result in the production of amines, ammonia, carbon dioxide, keto acids and fatty acids (Table 18.2). Coliform bacteria also have the ability to deaminate and decarboxylate amino acids.

Organic acids and buffering capacity

The buffering capacity (Bc) of plants, i.e. their ability to resist changes in *p*H, is an important factor in ensilage. Within the *p*H range 4 to 6, about

TABLE 18.2   Some fermentation pathways in ensilage

(A)   *Lactic acid bacteria*
  Homofermentative
    Glucose →2 Lactic acid
    Fructose →2 Lactic acid
    Pentose → Lactic acid + Acetic acid

  Heterofermentative
    Glucose → Lactic acid + Ethanol + CO_2
    3 Fructose → Lactic acid + 2 Mannitol + Acetic acid + CO_2
    2 Fructose + Glucose → Lactic acid + 2 Mannitol + Acetic acid + CO_2
    Pentose → Lactic acid + Acetic acid
(B)   *Clostridia*
  Saccharolytic
    2 Lactic acid → Butyric acid + $2CO_2$ + $2H_2$
  Proteolytic
    Deamination
    Leucine → Isovaleric acid + NH_3 + CO_2
    Lysine → Acetic acid + Butyric acid + $2NH_3$
    Serine → Pyruvic acid + NH_3
    Tryptophan → Indolepropionic acid + NH_3
  Decarboxylation
    Arginine → Putrescine + CO_2
    Glutamic acid → γ-aminobutyric acid + CO_2
    Histidine → Histamine + CO_2
    Lysine → Cadaverine + CO_2
    Tryptophan → Tryptamine + CO_2
  Oxidation/reduction (Stickland)
    Alanine + 2 Glycine → 3 Acetic acid + $3NH_3$ + CO_2

70 to 80 per cent of the Bc of herbage can be attributed to salts of organic acids, orthophosphates, sulphates, nitrates and chlorides, with only about 10 to 20 per cent attributable to plant proteins. Buffering capacity is frequently expressed in terms of mequiv alkali per kg DM required to change the pH of herbage macerates from 4 to 6. On this basis, ryegrasses have Bc values usually within the range 250 to 400. Legumes such as clover and lucerne are more highly buffered with Bc values of 500 to 600, a property which makes them more difficult to conserve satisfactorily as silage. During fermentation, because of the formation of lactate, acetate and other products, the Bc increases. In silages of low pH this increase may amount to 3 or 4 times the original value. Treatments which restrict fermentation such as wilting or the addition of chemical inhibitors reduce the development of buffering substances. In grasses, the main organic acids present in the original crop are citric and malic which, during ensilage, are fermented by the lactic acid bacteria to a number of products including lactic acid, acetic acid, formic acid, ethanol, butanediol and acetoin.

Pigments

The most obvious visible change during ensilage is that of the colour of the herbage. The light brown colour of silage is caused by the action of organic acids on chlorophyll, which is converted into the magnesium-free pigment phaeophytin.

Destruction of the provitamin A pigment, β-carotene, is related to temperature and the degree of oxidation. When these are both high, losses of β-carotene can be considerable. In well preserved silages, however, carotene losses are usually less than 30 per cent. In one study in which samples of 25 different silages were examined, β-carotene values ranged from 123 to 696 mg/kg DM with a mean value of 417, which compares favourably with amounts present in fresh herbage (see p. 390).

LOSSES OF NUTRIENTS DURING ENSILAGE

Field losses

With crops cut and ensiled the same day, nutrient losses are negligible and even over a 24 h wilting period, losses of not more than 1 or 2 per cent DM can be expected. Over periods of wilting longer than 48 h, considerable losses of nutrients can occur depending upon weather conditions. Dry matter losses as high as 6 per cent after 5 days and 10 per cent after 8 days wilting in the field have been reported. The main nutrients affected are the water soluble carbohydrates, and proteins which are hydrolysed to amino acids.

Oxidation losses

These result from the action of plant and microbial enzymes on substrates such as sugars in the presence of oxygen, leading to the formation of carbon dioxide and water. In a silo which has been rapidly filled and sealed, the oxygen trapped within the plant tissues is of little significance causing a dry matter loss of about 1 per cent only. Continuous exposure of herbage to oxygen, as sometimes occurs on the sides and upper surface of ensiled herbage, leads to the formation of inedible composted material. Measurements of this surface waste can be misleading since losses of up to 75 per cent DM can occur in its formation.

Fermentation losses

Although considerable biochemical changes occur during fermentation, especially to the soluble carbohydrates and proteins, overall dry matter and energy losses arising from the activities of lactic acid bacteria are low. Dry matter losses can be expected to be less than 5 per cent and gross energy losses, because of the formation of high energy compounds such as ethanol, are even less: In clostridial fermentations, because of the evolution of the gases carbon dioxide, hydrogen and ammonia, nutrient losses will be much higher than in lactate fermentations.

Effluent losses

In most silos, free drainage occurs and the liquid or effluent carries with it soluble nutrients. The amount of effluent produced depends largely upon the initial moisture content of the crop, but it will obviously be increased if the silo is left uncovered so that rain enters. Effluent contains sugars, soluble nitrogenous compounds, minerals and fermentation acids all of which are of high nutritional value. Crops ensiled with a dry matter content of 150 g/kg may result in effluent dry matter losses as high as 10 per cent, whereas crops wilted to about 300 g/kg DM produce little, if any, effluent.

CLASSIFICATION OF SILAGES

Lactate silage

In this, the commonest type of silage produced from unwilted grasses and whole cereal crops, lactic acid bacteria have dominated the fermentation. Lactate silages are characterised by having a low pH (c. 3.7–4.2), and a high concentration of lactic acid. In grass silages, lactic acid contents normally lie in the range 80–120 g/kg DM but may be higher if silages are made from crops rich in water-soluble carbohydrates. The lactic acid contents of maize silages are usually much lower (30–80 g/kg DM) than those of well-preserved grass silages because of the higher dry matter and lower buffering properties of the maize crop (Table 18.3).

Lactate silages usually contain small amounts of acetic acid and may also contain traces of propionic and butyric acids. Variable amounts of ethanol and mannitol derived from the activities of heterofermentative lactic acid bacteria and yeasts are present. The buffering capacities of these silages are high and the water soluble carbohydrates low (Table 18.3). The nitrogenous components are mainly in a non protein, soluble form and, because very little deamination of amino acids will have occurred, the free ammonia N values will be low, being usually less than 100 g/kg of total N. However, the high soluble non-protein N content of these silages, coupled with the low levels of soluble carbohydrate, can result in high ammonia concentrations in the rumen, which lead to poor utilisation of the silage N.

TABLE 18.3 Typical compositions of grass, lucerne and maize silages (From P. McDonald and R. A. Edwards, 1976, *Proc. Nutr. Soc.*, **35**, 201; M. Ohshima, P. McDonald and T. Acamovic, 1979, *J. Sci. Food Agric.*, **30**, 97; J. M. Wilkinson and R. H. Phipps, 1979, *J. agric. Sci. Camb.*, **92**, 485)

| | Silage type | | | | | | | |
| | Grass | | | | | Lucerne | | Maize |
	Lactate	Butyrate	Acetate	Wilted	Additive-treated*	Acetate	Additive-treated*	Lactate
pH	3.9	5.2	4.8	4.2	5.1	7.0	4.7	3.9
DM (g/kg)	190	170	176	308	212	131	145	285
Bc (mequiv/kg DM)	1120	—	1090	890	560	2530	787	—
DM composition								
Protein N (g/kg TN)	235	353	440	289	740	260	657	545
Ammonia N (g/kg TN)	78	246	128	83	30	292	21	63
Lactic acid (g/kg)	102	1	34	59	26	13	5	53
Acetic acid (g/kg)	36	24	97	24	10	114	8	26
Butyric acid (g/kg)	1	35	2	1	1	8	0.1	0
Water-soluble carbohydrates (g/kg)	10	6	3	48	133	0	64	16
Mannitol (g/kg)	41	—	2	36	—	—	—	—
Ethanol (g/kg)	12	—	8	6	4	—	—	<10

* Treated with formalin–formic acid (3 : 1 w/w mixture; 10 g/kg.)

Because of the extensive changes to the soluble carbohydrates, resulting in the formation of high energy compounds such as ethanol (gross energy = 29.8 MJ/kg), the gross energy concentrations of lactate silages increase. These changes are also reflected in the metabolisable energy values of the silages (Table 18.4). The digestibility of lactate silages is similar to that of the original crop. Methane production in the rumen is the same as that arising from animals on fresh grass diets, although the urinary energy losses might be slightly higher from animals consuming lactate silages than from those consuming grass. Very few net energy determinations of silages have been carried out, but from the limited data available for grass silages there is some evidence to suggest that the efficiency of utilisation of metabolisable energy for growth and fattening (k_f) is reduced during ensiling. One disadvantage of lactate silages is that when given *ad libitum* to ruminants, they promote lower dry matter intakes than comparable fresh or dried herbages. This reduction in intake is generally greater in sheep than in cattle. Although dry matter intake has been negatively correlated with total acidity and acetic acid content, the exact cause of low intake is unknown.

Acetate silage

Under certain ill-defined conditions, acetic acid producing bacteria may dominate the fermentation. Such acetate silages seem to be rare in temperate countries, but have been reported with tropical grass species such as *Setaria sphacelata*. Acetate silages contain high levels of acetic acid and relatively low levels of lactic acid (Table 18.3). Deamination of amino acids

TABLE 18.4 Nutritional value of some silages compared with original perennial ryegrass* (From E. Donaldson and R. A. Edwards, 1976, *J. Sci. Fd Agric.*, **27**, 536)

	Ryegrass	Lactate silage	Wilted silage	Additive-treated wilted silage[†]
pH	6.1	3.9	4.2	4.4
DM (g/kg DM)	175	186	316	336
Lactic acid (g/kg DM)	—	102	59	43
Water-soluble carbohydrates (g/kg DM)	140	10	47	151
DM digestibility	0.784	0.794	0.752	0.776
GE (MJ/kg DM)	18.5	20.7	18.7	19.1
ME (MJ/kg DM)	11.6	13.6	11.4	12.0
DM intake (g/kg/day)	11.2	8.5	9.7	12.3

* Trials carried out with sheep, access to silage limited to two periods of 2 h daily.
† Original wilted grass treated with formic acid, 3.3 g/kg

is usually extensive, and consequently ammonia levels in these silages are higher than those found in lactate silages.

Because of the negative correlation of acetic acid content with dry matter intake, it is reasonable to assume that the latter will be low in animals given these silages *ad libitum*.

Butyrate silages

These silages have undergone a clostridial fermentation. They usually have *p*H values within the range 5 to 6, and contain low concentrations of lactic acid and water-soluble carbohydrates. Butyric acid is usually the dominant fermentation product, although acetic acid contents are also frequently high. Because of the extensive breakdown of amino acids caused by clostridia, silages of this type will contain high concentrations of ammonia-N, usually in excess of 200 g/kg total N. Decarboxylation of amino acids to amines also occurs (Table 18.2). As a result of these changes, the subsequent utilisation by ruminants of the nitrogenous compounds in butyrate silages is likely to be low. The dry matter intake of ruminants given these silages is low, but whereas there is a close negative correlation between dry matter intake and the concentration of silage ammonia-N, the exact cause of these reduced intakes by animals is unknown.

Wilted silages

Wilting a crop prior to ensiling restricts fermentation increasingly as dry matter content increases. In such wilted silages, clostridial activity is minimal although some growth of lactic acid bacteria occurs, even in herbage wilted to dry matter contents as high as 500 g/kg. With very dry silages of this type, anaerobic storage in bunker silos is difficult and tower silos are preferred because there is less risk of air penetration. For bunker-type silos, a more normal aim is to prewilt the crop to a dry matter content of 280 to 320 g/kg. The composition of a typical wilted silage is shown in Table 18.3. Total fermentation acids and buffering capacity values are lower than in unwilted silages, and there are usually some residual water soluble carbohydrates. Wilting does not prevent proteolysis occurring, although deamination of amino acids is considerably reduced. Both gross energy and metabolisable energy concentrations of wilted silages will be similar to those of the parent material (Table 18.4).

Although wilting usually results in an increase in silage dry matter intake, any beneficial effects in terms of improved animal performance, compared with well-preserved, unwilted, silage diets are difficult to demonstrate. The

main advantages of wilting are that the risk of obtaining a butyrate silage
is decreased and the production of effluent is reduced.

Additive-treated silages

Silage additives can be classified into two main types: *stimulants*, such as
inoculants and sugars, which encourage the development of lactic acid
bacteria, and *inhibitors* such as acids and formalin, which partially or
completely inhibit microbial growth.

Stimulants. Most of the early work carried out on inoculants gave
variable and inconsistent results. In recent years a number of commercial
inoculants containing freeze-dried cultures of homofermentative lactic acid
bacteria have become available. Most of these contain *Lactobacillus plantarum* and often other suitable organisms such as *Pediococcus acidilactici*.
Successful control of fermentation, using these inoculants, depends upon a
number of factors including the inoculation rate, which should be at least
10^5 (but preferably 10^6) organisms/g fresh forage, and the presence of an
adequate level of fermentable carbohydrates. The rapid domination of the
fermentation by homolactic bacteria ensures the most efficient use of the
water-soluble carbohydrates and, when levels of these in the crop are
critical, increases the chances of producing a well-fermented lactate silage.

Molasses, which is a by-product of the sugar beet and sugar cane industries (see p. 433), was one of the earliest silage additives to be used as a
source of sugars. The by-product has a water-soluble carbohydrate content
of about 700 g/kg DM and the additive has been shown to increase the dry
matter and lactic acid contents, and to reduce the pH and ammonia levels
in treated silages.

Inhibitors. A large number of chemical compounds have been tested
as potential fermentation inhibitors, but relatively few have been accepted
for commercial use. One of the earliest was a mixture of mineral acids
proposed by A. I. Virtanen, the technique being referred to as the A.I.V.
process. The acids, usually hydrochloric and sulphuric, were added to the
herbage during ensiling in sufficient quantity to lower the pH value below
4.0. This process was for many years very popular in the Scandinavian countries and, if carried out effectively, is a very efficient method of conserving
nutrients. In recent years, however, formic acid has largely replaced mineral
acids in Scandinavia and this organic acid, which is less corrosive than
mineral acids, has also been accepted as an additive in many other countries. In the UK the commercial product most commonly used contains 85
per cent formic acid and is applied to the herbage, undiluted, from a

container attached to the forage harvester. The recommended application rate is 2.7 kg/tonne fresh crop, and when applied to grass at this level it reduces the *p*H value to about 4.8. Complete inhibition of microbial growth does not take place, some lactic acid fermentation occurring. To inhibit the lactic acid bacteria, an application rate of 2 to 3 times the recommended commercial rate is required, greater amounts being required for wetter crops. The beneficial effects of formic acid on the fermentation characteristics of difficult crops low in water-soluble carbohydrates, such as legumes and grasses, have been well established. Improvements in animal performance and dry matter intake have also been demonstrated.

More recently, attention has been focused on the use of formalin, a 40 per cent solution of formaldehyde in water, which is applied either on its own or more effectively with an acid such as sulphuric or formic. Typical results of a formalin/formic acid mixture (3:1) applied to ryegrass at the rate of 10 g/kg are shown in Table 18.3; the formaldehyde combines with the protein protecting it from hydrolysis by plant enzymes and micro-organisms in the silo and in the rumen. The protein is subsequently liberated under the acid conditions of the abomasum and is digested in the small intestine. The acid in the additive mixture acts as a fermentation inhibitor, preventing in particular the development of clostridial bacteria in the silage. Unfortunately, the level of formalin addition is critical. If applied at too high a concentration, then the normal microbial activity in the rumen will be affected and both digestibility and dry matter intake will be reduced. The optimum level of addition varies with crop species and protein content, but should not exceed 50 g formaldehyde/kg protein.

Deteriorated silages

The continuous infiltration of air during the storage period in the silo results in the growth of aerobic micro-organisms which break down the organic matter to form composted material unfit for animals. Such waste material is commonly found on the surface and sides of silage made in bunker and stack silos. This deterioration process will also occur during the feeding period when silage is exposed to air for varying periods of time. Initially the soluble components in the silage, such as organic acids, alcohols and sugars will be oxidised, but continuous exposure to air eventually leads to the destruction of the more stable components such as cell wall polysaccharides. The organisms responsible are initially yeasts and bacteria which are followed by moulds. The factors governing the rate of deterioration are unknown but the extent of dry matter breakdown in silage exposed to air over a 10-day period may range from virtually nil to over 30 per cent. Silages in which fermentation has been restricted by wilting or by the use of chemical additives, are more prone to aerobic deterioration than those in which either lactic acid bacteria or clostridia have been active. The presence of propionic, butyric and caproic acids in silages improves their stability in air.

Overheated silages

These silages are produced from overwilted material ensiled usually in bunker-type or stack silos without adequate consolidation. If the temperature in the mass exceeds 55°C, protein digestibility may be reduced. Overheated silages, which are dark brown or even black in colour, may be palatable to animals but are of low nutritional value because of excessive oxidation of soluble nutrients.

FURTHER READING

P. McDonald, 1981. *The Biochemistry of Silage*. John Wiley, Chichester.
M. J. Nash, 1985. *Crop Conservation and Storage*. Pergamon Press, Oxford.
J. A. F. Rook and P. C. Thomas (eds), 1982. *Silage for Milk Production*. Technical Bulletin No. 2, Hannah Research Institute, Ayr.
M. K. Woolford, 1984. *The Silage Fermentation*. Marcel Dekker, New York.

19

Hay, artificially dried forages, straws and chaff

HAY

The traditional method of conserving green crops is that of haymaking, the success of which until fairly recently was entirely dependent upon the chance selection of a period of fine weather. The introduction of rapid drying techniques using field machinery and barn drying equipment has, however, considerably improved the efficiency of the process, and reduced the need to be dependent upon the weather.

The aim in haymaking is to reduce the moisture content of the green crop to a level low enough to inhibit the action of plant and microbial enzymes. The moisture content of a green crop depends on many factors, but may range from about 650 to 850 g/kg, tending to fall as the plant matures. In order that a green crop may be stored satisfactorily in a stack or bale, the moisture content must be reduced to 150–200 g/kg. The custom of cutting the crop in a mature state when the moisture content is at its lowest is clearly a sensible procedure for rapid drying and maximum yield, but unfortunately the more mature the herbage, the poorer is the nutritive value (see Ch. 17).

Chemical changes and losses during drying

Chemical changes resulting in losses of valuable nutrients inevitably arise during the drying process. The magnitude of these losses depends to a large extent upon the speed of drying. In the field the loss of water from the swath is governed by the natural biological resistance of the leaf and swath to water loss, the prevailing weather conditions and swath micro-climate, and the mechanical treatment of the crop during harvesting and conditioning.

The losses of nutrients during haymaking arise from the action of plant and microbial enzymes, chemical oxidation, leaching, and mechanical damage.

Action of plant enzymes. In warm, dry, windy weather the wet herbage, if properly handled and mechanically agitated, will dry rapidly and losses arising from plant enzyme activity will be small. The main changes involve the soluble carbohydrates and nitrogenous components. In the early stages of the drying process changes in individual components of the water soluble carbohydrates occur such as the formation of fructose from hydrolysis of fructans. During extended periods of drying considerable losses of hexoses occur as a result of respiration and this loss leads to an increase in the concentration of other constituents in the plant, especially the cell wall components reflected in the fibre content. In the freshly cut crop, proteases present in the plant cells rapidly hydrolyse the proteins to peptides and amino acids, hydrolysis being followed by some degradation of specific amino acids. The effects of wilting ryegrass under ideal dry conditions and in a poor humid environment are compared in Table 19.1.

A number of devices and methods of treatment are used to speed up the drying process in the field and these include 'conditioners', such as crushers, rollers and crimpers which break the cellular structure of the plant and allow the air to penetrate the swath more easily. A more traditional method, which is still practised in some parts of the world, notably Switzerland, Italy, West Germany and Scandinavia, is to make hay on racks, frames or tripods. Table 19.2 shows a comparison in composition and nutritive value between hay made by 'tripoding' and the traditional 'field curing' method. The difference between these two methods is reflected in the crude fibre, digestible crude protein, digestible organic matter and metabolisable energy values.

Action of micro-organisms. If drying is prolonged because of bad weather conditions, changes brought about by the activity of bacteria and fungi may occur. Mouldy hay is unpalatable and may be harmful to farm animals and man because of the presence of mycotoxins. Such hay may also contain actinomycetes which are responsible for the allergic disease affecting man known as 'farmer's lung'.

Oxidation. When herbage is dried in the field, a certain amount of oxidation occurs. The visual effects of this can be seen in the pigments, many of which are destroyed. The provitamin, carotene, is an important compound affected and may be reduced from 150–200 mg/kg in the dry matter of the fresh herbage to as little as 2–20 mg/kg in the hay. Rapid drying of the crop by tripoding or barn drying conserves the carotene more

TABLE 19.1 Changes in nitrogenous components of ryegrass/clover during the early stages of field drying
(From M. C. Carpintero, A. R. Henderson & P. McDonald, 1979. *Grass and Forage Sci*, **34**, 311)

	Dry matter (DM) (g/kg)	Water soluble carbohydrates (g/kg DM)	Total N (g/kg DM)	N components (g/kg total N)		
				Protein N	Non-protein N	Ammonia N
Fresh grass	173	213	26.6	925	75	1.2
Wilted 6 h (Dry conditions)	349	215	28.2	876	124	1.1
Wilted 48 h (Dry conditions)	462	203	28.9	835	165	2.6
Wilted 48 h (Humid conditions)	199	211	29.9	753	247	2.6
Wilted 144 h (Humid conditions)	375	175	31.0	690	310	26.4

TABLE 19.2 Composition (g/kg DM) and nutritive value of perennial ryegrass and of hay made from it by two different methods in S. E. Scotland

	Fresh grass	Tripod hay	Field cured hay
Organic matter	932	908	925
Crude fibre	269	324	362
Crude protein	128	121	99
Digestible crude protein	81	72	47
Digestible organic matter	711	614	547
Metabolisable energy* (MJ/kg DM)	10.7	9.2	8.2

* Calculated from digestible organic matter.

efficiently, and losses as low as 18 per cent in barn dried hay have been reported. On the other hand sunlight has a beneficial effect on the vitamin D content of hay because of irradiation of the ergosterol present in the green plants.

Leaching. Losses due to leaching by rain mainly affect the crop after it has been partly dried. Leaching causes a loss of soluble minerals, sugars and nitrogenous constituents, resulting in a concentration of cell wall components which is reflected in a higher fibre content. Rain may prolong the enzyme action within the cells, thus causing greater losses of soluble nutrients, and may also encourage the growth of moulds.

Mechanical damage. During the drying process the leaves lose moisture more rapidly than the stems, so becoming brittle and easily shattered by handling. Excessive mechanical handling is liable to cause a loss of this leafy material, and since the leaves at the hay stage are richer in digestible nutrients than the stems, the resultant hay may be of low feeding value. There are a number of modern machines available which reduce the losses caused by leaf shattering. If the herbage is bruised or flattened, the drying rates of stems and leaves differ less. Baling the crop in the field at a moisture content of 300–400 g/kg, and subsequent drying by artificial ventilation, will reduce mechanical losses considerably.

Stage of growth

The stage of growth of the crop at the time of cutting is the most important factor determining the nutritive value of the conserved product. The later the date of cutting the larger will be the yield, the lower the digestibility and net energy value, and the lower the voluntary intake of dry matter by animals. It follows that if their drying conditions are similar, hays made from early-cut crops will be of higher nutritive value than hays made from mature crops.

Plant species

The differences in chemical composition between species have already been discussed in Chapter 17. Hays made from legumes are generally richer in protein and minerals than grass hay. Pure clover swards are not commonly grown for making into hay in the United Kingdom, although many 'grass' hays contain a certain amount of clover. Lucerne or alfalfa (*Medicago sativa*) is a very important legume which is grown as a hay crop in many countries. The value of lucerne hay lies in its relatively high content of crude protein, which may be as high as 200 g/kg DM if it is made from a crop cut in the early bloom stage.

TABLE 19.3   Composition and nutritive value of hays
(After S. J. Watson and M. J. Nash, 1960. *The Conservation of Grass and Forage Crops.* Oliver and Boyd, Edinburgh)

		Dry matter basis			
	No. of samples	Crude fibre (g/kg)	Crude protein (g/kg)	Digestible crude protein (g/kg)	Metabolisable* energy (MJ/kg)
Grasses					
Meadow	686	298	113	67	8.8
Mixed grass	68	301	114	63	8.6
Cocksfoot (orchard grass)	17	356	82	42	8.0
Fescue	22	315	90	48	8.6
Ryegrass	39	305	96	48	8.9
Timothy	218	341	77	36	8.2
Legumes					
Clover	284	319	143	89	8.6
Lucerne	474	322	165	118	8.3
Vetches	28	277	213	163	9.1
Soya bean	42	366	156	101	7.8
Cereals					
Barley	19	265	93	52	8.6
Oat	48	329	80	41	8.5
Wheat	20	268	82	44	7.8

* Calculated from TDN values

TABLE 19.4 Composition (g/kg) and nutritive value of the dry matter of 47 grass hays* made during 1963–65 in England and Wales
(From *A. D. A. S. Science Arm Report*, 1972, p. 95. MAFF, HMSO, London)

	Range	Mean	S.D.
Ash	57–117	80	±11
Crude fibre	274–412	335	±31
Crude protein	63–167	96	±22
Digestible crude protein+	21–115	51	—
Digestible organic matter	391–711	563	±61
Metabolisable energy (MJ/kg)	5.7–11.5	8.5	±1.1

* Mainly ryegrass but including some timothy/meadow fescue hays.
+ Estimated from crude protein.

Cereals are sometimes cut green and made into hay, and this usually takes place when the grain is at the 'milky' stage. The nutritive values of cereal hays cut at this stage of growth are similar to those of hays made from mature grass, although the protein content is generally lower. Table 19.3 shows the composition of a number of hays made from different species. These values give no indication of the range in nutritive value. If extremes are considered, it is possible to produce hay of excellent quality with a digestible crude protein content of 115 g/kg DM and an ME value in excess of 10 MJ/kg DM (see Table 19.4). On the other hand poor-quality hay made from mature herbage harvested under bad weather conditions may have a negative digestible crude protein content and an ME value below 7 MJ/kg DM; material of this type is no better in feeding value than oat straw.

Changes during storage

The chemical changes and losses associated with haymaking do not completely cease when hay is stored in the stack or barn. The stored crop may contain from 100 to 300 g/kg moisture. At the higher moisture levels chemical changes brought about by the action of plant enzymes and micro-organisms are likely to occur.

Respiration ceases at about 40 °C, but the action of thermophilic bacteria may go on up to about 72 °C. Above this temperature chemical oxidation can cause further heating. The heat tends to accumulate in hay stored in bulk, and eventually combustion may occur.

Prolonged heating during storage can have an adverse effect on the proteins of hay. New linkages are formed within and between peptide chains. Some of these linkages are resistant to hydrolysis by proteases which reduces the solubility and digestibility of the proteins.

Susceptibility of proteins to heat damage is greatly enhanced in the presence of sugars. Maillard reactions are the result of a complex series of chemical changes initiated by the condensation between the carbonyl group of a reducing sugar and the free amino group of an amino acid or protein. Temperature has an important effect on the reaction rate, the rate being 9000 times faster at 70 °C than at 10 °C. The amino acid lysine is particularly susceptible to reactions of this type. The products are colourless at first, but eventually turn brown; the dark brown colour of overheated hays, and other foods, can be attributed mainly to Maillard reactions.

Losses of carotene during storage depend to a large extent on the temperature. Below 5 °C little or no loss is likely to occur, whereas in warm weather losses may be considerable.

The changes that take place during storage are likely to increase the proportion of cell wall constituents and reduce nutritive value.

Overall losses

The overall losses during haymaking can be appreciable under poor weather conditions. In one recent study on 6 commercial farms carried out over a 3-year-period in the N.E. of England by the Agricultural Development and Advisory Service (A.D.A.S.), losses of nutrients were measured between harvesting and feeding. Total dry matter losses averaged 19.3 per cent, made up of 13.7 per cent field loss and 5.6 per cent loss in the bale. The losses of digestible organic matter and digestible crude protein were both about 27 per cent.

Hay preservatives

The main objective in using hay preservatives is to allow hay to be stored at moisture levels which, in the absence of the preservative, would result in severe deterioration through moulding. A number of compounds have been tested, but of these, propionic acid and its less volatile derivative ammonium bispropionate, have received most attention.

It has been found that hays with moisture contents as high as 400–500 g/kg can, after propionate treatment, be stored satisfactorily, provided the additive is both applied in sufficient quantity and distributed uniformly.

More recently, the success of anhydrous ammonia treatment of straw (see p. 427) has stimulated studies on treatment of hay with this gas. Anhydrous ammonia injected into plastic-covered stacks of bales of moist hay has increased the stability, under aerobic and anaerobic conditions, and has improved the nutritional value of the hay.

The process of artificial drying is a very efficient, though expensive, method of conserving forage crops. In northern Europe, grass and grass-clover mixtures are the commonest crops dried by this method, whereas in the USA lucerne (alfalfa) is the primary crop that is dehydrated.

The drying is brought about by allowing the herbage to meet gases at a high temperature which varies with the type of drier used. In the 'low-temperature' type of equipment, the hot gases are usually at a temperature of about 150 °C and the drying time varies from about 20 to 50 minutes depending upon the drier design and the moisture content of the crop. With 'high-temperature' driers, the temperature of the gases is initially within the range of 500–1000 °C and the time taken to pass through the drier varies from about 0.5 to 2 minutes.

In both processes, the temperature and time of drying are very carefully controlled so that the forage is never completely desiccated, and the final product usually contains about 50–100 g/kg of moisture. As long as some moisture remains in the material, the temperature of the herbage is unlikely to exceed 100 °C. It is obvious, however, that if the material is left in contact with the hot gases too long it will be charred or even completely incinerated.

After drying, the forage may be processed to suit the class of animal for which it is intended. For pigs and poultry, it is usually hammer milled and stored either as meal in sacks or as pellets. For ruminants the dried forage can be used as a 'long roughage' or more commonly packaged into different forms described as pellets, cobs or wafers. A pellet is a package made in a rotary-die press from milled dried forage; a cob is made in a rotary-die press from chopped and dried forage, while a wafer is usually made in a piston-type machine from either milled or chopped dried forage.

More widespread use of artificial drying of forage crops is limited by the high initial capital and high running costs of the process.

Nutritive value

As a conservation technique, artificial drying is extremely efficient. Dry matter losses from mechanical handling and drying are together unlikely to exceed 10 per cent, and the nutritive value of the dried product is therefore close to that of the fresh crop.

Some oxidation of carotene may occur especially during long periods of storage of dried forage exposed to light and air; dried grass meal can lose as much as half its carotene during 7 months' storage under ordinary commercial conditions. A high quality meal should have a carotene content of about 250 mg/kg, although under exceptional conditions carotene contents as high as 450 mg/kg have been obtaind. Because irradiation of sterols cannot take place during the rapid drying process, the vitamin D content of dried forage will be very low. Originally most of the dried material produced from very young leafy herbage went into pig and poultry rations. For this market, which is still an important one, the product is sold mainly for its content of protein, carotene and xanthophyll (the pigment mainly responsible for egg yolk colour).

Dried grass made from young herbage is also used in the diets of ruminant animals, mainly as a replacement for the cereal and protein concentrates given with silage and hay. A particular feature of dried grass/silage combinations is the high total dry matter intake which can be obtained. Although there is some evidence to indicate that the apparent digestibility of the crude protein might be reduced slightly during drying, the disadvantages of this are more than offset by the increased amino acid supply to the animal because of the larger amount of protein which escapes degradation in the rumen and is digested in the small intestine. In a recent study with sheep the proportion of the total amino acid nitrogen intake that was apparently absorbed in the small intestine, was 0.41 for fresh grass and 0.51 for the same grass after drying.

In the USA considerable amounts of lucerne (alfalfa) are artificially dried and sold as a high-vitamin feed supplement for broilers. As the production of this is seasonal large volumes of dried meal must be stored for periods of six months or more, and unless precautions are taken losses of carotene, xanthophyll and vitamin E are liable to occur as a result of oxidation. Since the rate of loss is a function of temperature, large volumes were stored under refrigeration in the past. More recently the dried product has been stored satisfactorily under inert gas, which virtually eliminates oxidative losses up to the time of removal from storage. Many manufacturers also add antioxidants which protect the product from the time it is removed from the inert gas until it is used.

STRAWS AND CHAFF

Straws consist of the stems and leaves of plants after the removal of the ripe seeds by threshing, and are produced from most cereal crops and from some legumes. Chaff consists of the husk or glumes of the seed which are separated from the grain during threshing. These products are extremely fibrous, rich in lignin and of extremely low nutritive value. They should not be used as pig or poultry foods.

On a world scale, maize, wheat and rice are the predominant straw crops. In the UK barley and oat straws have traditionally been the only cereal straws given to ruminants.

Barley and oat straw

Of the cereal straws, oat straw used to be popular in many areas of the UK as a bulky food for store cattle, along with roots and concentrates, and in limited quantities as a source of fibre for dairy cows. With much of the oat crop combine-harvested the present cultivars tend to be short-strawed types and because of this, and the declining interest in oats as a grain crop, the amount of oat straw produced is limited. With the increasing use of barley grain as a major concentrate food for farm animals, especially in Northern Europe, large quantities of barley straw are available and attention has concentrated in recent years on methods of trying to improve the nutritional value of this low-grade material.

The composition of both barley and oat straw may vary, although this is influenced more by stage of maturity of the crop at harvesting and environment than by the cultivar grown. The crude protein content of the dry matter of both straws is low, usually between 20 and 50 g/kg with the higher values obtained in crops grown under cold and wet conditions where they do not mature completely. The apparent digestibility of the crude protein is also very low and will be negative if the metabolic nitrogen excreted is greater than the nitrogen absorbed from the gut. This is usually the case with roughages containing less than 30 to 40 g crude protein/kg DM. A major component of the dry matter is the fibre which contains a relatively high proportion of lignin. In a chemical examination of the dry matter of barley straw, it was shown that this consisted of about 400–450 g/kg of cellulose, 300–500 g/kg of hemicellulose and 80–120 g/kg of lignin.

The digestibility of the organic matter of these straws rarely exceeds 0.5 and the metabolisable energy value is about 7 MJ/kg DM or in the case of winter barley cultivars, less than this. Of the ash fractions, silica is the main component and straws generally are poor sources of essential mineral elements as can be seen from the results of a comparison between hays and barley straws shown in Table 19.5.

Apart from the low digestibility of these cereal straws, a major disadvantage is the low intake obtained when they are given to ruminant animals. Whereas a cow will consume up to 10 kg of medium-quality hay, it will eat only about 5 kg of straw. Improvements in both digestibility and intake can be obtained by the addition of nitrogen in the form of protein or urea.

Maize straw

Maize straw, or corn stover, has a higher nutrient content and is more digestible than most other straws. It has a crude protein content of about

TABLE 19.5   Some mineral contents of hay and barley straw
obtained from 50 farms in S.E. Scotland 1964–66
(From E. J. Mackenzie and D. Purves, 1967. *Edinb. Sch. Agric. Exp. Work.* p. 23)

Hay		Barley straw	
Range	Mean	Range	Mean
g/kg DM			
Ca 3.0–6.3	4.6	1.5–4.5	3.1
Mg 0.6–1.4	1.1	0.3–0.6	0.5
Na 0.2–1.9	1.0	0.1–1.0	0.5
mg/kg DM			
Cu 1.5– 10.0	6.0	0.6–4.0	2.4
Mn 30 – 150	80	1.8–22.0	12.1
Fe 30 – 120	106	18–170	78

60 g/kg DM and a metabolisable energy value of about 9 MJ/kg DM. In North America, corn stover is frequently used as a major part of the diet for dry, pregnant beef cows. The animals may be turned into the cornfields after the grain has been harvested or the stover may be chopped, ensiled and fed in a similar way to maize silage. Alternatively the stover, after drying in the field, can be stacked or harvested as large round bales.

Rice straw

In many of the intensive rice-growing areas of the world, particularly Asia, this straw is used as a food for farm animals. Its protein content and metabolisable energy value are similar to those of spring barley straw. It has an exceptionally high ash content, about 170 g/kg DM, which consists mainly of silica. The lignin content of this straw, about 60–70 g/kg DM is however, lower than that of other cereal straws. In contrast to other straws, the stems are more digestible than the leaves.

Wheat and rye straws

Until recently, wheat and rye straws were considered to be so poor in nutritional value that their use as foods for farm animals was not recommended. However, recent research has demonstrated that the digestibility of most cereal straws can be markedly improved with alkali treatment (see below).

Legume straws

The straws of beans and peas are richer in protein, calcium and magnesium than the cereal straws, and if properly harvested are useful roughage foods for ruminant animals. Because of their thick fibrous stems they are more difficult to dry than cereal straws and frequently become mouldy on storage.

Alkali treatment of straws

When straw is treated with alkali, the ester linkages between lignin and the cell wall polysaccharides, cellulose and hemicelluloses, are hydrolysed thereby causing the carbohydrates to become more available to the micro-organisms in the rumen. In the early process (the Beckman process) widely used in Germany in the First World War, straw was soaked for one to two days in a dilute solution (15–30 g/l) of sodium hydroxide, then washed exhaustively to remove residual alkali. The process increased the digestibility of the straw, but a considerable proportion of the soluble nutrients was lost in the washings. More recently attention has concentrated on the so called 'dry process' in which the chopped, or milled, straw is treated with a relatively small volume (100–400 l/t) of concentrated sodium hydroxide (200–400 g/l). In this process, in which the excess sodium is not removed before feeding, the digestibility of the straw dry matter is increased from about 0.4 to 0.5–0.7.

An alternative approach is to use anhydrous ammonia or solutions of ammonia in water (ammonium hydroxide). Ammonia, if applied to straw under appropriate conditions, improves digestibility to the same extent as does sodium hydroxide. The use of ammonia has the added advantages that it increases the crude protein content of the straw, it acts as a fungicide and it avoids the problems arising from alkali residues found with sodium hydroxide. Any excess ammonia quickly volatilizes when the treated material is exposed to air. However, this also is a disadvantage as the straw has to be enclosed in an airtight container. The treatment time varies from several weeks at low ambient temperature to only a few hours at high temperature. The recommended application rate is about 30–35 kg anhydrous ammonia per tonne of straw. When used at this level, ammonia increases the crude protein content of the treated material by about 50 g/kg DM. Table 19.6 summarises the results of 44 experiments in which animals were fed on diets containing a high proportion of poor-quality roughages, mainly straws, either treated or untreated with alkali.

A cheaper, and safer form of ammonia is urea. Urea decomposes to ammonia when acted upon by the enzyme urease:

$$NH_2 - CO - NH_2 + H_2O \rightarrow 2NH_3 + CO_2$$

TABLE 19.6 Performance of cattle and sheep fed on diets containing treated and untreated roughages*
(From J. F. D. Greenhalgh, 1983. *Agricultural Progress*, **58**, 11)

Species	Alkali:	NaOH		NH₃	
	Treatment:	−	+	−	+
Cattle	No of experiments	17		10	
	Roughage in diet (%)	64		61	
	Digestibility†	0.56	0.64	0.58	0.63
	Intake (kg/day)†	7.2	8.1	6.8	7.8
	Liveweight gain (kg/day)	0.62	0.82	0.40	0.71
Sheep	No. of experiments	10		7	
	Roughage in diet (%)	66		65	
	Digestibility†	0.57	0.65	0.52	0.62
	Intake (g/day)†	994	1259	1156	1147
	Liveweight gain (g/day)	39	126	73	99

* The roughages used were mainly wheat and barley straws but included rice straw, and also maize by-products such as cobs and stalks. Some of the roughages were ground and pelleted.
† Of dry matter in total diet.

The straw should contain sufficient of the enzyme to bring about the above reaction. In a recent study carried out in Bangladesh using rice straw stored in earthen pits, the addition of 30 kg urea per tonne of straw increased the crude protein content from 29 to 59 g/kg DM and the organic matter digestibility from 0.45 to 0.54 over a 20-day period. However, urea has yet to prove as consistently effective in improving digestibility as ammonia or sodium hydroxide.

Chaff

Cereal chaff is similar to straw in being a fibrous food, but is generally more digestible and richer in protein content. The most valuable is oat chaff. Care is required in feeding animals on barley chaff because of the presence of the awns, whose serrated edges may cause irritation.

FURTHER READING

M. G. Jackson, 1977. Review article: the alkali treatment of straws. *Animal Feed Science and Technology*, **2**, 105–30.
M. J. Nash, 1985. *Crop Conservation and Storage*. Pergamon Press, Oxford.

A. R. Staniforth, 1979. *Cereal Straw*. Clarendon Press, Oxford.

J. T. Sullivan, 1973. Drying and storing herbage as hay. In *Chemistry and Biochemistry of Herbage,* Vol. 3, G. W. Butler and R. W. Bailey (eds). Academic Press, London.

F. Sundstøl and E. Owen (eds), 1984. *Straw and Other Fibrous By-products as Feed.* Elsevier, Amsterdam.

20

Roots and tubers

The roots include turnips, swedes, mangels, fodder beet, carrots and parsnips. Sugar beet is also an important root crop, but is grown primarily for its sugar content and is normally not given to animals as such. However the two by-products from the sugar extraction industry, sugar beet pulp and molasses, are important and nutritionally valuable animal foods.

The main tubers are potatoes, Jerusalem artichokes, cassava and sweet potatoes; the last two being tropical crops.

ROOTS

The main characteristics of roots are their high moisture content (750–940 g/kg) and low crude fibre content (40–130 g/kg DM). The organic matter of roots consists mainly of sugars (Table 20.1) and is of high digestibility (*c.* 0.80–0.87). Roots are generally low in crude protein content

TABLE 20.1   Ranges of major constituents of roots

	Dry matter (g/kg)	Components of dry matter (g/kg)		
		Crude protein	Crude fibre	Sugars
Turnips	60–100	70–130	50–130	500–610
Swedes	80–130	70–120	50–130	500–630
Mangels	90–150	80–100	40–60	550–700
Fodder beet	140–220	50–80	40–60	600–730
Sugar beet	200–250	40–60	40–60	650–750

although like most other crops this component can be influenced by the application of nitrogenous fertilisers.

The composition is influenced by season and the variation may be quite large—low DM roots are produced in a wet season and relatively high DM roots in a hot dry season. The composition also varies with size, the large root containing lower DM and crude fibre contents, and being of higher digestibility than the small root. In the past, root crops have been considered as an alternative to silage in ruminant diets but their value as cereal replacements is now recognised. Roots are not a popular food for pigs and poultry because of their bulky nature, although those which have higher DM contents, such as fodder beet, are given to pigs. Roots are poor sources of vitamins, with the exception of carrots which are rich in the provitamin, β-carotene.

Roots are often stored in clamps during winter; during this period losses of dry matter of up to 10 per cent are not uncommon.

Swedes and turnips

Swedes (*Brassica napus*), which were introduced into Britain from Sweden about 200 years ago, and turnips (*Brassica campestris*) are chemically very similar although turnips generally contain less DM than swedes (Table 20.1). Of the two types of turnips that are grown, the yellow-fleshed cultivars are of higher DM content than the white-fleshed cultivars. The ME value of swedes is usually higher than that of turnips, i.e. about 13.5 and 11 MJ/kg DM respectively, (Table 20.2). The main sugar present in these roots is glucose.

Both swedes and turnips are liable to taint milk if given to dairy cows at or just before milking time. The volatile compound responsible for the taint is absorbed from the air by the milk and is not passed through the cow.

Mangels, fodder beet and sugar beet

Mangels, fodder beet and sugar beet are all members of the same species, *Beta vulgaris*, and for convenience they are generally classified according to their dry matter content. Mangels are the lowest in DM content, richest in crude protein and lowest in sugar content of the three types. Fodder beet can be regarded as lying in between mangels and sugar beet in terms of DM and sugar content, while sugar beet is richest in DM and sugar content, though poorest in crude protein. On a DM basis the ME values range from about 12 to 14 MJ/kg, the higher values occurring in sugar beet. The main sugar present in these roots is sucrose.

TABLE 20.2   Typical composition and nutritional values of roots, root by-products and tubers

| | Dry matter basis | | | | |
	Dry matter (g/kg)	Organic matter (g/kg)	Crude protein (g/kg)	Crude fibre (g/kg)	Rumen degradability* of protein	ME* (MJ/kg)
Roots						
Swedes	100	942	90	100	0.85	13.5
Turnips	80	922	122	111	0.85	11.2
Mangels	110	933	100	58	0.85	12.4
Fodder beet	160	926	60	45	0.85	12.5
Sugar beet	230	970	48	48	—	13.7
Root by-products						
Molasses	750	931	40	0	0.80	12.0
Sugar beet pulp (molassed)	860	918	92	144	0.70	12.5
Tubers						
Cassava	880	970	30	43	0.80	12.8
Potatoes	210	957	110	38	0.85	13.3
Jerusalem artichokes	200	945	75	35	—	13.2
Sweet potatoes	320	966	39	38	—	12.7

* For ruminants.

Mangels. 'Low dry matter' mangels have a DM content between 90 and 120 g/kg. 'Medium dry matter' mangels have a DM content of 120 to 150 g/kg. This group is usually smaller in size than the low dry matter group, but usually develops fairly large tops.

It is customary to store mangels for a few weeks after lifting, since freshly lifted mangels may have a slightly purgative effect. The toxic effect is associated with the nitrate present, which on storage is converted into asparagine. Unlike turnips and swedes, mangels do not cause milk taints when given to dairy cows.

Fodder beet. 'Medium dry matter' fodder beet contains from 140 to 180 g/kg of dry matter, whereas the 'high dry matter' varieties may contain up to 220 g/kg.

Fodder beet is a popular food in Denmark and the Netherlands for dairy cattle and young ruminants. Care is required in feeding cattle on high dry matter fodder beet, since excessive intakes may cause digestive upsets, hypocalcaemia and even death. The digestive disturbances are probably associated with the high sugar content of the root.

The use of fodder beet as the bulk ration for feeding pigs has given satisfactory results, but experiments have shown that the fattening period is slightly longer than when sugar beet is used. The digestibility of the organic matter of fodder beet is very high (about 0.90).

Sugar beet. Most sugar beet is grown for commercial sugar production, though it is sometimes given to animals, especially cows and pigs. Because of its hardness the beet should be pulped or chopped before feeding.

After extraction of the sugar at the sugar beet factory, two valuable by-products are obtained which are given to farm animals. These are sugar beet pulp and molasses.

Sugar beet pulp. The sugar beet on arrival at the factory is washed, sliced and soaked in water, which removes most of the sugars. After extraction of the sugar the residue is called sugar beet pulp. The water content of this product is 800–850 g/kg and the pulp may be sold in the fresh state for feeding farm animals, but because of transport difficulties it is frequently dried to a moisture content of 100 g/kg. Since the extraction process removes the water-soluble nutrients, the dried residue consists mainly of cell wall polysaccharides, and consequently the crude fibre content is relatively high (about 200 g/kg); the crude protein and phosphorus contents are low, the former being about 100 g/kg. Beet pulp is extensively used as a food for dairy cows and is also given to fattening cattle and sheep. It is not a popular food for pigs, especially fattening pigs, because of its fibrous nature, and is not suitable for poultry.

Beet molasses. After crystallisation and separation of the sugar from the water extract, a thick black liquid termed molasses remains. This product contains 700–750 g/kg of DM, of which about 500 g consists of sugars. The molasses dry matter has a crude protein content of only 20–40 g/kg, most of this being in the form of non-protein nitrogenous compounds, including the amine, betaine, which is responsible for the 'fishy' aroma associated with the extraction process.

Molasses is a laxative food and is normally given to animals in small quantities. Usually molasses is added to the beet pulp, which is then marketed as 'dried molassed beet pulp'. A variety of products containing molasses are available. Absorbents used with molasses include bran, brewers' grains, malt culms, spent hops and sphagnum moss. The nutritive value of these molassed products will depend upon the absorbent used.

Molasses is used, generally at levels of 5–10 per cent, in the manufacture of compound cubes and pellets. The molasses not only improves the palatability of the product but also acts as a binding agent. Since molasses is a rich and relatively cheap source of soluble sugars it is sometimes used as an additive in silage making.

Carrots and parsnips

Carrots (*Daucus carota*) are not grown on a large scale for feeding to farm animals, largely for economic reasons, but they are a valuable food for all classes of farm animals, being particularly favoured as a food for horses. Carrots have a dry matter content of about 110–130 g/kg and an ME value of 12.8 MJ/kg DM.

Parsnips (*Pastinaca sativa*) are slightly richer in dry matter than carrots but the dry matter has a similar ME value. The carrot is rich in carotene, and this provitamin is also present in parsnips.

TUBERS

Tubers differ from the root crops in containing either starch or fructan instead of sucrose or glucose as the main storage carbohydrate. They have higher dry matter and lower crude fibre contents (Table 20.2) and consequently are more suitable than roots for feeding to pigs and poultry.

Potatoes

In potatoes (*Solanum tuberosum*) the main component is starch. The starch content of the dry matter is about 700 g/kg (see Table 20.3); this carbohydrate is present in the form of granules which vary in size depending upon the variety. The sugar content in the dry matter of mature, freshly lifted potatoes rarely exceeds 50 g/kg although values in excess of this figure in stored potatoes have been obtained. The amount present is affected by the temperature of storage and values as high as 300 g/kg have been reported for potatoes stored at −1 °C.

The crude protein content of the dry matter is approximately 110 g/kg, about half of this being in the form of non-protein nitrogenous compounds. One of these compounds is the alkaloid solanidine, which occurs free and also in combination as the glyco-alkaloids, chaconine and solanine. Solanidine and its derivatives are toxic to animals, causing gastroenteritis. The alkaloid levels may be high in potatoes exposed to light. Associated with light exposure is greening due to the production of chlorophyll. Green potatoes should be regarded as suspect, although removal of the eye and peel, in which the solanidine is concentrated, will reduce the toxicity. Young shoots are also likely to be rich in solanidine and these should be removed and discarded before feeding. Immature potatoes have been found to

TABLE 20.3 Approximate composition (g/kg) of the potato tuber*
(From W. G. Burton, 1966. *The Potato*, p. 146. Veenman and Zonen,
Wageningen)

	Range	Normal value (approx)
Dry matter	180–280	230
Components of dry matter		
Starch	600–800	700
Reducing sugars	2.5–30	5–20
Sucrose	2.5–15	5–10
Citric acid	5–70	20
Total N	10–20	10–20
Protein N	5–10	5–10
Fat	1–10	3–5
Fibre	10–100	20–40
Ash	40–60	40–60

* Values are for mature unstored tubers

contain more solanidine than mature tubers. The toxic risk is reduced considerably if potatoes are steamed or otherwise cooked, the water in which the tubers have been boiled being discarded.

The fibre content of potatoes is very low which makes them particularly suitable for pigs and poultry. However, the protein in uncooked potatoes is frequently poorly digested by these animals, and protein digestibility coefficients as low as 0.23 have been reported for pigs. In similar trials with cooked potatoes, digestibility coefficients for protein generally exceed 0.70. This depressing effect on protein digestibility has been attributed to the presence in raw potatoes of a protease inhibitor which is destroyed on heating. It is normal practice to cook potatoes for pigs and poultry, although cooking is unnecessary for ruminants, presumably because the protease inhibitor is destroyed in the rumen. For pigs and poultry the ME value of cooked potatoes is similar to that of maize, about 14–15 MJ/kg DM.

Potatoes are a poor source of minerals, except of the abundant element potassium, the calcium content being particularly low. The phosphorus content is rather higher since this element is an integral part of the potato starch molecule, but some 20 per cent of it is in the form of phytates (see p. 95).

During the storage of potatoes considerable changes in composition may occur. The main change is a conversion of some of the starch to sugar and the oxidation of this sugar, with the production of carbon dioxide during respiration. The respiration rate increases with an increase in temperature. There may also be a loss of water from the tubers during storage.

Dried potatoes. The difficulty of storing potatoes satisfactorily for any prolonged period of time has led to a number of processing methods. Several methods of drying are used. In one method the cooked potatoes are passed through heated rollers to produce dried potato flakes. In another method sliced tubers are dried direct in flue gases; the resultant potato slices are frequently ground to a meal before marketing. The products are valuable concentrate foods for all classes of animals.

Cassava

Cassava (*Manihot esculenta*), also known as manioc, is a tropical shrubby perennial plant which produces tubers at the base of the stem. The chemical composition of these tubers varies with maturity, cultivar and growing conditions. Cassava tubers are used for the production of tapioca starch for human consumption although the tuber is also given to cattle, pigs and poultry. Its ME value is similar to that of potatoes but it has higher dry matter and lower crude protein contents (Table 20.2).

Cassava plants and tubers are to a certain degree poisonous since they contain varying proportions of two cyanogenetic glucosides (linamarin and lotaustralin), which readily break down to give hydrocyanic acid (see p. 16). In all cases care must be taken in the use of the tuber and wherever the plant is grown indigenous methods of removing the glucosides have been devised. Such treatments include boiling, or grating and squeezing, or grinding to a powder and then pressing.

Large amounts of dehydrated cassava meal have recently become available for feeding to farm animals. The meal can be used as a partial cereal grain replacer, provided the protein deficiency is rectified.

Cassava pomace is the residue from the extraction of starch from cassava tubers. Because of its high crude fibre content (*c*. 270 g/kg DM) its use in the diets of non-ruminant animals should be restricted.

Sweet potatoes

The sweet potato (*Ipomoea batatas*) is a very important tropical plant whose tubers are widely grown for human consumption and as a commercial source of starch. The tubers are of similar nutritional value to ordinary potatoes although of much higher dry matter and lower crude protein contents (Table 20.2).

Jerusalem artichoke

The Jerusalem artichoke (*Helianthus tuberosus*) is a tuber similar to the potato in dry matter and fibre content, but differs in containing the fructan inulin as the main storage carbohydrate instead of starch. The tubers are of high digestibility (Table 20.2) but do not store well and have never been a very popular food for farm animals.

FURTHER READING

W. P. Barber and C. R. Lonsdale, 1980. By-products from cereal, sugarbeet and potato processing, in *By-products and Wastes in Animal Feeding*. British Society Animal Production Occasional Symposium Publication No 3, Reading.

W. G. Burton, 1966. *The Potato*. Veenman and Zonen, Wageningen, the Netherlands.

B. Gohl, 1981. *Tropical Feeds*. FAO, Rome.

J. F. D. Greenhalgh, I. H. McNaughton and R. F. Thow (eds) 1977. *Brassica Fodder Crops*. Scottish Agricultural Development Council, Edinburgh.

M. J. Nash, 1985. *Crop Conservation and Storage*. Pergamon Press, Oxford.

21

Cereal grains and cereal by-products

The name 'cereal' is given to those members of the Gramineae which are cultivated for their seeds. Cereal grains are essentially carbohydrate concentrates, the main component of the dry matter being starch which is concentrated in the endosperm (Fig. 21.1). The dry matter content of the grain depends on the harvesting method and storage conditions but is generally within the range of 800–900 g/kg.

Of the nitrogenous components 85–90 per cent are in the form of proteins. The proteins occur in all tissues of cereal grains, but higher concentrations are found in the embryo and aleurone layer than in the starchy endosperm, pericarp and testa. Within the endosperm, the concen-

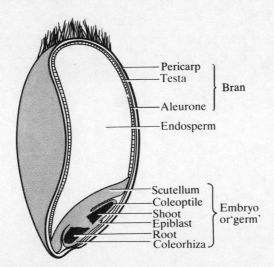

Fig. 21.1.   Longitudinal section of caryopsis (grain) of wheat.

tration of protein increases from the centre to the periphery. The total content of protein in the grain is very variable; expressed as crude protein it normally ranges from 80 to 120 g/kg DM, although some cultivars of wheat contain as much as 220 g/kg DM. Cereal proteins are deficient in certain indispensable amino acids, particularly lysine and methionine. It has been shown that the value of cereal proteins for promoting growth in young chicks is in the order, oats > barley > maize or wheat. The high relative value of oat protein for growth has been attributed to its slightly higher lysine content. This is demonstrated in Fig. 21.2 which compares the main limiting amino acid components of a number of cereal grains.

The lipid content of cereal grains varies with species. Wheat, barley, rye and rice contain 10–30 g/kg DM, sorghum 30–40 g/kg DM and maize and oats 40–60 g/kg DM. The embryo or 'germ' contains more oil than the endosperm; in wheat, for example, the embryo has 100–170 g/kg DM of oil while the endosperm contains only 10–20 g/kg DM. The embryo of rice is

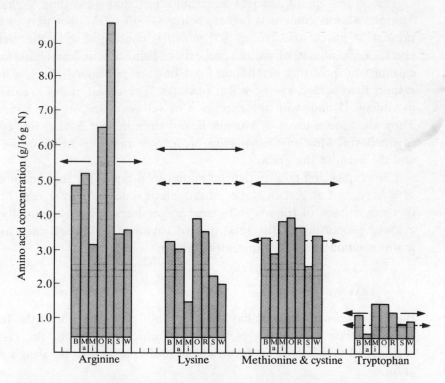

Fig. 21.2.   Main limiting indispensable amino acids of cereal grain proteins (g/16 g N).
(Straight lines indicate requirements for chicks: dotted lines for growing pigs.)
B = Barley   Ma = Maize   Mi = Millet   O = Oats   R = Rice
S = Sorghum   W = Wheat

exceptionally rich in oil, containing as much as 350 g/kg DM. Cereal oils are unsaturated, the main acids being linoleic and oleic, and because of this they tend to become rancid quickly, and also produce a soft body fat in pigs and poultry.

The crude fibre content of the harvested grain is highest in those such as oats or rice which contain a husk or hull formed from the fused glumes (palea and lemma), and is lowest in the 'naked' grains, wheat and maize. The husk has a diluent effect on the grain as a whole and reduces the energy value proportionally. Of the grains as harvested, oats have the lowest metabolisable energy value and maize has the highest, the respective values (MJ/kg DM) for poultry being 11 and 15 and for ruminants 12 and 14.

Starch occurs in the endosperm of the grain in the form of granules, whose size and shape vary with different species. Cereal starches consist of about 25 per cent amylose and 75 per cent amylopectin, although 'waxy' starches contain greater amounts of amylopectin.

The cereals are all deficient in calcium, containing less than 1 g/kg DM. The phosphorus content is higher, being 3–5 g/kg DM, but part of this is present as phytic acid (see p. 95) which is concentrated in the aleurone layer. Cereal phytates have the property of being able to immobilise dietary calcium and probably magnesium; oat phytates are more effective in this respect than barley, rye or wheat phytates. The cereal grains are deficient in vitamin D and, with the exception of yellow maize, in provitamins A. They are good sources of vitamin E and thiamin, but have a low content of riboflavin. Most of the vitamins are concentrated in the aleurone layer and the germ of the grain.

Calves, pigs and poultry depend upon cereal grains for their main source of energy, and at certain stages of growth as much as 90 per cent of their diet may consist of cereals and cereal by-products. Cereals generally form a lower proportion of the total diet of ruminants, although they are the major component of the concentrate ration.

BARLEY

Barley (*Hordeum sativum*) has always been a popular grain in the feeding of farm animals, especially pigs. In most varieties of barley the kernel is surrounded by a hull which forms about 10–14 per cent of the weight of the grain.

The metabolisable energy value (MJ/kg DM) is about 13.0 for ruminants, 13.7 for pigs and 12.5 for poultry. The crude protein content of barley grain ranges from about 60 to 160 g/kg DM with an average value of about 110 g/kg DM. As with all cereal grains the protein is of low quality being particularly deficient in the amino acid lysine. Recently high lysine mutants

TABLE 21.1 Composition and nutritive value of whole grain barley samples of parent variety NP–113, and of mutant varieties Notch 1 and Notch 2 (After S. P. Balaravi *et al.*, 1976. *J. Sci. Fd Agric.*, **27**, 545)

	NP 113	Notch 1	Notch 2
Protein (g/kg)	117	157	146
Lysine (g/16 g N)	3.88	4.00	3.96
Starch (g/kg)	662	396	414
Crude fibre (g/kg)	70	104	128
BV*	0.76	0.86	0.88
NPU*	0.66	0.68	0.73

* With rats

of barley have been produced by plant breeders and the superior nutritional value of two such mutants—Notch 1 and Notch 2—is shown in Table 21.1. Unfortunately with many of these mutants the yields of grain are much lower (about 30 per cent) than from parent varieties, and the starch contents may be reduced.

The lipid content of barley grain is low, usually less than 25 g/kg DM. The range in dry matter composition of 179 samples of barley grain harvested in Wales is given in Table 21.2.

In many parts of the world, and in particular in the UK, barley forms the main concentrate in the diets of pigs and ruminants. In the 'barley beef' system of cattle feeding, beef cattle are fattened on concentrate diets consisting of about 85 per cent bruised barley without the use of roughages. In this process the barley is usually treated so that the husk is kept intact and at the same time the endosperm is exposed, the best results being obtained by rolling grain at a moisture content of 160–180 g/kg. Storage of high-moisture barley of this type can present a problem because of the possibility of mould growth. Satisfactory preservation of the moist grain can be obtained if it is stored anaerobically. An additional or alternative safeguard is to treat the grain with a mould inhibitor such as propionic acid. Certain hazards, such as bloat, can be encountered with high-concentrate diets given to ruminants and it is necessary to introduce this type of feeding gradually over a period of time. It is important that a protein concentrate with added vitamins A and D and minerals be used to supplement high-cereal diets of this type.

Barley should always have the awns removed before they are offered to poultry, otherwise digestive upsets may occur.

TABLE 21.2 Dry matter composition 171 oat and 179 barley grain samples grown in Wales 1961–63 (After D. E. Morgan, 1967 and 1968. *J. Sci. Fd Agric.*, **18**, 21 and **19**, 393)

	Oats			Barley		
	Range	Mean	Coeff. of variation*	Range	Mean	Coeff. of variation*
Proximate constituents (g/kg)						
Crude protein	72–145	107	13.4	66–153	108	15.7
Crude fibre	80–179	125	13.6	38–73	56	12.5
Ether extract	9–80	52	20.2	11–32	19	15.8
Ash	22–41	31	8.7	17–42	25	12.4
Major mineral elements (g/kg)						
Ca	0.7–1.8	1.1	18.2	0.5–1.6	0.8	25.6
Mg	1.0–1.8	1.3	13.1	0.9–1.6	1.2	8.3
K	3.1–6.5	4.7	17.0	3.5–6.3	4.9	12.2
Na	0.04–0.6	0.2	47.8	0.06–0.4	0.2	41.2
P	2.9–5.9	3.8	10.5	2.6–5.2	3.8	11.8
Cl	0.4–1.8	0.9	33.3	0.8–2.2	1.4	21.4
Trace elements (mg/kg)						
Cu	3.0–8.2	4.7	17.0	3.5–19.8	6.6	27.3
Co	0.02–0.17	0.05	53.0	0.02–0.18	0.07	48.6
Mn	22–79	45	28.9	5–47	16	31.3
Zn	21–70	37	27.0	19–77	37	77.0
Energy (MJ/kg)						
Metabolisable energy for poultry	9.5–14.4	12.1	7.2	12.9–15.0	14.1	2.6

* Standard deviation as % of mean

Barley by-products

By-products of the brewing industry. In brewing, barley is first soaked and allowed to germinate. During this process, which is allowed to continue for about 6 days, there is development of a complete enzyme system for hydrolysing starch to dextrins and maltose. Although the enzymatic reactions have been initiated in this germination or 'malting' process, the main conversion of the starch in the grain to maltose and other sugars takes place during the next process, described as 'mashing'. After germination but before mashing, the grain or malt is dried, care being taken not to inactivate the enzymes. The sprouts are removed and are sold as malt culms or coombs. The dried malt is crushed, and small amounts of other cereals such as maize or rice may be added. Water is sprayed on to the mixture and the temperature of the mash increased to about 65 °C.

The object of mashing is to provide suitable conditions for the action of enzymes on the proteins and starch, the latter being converted to dextrins, maltose and small amounts of other sugars. After the mashing process is completed the sugary liquid or 'wort' is drained off, leaving 'brewers' grains' as a residue. Brewers' grains are sold wet or dried as food for farm animals.

The wort is next boiled with hops, which give it a characteristic flavour and aroma; the hops are then filtered off and after drying are sold as spent hops. The wort is then fermented in an open vessel with yeast for a number of days, during which time most of the sugars are converted to alcohol and carbon dioxide. The yeast is filtered off, dried, and sold as brewers' yeast.

The by-products obtained from the brewing process are therefore: malt culms, brewers' grains, spent hops and brewers' yeast.

Malt culms. Malt culms consist of the plumule and radicle of barley, and are relatively rich in crude protein (about 270 g/kg DM). They are not a high-energy food, however, and because of their fibrous nature their use is generally restricted to the feeding of ruminants and horses. Malt culms absorb water readily, and since they tend to swell in the stomach are usually moistened before they are given to animals. A further disadvantage is that they have a bitter taste and diets containing large amounts of this product may be rejected by animals as unpalatable.

Brewers' grains. Brewers' grains or 'draff' consist of the insoluble residue left after removal of the wort. In addition to the insoluble barley residue this product may contain maize and rice residues and, because of this, the composition of the product can be very variable as is illustrated in Table 21.3.

The fresh brewers' grains contain about 700–760 g water/kg and may be

TABLE 21.3   The nutritional value of fresh brewers' grains*
(From W. P. Barber and C. R. Lonsdale, 1980. Occasional Publication No. 3.
British Society of Animal Production, Reading, pp 61–9)

	Mean	Range
Dry matter (g/kg)	263	244–300
Crude protein (g/kg DM)	234	184–262
Crude fibre (g/kg DM)	176	155–204
Ether extract (g/kg DM)	77	61–99
Total ash (g/kg DM)	41	36–45
Digestible organic matter (g/kg DM)†	594	552–643
Metabolisable energy (MJ/kg DM)‡	11.2	10.5–12.0
Digestible crude protein (g/kg DM)‡	185	139–213

* Results for seven samples selected from widely different sources in the UK.
† *In vitro.*
‡ Measured in sheep.

given to cattle, sheep and horses in this fresh state or alternatively preserved
as silage. The wet product is sometimes dried to about 100 g water/kg and
sold as dried brewers' grains. The rumen degradability of the protein of the
dried product is about 0.6 compared with about 0.8 in the original barley.
Brewers' grains have always been a popular food for dairy cows, but they
are of little value to poultry and are not very suitable for pigs except in small
amounts.

Spent hops. Dried spent hops are a fibrous product and can be
compared to poor hay in nutritive value, but are less palatable, probably
because of their bitter flavour. This product is rarely used as a food for
animals today, most of it being sold for use as fertiliser.

Dried brewers' yeast. Dried yeast is a rich protein concentrate
containing about 420 g crude protein/kg. It is highly digestible and may be
used for all classes of farm animals. The protein is of fairly high nutritive
value and is specially favoured for feeding pigs and poultry. It is a valuable
source of many of the B group of vitamins, is relatively rich in phosphorus
but has a low calcium content. Other yeasts are now available as protein
concentrates and these will be described in Chapter 22.

Distillers' grains and distillers' solubles. In distilling, the soluble
materials may be extracted, as in brewing, or the whole mass fermented,
the alcohol then being distilled off. The residue after filtration is sold as wet
or dried distillers' grains. In Scotland, whisky distilleries are either 'malt'
or 'grain' types. The former use barley malt alone whereas the latter use

a mixture of cereals which may include barley, maize, wheat and oats. The composition of the distillers' grains obviously will depend on the starting materials and can vary widely. Malt distillers' grains are less variable in composition and contain, about 200–280 g crude protein/kg DM, 160–210 g crude fibre/kg DM and 50–80 g/kg DM of lipid material. Distillers' grains are low in soluble minerals, sodium and potassium, and also in calcium. Some recent studies have shown that with ruminants the digestibility of the organic matter of malt distillers' grains can be significantly improved (from 0.59 to 0.65) by the addition of calcium carbonate.

After distilling off the alcohol, the liquor ('spent wash' in grain distilleries and 'pot ale' in malt distilleries) remaining in the whisky still is evaporated and then spray dried to produce a light brown powder of variable composition known as 'distillers' solubles'. These are often mixed with the distillers' grains and dried together to yield a material sold as 'distillers' dried grains with solubles' or 'dark grains'. A cruder preparation of pot ale is the condensed form 'pot ale syrup' which contains about 300–500 g DM/kg and 350 g crude protein/kg DM. This may be used as a protein supplement for ruminants, but should be given with caution to sheep because of its high content of copper, present as a contaminant from copper stills.

By-products of the pearl barley industry. In the preparation of pearl barley for human consumption, the bran coat is removed and the kernel is polished to produce a white shiny grain. During this process three by-products, described as coarse, medium and fine dust, are produced and these are frequently mixed and sold as barley feed. Barley feed contains about 140 g crude protein/kg DM and about 100 g crude fibre/kg DM. The amount of this product available in the UK is very small.

MAIZE

A number of different types of maize (*Zea mays*) exist and the grain appears in a variety of colours, yellow, white or red. Yellow maize contains a pigment, cryptoxanthin, which is a precursor of vitamin A. In the USA, where it is known as corn, large amounts of this cereal are grown; the yellow varieties being preferred for animal feeding. The pigmented grain tends to colour the carcass fat, which in the UK is not considered desirable, so that white maize varieties are generally preferred here for feeding fattening animals.

Maize, like the other cereal grains, has certain limitations as a food for farm animals. Though an excellent source of digestible energy it is low in protein, and the proteins present are of poor quality (Fig. 21.2). Maize contains about 730 g starch/kg DM is very low in fibre and has a high metabolisable energy value.

The crude protein content of maize is very variable and generally ranges from about 90 to 140 g/kg DM, although varieties have been developed recently containing even higher amounts. In the USA the tendency has been to develop hybrid varieties of lower protein content.

The maize kernel contains two main types of protein. Zein, occurring in the endosperm, is quantitatively the most important, but this protein is deficient in the indispensable amino acids, tryptophan and lysine (see Fig. 21.2). The other protein, maize glutelin, occurring in lesser amounts in the endosperm and also in the germ, is a better source of these two amino acids. Recently plant breeders have produced new varieties of maize with amino acid components different from those present in normal maize. One such variety is *Opaque-2* which has a high lysine content. The difference between this variety and normal maize is primarily attributed to the zein: glutelin ratio. *Opaque-2* has been reported to be nutritionally superior to normal maize for the rat, pig, man and chick, but only in methionine-supplemented diets. A newer variety, *Floury-2*, has both increased methionine and lysine contents and has been shown in studies with chicks to be superior to *Opaque-2* maize in diets not supplemented with methionine. The oil of maize 40–60 g/kg DM contains a high proportion of unsaturated fatty acids and tends to produce a soft body fat.

Maize by-products

In the manufacture of starch and glucose from maize, a number of by-products are obtained which are suitable for feeding farm animals.

The cleaned maize is soaked in a dilute acid solution and is then coarsely ground. The maize germ floats to the surface and is removed for further processing. The de-germed grain is then finely ground and the bran is separated by wet screening. The remaining liquid consists of a suspension of starch and protein (gluten), which are separated by centrifugation. The process gives rise to three by-products—germ, bran and gluten.

The germ is very rich in oil, most of which may be extracted before producing the germ meal. Maize gluten (prairie) meal has a very high protein content (up to about 700 g/kg DM). The three by-products are frequently mixed together and sold as maize gluten feed. This food has a variable protein content, normally in the range 220–290 g/kg DM, of which about 0.6 is degraded in the rumen. It has a crude fibre content of about 100 g/kg DM and metabolisable energy values of about 8.0, 10.5 and 12.5 MJ/kg DM for poultry, pigs and ruminants respectively.

The oat (*Avena sativa*) has always been a favourite cereal for ruminant animals and horses, but has been less popular in pig and poultry feeding because of its comparatively high fibre content and low energy value.

The nutritive value of oats depends to a large extent on the proportion of kernel (groat) to hull. The proportion of hull in the whole grain depends upon the variety, environment and season, and can vary from 23 to 35 per cent (average 27 per cent). Oats of high hull content are richer in crude fibre and have a lower metabolisable energy value than low-hulled oats.

The crude protein content, which ranges from 70 to 150 g/kg DM, is increased by the application of nitrogenous fertilisers. Oat proteins are of poor quality and are deficient in the essential amino acids methionine, histidine and tryptophan, the amount of each of these acids in oat protein being generally below 20 g/kg. The lysine content is also low but is slightly higher than that of the other cereal proteins. Glutamic acid is the most abundant amino acid of oat protein, which may contain up to 200 g/kg.

The oil content of oats is high compared with that of most of the other cereal grains, and about 60 per cent of it is present in the endosperm. As mentioned earlier, the oil is rich in unsaturated fatty acids and has a softening effect on the body fat. The range in dry matter composition of 171 samples of oat grain harvested in Wales is shown in Table 21.2.

Oat by-products

During the commercial preparation of oatmeal for human consumption, a number of by-products are obtained which are available for animal feeding. When the oats are received at the mill they contain a number of foreign grains, mainly other cereals and weed seeds, which are removed as cockle before processing. The cleaned oats are then stabilised by steaming to inactivate the enzyme lipase which is located almost entirely in the pericarp of the kernel. After stabilisation, the oats are kiln dried before passing on to the huller, which removes the husks. The kernels are then brushed or scoured to detach the fine hairs that cover much of their surface.

The main by-products of oatmeal milling are oat husks or hulls, oat dust and meal seeds. The hulls form the main by-product, about 70 per cent of the total and the commercial product consists of the true husks with a variable proportion, up to 10 per cent, of kernel material. Oat hulls are of very low feeding value, being little better than oat straw. Their crude protein content is so low (about 30 g/kg DM) that in digestibility studies negative digestibility coefficients for nitrogen are likely to be obtained, owing to the relatively high amount of metabolic nitrogen excreted compared with that digested from them. The crude fibre content is usually between 350 and 380 g/kg DM which makes the by-product valueless as food for animals other than ruminants.

Oat dust is rich in kernel material and includes the kernel hairs removed from the grain during brushing. It has a protein content of about 100 g/kg DM. Meal seeds consist of slivers of husk and fragments of kernels in approximately equal proportions.

Oat hulls may be combined with oat dust in the proportion in which they come from the mill (4 to 1) to produce a product sold as 'oat feed'. This material is rather better in feeding value than the hulls alone. In the United Kingdom oat feed should not, by legal definition, contain more than 270 g crude fibre/kg. An alternative use for the hulls is in the brewing industry, where they are often added to the malt to assist in the drainage of wort from the mash tun.

The dehusked oats themselves (kernels or groats) are of high nutritive value containing about 180 g crude protein/kg DM and less than 30 g crude fibre/kg DM. The groats are generally too expensive to give to farm animals, and are ground into oatmeal after removal of the tips. The tips are mixed with any residues which accumulate during the flow of the oats during milling and the product is designated 'flowmeal'. Flowmeal can be a very valuable food since it may contain the germ; most of this by-product, however, is absorbed by the compound trade.

WHEAT

Grain of wheat (*Triticum aestivum*) is very variable in composition. The crude protein content, for example, may range from 60 to 220 g/kg DM, though it is normally between 80 and 140 g/kg DM. Climate and soil fertility as well as variety influence the protein content. The amount and properties of the proteins present in wheat are very important in deciding the quality of the grain for flour production. The most important proteins present in the endosperm are a prolamin (gliadin) and a glutelin (glutenin). The mixture of proteins present in the endosperm is often referred to as 'gluten'. The amino acid composition of these two proteins differs, glutenin containing about three times as much lysine as that present in gliadin. The main amino acids present in wheat gluten are the dispensable acids glutamic acid (330 g/kg) and proline (120 g/kg). Wheat glutens vary in properties and it is mainly the properties of the gluten which decide whether the flour is suitable for bread or biscuit making. All glutens possess the property of elasticity. Strong glutens are preferred for bread making, and form a dough which traps the gases produced during yeast fermentation.

This property of gluten is considered to be the main reason why finely ground wheat is unpalatable when given in any quantity to animals. Wheat, especially if finely milled, forms a pasty mass in the mouth and this may lead to digestive upsets. Poultry are less susceptible, although wheat with a high gluten content should not be given since a doughy mass may accumulate in the crop. Newly harvested wheat is apparently more harmful in this respect than wheat which has been stored for some time.

Wheat by-products

The wheat grain consists of about 82 per cent endosperm, 15 per cent bran or seed coat and 3 per cent germ. In modern flour milling the object is to separate the endosperm from the bran and germ. The wheat after careful cleaning and conditioning is blended into a suitable mix (grist) depending upon the type of flour required, and is passed through a series of rollers arranged in pairs. The first pair have a tearing action and release the bran coat from the endosperm. The rollers gradually break up the kernels, and at the end of the various stages the flour is removed by sieving. The proportion of flour obtained from the original grain is known as the extraction rate. The mechanical limitations of milling are such that in practice about 75 per cent is the limit of white flour extraction; higher extraction rates result in the inclusion of bran and germ with the flour. In the UK, wholemeal and brown flour are frequently made by adding all, or some of, the milling by-products respectively to the straight-run white flour. Alternatively, the whole grain may be ground between stones to form a coarse wholemeal.

In the production of white flour, the extraction rate varies in different countries but in the UK is about 74 per cent. The remaining 26 per cent constitutes the residues or 'offals'. Before roller milling replaced stone milling, many different grades of wheat offals were sold. The names of these varied in different parts of the country and even from mill to mill. Some names simply indicated the quality of the by-product or the stage of the process at which they arose, for example middlings and thirds. In modern roller milling, the offals may be sold complete as straight-run wheat feed or as three separate products—germ, fine wheat feed (shorts in the USA; pollard in Australia) and coarse wheat feed or bran.

The germ or embryo is very rich in protein (*c.* 250 g/kg DM), low in fibre and an excellent source of thiamin and vitamin E. It may be collected separately or may be allowed to flow on to the fine wheat feed by-product.

Fine wheat feed varies considerably in composition depending on the original grist and the extraction rate. The crude protein content is generally within the range 160–210 g/kg DM and the crude fibre content about

40–100 g/kg DM. Fine wheat feed can be safely used for all classes of farm animals and levels up to 45 per cent have been used satisfactorily in diets for bacon pigs.

Coarse wheat feed, or bran, contains more fibre and less protein than fine wheat feed and has always been a popular food for horses. It is not considered to be a suitable food for pigs and poultry because of its high fibre content. However, very little bran is now available for feeding animals as most of it is used in the preparation of breakfast cereals.

RICE

Rice (*Oryza sativa*), the main cereal crop of eastern and southern Asia, requires a sub-tropical or warm temperate climate and little is grown in Europe north of latitude 49°.

Rice, when threshed, has a thick fibrous husk or hull like that of oats, and in this state is known as rough rice. The hull amounts to some 20 per cent of the total weight and is rich in silica. The hull is easily removed to leave a product known as brown rice. Brown rice is still invested in the bran, which may be removed with the aleurone layer and the germ by skinning and polishing, thus producing polished rice.

Rough rice may be used as a food for ruminants and horses, but brown rice is preferable for pigs and poultry and compares favourably with maize in protein and energy value. Most rice, however, is used for human consumption and little is available in the United Kingdom for farm animals.

The two main by-products obtained from rice milling are the hulls and rice meal. The hulls are high in fibre content and can contain up to 210 g/kg DM of silica. They also have sharp edges which may irritate the intestine, and should never be given to animals. Rice meal or rice bran comprises the pericarp, the aleurone layer, the germ and some of the endosperm, and is a valuable product containing about 120–145 g crude protein/kg DM and 110–180 g oil/kg DM. The oil is particularly unsaturated and may become rancid very quickly; if it is removed a product of better keeping quality is obtained. The amounts of oil, crude protein and crude fibre must be declared in rice meal sold in the United Kingdom.

In the preparation of starch from rice, a product known as rice sludge or rice slump is left as a residue. The dried product has a crude protein content of about 280 g/kg DM and low crude fibre and oil contents, and is suitable for ruminants and pigs.

RYE

The use of rye (*Secale cereale*) in the United Kingdom is relatively small and little is grown for feeding farm animals. Rye grain is very similar to wheat in composition although rye protein has higher lysine and lower tryptophan contents than wheat protein. It is regarded as being the least palatable of the cereal grains. It is also liable to cause digestive upsets and should always be given with care and in restricted amounts.

Rye contaminated with ergot (*Claviceps purpurea*) may be dangerous to animals. This fungus contains a mixture of alkaloids which, if consumed by pregnant animals, may cause abortion. Like wheat, rye should be crushed or coarsely ground for feeding to animals. Rye is not commonly given to poultry. Studies with chicks have shown that rye contains at least two detrimental factors, an appetite-depressing factor located primarily in the bran, and a growth-depressing factor found in all parts of the grain.

Most of the rye grown in the United Kingdom is used for the production of rye breads and speciality products for human consumption. Some is used for brewing and distilling. The offals from the production of rye malt are rye bran and rye malt culms, but these are available in such small amounts in the United Kingdom as to be of little importance.

TRITICALE

Triticale is a hybrid cereal derived from crossing wheat with rye. Its name is derived from a combination of the two generic terms for the parent cereals (*Triticum* and *Secale*). The objective in crossing the two cereals was to combine the desirable characteristics of wheat such as grain quality, productivity and disease resistance with the vigour and hardiness of rye.

Triticale is grown commercially in the USSR, USA, Europe and South America mainly for animal feeding. Its composition is very variable; Hungarian strains for example, can range in crude protein content from 110 to 185 g/kg DM. Recent strains of triticale are at least equal in protein content to wheat and the quality of protein in the hybrid is better than that in wheat because of its higher proportion of lysine and sulphur-containing amino acids. As with rye, triticale is subject to ergot infestation. Studies using this hybrid have demonstrated poor palatability in pigs and, when compared with sorghum diets, increased liver abcesses in steers. It is generally recommended that triticale be limited to 50 per cent of the grain in the diets of farm animals.

MILLET

The name 'millet' is frequently applied to several species of cereals which produce small grains and are widely cultivated in the tropics and warm temperate regions of the world.

The most important members of this group include *Pennisetum americanum* (pearl or bulrush millet), *Panicum miliaceum* (proso or broomcorn millet), *Setaria italica* (foxtail or Italian millet), *Eleusine coracana* (finger or birdsfoot millet), *Paspalum scorbiculatum* (kodo or ditch millet) and *Echinochloa crusgalli* (Japanese or barnyard millet).

The composition of millet is very variable, the crude protein content being generally within the range 100–120 g/kg DM, the ether extract 20–50 g/kg DM and the crude fibre 20–90 kg DM.

Millet has a nutritive value very similar to that of oats, and contains a high percentage of indigestible fibre owing to the presence of hulls which are not removed by ordinary harvesting methods. Millet is a small seed and is usually ground for feeding to animals other than poultry.

SORGHUM

Sorghum (*Sorghum bicolor*) is the main food grain in Africa and parts of India and China. This cereal is also grown in the southern parts of the United States, as it is more drought-resistant than maize.

The kernel of sorghum is very similar to that of maize, although smaller in size. It generally contains rather more protein but less oil than maize. Whole sorghum grains can be given to sheep, pigs and poultry but are usually ground for cattle.

CEREAL PROCESSING

The processing of cereals using simple techniques, such as rolling or grinding, has been common practice for many years. More recently a range of other techniques have become available and these can be classified into two basic types—'hot processes' in which heat is either applied or created during the treatment process, and 'cold processes' in which the temperature of the grain is not increased significantly. The hot treatments include steam-flaking, micronisation, roasting and hot pelleting. Steam flaking is often carried out on maize by cooking the grain first with steam, then passing it through rollers to produce a thin flake which is then dried. Flaked maize is considered to be more acceptable to animals and is of slightly higher digestibility than the unprocessed grain. Steaming and flaking is also known to increase the proportion of propionic acid in the volatile fatty acids in the rumen. The term micronisation, in the context of grain processing, is used specifically to describe cooking by radiant heat followed by crushing in a roller mill. In this process the starch granules swell, fracture and gelatinise thus making them more available to enzyme attack in the digestive tract. For poultry, hot (steam) pelleting appears to be superior to cold pelleting as measured by growth rate and feed conversion efficiency. Steam processed or pressure-cooked sorghum grains also appear to be better utilised than unprocessed sorghum by chicks.

The 'cold processes' include grinding, rolling, cracking or crimping, cold pelleting, and addition of organic acids or alkalis. Grinding of cereals is essential for maximum performance of poultry kept under intensive conditions. Similarly pigs receiving ground barley generally perform better than those given crimped barley. With ruminants it is generally accepted that for cattle, barley grain should be coarse ground or rolled. With sheep, however, because they masticate their feed so well, there is generally no advantage in processing grains. This is illustrated in Table 21.4 which shows the results of a study with early weaned lambs given whole grains or ground and pelleted grains. There were no marked effects of grinding and pelleting on liveweight gain and food conversion efficiency, although there were differences between cereals. Although processing had no effect on nitrogen digestibility it significantly depressed the organic matter digestibility of barley and increased that of wheat.

Organic acids, such as propionic acid, are sometimes added to high moisture grain, especially barley, as a mould inhibitor. Unless the acid is effectively distributed, patches of mouldy grain may result which may present a health hazard. Certain *Fusarium* spp have been associated with such mouldy grain and these are known to produce metabolites such as zearalenone which can cause vulvovaginitis and the characteristic *splay-leg* syndrome in pigs.

Chemical treatment with sodium hydroxide solution has recently been introduced as an alternative to mechanical treatment (e.g. rolling) of barley and other cereal grains. The intention is to soften the husk but not to expose

TABLE 21.4 Performance and digestibility of early weaned lambs given four cereals (From E. R. Ørskov, C. Fraser and J. G. Gordon, 1974. *Br. J. Nutr.* **32**, 59, and E. R. Ørskov, C. Fraser and I. McHattie, 1974, *Anim. Prod*, **18**, 85)

Cereal	Processing	Liveweight gain (g/day)	Feed conversion efficiency (kg feed/kg gain)	Digestibility Organic matter	Nitrogen
Barley	Whole, loose	340	2.75	0.81	0.72
	Ground, pelleted	347	2.79	0.77	0.66
Maize	Whole, loose	345	2.52	0.84	0.75
	Ground, pelleted	346	2.62	0.82	0.69
Oats	Whole, loose	241	3.07	0.70	0.78
	Ground, pelleted	238	3.33	0.68	0.77
Wheat	Whole, loose	303	2.97	0.83	0.71
	Ground, pelleted	323	2.56	0.87	0.76

the endosperm to rapid fermentation in the rumen. In practice it has proved difficult to achieve these objectives by alkali treatment, and in cattle in particular, the digestibility of alkali-treated whole grain is often lower than that of rolled or ground grain.

FURTHER READING

D. C. Church, 1984. *Livestock Feeds and Feeding*, 2nd edn. O and B Books, Corvallis, Oregon.

C. M. Duffus and J. C. Slaughter, 1980. *Seeds and Their Uses*. John Wiley and Sons, Chichester.

B Göhl, 1981. *Tropical Feeds*. FAO, Rome.

R. C. Hoseney, E. Varriano-Marston and D. A. V. Dendy, 1981. Sorghum and millets, in *Advances in Cereal Science and Technology*, Vol. 4, Y. Pomeranz (ed). American Association of Cereal Chemists, St Paul, Minnesota.

N. L. Kent, 1983. *Technology of Cereals*, 3rd edn. Pergamon Press, Oxford.

E. L. Ørskov, 1981. Recent advances in the understanding of cereal processing for ruminants, in *Recent Developments in Ruminant Nutrition*, W. Haresign and D. J. A. Cole (eds). Butterworths, London.

22

Protein concentrates

OILSEED CAKES AND MEALS

Oilseed cakes and meals are the residues remaining after the removal of the greater part of the oil from oilseeds. Most of these are of tropical origin; they include groundnut, cottonseed, linseed and soya bean. The residues are rich in protein (200–500 g/kg) and most are valuable foods for farm animals. Some seeds, such as castor bean, yield residues unsuitable for animal feeding because they contain toxic substances.

Two main processes are used for removing oil from oilseeds. One employs pressure to force out the oil, while the other uses an organic solvent, usually hexane but occasionally trichlorethylene, to dissolve the oil from the seed. Some seeds such as groundnut, cottonseed and sunflower have a thick coat or husk, which is rich in fibre and of low digestibility and lowers the nutritive value of the material. It may be completely or partially removed by cracking and riddling, a process known as decortication. The effect of decortication of cottonseed upon the nutritive value of the cake derived from it is shown in Table 22.1; removal of the husk lowers the crude fibre content and has an important effect in improving the apparent digestibility of the other constituents. As a result the nutritive value of the decorticated cake is raised significantly above that of the undercorticated. The latter is only suitable for feeding adult ruminants, for whom it may have a role in maintaining the crude fibre level. Undecorticated cakes are rarely produced nowadays.

The seed from which oil is to be removed is cracked and crushed to produce flakes about 0.25 mm thick, which are cooked at temperatures up to 104 °C for 15 to 20 minutes. The temperature is then raised to about 110 to 115 °C until the moisture content is reduced to about 30 g/kg. The

TABLE 22.1   Composition and nutritive value of cottonseed cakes

| | Composition (g/kg) | | | | | |
	Dry matter	Crude protein	Ether extract	N-free extractives	Crude fibre	Ash
Undecorticated	880	231	55	400	248	66
Decorticated	900	457	89	293	87	74

| | Digestibility | | | | |
	Crude protein	Ether extract	N-free extractives	Crude fibre	Metabolisable energy (MJ/kg DM)
Undecorticated	0.77	0.94	0.54	0.20	8.5
Decorticated	0.86	0.94	0.67	0.28	12.3

material is then passed through a perforated horizontal cylinder in which revolves a screw of variable pitch. Pressures up to 40 MN/m² are attained. The residue from screw pressing usually has an oil content between 25 and 40 g/kg. The cylindrical presses used for extraction are called expellers and the method of extraction is usually referred to as the expeller process.

Only material with an oil content of less than 350 g/kg is suitable for *solvent extraction*. If material of higher oil content is to be so treated, it first undergoes a modified screw pressing to lower the oil content to a suitable level. The first stage in solvent extraction is flaking; after this the solvent is allowed to percolate through the flakes, or a process of steeping may be used. The oil content of the residual material is usually below 10 g/kg and it still contains some solvent, which is removed by heating. Some meals may benefit from being heated, and advantage is taken of the evaporation of the solvent to do this; soya bean meal, for example, is toasted at this stage in its production.

About 950 g/kg of the nitrogen in oilseed meals is present as true protein. It usually has a digestibility of 0.75 to 0.90 and is of good quality. When biological value is used as the criterion for judging protein quality, that of the oilseed proteins is considerably higher than that of the cereals (Table 22.2). Some of them approach animal proteins like fish meal and meat meal in quality though, as a class they are not as good. Certainly they are of poorer quality than the better animal proteins such as those of milk and eggs. The figures for protein efficiency ratio and gross protein value confirm the good quality of oilseed proteins, but their chemical scores are low. This means that they have a badly balanced amino acid constitution, having a large deficit of at least one indispensable amino acid. In general, oilseed

TABLE 22.2 Nutritive value of some food proteins

Source	Biological value (rat)	Chemical score	Protein efficiency ratio (rat)	Gross protein value (chick)
Oats	0.65	0.46	—	—
Wheat	0.67	0.37	1.5	—
Maize	0.55	0.28	1.2	—
Cottonseed meal	0.80	0.37	2.0	0.77
Groundnut meal	0.58	0.24	1.7	0.48
Soya bean meal	0.75	0.49	2.3	0.79
White-fish meal	0.77	—	—	1.02
Milk	0.85	0.69	2.8	0.90
Whole egg	0.95	1.00	3.8	—

proteins have a low cystine and methionine content, and a variable but usually low lysine content. As a result they cannot provide adequate supplementation of the cereal proteins with which they are commonly used, and should be used in conjunction with an animal protein when given to simple-stomached animals. The quality of the protein in a particular oilseed is relatively constant, but that of the cake or meal derived from it may vary, depending upon the conditions used for the removal of oil. The high temperatures and pressures of the expeller process may result in a lowering of digestibility and in denaturation of the protein, with a consequent lowering of its nutritive value. For ruminant animals such a denaturation may be beneficial owing to an associated reduction in degradability. The high temperatures and pressures also allow control of deleterious substances such as gossypol and goitrin. Solvent extraction does not involve pressing, temperatures are comparatively low, and the protein value of the meals is almost the same as the original seed.

The oilseed cakes may make a significant contribution to the energy content of the diet, particularly where the oil content is high. This will depend upon the process employed and its efficiency. Expeller soya bean meal may have an oil content of 66 g/kg DM and a metabolisable energy concentration of 14 MJ/kg DM, compared with 17 g oil and 12.3 MJ metabolisable energy per kg DM for solvent extracted meal. Digestive disturbances, however, may result from uncontrolled use of cakes rich in oil, and if the oil is unsaturated, milk or body fat may be soft and carcass quality lowered.

The oilseed meals usually have a high phosphorus content, which tends to aggravate their generally low calcium content. They may provide useful amounts of the B-vitamins, but are poor sources of carotene and vitamin E.

Soya bean meal

Soya beans contain from 160 to 210 g oil/kg and are normally solvent-extracted; the residual meal has an oil content of about 10 g/kg. The meal is generally regarded as one of the best sources of protein available to animals. The protein contains all the indispensable amino acids, but the concentrations of cystine and methionine are sub-optimal. Methionine is the first limiting amino acid and may be particularly important in high-energy diets.

Soya bean meal contains a number of toxic, stimulatory and inhibitory substances, including allergenic, goitrogenic and anticoagulant factors. Of particular importance in nutrition are the protease inhibitors of which six have been identified. Two of these, the Kunitz anti-trypsin factor and the Bowman-Birk chymotrypsin inhibitor, are of practical significance. The protease inhibitors are partly responsible for the growth-retarding property of raw soya beans or unheated soya bean meal. This has been attributed to inhibition of protein digestion, but there is evidence that pancreatic hyperactivity, resulting in destruction of methionine, is also involved. Another substance contributing to the growth retardation is a haemagglutinin which is capable of agglutinating red blood cells in rats, rabbits and human beings but not in sheep and calves. The inhibitors are inactivated by heating, which accounts for the preference shown for toasted meals for simple-stomached animals. For ruminant animals the inhibitors are not important and toasting is unnecessary. The process of toasting must be carefully controlled, since overheating will reduce the availability of lysine and arginine and reduce the value of the protein.

Provided the meal has been properly prepared, it forms a very valuable food for all farm animals. However, if soya bean meal is used as the major protein food for simple-stomached animals, certain problems arise. The meal is a poor source of B-vitamins, and these must be provided either as a supplement or in the form of an animal protein such as fish meal. If such supplementation is not practised, sows may produce weak litters which grow slowly because of reduced milk yields, and older pigs show uncoordination and failure to walk. On such diets, breeding hens produce eggs of poor hatchability giving chicks of poor quality; such chicks may also have an increased susceptibility to haemorrhages owing to a shortage of vitamin K. Soya bean meal is a better source of calcium and phosphorus than the cereal grains, but where it replaces animal protein foods, adjustments must be made in the diet, particularly for rapidly growing animals and laying hens. As long as adequate supplementation is practised it may form up to 400 kg/t of poultry diets and 250 kg/t of those of pigs.

Soya bean meal contains about 1 g/kg of genistein, which has oestrogenic properties and a potency of 4.44×10^{-6} times that of diethylstilboestrol. The effect of this constituent on growth rate has not been elucidated.

The oil contained in the bean has a laxative effect and may result in the production of a soft body fat. The extracted meal does not contain sufficient oil to cause this problem, but it should be borne in mind in view of the increasing tendency to use full fat soya bean products in dietary formulations, especially those for pigs. These products may be produced by batch pressure cooking or extrusion. The extruded product has a higher metabolisable energy content, but this advantage is nullified if the products are ground and pelleted.

Groundnut meal

The seeds of the groundnut are borne in pods, usually in pairs or threes. The seeds contain 250 to 300 g/kg of crude protein and 350 to 600 g/kg of lipid material. The pod or husk is largely fibrous. Groundnut meal is now usually made from the kernels and only occasionally is the whole pod used as the source, when an undecorticated meal is produced. The most common method of extraction is screw pressing, giving a residual meal with 50 to 100 g/kg of oil. Lower oil levels can only be achieved by solvent extraction, but this has to be preceded by screw pressing to reduce the initially high oil content. The composition of the meal will depend upon the raw material and the method of extraction used.

The protein of groundnut meal has sub-optimal amounts of cystine and methionine, although the first limiting amino acid is lysine. Where the meal is used in high-cereal diets, adequate supplementation with animal protein is necessary. This also ensures that the deficiencies of vitamin B_{12} and calcium are made good. Such supplementation is particularly important in fast-growing animals such as pigs and poultry. The palatability of the meal for pigs is high, but it should not form more than 25 per cent of the diet as it tends to produce a soft body fat and may have a troublesome laxative action. This also limits its use for lactating cows, for whom it otherwise forms an excellent and acceptable protein source. It has been reported that both a growth factor and an antitrypsin factor occur in groundnut meal. The latter has antiplasmin activity and so shortens bleeding time. It is destroyed by heating.

In 1961 reports appeared which implicated certain batches of groundnut meal in the poisoning of turkey poults and ducklings. The toxic factor was shown to be a metabolite of the fungus *Aspergillus flavus* and was named aflatoxin. This is now known to be a mixture of four compounds designated aflatoxins B_1, G_1, B_2 and G_2 of which B_1 is the most toxic. There are considerable species differences in the susceptibility to these toxins; turkey poults and ducklings are highly susceptible, calves and pigs are susceptible while mice and sheep are classed as resistant. Young animals are more susceptible than adults of the same species. A common feature in affected animals is liver damage with marked bile duct proliferation, liver necrosis, and in many cases, hepatic tumours. In fact, aflatoxin has been shown to be a potent liver toxin and a very active carcinogen. There are several recorded cases of deaths in calves under six months of age when fed on contaminated groundnut meal. Older cattle are more resistant but cases of death in store cattle and loss of appetite and reduced milk yield in cows have been reported. No clinical cases of poisoning in sheep have been reported.

Deaths have been reported in experiments where six-month-old steers were given 1 mg/kg of aflatoxin B_1 in their diet for a period of 133 days, while liveweight gains were generally significantly reduced. Feeding of 0.2 mg/kg of aflatoxin B_1 in the diets of Ayrshire calves significantly reduced liveweight gains. Experiments on the inclusion of 150 to 200 kg/t of toxic groundnut in dairy cow rations have shown significant falls in milk yield. A metabolite known as aflatoxin M has been shown to be present in the milk of cows fed on toxic meals, and to cause liver damage in ducklings. Concentrations of aflatoxin in milk may rise to dangerous levels if highly contaminated meal is given and a fall in yield may result. The effect of aflatoxin on man has not been clearly established.

Aflatoxins are relatively stable to heat and methods of eliminating them from meals are elaborate. The best method of control is suitable storage to prevent mould growth. In the UK prescribed limits for aflatoxins in foods for animals are laid down by law. Straight foods and complete foods for cattle, sheep and goats (except dairy cattle, calves, lambs and kids) must contain less than 0.05 mg/kg, complete foods for pigs and poultry (except piglets and chicks), and complementary foods for dairy cattle and other complete foods, less than 0.01 mg/kg.

Cottonseed meal

The protein of cottonseed meal is of good quality, but has the common disadvantage of oil seed residues of having a low content of cystine, methionine and lysine, the last named being the first limiting amino acid. The calcium content is low and as the calcium to phosphorus ratio is about 1:6 deficiencies of calcium may easily arise. It is a good though variable source of thiamin, but is a poor source of carotene.

Where cottonseed meal is used as a protein source for young, pregnant or nursing pigs, or young and laying poultry, it must be supplemented with fish meal or meat and bone meal to make good a shortage of indispensable amino acids and calcium. A supplement of vitamins A and D should also be provided. Pigs and poultry do not readily accept the meal, largely owing to its dry dusty nature. No such difficulty is encountered with lactating cows, but complications may arise where large amounts are given, since the milk fat tends to become hard and firm and butter made from such fat is often difficult to churn and tends to develop tallowy taints. Another factor to be considered in feeding with cottonseed meal is that it has a costive action, though this is not normally a problem and may indeed be beneficial in diets containing large amounts of laxative constituents.

Cotton seeds may contain from 0.3 to 20 g/kg DM of a yellow pigment known as gossypol and concentrations of 4 to 17 g/kg DM have been quoted for the kernels. Gossypol is a polyphenolic aldehyde which is an antioxidant and polymerisation inhibitor and is toxic to simple-stomached animals. The general symptoms of gossypol toxicity are depressed appetite and loss of weight; death usually results from circulatory failure. Although acute toxicity is low, ingestion of small amounts over a prolonged period can be lethal. It is important to distinguish between free (soluble in 70–30 v/v aqueous acetone) and bound gossypol since only the former is considered to be physiologically active.

The free gossypol content of cottonseed meal decreases during processing and varies according to the methods used. Screw pressed materials have 200–500 mg free gossypol/kg, prepressed solvent extracted meals 200–700 and solvent extracted 1000–5000 mg/kg. Processing conditions have to be carefully controlled to prevent loss of protein quality owing to binding of gossypol to lysine at high temperatures. Fortunately the shearing effect of the screw press in the expeller process is an efficient gossypol inactivator at temperatures which do not reduce protein quality.

It is generally considered that pig and poultry diets should not contain more than 100 mg free gossypol/kg and that inclusions of cottonseed meal should be between 50 and 100 kg/t. Particular care is required with laying hens since comparatively low levels of the meal may cause an olive green discoloration of the yolk in storage. An associated pink discoloration of the albumen is now considered to be due to cyclopropenoids and not gossypol as was once thought. Treatment with ferrous sulphate can ameliorate the effects of gossypol with doses ranging from 1 to 4 parts to 1 part gossypol. Ruminant animals do not show ill-effects even when they consume large quantities of cottonseed meal.

In the UK the free gossypol content of foods is strictly controlled by law. Straight foods except cotton cake or meal must not contain more than 20 mg/kg and the same limit applies to complete foods for laying hens and piglets. For poultry and calves the limit is 100, for pigs 60 and for cattle, sheep and goats 500. Cotton cakes and meals are allowed up to 1200 mg free gossypol/kg.

Coconut meal

The oil content of coconut meal varies from 25 to 65 g/kg, the higher-oil meals being very useful in the preparation of high-energy diets. They have the disadvantage, however, of a susceptibility to develop rancidity in storage. The protein is low in lysine and histidine and this, together with the generally high fibre content of about 120 g/kg, limits the use of the meal for simple-stomached animals. It is usually recommended that it should form less than 25 kg/t of pig diets and less than 50 kg/t of poultry diets. Where low-fibre coconut meals are available for simple-stomached animals, they have to be supplemented with animal proteins to make good their amino acid deficiencies. Neither protein quality nor fibre content is limiting for ruminant animals, and coconut meal provides an acceptable and very useful protein supplement. In diets for dairy cows, it is claimed to increase milk fat content. Recent work has shown increased butterfat contents in milks of cows given supplements of coconut meal, but the basal diet had a fat content of only about 10 g/kg. The milk fats produced on diets containing considerable amounts of coconut meal are firm and excellent for butter making.

Coconut meal has the valuable property of absorbing up to half its own weight of molasses, and as a result is popular in compounding.

Palm kernel meal

This food has a comparatively low content of protein which is however of high quality, the first limiting amino acid being methionine. The ratio of calcium to phosphorus is more favourable than in many other oilseed residues. The meal is dry and gritty, especially the solvent-extracted product, and is not readily eaten; it is therefore used in mixtures with more acceptable foods. Attempts to use it mixed with molasses, as molassed palm kernel cake have not been successful. It is used chiefly for dairy cows, for whom it has a reputation for increasing the fat content of the milk. Palm kernel meal has been described as being balanced for milk production, but in fact contains too high a proportion of protein to energy.

Despite the high quality of the protein and the relatively satisfactory calcium and phosphorus balance, palm kernel meal is not used widely in pig and poultry diets. This is due partly to its unpalatability and partly to its high fibre content of 150 g/kg which reduces its apparent digestibility for such animals. The highest level of palm kernel meal recommended in the diet of simple-stomached animals is about 200 kg/t.

Linseed meal

Linseed meal is unique among the oilseed residues in that it contains from 30 to 100 g/kg of mucilage. This is almost completely indigestible by non-ruminant animals, but can be broken down by the microbial population of the rumen. It is readily dispersible in water, forming a viscous slime. Immature linseed contains a small amount of a cyanogenetic glycoside, linamarin, and an associated enzyme, linase, which is capable of hydrolysing it with the evolution of hydrogen cyanide. Low-temperature removal of oil may result in a meal in which unchanged linamarin and linase persist; such meals have proved toxic when given as a gruel. Normal processing conditions however destroy linase and most of the linamarin, and the resultant meals are quite safe. In the dry state, meal containing linase and linamarin is a safe food. The pH of the stomach contents of the pig is sufficiently low to inactivate linase. In ruminants the hydrogen cyanide formed by linase action is absorbed into the blood very slowly, and this, coupled with its rapid detoxication in the liver and excretion via the kidney and lungs, ensures that it never reaches toxic levels in the blood.

It has been reported that linseed meal has a protective action against selenium poisoning.

The protein of linseed meal is not of such good quality as that of soya bean or cottonseed meals, having a lower methionine and lysine content. Linseed meal has only a moderate calcium content but is rich in phosphorus, part of which is present as phytate. It is a useful source of thiamin, riboflavin, nicotinamide, pantothenic acid and choline.

Linseed meal has a very good reputation as a food for ruminant animals which is not easy to justify on the basis of its proximate analysis. Part of the reputation may be the result of the ability of the mucilage to absorb large amounts of water, which results in an increase in the bulk of linseed meal in the rumen; this may increase the retention time in the rumen and give a better opportunity for microbial digestion. The lubricating character of the mucilage also protects the gut wall against mechanical damage and, together with the bulkiness, regulates excretion, preventing constipation without causing looseness. Linseed meal given to fattening animals results in rapid gains compared with other vegetable protein supplements making the same protein contribution, and cattle attain a very good sleek appearance, though the body fat may be soft. The meal is readily eaten by dairy cows but tends to produce a soft milk fat.

Linseed meal is an excellent protein food for pigs, provided it is given with an animal protein supplement to make good its deficiency of methionine, lysine and calcium. This is particularly important with diets containing large amounts of maize.

Linseed meal is not a satisfactory food for inclusion in poultry diets. Retardation of chick growth has been reported on diets containing 50 kg/t of linseed meal, and deaths in turkey poults at a 100 kg level. These adverse effects can be avoided by autoclaving the meal or by increasing the levels of vitamin B_6 in the diet. Some workers consider the adverse effects of the meal to be due to the mucilage, since this collects as a gummy mass on the beak, causing necrosis and malformation and reducing the bird's ability to eat. Pelleting or coarse granulation can overcome this trouble. If linseed meal has to be included in poultry diets the level should not exceed 30 kg/t.

In the UK, linseed cake or meal must by law, contain less than 350 mg hydrocyanic acid/kg.

Rapeseed meal

The 1981 world production of rapeseed was estimated at about 11.5 million tonnes, making it the fifth in order of importance of the oil-producing seed crops. The current picture is of increasing production, particularly in Western Europe where its production is actively encouraged by the EEC. Extraction of the oil, usually by a prepress solvent extraction procedure, leaves a residue containing about 400 g protein/kg DM. It contains more fibre (140 g/kg DM) than soya bean meal and its metabolisable energy value is lower; about 8.2 MJ/kg DM for poultry and 12.0 for pigs and ruminants. Both protein content and digestibility are lower than for soya bean meal but the balance of indispensable amino acids compares favourably, with less lysine but more methionine. The balance of calcium and phosphorus is satisfactory and it contains a higher phosphorus content than other oilseed residues. The use of rapeseed meals produced from rape (*Brassica napus*) grown in Europe is restricted, particularly for pigs and poultry, by the presence of thiocyanates and 5-vinyloxazolidine-2-thione. Both are goitrogenic but the latter, known as goitrin, is much more powerfully so. They are produced from glucosinolates (thioglucosides) present in the seed by the action of an accompanying thioglucosidase, sometimes referred to as myrosinase.

In other parts of the world, notably Canada, *B. campestris* is preferred and careful selection has resulted in the production of 'double zero' varieties having very low contents of glucosinolates and erucic acid. The latter has caused heart lesions in experimental animals, but is unlikely to be a problem to farm animals since it partitions with the oil during extraction. The Canadian meals are produced from low glucosinolate varieties and are

TABLE 22.3 Best estimates of permissible levels of inclusion of rapeseed meal in pig diets

	HGRSM (kg/t)	LGRSM (kg/t)
Starting pigs (<20 kg)	40	120
Growing pigs (20–60 kg)	50	120
Finishing pigs (>60 kg)	80	150
Gilts	0	120
Sows	30	120

known as Canola meals. They are frequently referred to as low glucosinolate meals (LGRSM) having less than one-eighth of the glucosinolate content of the high glucosinolate meals (HGRSM) produced in Europe.

A certain measure of control of the goitrogenic activity of meals is achieved by manipulating the pretreatment of the seed before extraction, so as to ensure the earliest possible destruction of myrosinase. Such control is partial only, since bacterial thioglucosidases produced in the gut may hydrolyse residual glucosinolate in the meal.

Evidence on permissible levels of inclusion of rapeseed meals in the diet is conflicting, but those for high and low glucosinolate meals appear to differ markedly, as illustrated in Table 22.3. The highly variable responses to dietary inclusions of rapeseed meal requires that such figures are applied with caution in practice.

For calves of less than 100 kg, levels of HGRSM in the diet should be limited to 100 kg/t whereas LGRSM can serve as the sole source of supplemental protein for such animals. For more mature animals it would appear that HGRSM can be used as the major supplemental protein and LGRSM as the sole supplemental protein. There is some evidence that the consumption of HGRSM may have beneficial effects on performance, owing to a possible lowering of basal metabolic rate.

For lactating cows HGRSM may form up to 100 kg/t of the concentrate and LGRSM may be used as the sole protein source. Currently, rapeseed meals make up about 15 per cent of the supplemental protein sources used in the UK, and is second only to soya bean meal in importance.

In the UK whole foods for poultry must, by law, contain less than 1000 mg vinylthioxazolidine/kg except those for laying hens which must contain less than 500 mg/kg. Levels of isothiocyanates in whole foods also are strictly prescribed and rape cakes or meals must contain less than 4000 mg allylthiocyanate/kg.

Sunflower seed meal

The meal is produced when the oil is removed from the seed by hydraulic pressure or solvent extraction. The hulls are usually partially, rather than completely removed, but the resulting high-fibre meals (up to 380 g/kg) are readily accepted by animals, provided they are finely ground.

Solvent extracted meals contain, on average, about 200 g crude fibre and 390 g crude protein/kg and have metabolisable energy contents of 8.1, 10.4 and 11.2 MJ/kg DM for poultry, cattle and sheep respectively. For pigs, a digestible energy content of 10.6 MJ/kg DM would be appropriate. Expeller meals have higher fat and lower crude fibre and crude protein contents, and have a metabolisable energy content of about 13 MJ/kg DM for cattle.

The meals are useful sources of protein which is low in lysine but has about twice as much methionine as does soya protein. Maximum rates of inclusion in diets are 200 kg/t for ruminants, 25 kg/t for growing pigs, 50 kg/t for finishing pigs and 100 kg/t for sows. They are not recommended for young pigs. For adult poultry, sunflower meals may be included at 100 g/t of the diet, but they are not recommended for young birds.

The meal is palatable but is laxative and has a very short shelf life.

Sesame seed meal

The meals currently available may be produced by hydraulic pressing or solvent extraction. The former has the lower protein content (about 400 g/kg DM compared with 500 g/kg DM for the solvent extracted material), but has an oil content of more than 100 g/kg DM compared with 20 g/kg DM for the extracted meal, and makes a significantly greater contribution to the energy of the diet.

The protein is rich in leucine, arginine and methionine, but is relatively low in lysine. The meal therefore needs to be combined with foods rich in lysine when fed to pigs and poultry. The protein has a degradability of 0.65 to 0.75 depending upon the rate of passage through the rumen.

The residual oil of the cake or meal is highly unsaturated and may result in soft body and milk fat if consumed in excessive amounts, and may also impart a disagreeable flavour to milk. The oil rapidly becomes rancid and unpalatable and such meals have been implicated in cases of vitamin E deficiency. The meal has a high content of phytic acid which makes much of its phosphorus unavailable: rations containing the meal may also need extra supplementation with calcium.

The hulls of sesame seeds contain oxalates and it is essential that meals should be completely decorticated in order to avoid toxicities.

Meals in good condition are palatable but have a laxative action. The diets of young ruminants should not contain more than 50 kg sesame meal/t, whereas the maximum inclusion rate for adults is 100–150 kg/t. The meal should not be fed to young pigs or poultry, but may be included at up to 50 kg/t for adults of these species.

LEGUMINOUS SEEDS

The Leguminosae is a large family of plants with about 12 000 recognised species. Four tribes within the family are of particular importance since they include all the common peas and beans. The Hedysareae contains the groundnut; the Vicieae contains the genera *Vicia, Cicer, Pisum, Lens* and *Lathyrus*; the Genisteae contains the genus *Lupinus*; the Phaseoleae contains the genera *Phaseolus, Dolichos* and *Glycine*.

Many leguminous plants are toxic to animals. Species within the genus *Lathyrus* such as the Indian pea (*L. sativus*) cause skeletal lesions, retardation of sexual development and various degrees of paralysis, which sometimes proves fatal when the larynx is affected. The condition is known as lathyrism and is caused by β-aminopropionitrile present in the seed. *Vicia faba* (broad beans) may cause the condition in man known as favism. This is characterised by haemolytic anaemia and occurs in individuals with a genetic deficiency of glucose-6-phosphate dehydrogenase in their erythrocytes. *Phaseolus lunatus* (Lima bean, Java bean) contains the cyanogenetic glucoside phaseolunatin, which is extremely toxic when hydrolysed. The glucoside is present, in small amounts only, in cultivated varieties of *P. lunatus* such as the 'butter bean'. Unfamiliar leguminous seeds should be fed with caution until they have been proved to be safe.

Beans

Beans belong mainly to the Vicieae and Phaseoleae and are used as food for humans and animals all over the world. The chief member of the Vicieae is *Vicia faba* known as the broad bean, horse bean and Windsor bean. The most numerous genus in the Phaseoleae is the *Phaseolus*, and the best known species is *P. vulgaris* with a number of varieties known as kidney, field, garden and haricot beans. There are a large number of others of the species important locally as sources of food, as are several other genera such as *Vigna, Dolichus* and *Canavalia*. Nutritionally, the various species are very similar, being good sources of protein with a high lysine content, good sources of energy and phosphorus but with a low calcium content. Beans have little or no carotene or vitamin C but may contain significant amounts of thiamin, niacin and riboflavin.

There are a large number of varieties of field bean, which fall into two classes, winter and spring. The winter varieties outyield the spring varieties, the yield levels being about 3.4 and 3.0 tonnes/hectare respectively. However, spring varieties usually have a higher protein content than winter varieties, about 270 compared with 230 g/kg and a lower fibre content, about 68 compared with 78 g/kg. These are higher than in the common cereals except for oats. The ether extract content of both winter and spring varieties of beans is low, about 13 g/kg, but has a high proportion of the highly unsaturated linoleic and linolenic acids. The mineral composition of beans is similar to that of the cereals and oil seed residues, with high phosphorus and low calcium contents. They contain little or no sodium or chlorine and are poor sources of manganese.

Beans are regarded primarily as sources of protein, which is of relatively good quality. This is a reflection of the amino acid composition which is characterised by a high lysine content, similar to that of fish meal protein, but by a low level of cystine and methionine, which is lower than in the common animal and vegetable protein sources.

As well as being a good source of protein, beans make a significant contribution to the energy economy of the animal, having a metabolisable energy content of 13.5 MJ/kg DM for ruminants, 12.0 MJ/kg DM for poultry and 13.3 MJ/kg DM for pigs.

Beans are used in the diets of all the major classes of farm animals. Levels in the diets of calves up to 3 months of age are usually of the order of 150 kg/t but can be increased considerably thereafter, and mixtures containing 250 kg/t have been used quite satisfactorily for intensively fed steers. The beans are usually cracked, kibbled or coarsely ground for feeding but it would appear that whole beans are quite satisfactory for older ruminants, who rapidly adapt to chewing them. Dairy cow concentrates may contain 150–200 kg/t of beans and recent work has shown that levels of up to 350 kg/t may be used with no loss of yield. For pigs, beans are usually ground to pass a 3 mm screen and are used in sow, weaner and fattening diets but it is not usual to include them in creep feeds. Although there is no objective evidence in its support, the view is widely held that newly harvested beans should be allowed to mature for several weeks before being given to pigs. The usual rate of inclusion of beans is 50 to 150 kg/t and should not exceed 200 kg/t. Cystine and methionine are important in the diets of poultry owing to the demands of feathering and beans would be expected to be of limited value. Traditionally they have been little used in poultry diets but present evidence indicates that field beans can be used to replace soya bean meal providing adequate methionine supplementation is practised.

Many species of beans have been shown to contain antitryptic factors but the field bean as grown in the UK has not been shown to have any significant antitryptic activity.

Peas

The peas grown as a source of protein for animals in the UK (*Pisum sativum*) belong to the Vicieae. Other species, such as the chick pea (*Cicer arietinum*) in India, are important locally. Peas are basically similar to beans but have lower contents of crude protein (260 g/kg DM) and crude fibre (<60 g/kg DM). The oil content is slightly higher than that of beans but the degree of saturation is similar. Like beans, peas are regarded primarily as a source of protein. That of peas has the better balance of amino acids having higher contents of lysine, methionine and cystine. Methionine is, however, still the main limiting amino acid. Peas make a significant contribution to the energy intake of the animal with a metabolisable energy content of 13.4 MJ/kg DM for ruminants, 12.7 MJ/kg DM for poultry and a digestible energy content of 15.0 MJ/kg DM for pigs.

The maximum rate of inclusion of peas in ruminant diets may be as high as 400 kg/t, but problems of mixing and cubing limit their inclusion in pelleted foods to a maximum of 20 kg/t. They are particularly useful in being able to replace soya bean meal in pig and poultry diets, whereas beans are largely confined to ruminant diets.

The growing appreciation of the value of peas is reflected in the increase in European production from 540 000 t in 1982 to 1 160 000 in 1984. This compares with figures for beans of 405 and 338 000 t in 1982 and 1984 respectively.

Lupin seed meal

Lupin seed meal is made by grinding the whole seeds. It is a useful source of protein which can be grown in Europe. There are three species of lupins distinguished by the colour of the flowers. Those of *Lupinus albus* are white, those of *L. angustifolius* blue and those of *L. luteus* yellow. Within species there are sweet and bitter varieties. The latter contain from 10 to 20 g/kg of toxic alkaloids such as lupinin and angustifolin and should not be offered to animals: even sweet varieties may contain low levels of alkaloids. For safety the alkaloid content must be less than 0.6 g/kg.

The seed coat is fibrous and its inclusion in the meal adversely affects its digestibility, especially for young monogastric animals. White varieties have the lowest fibre and highest oil and protein contents. Meals made from them have higher values for pigs and poultry than do the blue and yellow varieties. A typical meal will have a metabolisable energy content for poultry of 11.5 MJ/kg DM, for ruminants 13.2 MJ/kg DM, and a digestible energy content for pigs of 17.3 MJ/kg DM. The amino acid pattern is not well balanced, and diets with significant quantities of the meal may require supplementation with methionine, which is the first limiting amino acid. Maximum levels of inclusion are 150 kg/t for ruminant diets, 100 kg/t for those of adult poultry and pigs, and 50 kg/t for growing pigs and broilers.

Because of rapid oxidation of the oil, the meal has to be used immediately or an antioxidant incorporated.

ANIMAL PROTEIN CONCENTRATES

These materials are given to animals in much smaller amounts than the oil seed derivaties so far discussed, since they are not used primarily as sources of protein *per se* but to make good deficiencies of certain indispensable amino acids from which non-ruminant animals may suffer when they are fed on all-vegetable protein diets. In addition they often make a significant contribution to the animals' mineral nutrition, as well as supplying various vitamins of the B-complex. A further reason why these products are given in limited quantities to farm animals is that they are expensive, which makes their large-scale use uneconomic.

Fish meal

In the UK, fish meal is legally defined as . . . the product obtained by drying and grinding whole fish, or parts thereof, of various species. Concentrated press liquid may be added. Products whose chloride content expressed as NaCl is less than 2 per cent may be referred to as 'low in salt'. (Feeding Stuffs Regulations 1986).

White fish meal, although still available, has no longer a legal identity or definition. For the purposes of trading and to clarify questions of quality the International Association of Fish Meal Manufacturers have proposed that fish meals should be classified as follows:

1. High protein meals having more than 680 g protein/kg and less than 90 g oil/kg. Most herring meals would fall into this class.
2. Regular protein meals with between 640 and 679 g protein/kg. Such meals, including Capelin and South American meals, could have as much as 130 g oil/kg.
3. Regular protein, low oil meals having 640 to 679 g protein/kg and less than 60 g oil/kg. Meals containing significant amounts of offal would be in this class.
4. Standard protein meals with 600–639 g protein/kg. These would include the American Menhaden meals.

Fish meals are produced in two ways. The first is by drying in steam jacketed vessels, which may be either a batch process carried out under vacuum or a continuous process not employing reduced pressure. In both cases heating is carried out in steam-jacketed vessels. In the flame-drying process the meal is dried in a revolving drum by hot air from a furnace at one end of the drum. Flame drying is a more drastic process than steam drying, and this may affect the quality of the protein.

In well-produced meals the protein has a digestibility of between 0.93 and 0.95. Meals which are heated too strongly in processing have values as low as 0.60. The quality of protein in fish meal is high though variable as indicated by values of 0.36 and 0.82 quoted as the biological value for rats. Processing, particularly degree and length of time of heating, is probably the major determinant of protein quality as shown by the figures for available lysine quoted in Table 22.4. Fish meal protein has a high content of lysine, methionine and tryptophan and is a valuable supplement to cereal-based diets, particularly where they contain much maize. Fish meals have a high mineral content, about 210 g/kg, which is of value nutritionally since it contains a high proportion of calcium (80 g/kg) and phosphorus (35 g/kg) and also a number of desirable trace minerals including manganese, iron and iodine. They are a good source of vitamins of the B-complex, particularly choline, B_{12} and riboflavin, and have an enhanced nutritional value because of their content of growth factors known collectively as the Animal Protein Factor (APF).

Fish meals find their greatest use with simple-stomached animals. They are used mostly in diets for young animals, whose demand for protein and the indispensable amino acids is particularly high and for whom the growth-promoting effects of APF are valuable. Such diets may include up to 150 kg/t of fish meal. With older animals, who need less protein, the level of fish meal in the diet is brought down to about 50 kg/t and it may be eliminated entirely from diets for those in the last stages of fattening. This is partly for economic reasons, since the protein needs of such animals are small, and partly to remove any possibility of a fishy taint in the finished carcass. This possibility must also be carefully considered with animals producing milk and eggs, which are vulnerable to taint development. Fully ruminant animals are able to obtain amino acids and B vitamins by micro-

TABLE 22.4 Effect of various heat treatments on the available lysine contents of fish meals

Treatment	Available lysine (g/kg CP)
Freeze dried	86
Oven dried at:	
105 °C for 6 h	83
170 °C for 6 h	69

bial synthesis and the importance of fish meal for such animals is as a source of undegradable protein. This is particularly important for actively growing, pregnant and lactating animals. Rates of inclusion in the diet are usually about 50 kg/t. Recent research indicates that the energy value of fish meal for ruminants has traditionally been underestimated. Thus a metabolisable energy content of 14 MJ/kg DM would appear to be more realistic than the generally accepted value of 11.1.

The nitrite content of fish meals sold in the UK is strictly controlled by legislation. They may not contain more than 60 mg/kg of food referred to a moisture content of 120 g/kg and stated as sodium nitrite. This is equivalent to 14 mg nitrite-nitrogen/kg DM.

Meatmeal

In the UK meat meal is legally defined as . . . the product obtained by drying and grinding carcasses and parts of carcasses of warm-blooded land animals, if need be with the fat removed by an appropriate process. It should be virtually free of hair, bristle, feathers, horn, hoof and skin and of the contents of the stomach and viscera. It shall be technically free of organic solvents.

Products with a fat content of more than 11 per cent should be described as 'rich in fat'. (Feeding Stuffs Regulations 1986).

Meat and bone meal is defined as . . . the product obtained by drying and grinding meat pieces containing a high proportion of bone from warm-blooded land animals. The product should be subtantially free of hair, bristle, feathers, horn, hoof, skin and blood and of the contents of the stomach and viscera. It shall be technically free of organic solvents. (Feeding Stuffs Regulations 1986).

The meals may be produced by dry rendering, in which the material is heated in steam-jacketed cookers and the fat which separates from the dehydrated product is allowed to drain away. More fat is removed under pressure, and the residue is ground to give the final product. In the wet rendering process the material is heated by live steam after water has been added. Fat separates and is skimmed off, the residue is allowed to settle and the supernatant liquor is drained off. This is known as 'stick' and contains a considerable amount of protein. The residue is pressed to remove fat, dried and ground.

Meat meal generally contains from 600 to 700 g/kg of protein compared with about 450 to 550 g/kg for meat and bone meal. The fat content is variable, ranging from 30 to 130 g/kg, but is normally about 90 g/kg. Meat and bone meal contains more ash than meat meal and is an excellent source of calcium, phosphorus and manganese. Both meals are good sources of vitamins of the B-complex, especially riboflavin, choline, nicotinamide and B_{12}.

The protein of these meat by-products is of good quality (BV approximately 0.67 for adult man) and is particularly useful as a lysine supplement. Unfortunately it is a poor source of methionine and tryptophan. Various unidentified beneficial factors have been claimed to be present in meat meals, among them the enteric growth factor from the intestinal tract of swine, the *Ackerman* factor and a growth factor located in the ash.

Meat products are more valuable for simple-stomached than for ruminant animals, since the latter do not require a dietary supply of high quality protein. The low methionine and tryptophan levels of the meals affect their value, however, because they cannot adequately make good the deficiencies of these amino acids in the high-cereal diets of pigs and poultry. This is particularly so where high proportions of maize are given, maize being particularly low in tryptophan. Usually meat meal is given in conjunction with another animal protein or with a vegetable protein to make good its low content of methionine and tryptophan. Both meat meal and meat and bone meal are eaten readily by pigs and poultry, and may be given at levels of up to 150 kg/t of the diet for laying hens and young pigs; for fattening pigs the level is usually kept below 100 kg/t. As well as being less beneficial to ruminant than to simple-stomached animals, these products are not readily acceptable to ruminants and must be introduced into their diets gradually. Considerable care is required in storing the meat products to prevent the development of rancidity and loss of vitamin potency.

In the past these meals were produced under conditions which frequently were unsatisfactory and, unless great care was taken to ensure a final sterilisation by heating to 100°C for one hour, their use often presented a health hazard. Modern processing conditions and the demands of the Protein Processing Order should result in meals guaranteed to be free of pathogens. This should allow their use in greater amounts and in a wider variety of rations.

Blood meal

In the UK, blood meal is defined as . . . the product obtained by drying the blood of slaughtered animals and poultry. This product should be substantially free of foreign matter. (Feeding Stuffs Regulations 1986).

It is manufactured by passing live steam through the blood until the temperature reaches 100 °C. This ensures efficient sterilisation and causes the blood to clot. It is then drained, pressed to express occluded serum, dried by steam heating and ground.

Blood meal is a dark chocolate-coloured powder with a characteristic smell. It contains about 800 g/kg of protein, small amounts of ash and oil and about 100 g/kg of water, and is important nutritionally only as a source of protein. Blood meal is one of the richest sources of lysine, a rich source of arginine, methionine, cystine and leucine, but is very poor in isoleucine and contains less glycine than either fish, meat or meat and bone meals. Owing to the poor balance of amino acids its biological value is low and in addition it has a low digestibility. It has the advantage, in certain situations, that its protein has a very low degradability (*c*. 0.20).

The meal is unpalatable and its use has resulted in reduced growth rates in poultry and it is not recommended for young stock. For older birds, rates of inclusion are limited to about 10 to 20 kg/t of diet. It should not be used in creep foods for pigs. Normal levels of inclusion for older animals are of the order of 50 kg/t of diet and it is usually used together with a high-quality protein source. At levels over 100 kg/t of the diet it tends to cause scouring and, overall, is best regarded as a food for boosting dietary lysine levels.

Milk products

Whole milk from cows contains about 875 g/kg of water and 125 g/kg of dry matter, usually referred to as the total solids. Of this about 37.5 g/kg is fat, the remainder, forming the solids-not-fat (SNF), consists of protein (33 g/kg) lactose (47 g/kg) and ash (7.5 g/kg).

Most of the fat is neutral triacylglycerol having a characteristically high proportion of fatty acids of low molecular weight and forming an excellent source of energy. It has about $2\frac{1}{4}$ times the energy value of the milk sugar or lactose. The crude protein fraction of milk is complex, about 5 per cent of the nitrogen being non-protein. Casein, the chief milk protein, contains about 78 per cent of the total nitrogen, and is of excellent quality, but has a slight deficiency of the sulphur-containing amino acids, cystine and methionine. Fortunately the β-lactoglobulin is rich in these acids, so that the combined milk proteins have a biological value of about 0.85. The most economical use of the protein is to supplement poor-quality proteins like those of the cereals, for which purpose it is better than either the meat or fish products. When, however, milk products are used to replace fish meal or meat and bone meal, the diet must be supplemented with inorganic elements, particularly calcium and phosphorus, since the ash content of milk is so low. Milk has a low magnesium content and is seriously deficient in iron. Normally milk is a good source of vitamin A but poor in vitamins D and E. It is a good source of thiamin and riboflavin and contains small amounts of vitamin B_{12}.

Whole milk is fed to suckled calves and lambs, and to young dairy and bull calves, and animals being prepared for competition. Two milk by-products are widely used, however, and are valuable foods for farm animals.

Skim milk. This is the residue after the cream has been separated from milk by centrifugal force. The fat content is very low, below 10 g/kg,

TABLE 22.5 Effect of processing on the nutritive value of skim milk

	Digestibility of protein	Biological value of protein	Available lysine (g/kg CP)
Spray-dried	0.96	0.89	81
Roller-dried	0.92	0.82	59

and the gross energy of the by-product is much reduced, about 1.5 MJ/kg as compared with 3.1 MJ/kg for whole milk. Removal of the fat in the cream also means that skim milk has little or none of the fat-soluble vitamins. However, it does result in a concentration of the SNF constituents. Skim milk finds its main use as a protein supplement in the diets of simple-stomached animals and is rarely used for ruminant animals; it is particularly effective in making good the amino acid deficiencies of the largely cereal diets of young pigs and poultry. For pigs it is usually given in the liquid state, and limited to a *per capita* consumption of 2.8 l (3.0 kg) to 3.4 l (3.6 kg)/day. Where the price is suitable it may be given *ad lib.*, and up to 23 l (24 kg) per pig per day may be consumed along with about 1 kg of meal. Scouring may occur at these levels but can be avoided with reasonable care. Liquid skim milk must always be given in the same state, either fresh or sour, if digestive troubles are to be avoided. It may be preserved by adding 1.5 l of formalin to 1000 l of skim milk. For feeding to poultry, skim milk is normally used as a powder, and may form up to 150 kg/t of the diet. It contains about 350 g/kg of protein, the quality of which varies according to the manufacturing process used: roller-dried skim milk is subjected to a higher drying temperature than the spray-dried product and has a lower digestibility and biological value (Table 22.5). For poultry, skim milk protein has the disadvantage of a low cystine content.

Whey. When milk is treated with rennet in the process of cheese-making, casein is precipitated and carries down with it most of the fat and about half the calcium and phosphorus. The remaining serum is known as whey, which as a result of the partition of the milk constituents in the rennet coagulation, is a poorer source of energy (1.1 MJ/kg), fat-soluble vitamins, calcium and phosphorus. Quantitatively it is a poorer source of protein than milk, but most of the protein is β-lactoglobulin and of very good quality. Whey is usually given *ad lib.* to pigs in the liquid state. Dried skim milk and whey find their main use as constituents of proprietary milk replacers for young calves.

SINGLE-CELL PROTEIN

In recent years there has been considerable interest in exploiting microbial fermentation for the production of protein. Single-cell organisms such as yeasts and bacteria grow very quickly and can double their cell mass, even in large-scale industrial fermentors, in three to four hours. A range of nutrient substrates can be used including cereal grains, sugar beet, sugar cane and its by-products, hydrolysates from wood and plants, and waste products from food manufacture. Bacteria such as *Pseudomonas* sp. can be grown on more unconventional materials such as methanol, ethanol, alkanes, aldehydes and organic acids. The figures in Table 22.6 show that the material obtained by culturing different organisms varies considerably in composition. The protein content of bacteria is higher than that of yeasts and contains higher concentrations of the sulphur-containing amino acids but a lower concentration of lysine. Single-cell protein (SCP) contains unusually high levels of nucleic acids ranging from 50 to 120 g/kg DM in yeasts and 80 to 160 g/kg DM in bacteria. While some of the purine and pyrimidine bases in these acids can be used for nucleic acid biosynthesis, large amounts of uric acid or allantoin, the end products of nucleic acid catabolism, are excreted in the urine of animals consuming SCP. The oil content of yeasts and bacteria varies from 25 to 236 g/kg DM and the oils themselves are rich in unsaturated fatty acids. Athough SCP does contain a crude fibre fraction, and this can be quite high in some yeasts, it is not composed of cellulose, hemicelluloses and lignin as in foods of plant origin; it consists chiefly of glucans, mannans and chitin.

Studies with pigs have yielded digestibility coefficients for energy varying from 0.7 to 0.90 for yeasts grown on whey and *n*-paraffins respectively, and a value of 0.80 was obtained for a bacterium grown on methanol. The inclusion of SCP in pig diets at levels up to 150 kg/t has given performances comparable with diets containing soya bean meal or fishmeal. Similarly encouraging results have been obtained with calves although it is usually recommended that the maximum level of inclusion of alkane yeasts in calf milk substitutes should be 80 kg/t.

In the case of poultry, dietary SCP concentrations of 20 to 50 kg/t have proved optimal for broilers while 100 kg/t has been suggested for diets for laying hens.

PROTEIN DEGRADABILITY

Current concepts of ruminant protein nutrition make it highly desirable that, in assessing the value of protein supplements, their susceptibility to breakdown in the rumen should be known. Unfortunately, little definitive information is available. Table 22.7 presents some experimental data and suggests degradability values for use in ration formulation. It is most important not to assign too great an accuracy to these estimates.

TABLE 22.6 Chemical composition of SCP grown on different substrates
(After E. Schulz and H. J. Oslage, 1976. *Animal Feed Science and Technology* **1**, 9.) (g/kg DM)

Substrates used	Micro-organism	DM (g/kg)	Organic matter	Crude protein	Crude fat	Crude fibre	Ash
Gas oil	*Candida lipolytica*	916	914	678	25	44	86
Gas oil	*Candida lipolytica*	903	917	494	132	41	84
n-Paraffin	*Candida lipolytica*	932	934	644	92	47	66
n-Paraffin	*Candida lipolytica*	914	933	480	236	47	67
n-Alkanes	*Pichia guillerm*	971	941	501	122.	76	59
Whey (lactic acid)	*Candida pseudotropicalis*	900	900	640	56	50	100
Methanol	*Candida boidinii*	938	939	388	77	107	61
Methanol	*Pseudomonas methylica*	967	903	819	79	5	97
Sulphite liquor	*Candida utilis*	917	925	553	79	13	75
Molasses	*Saccharomyces cerevisiae*	908	932	515	63	18	68
Extract of malt	*Saccharomyces carlsbergensis*	899	926	458	31	11	74

TABLE 22.7   Rumen degradability of some proteins

Source	Range of degradability reported	Suggested values for ration formulation
Soya bean meal	0.39–0.86	0.60
Groundnut meal	0.55–0.85	0.80
Cottonseed meal	—	0.60
Coconut meal	0.43	0.45
Linseed meal	0.63	0.60
Rapeseed meal	0.60–0.78	0.75
Sunflower meal	0.77–0.84	0.80
Sesame meal	—	0.60
Bean meal	0.47–0.55	0.55
Fish meal	0.16–0.71	0.40
Meat and bone meal	0.40	0.40

NON-PROTEIN NITROGEN COMPOUNDS AS PROTEIN SOURCES

Non-protein nitrogen compounds are recognised as useful sources of nitrogen for ruminant animals, and simple nitrogenous compounds, such as ammonium salts of organic acids, can be utilised to a very limited extent by pigs and poultry. Commercially, non-protein nitrogen compounds are important for ruminants only. Their use depends upon the ability of the rumen micro-organisms to use them in the synthesis of their own cellular tissues (Ch. 8), and they are thus able to satisfy the microbial portion of the animal's demand for nitrogen and, by way of the microbial protein, at least part of its nitrogen demand at tissue level. The compounds investigated include urea, ammonium salts of organic acids, inorganic ammonium salts and various amides as well as thiourea, hydrazine, and biuret. Studies *in vitro* have shown that ammonium acetate, ammonium succinate, acetamide and diammonium phosphate are better substrates for microbial protein synthesis than urea; but from considerations of price, convenience, palatability and toxicity, urea has been the most widely used and investigated non-protein nitrogen compound in farm animal foods.

Urea

Urea is a white, crystalline, deliquescent solid with the following formula:

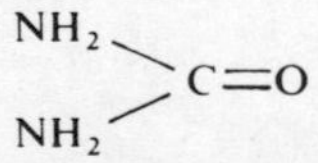

$$\begin{array}{c} NH_2 \\ \diagdown \\ \diagup \\ NH_2 \end{array} C{=}O$$

Pure urea has a nitrogen content of 466 g/kg, which is equivalent to a crude protein content of $466 \times 6.25 = 2913$ g/kg. 'Feed urea' contains an inert conditioner to keep it flowing freely, which reduces its nitrogen content to 464 g/kg, equivalent to 2900 g/kg crude protein. Urea is hydrolysed by the urease activity of the rumen micro-organisms with the production of ammonia:

$$\begin{array}{c} NH_2 \\ \diagdown \\ C=O + H_2O \rightarrow 2NH_3 + CO_2 \\ \diagup \\ NH_2 \end{array}$$

The ease and speed with which this reaction occurs when urea enters the rumen, gives rise to two major problems owing to excessive absorption of ammonia from the rumen. Thus wastage of nitrogen may occur, and there may be a danger of ammonia toxicity. This is diagnosed by muscular twitching, ataxia, excessive salivation, tetany, bloat and respiration defects (both rapid shallow, and slow deep, breathing have been reported). Dietary levels of urea vary in their effects and it is not possible to give accurate safety limits for any particular animal. Thus, sheep receiving as little as 8.5 g/day have died, whereas others have consumed up to 100 g/ day without ill effects. Toxic symptoms appear when the ammonia level of peripheral blood exceeds 10 mg/kg and the lethal level is about 30 mg/kg. Such levels are usually associated with ruminal ammonia concentrations of 800 mg/kg, the actual level depending on pH. Ammonia, which is the actual toxic agent in urea poisoning, is most toxic at high ruminal pH owing to the increased permeability of the rumen wall to unionised ammonia compared with the ammonium ion which predominates at low pH.

Urea should be given in such a way as to slow down its rate of breakdown and encourage ammonia utilisation for protein synthesis. Urea is most effective when given as a supplement to diets of low protein content, particularly if the protein is resistant to microbial breakdown. The diet should also contain a source of readily available energy so that microbial protein synthesis is enchanced and wastage reduced. At the same time, the entry of readily available carbohydrate into the rumen will bring about a rapid fall in rumen pH and so reduce the likelihood of toxicity. Many of the difficulties encountered with urea would be avoided if the amount and size of meals were restricted. To avoid the danger of toxicity, not more than one-third of the dietary nitrogen should be provided as urea, and where possible this should be in the form of frequent and small intakes.

Studies of the degradability of food proteins in the rumen have now provided a valuable guide to the use of urea; in general, urea, like other non-protein nitrogen sources, will not be used efficiently by the ruminant animal unless its diet contains insufficient degradable protein to satisfy the needs of its rumen micro-organisms.

Although urea provides an acceptable protein source there is evidence that where it forms a major part of dietary nitrogen, deficiencies of the sulphur-containing amino acids may occur. In such cases supplementation of the diet with a sulphur source may be necessary. An allowance of 0.13 g anhydrous sodium sulphate/g urea is generally considered to be optimal. Urea provides no energy, minerals or vitamins for the animal, and where it is used to replace conventional protein sources care must be taken to ensure that satisfactory dietary levels of these nutrients are maintained by adequate supplementation.

Urea is available in proprietary foods in several forms. It may be included in solid blocks which also provide vitamin and mineral supplementation and contain a readily available source of energy, usually starch. Animals are allowed free access to the blocks, intake being restricted by the blocks having to be licked and by their high salt content. There is some danger of excessive urea intakes if the block should crumble or if there should be a readily available source of water allowing the animal to cope with the high salt intakes. Solutions of urea, containing molasses as the energy source and carrying a variable amount of mineral and vitamin supplementation, are also in use. Like the blocks they contain 50–60 g urea/kg and about 250 g sugar/kg and are supplied in special feeders in which the animal licks a ball floating in the solution; the animal thus has no direct access to the solution. Many proprietary concentrate foods now contain urea at levels ranging from 10–30 g/kg. Protein supplements for balancing cereals may contain up to 100 g/kg urea and there are some highly concentrated products with as much as 500 g/kg. Where urea is included in the concentrate diet thorough mixing is essential to prevent localised concentrations which may have toxic effects. A recent development in the feeding of urea has been its addition, to the roughage or concentrate portion of the ration, as a solution which also supplies some mineral and vitamin supplementation. Such solutions contain about 350 g urea/kg and 100 g molasses/kg. The efficiency of capture of urea nitrogen by rumen micro-organisms is about 0.8 of that of protein nitrogen. This means that the rumen degradable protein requirement must be increased by about 0.25 of the calculated value, if it is to be supplied by urea. Looked at in another way, about 0.44 of dietary urea nitrogen arrives as amino acid at the small intestine compared with 0.54 of dietary rumen degradable protein.

Urea may be used for all classes of ruminants but is less effective in animals in which the rumen is not fully functional. Low intensity, ranch-like conditions, in which animals are fed on diets containing poor quality protein in low concentration, are very suitable for the use of block or liquid feeds. Under such conditions, raising the nitrogen content of the diet may increase digestibility and intake of roughage owing to stimulation of microbial activity. Beef cattle and sheep consuming small amounts of concentrates are able to utilise urea efficiently, as are animals receiving even large quantities of concentrate where feeding *ad libitum* results in small and frequent intakes of food. Low-yielding dairy cows also use urea-containing concentrates efficiently owing to their low concentrate intakes, but cows of moderate and high yield, which may receive large concentrate meals at milking, do not. There is much evidence for reduced performance by such animals on urea-containing diets.

When straws and other low-quality roughages are treated with ammonia (see p. 427), about 0.3–0.5 of the ammonia is retained by the roughage and may be utilised by the rumen micro-organisms in the same way as ammonia derived from urea.

Biuret

Biuret is produced by heating urea. It is a colourless, crystalline compound with the following formula:

$$NH_2.CO.NH.CO.NH_2$$

It contains 408 g nitrogen/kg, equivalent to 2550 g crude protein/kg. Biuret is utilised by ruminants but a considerable period of adaptation is required. Adaptation is speeded up by inoculation with rumen liquor from an adapted rumen. Biuret nitrogen is not as efficiently utilised as that of urea, and it is very much more expensive. It has the considerable advantage that it is not toxic even at levels very much higher than those likely to be found in foods.

Compound foods containing urea, biuret, urea phosphate or diureidoiso-butane must by law carry a declaration of

(a) the name of the material,
(b) the amount present,
(c) the proportion of nitrogen, expressed as protein equivalent provided by the non-protein nitrogenous content of the food,
(d) directions for use specifying the animals for which the food is intended and the maximum level of non-protein nitrogen which must not be exceeded in the daily ration.

Poultry waste

In the UK poultry waste is legally defined as . . . the waste from intensive poultry units which consists principally of excreta, with or without litter, and which has been suitably treated for use as a feeding stuff (Feeding Stuffs Regulations 1986).

Despite aesthetic objections dried poultry excreta have been successfully used for feeding ruminants. Poultry manures vary considerably in composition, depending upon their origins. That from caged layers has a lower fibre content than broiler litter which has a base of the straw, wood chips or sawdust used as bedding. Broiler litters also vary in composition depending upon the number of batches put through between changes of bedding. Both types of waste have a high ash content particularly that of layers, which is usually about 280 g/kg DM. Digestibility is low and, although metabolisable energy values of 6 to 9 have been quoted, 7.5 MJ/kg DM is probably an acceptable figure. Protein (N × 6.25) contents are variable, between 250 and 350 g/kg DM, with a digestibility of about 0.65. Most of the nitrogen (600 g/kg at least) is present as non-protein compounds, mostly urates, which must first be converted into urea and then ammonia in order to be utilised by the animal. The conversion to urea is usually a slow process and wastage and the danger of toxicity are both less than for foods containing urea itself. Layer wastes are excellent sources of calcium (about 65 g/kg DM) but the ratio of calcium to phosphorus is rather wide at 3:1; broiler litters have less calcium with ratios closer to 1:1.

Dietary inclusion rates of up to 250 kg/t have been used for dairy cows and up to 400 kg/t for fattening cattle and have supported very acceptable levels of performance. Thus dairy cows, given 110 kg waste per tonne of diet to replace half the soya bean of a control diet, yielded 20 kg milk, the same as the control, but gained only 0.58 kg/day compared with 0.95 kg for the controls. With fattening steers, concentrate diets containing wastes have supported gains of the order of 1 kg/day but it has been estimated that for every inclusion of 100 kg excreta/t of diet, liveweight gains are reduced by about 40 g/kg.

One of the major anxieties constraining the use of poultry wastes in animal diets has been the fear of health hazards arising from the presence of pathogens such as *Salmonella*, and the presence of pesticide and drug residues. The heat treatment involved in drying, and the ensiling procedures used for storing the materials, appear to offer satisfactory control of pathogens, and pesticides have not yet been a problem. Drug residues may be a hazard but this can be overcome by having a withdrawal period of three weeks before slaughter.

The best method of treating poultry wastes for use as animal foods is by drying but this is costly, and ensiling, either alone, with forage, or with barley meal and malt have all proved satisfactory.

Poultry wastes must, by law, carry a declaration of the amount of protein equivalent of uric acid, if 1 per cent or greater, and of calcium if in excess of 2 per cent.

FURTHER READING

F. X. Aherne and J. J. Kennelly, 1982. Oilseed meals for livestock feeding, in *Recent Advances in Animal Nutrition*. W. Haresign (ed.). Butterworths, London.

A. M. Altschul, 1958. *Processed Plant Protein Foodstuffs*. Academic Press, New York.

M. H. Briggs (ed.), 1967. *Urea as a Protein Supplement*. Pergamon Press, London.

M. T. Gillies, 1978. *Animal Feeds from Waste Materials*. Noyes Data, N. J., U.S.A.

I. E. Liener, 1969. *Toxic Constituents of Plant Foodstuffs*. Academic Press, New York.

National Academy of Sciences, 1976. *Urea and other Non-protein Nitrogenous Compounds in Animal Nutrition*. NAS Publishing and Printing Office, Washington DC.

B. H. Schneider, 1947. *Feeds of the World, Their Digestibility and Composition*. Agri. Exp. Stn., West Virginia University.

APPENDIX

LIST OF TABLES

1	Nutritive value of foods for ruminants	485
2	Nutritive value of foods for pigs	488
3	Nutritive value of foods for poultry	490
4	Amino acid composition of foods	492
5	Vitamin potency of foods	494
6	Mineral contents of foods	496
7	Feeding standards for lactating and pregnant cattle	499
8	Feeding standards for growing cattle	506
9	Feeding standards for pregnant ewes	510
10	Feeding standards for lactating ewes	512
11	Feeding standards for growing lambs	515
12	Dietary allowances of trace elements for ruminants	518
13	Feeding standards for pigs	519
14	Feeding standards for poultry	521
15	Water allowances of farm animals	523
16	Values for metabolic liveweight ($W^{0.75}$) for weights at 10 kg intervals to 590 kg	524

NOTE ON THE USE OF THE TABLES

The data given in these Tables have been compiled from a number of sources, a full list of which is given at the end of the Appendix. *Absence of figures does not imply a zero*, but merely that the information was not given in these sources.

Composition, Tables 1–6

The composition of a particular food is variable, and figures given in these tables should be regarded only as representative examples and not constant values. For more comprehensive data readers should consult the references.

Feeding standards, Tables 7–13

The scientific feeding of farm animals is based on standards expressed in terms of either 'nutrient requirements' or 'nutrient allowances'. These terms are defined in Chapter 14. The figures in the tables are expressed as nutrient allowances since they include margins of safety, when necessary, for variations between animals. It has not been possible to include every class of farm animal in these tables and only a respresentative selection is given. For more detailed information readers should consult the relevant references.

ABBREVIATIONS USED IN THE TABLES

APL = Animal production level, CF = Crude fibre, CP = Crude protein, DCP = Digestible crude protein, DE = Digestible energy, dg = degradability of protein in the rumen, DM = Dry matter, DMI = Dry matter intake, EE = Ether extract, E_{mp} = Net energy allowance for maintenance and production, ME = Metabolisable energy, M/D = ME concentration in the dry matter (MJ/kg), NE = Net energy, RDP = Rumen degradable protein, UDP = Undegradable protein in the rumen, W = Liveweight.

TABLE 1 Nutritive value of foods for ruminants[5,9,10,13,14,15]

Food	DM basis						
	DM (g/kg)	CF (g/kg)	EE (g/kg)	Ash (g/kg)	CP (g/kg)	dg	ME (MJ/kg)
Green crops							
Barley in flower	250	316	16	64	68	—	10.0
Cabbage	150	160	47	107	160	0.90	10.8
Clover, red, early flowering	190	274	37	84	179	—	10.2
Clover, white, early flowering	190	232	42	116	237	—	9.0
Grass, close grazing	200	130	55	105	265	0.80	12.1
Grass, extensive grazing	200	200	40	100	175	0.80	11.0
Kale	140	179	36	136	157	0.90	11.0
Lucerne, early flowering	240	300	17	100	171	—	8.2
Maize	190	289	26	63	89	—	8.8
Rape	140	250	57	93	200	0.85	9.5
Sugar beet tops	160	100	31	212	125	—	9.9
Sugar cane	279	312	22	57	97	—	8.9
Swede tops	120	125	42	183	192	0.85	9.2

Table 1   contd.

Food	DM basis						
	DM (g/kg)	CF (g/kg)	EE (g/kg)	Ash (g/kg)	CP (g/kg)	dg	ME (MJ/kg)
Silages							
Barley, whole crop	324	248	15	153	64	0.85	8.7
Grass, young	250	270	52	91	186	0.85	11.6
Grass, mature	294	340	52	110	125	0.85	9.9
Lucerne	250	296	84	100	168	—	8.5
Maize, whole crop	210	233	57	62	110	—	10.8
Potato	270	26	19	52	81	—	11.8
Hays							
Clover, red	850	266	39	84	184	—	9.6
Grass, poor quality	800	380	16	70	55	0.70	7.0
Grass, good quality	900	298	18	82	110	0.60	9.5
Lucerne, early flowering	850	302	13	95	225	—	8.2
Dried grasses							
Leafy	900	213	38	102	187	0.65	10.6
Early flower	900	258	28	107	154	—	9.7
Straws							
Barley	860	394	21	53	38	0.75	7.3
Bean	860	501	9	53	52	—	7.4
Oat	860	394	22	57	34	0.75	6.7
Pea	860	410	19	77	105	—	6.5
Rye	860	429	19	30	37	—	6.2
Wheat	860	417	15	71	34	—	5.6
Roots and tubers							
Artichoke, Jerusalem	200	35	10	55	75	—	13.2
Cassava	370	43	9	30	35	—	12.8
Mangels	120	58	8	67	83	0.85	12.4
Potatoes	210	38	5	43	90	0.85	12.5
Sugar beet pulp, dried	900	203	7	34	99	0.70	12.7
Sugar beet molasses	750	0	0	69	47	0.80	12.9
Swedes	120	100	17	58	108	0.85	12.8
Sweet potato	320	38	16	34	39	—	12.7
Turnips	90	111	22	78	122	0.85	11.2
Cereals and by-products							
Barley	860	53	17	26	108	0.80	13.0
Barley, brewers' grains, dried	900	169	71	43	204	0.60	10.3
Barley, malt culms	900	156	22	80	271	—	11.2
Brewers' yeast, dried	900	2	11	102	443	—	11.7
Maize	860	24	42	13	98	0.65	14.2
Maize, flaked	900	17	49	10	110	0.65	15.0
Maize gluten feed	900	39	38	28	262	0.60	13.5
Millet	860	93	44	44	121	—	11.3
Oats	860	121	49	33	109	0.80	12.0
Oat husks	900	351	11	42	21	0.65	4.9
Rice, polished	860	17	5	9	77	—	15.0

Table 1 contd.

Food	DM basis						
	DM (g/kg)	*CF* (g/kg)	*EE* (g/kg)	*Ash* (g/kg)	*CP* (g/kg)	*dg*	*ME* (MJ/kg)
Rye	860	26	19	21	124	—	14.0
Sorghum	860	21	43	27	108	—	13.4
Wheat	860	26	19	21	124	0.80	13.5
Wheat feed	880	74	45	50	178	0.75	12.0
Wheat bran	880	114	45	67	170	0.75	10.1
Oilseed by-products							
Coconut cake meal	900	153	76	72	220	0.45	12.7
Cottonseed cake, undec.	900	248	54	66	231	—	8.5
Cottonseed cake, dec.	900	87	89	74	457	0.60	12.3
Groundnut meal, undec.	900	273	21	47	343	—	9.2
Groundnut meal, dec.	900	88	8	63	552	0.80	11.7
Linseed meal	900	102	36	73	404	0.60	11.9
Palm kernel meal	900	167	10	44	227	—	12.2
Soyabean meal	900	58	17	62	503	0.60	12.3
Sunflower seed cake, undec.	900	323	80	80	206	—	9.5
Sunflower seed cake, dec.	900	134	152	74	413	0.80	13.3
Leguminous seeds							
Field beans	860	80	15	40	314	0.55	12.8
Gram	860	57	13	57	263	—	12.4
Peas	860	63	19	33	262	—	13.4
Animal by-products							
Fish meal	900	9	64	194	736	0.40	14.5
Meat meal	900	0	148	42	810	—	16.3
Meat and bone meal	900	0	50	62	597	0.40	9.7
Milk, cows, whole	128	0	305	55	266	—	20.2
Milk, skim	100	0	70	80	350	—	15.3
Milk, whey	66	0	30	106	106	—	14.5

TABLE 2　Nutritive value of foods for pigs[7,8]

Food	Fresh basis							DM basis	
	DM (g/kg)	CP (g/kg)	CF (g/kg)	EE (g/kg)	Ash (g/kg)	DCP (g/kg)	DE (MJ/kg)	DCP (g/kg)	DE (MJ/kg)
Green crops									
Pasture grass, closely grazed	200	52	34	8	17	35	2.2	175	11.0
Pasture grass, rotational grazing	200	34	39	6	16	19	2.1	95	10.3
Kale, marrow-stem, minced	140	22	25	5	19	14	1.6	100	11.4
Lucerne meal, dried	900	197	223	46	91	116	8.5	129	9.4
Roots and tubers									
Cassava, dried	900	25	78	3	31	17	13.4	19	14.9
Mangels	130	10	8	1	9	8	2.0	62	15.4
Potato meal	900	88	21	5	36	35	14.2	39	15.8
Swedes	120	15	12	2	13	9	1.7	75	14.2
Turnips	90	11	10	2	7	9	1.4	100	15.6
Cereals and by-products									
Barley	860	93	46	15	22	76	12.9	88	15.0
Barley, brewers' yeast, dried	940	502	11	6	88	487	17.6	518	18.7
Maize	860	84	21	36	11	67	14.5	78	16.9
Maize, flaked	860	95	14	42	9	90	15.6	105	18.1
Oats	860	94	104	42	28	73	11.4	85	13.3
Rice	860	66	15	4	8	57	15.3	66	17.8
Rye	860	114	19	17	20	92	13.9	107	16.2
Sorghum	860	93	18	37	23	72	14.3	84	16.6
Wheat	860	107	22	16	18	98	14.0	114	16.3

Table 2 contd.

Food	Fresh basis							DM basis	
	DM (g/kg)	CP (g/kg)	CF (g/kg)	EE (g/kg)	Ash (g/kg)	DCP (g/kg)	DE (MJ/kg)	DCP (g/kg)	DE (MJ/kg)
Oilseed by-products									
Coconut meal	870	232	119	75	53	169	13.4	194	15.4
Cottonseed meal, dec.	930	408	112	68	67	335	12.9	360	13.9
Groundnut meal, dec.	900	497	79	7	57	462	13.7	513	15.2
Palm kernel meal	900	204	150	9	40	122	10.4	136	11.6
Rapeseed meal	900	360	137	26	72	310	11.8	344	13.1
Soyabean meal	900	453	52	15	56	390	13.9	433	15.4
Sunflower meal	900	381	120	23	65	310	9.0	344	10.0
Animal by-products									
Fish meal	900	662	8	58	175	586	15.1	662	16.8
Herring meal	900	686	0	82	92	644	18.0	715	20.0
Meat meal, low fat	900	645	0	28	188	568	12.1	631	13.4
Meat and bone meal	900	474	0	40	371	422	9.6	469	10.7
Milk, cows, whole	124	32	0	38	8	30	3.0	242	24.2
Milk, separated, fresh	90	34	0	1	8	33	1.5	367	16.7
Whey, fresh	66	9	0	3	6	9	1.1	136	16.7

TABLE 3 Nutritive value of foods for poultry[6]

| | Fresh basis | | | | | | DM basis | |
	DM g/kg	CP g/kg	EE g/kg	Ash g/kg	DCP g/kg	ME MJ/kg	DCP g/kg	ME MJ/kg
Green crops and tubers								
Dried grass	921	178	37	77	156	5.82	169	6.32
Dried lucerne	887	145	27	73	123	4.60	139	9.19
Potato meal	913	87	2	32	63	12.1	69	13.3
Cereals and by-products								
Barley	891	113	15	27	90	11.1	101	12.5
Malt distillers' dried solubles	949	268	2	172	—	6.82	—	7.19
Brewers' yeast, dried	867	425	21	89	374	11.0	431	12.7
Maize	882	82	32	12	67	13.2	76	15.0
Maize gluten feed	897	250	19	53	223	9.75	249	10.9
Millet	856	119	39	29	82	12.0	96	14.0
Oats	876	100	49	27	85	11.1	97	12.7
Rice, brown	907	101	21	8	84	15.0	93	16.5
Rye	846	85	11	19	67	12.1	79	14.3
Sorghum (milo)	867	107	29	18	84	13.0	97	15.0
Wheat	891	104	14	18	88	12.2	99	13.7
Wheat germ meal	889	248	73	43	198	11.1	223	12.5
Wheat middlings, coarse	874	149	39	42	127	9.75	145	11.2
Wheat middlings, fine	875	177	52	32	150	11.8	171	13.5

Table 3 contd.

	Fresh basis						DM basis	
	DM g/kg	CP g/kg	EE g/kg	Ash g/kg	DCP g/kg	ME MJ/kg	DCP g/kg	ME MJ/kg
Oilseed by-products								
Coconut meal	887	195	67	64	109	6.90	123	7.78
Cottonseed meal, dec.	901	378	61	67	280	10.9	311	12.1
Groundnut meal, dec.	912	454	51	64	408	13.2	447	14.5
Linseed meal	888	341	63	53	300	8.66	338	9.75
Palm kernel meal	900	190	20	40	171	6.74	190	7.49
Soya bean meal	873	499	15	47	428	10.7	490	12.3
Sunflower seed meal, dec.	916	321	27	64	248	8.83	270	9.6
Leguminous seeds								
Bean meal	866	250	13	39	211	10.4	244	12.0
Pea meal	871	271	17	28	206	11.1	237	12.7
Animal by-products								
Blood meal	868	800	8	35	720	13.0	829	15.0
Fish meal	910	655	42	215	590	11.5	648	12.6
Herring meal	905	740	70	95	666	13.4	736	14.8
Meat meal	902	722	132	38	650	15.7	721	17.4
Meat and bone meal	935	515	112	275	412	11.0	441	11.8
Milk, dried skim	934	340	9	80	275	12.3	294	13.2
Milk, dried whey	937	125	7	85	101	12.0	108	12.8

TABLE 4   Amino acid composition of foods (g/kg) (fresh basis)[11,12]

Food	DM (g/kg)	Nitrogen (g/kg)	Arginine	Cystine	Glycine	Histidine	Isoleucine	Leucine	Lysine	Methionine	Phenylalanine	Serine	Threonine	Tryptophan	Tyrosine	Valine
Green crops																
Dried grass	897	23.5	7.6	3.4	7.5	2.9	5.8	10.9	7.1	3.0	7.1	6.1	6.5	1.2	4.8	4.9
Dried lucerne	—	35.7	10.9	5.0	10.2	4.7	9.3	16.1	11.7	2.8	10.5	9.0	9.2	1.6	8.2	11.3
Cereals and by-products																
Barley	856	15.6	5.4	4.3	4.1	4.1	3.5	6.9	3.8	2.1	5.0	4.3	3.4	1.0	3.4	5.1
Brewers' yeast, dried	930	71.0	21.9	5.0	21.9	10.7	21.4	31.9	32.3	7.0	18.1	—	20.6	4.9	14.9	23.2
Distillers' dark grains	900	39.4	10.1	8.8	10.6	4.5	8.5	15.5	9.6	4.3	8.9	9.3	8.8	2.1	7.5	11.7
Distillers' solubles	—	42.9	3.8	4.8	12.9	4.0	8.0	13.0	6.8	3.4	7.7	6.4	6.0	3.6	8.5	12.8
Maize	852	13.5	4.3	3.8	3.3	2.6	3.0	11.1	2.5	2.3	4.5	4.3	3.2	0.4	3.9	4.3
Maize gluten meal	—	106.2	24.1	25.2	17.4	14.0	28.4	117.7	10.8	24.5	41.0	37.7	24.0	2.6	34.7	33.0
Oats	869	16.8	7.0	8.0	5.7	2.3	3.7	7.3	4.5	2.6	5.1	5.7	3.7	0.7	4.1	5.1
Rice, polished, broken (Brewers' rice)	890	13.9	6.2	0.8	6.3	1.7	3.5	5.2	2.4	1.5	3.6	13.6	2.9	1.3	4.1	5.0
Sorghum	870	14.1	3.4	1.6	3.5	1.9	4.2	11.8	2.1	1.6	4.2	3.9	2.9	1.0	3.8	5.3
Wheat	858	16.2	5.2	4.5	4.1	2.5	3.5	7.1	3.1	2.1	4.8	4.8	3.1	1.2	3.3	4.5
Wheat feed	858	22.6	10.2	7.8	7.8	4.0	4.7	9.5	6.4	3.2	6.1	6.5	4.0	2.2	4.6	7.1
Oilseed by-products																
Cottonseed meal	900	66.2	45.9	6.4	17.0	11.0	13.3	24.1	17.1	5.2	22.2	—	13.2	4.7	10.2	18.9
Groundnut meal	897	75.5	57.0	11.2	26.4	11.3	15.7	29.9	16.4	5.6	25.1	23.1	13.5	3.0	19.9	20.7
Lupinseed meal	—	60.8	42.7	12.4	13.6	8.5	16.5	26.8	17.0	3.0	13.1	17.4	12.1	1.8	18.5	14.6
Rapeseed meal	899	50.0	23.2	15.1	18.5	9.9	14.2	25.9	21.5	7.9	14.3	16.3	16.8	1.7	11.5	18.6
Soyabean meal	861	70.9	35.3	12.0	19.5	12.6	20.3	35.0	28.5	7.9	23.0	23.5	17.9	5.5	17.7	22.2
Sunflower seed meal	—	44.5	23.1	9.2	15.6	7.2	11.6	18.5	10.1	7.6	13.4	11.9	10.4	1.4	8.1	14.3

Table 4 contd.

Food	DM (g/kg)	Nitrogen (g/kg)	Arginine	Cystine	Glycine	Histidine	Isoleucine	Leucine	Lysine	Methionine	Phenylalanine	Serine	Threonine	Tryptophan	Tyrosine	Valine
Leguminous seeds																
Beans (*Vicia faba*)	—	39.8	22.2	7.8	10.5	6.1	9.7	18.3	15.8	1.8	10.1	11.7	9.1	1.6	8.9	11.2
Peas (*Pisum sativum*)	—	31.4	17.1	6.0	8.7	5.3	8.2	14.4	15.2	2.5	8.9	9.5	8.0	0.9	7.2	9.2
Animal by-products																
Fishmeal	918	100.0	40.5	6.4	50.6	14.1	26.1	44.6	48.2	15.2	28.8	28.3	24.9	6.9	21.4	30.7
Meat and bone meal	957	73.3	32.4	5.4	70.6	7.7	11.6	26.1	22.0	6.5	14.6	16.0	14.3	2.1	9.6	19.6
Whey dried	930	19.2	3.4	3.0	3.0	1.8	8.2	11.9	9.7	1.9	3.3	3.2	8.9	1.9	2.5	6.8

TABLE 5 Vitamin potency of foods (fresh basis)[6,11,12]

Food	Vitamin A potency* i.u./g	Vitamin E i.u./kg	Thiamin mg/kg	Riboflavin mg/kg	Nicotinic acid mg/kg	Pantothenic acid mg/kg	Vitamin B_6 mg/kg	Vitamin B_{12} mg/kg	Choline mg/kg
Green crops									
Grass, dried	328	150	—	15.5	74	—	—	—	890
Lucerne, dried	267	200	—	16.6	43	—	—	0.003	1110
Cereals and by-products									
Barley	0.7	20	1.9	1.8	55	8	3.0	—	990
Brewers' yeast, dried	—	—	91.8	37.0	448	109	42.8	—	3984
Maize	5.0	22	3.5	1.0	24	4	7.0	—	620
Oats	0.6	20	6.0	1.1	12	—	1.0	—	946
Rice	—	12	—	0.4	15	—	—	—	780
Rye	0.2	17	3.6	1.6	19	8	2.6	—	419
Sorghum	0.7	12	4.0	1.1	41	12	3.2	—	450
Wheat	0.4	13	4.5	1.4	48	10	3.4	—	1090
Wheat, fine middlings	0.5	20	—	2.2	100	—	—	—	1110
Wheat, coarse middlings	0.4	57	—	2.4	95	—	—	—	1170

Table 5 contd.

Food	Vitamin A potency* i.u./g	Vitamin E i.u./kg	Thiamin mg/kg	Riboflavin mg/kg	Nicotinic acid mg/kg	Pantothenic acid mg/kg	Vitamin B_6 mg/kg	Vitamin B_{12} mg/kg	Choline mg/kg
Oilseed by-products									
Coconut meal	—	16	—	3.3	27	—	—	—	1110
Cottonseed meal (dec. exp)	0.3	39	6.4	5.1	38	10	5.3	—	2753
Groundnut meal (dec. extr)	—	3	5.7	11.0	170	53	10.0	—	2396
Groundnut meal (dec. exp)	0.3	3	7.1	5.2	166	47	10.0	—	1655
Linseed meal (extr)	0.4	—	—	3.5	40	—	—	—	1660
Soya bean meal (extr)	—	2	4.5	2.9	29	16	6.0	—	2794
Animal by-products									
Fish meal	—	8	2.1	6.0	49	10	4.1	0.081	5180
Meat meal	—	1	0.2	5.5	57	5	3.0	0.068	2077
Milk, dried skim	0.3	1	—	21.0	12	—	—	0.055	1060

* For chicks. The values for pigs and ruminants are about half those quoted for plant products.

TABLE 6   Mineral contents of foods (DM basis)[5,12]

Food	Calcium (g/kg)	Phosphorus (g/kg)	Magnesium (g/kg)	Sodium (g/kg)	Copper (mg/kg)	Manganese (mg/kg)	Zinc (mg/kg)	Cobalt (mg/kg)	Selenium (mg/kg)
Green crops									
Grass, close grazing	5.0	3.5	1.7	1.9	8.0	—	—	0.10	0.05
Grass, extensive grazing	4.8	2.8	1.7	1.7	7.0	16	5.0	0.08	0.04
Kale	21.0	3.2	2.5	2.0	4.5	38	—	0.10	0.05
Lucerne, late vegetative	21.9	3.3	2.7	2.1	11.0	41	—	0.17	—
Turnip tops	24.2	3.1	2.8	3.1	8.0	—	—	0.08	0.06
Silages									
Cereal, vegetative	4.0	2.7	1.0	1.8	6.0	80	25	0.07	0.06
Grass, early	8.0	4.0	3.0	3.0	11.0	90	25	—	0.10
Grass, mature	3.0	2.0	0.9	1.0	3.0	94	30	0.05	0.02
Hays									
Clover	15.3	2.5	4.3	1.9	11.0	73	17	0.16	—
Grass, poor quality	2.5	1.5	0.8	1.0	2.0	70	17	0.05	0.01
Grass, good quality	7.0	3.5	2.5	2.5	9.0	100	21	0.20	0.07
Lucerne, mature	11.3	1.8	2.7	0.8	14.0	44	24	0.09	—
Straws									
Barley	4.5	0.7	0.8	1.1	3.2	84	16	0.04	0.04
Oat	4.0	0.7	1.3	3.7	4.0	69	29	0.04	0.02

Table 6 contd.

Food	Calcium (g/kg)	Phosphorus (g/kg)	Magnesium (g/kg)	Sodium (g/kg)	Copper (mg/kg)	Manganese (mg/kg)	Zinc (mg/kg)	Cobalt (mg/kg)	Selenium (mg/kg)
Roots and tubers									
Cassava, dried	2.0	1.0	—	0.2	—	20	—	—	—
Mangels	2.9	2.1	5.3	9.9	9.4	—	—	0.09	0.03
Potatoes	1.0	2.1	1.0	0.5	4.5	42	28	0.06	0.03
Sugar beet pulp, molassed, dried	5.7	0.8	2.4	2.5	11.0	51	32	0.10	0.02
Swedes	3.6	3.2	1.2	2.6	3.8	21	19	0.07	0.03
Turnips	5.0	3.6	1.4	2.2	2.7	35	36	0.04	0.03
Cereals and by-products									
Barley	0.5	4.0	1.3	0.2	4.8	18	19	0.04	0.02
Brewers' grains, dried	3.2	7.8	1.8	0.4	25.0	50	—	0.03	—
Brewers' yeast	1.3	15.1	2.5	0.8	35.3	6	42	—	—
Distillers' grains, malt	1.7	3.7	1.4	0.9	10.0	—	—	0.02	0.02
Maize	0.3	2.7	1.1	0.2	2.5	6	16	0.02	0.02
Maize gluten meal	1.6	5.0	0.6	1.0	30.0	8	190	0.08	—
Millet	0.6	3.1	1.8	0.4	24.4	32	16	0.04	—
Oats	0.8	3.7	1.3	0.2	3.6	42	41	0.04	0.03
Oat feed	1.5	2.9	1.0	0.2	3.9	—	—	0.04	0.03
Rice	0.7	3.2	1.5	0.6	3.0	20	17	0.05	—
Rye	0.7	3.7	1.4	0.3	8.0	66	36	—	—

Table 6 contd.

Food	Calcium (g/kg)	Phosphorus (g/kg)	Magnesium (g/kg)	Sodium (g/kg)	Copper (mg/kg)	Manganese (mg/kg)	Zinc (mg/kg)	Cobalt (mg/kg)	Selenium (mg/kg)
Sorghum	0.5	3.5	1.9	0.4	10.8	16	15	0.14	—
Wheat	0.5	3.5	1.2	0.1	5.0	42	50	0.05	0.02
Wheat bran	1.6	13.6	5.0	0.4	12.9	143	189	0.03	0.40
Wheat feed	1.1	8.0	3.3	0.4	17.5	—	—	0.03	0.04
Oilseeds and by-products									
Coconut meal	2.3	6.6	2.8	0.4	20.4	59	—	0.14	—
Cottonseed meal, dec.	1.9	12.4	5.0	0.6	16.0	25	79	0.05	—
Groundnut meal, dec.	2.9	6.8	1.7	0.8	17.0	29	22	0.12	—
Linseed meal	4.1	8.6	5.8	0.7	25.0	42	—	0.55	0.91
Soyabean meal	3.5	6.8	3.0	0.4	25.0	32	61	0.20	0.55
Leguminous seeds									
Beans	1.0	5.5	2.0	0.1	14.0	16	46	0.20	—
Peas	1.5	4.4	1.4	0.5	—	—	33	—	—
Animal by-products									
Fishmeal	79.0	44.0	3.6	4.5	9.0	21	119	0.14	2.00
Meat and bone meal	120	58.0	2.5	7.2	24.0	—	—	0.20	0.20
Whey, dried	9.2	8.2	1.4	7.0	50.0	6	3	0.13	—

TABLE 7 Feeding standards for lactating and pregnant cattle[2,3]
7.1 *Daily allowances for cows producing milk of 38 g fat and 34 g protein/kg and weighing 500 kg*
($q_m = 0.55$)

	Milk yield (kg/day)						Month of pregnancy	
	0	5	10	15	20	25	8	9
Liveweight change (kg/day)	0	+0.6	+0.4	+0.25	0	−0.2		
DMI (kg)	9	11	13	15	16	16	10	10
ME (MJ)	56	112	130	150	166	186	72	85
RDP(g)	465	934	1 082	1 253	1 383	1 553	600	709
UDP (g)	2	15	95	174	256	335	0	0
E_{mp} (MJ)	39	71	82	94	103	114	41	43
APL	1.00	1.84	2.11	2.42	2.65	2.95	1.06	1.10
Ca (g)	19	32	45	58	71	85	30	30
P (g)	16	26	36	45	55	65	23	23
Mg (g)	9	13	17	21	25	29	12	12
Na(g)	4	8	11	14	18	21	6	6
Vit. A (i.u.)	←				50 000	→		
Vit. D (i.u.)	←				5 000	→		
Vit. E(i.u.)	←				300	→		

Table 7 contd.
7.2 Daily allowances for cows producing milk of 38 g fat and 34 g protein/kg and weighing 500 kg ($q_m = 0.60$)

Milk yield (kg/day)	0	5	10	15	20	25	Month of pregnancy	
							8	9
Liveweight change (kg/day)	0	+0.6	+0.4	+0.25	0	−0.2		
DMI (kg)	9	11	13	15	16	16	10	10
ME (MJ)	54	109	126	146	161	181	70	83
RDP (g)	450	909	1 051	1 218	1 343	1 510	584	692
UDP (g)	16	35	122	201	289	369	0	0
E_{mp} (MJ)	37	72	82	94	103	114	41	43
APL	1.00	1.86	2.12	2.42	2.65	2.95	1.06	1.10
Ca (g)	19	32	45	58	71	85	30	30
P (g)	16	26	36	45	55	65	23	23
Mg (g)	9	13	17	21	25	29	12	12
Na (g)	4	8	11	14	18	21	6	6
Vit. A (i.u.)					50 000			→
Vit. D (i.u.)					5 000			→
Vit. E (i.u.)					300			→

Table 7 contd.

7.3 Daily allowances for cows producing milk of 38 g fat and 32 g protein/kg and weighing 500 kg
($q_{\mathrm{m}} = 0.65$)

Milk yield (kg/day)	0	5	10	15	20	25	30	Month of pregnancy	
								8	9
Liveweight change (kg/day)	0	+0.6	+0.4	+0.25	0	−0.2	−0.4		
DMI (kg)	9	11	13	15	16	16	16	10	10
ME (MJ)	53	106	123	142	157	176	196	69	82
RDP (g)	443	886	1 026	1 188	1 311	1 471	1 634	575	684
UDP (g)	22	55	142	230	316	400	483	0	0
E_{mp} (MJ)	39	71	82	94	103	114	126	41	43
APL	1.00	1.84	2.11	2.42	2.65	2.95	3.24	1.06	1.10
Ca (g)	19	32	45	58	71	85	98	30	30
P (g)	16	26	36	45	55	65	75	23	23
Mg (g)	9	13	17	21	25	29	33	12	12
Na (g)	4	8	11	14	18	21	24	6	6
Vit. A (i.u.)					50 000				
Vit. D (i.u.)					5 000				
Vit. E (i.u.)					300				

Table 7 contd.
7.4 *Daily allowances for cows producing milk of 38 g fat and 34 g protein/kg and weighing 650 kg*
$(q_m = 0.55)$

Milk yield (kg/day)	0	5	10	15	20	Month of pregnancy	
						8	9
Liveweight change (kg/day)	0	+0.6	+0.4	+0.25	0		
DMI (kg)	10.5	13	15	17	17.5	12	12
ME (MJ)	67	123	141	161	177	88	105
RDP (g)	560	1 028	1 175	1 344	1 473	734	876
UDP(g)	11	23	104	182	265	0	0
E$_{mp}$ (MJ)	47	79	90	102	111	50	52
APL	1.00	1.69	1.92	2.18	2.37	1.06	1.11
Ca (g)	24	37	51	64	77	39	39
P (g)	19	29	39	49	59	27	27
Mg (g)	12	16	20	24	28	15	15
Na(g)	5	9	12	16	19	8	8
Vit. A (i.u.)				65 000			
Vit. D (i.u.)				6 500			
Vit. E (i.u.)				385			

Table 7 contd.
7.5 Daily allowances for cows producing milk of 38 g fat and 34 g protein/kg and weighing 650 kg
$(q_m = 0.60)$

Milk yield (kg/day)	0	5	10	15	20	25	Month of pregnancy	
							8	9
Liveweight change (kg/day)	0	+0.6	+0.4	+0.25	0	−0.2		
DMI (kg)	10.5	13	15	17	17.5	17.5	12	12
ME (MJ)	66	120	137	157	172	191	87	104
RDP (g)	550	1 000	1 143	1 309	1 434	1 597	726	867
UDP (g)	17	43	130	209	298	384	0	0
E_{mp} (MJ)	47	80	90	102	111	122	50	52
APL	1.00	1.71	1.93	2.19	2.37	2.62	1.06	1.11
Ca (g)	24	37	51	64	77	90	39	39
P (g)	19	29	39	49	59	69	27	27
Mg (g)	12	16	20	24	28	31	15	15
Na (g)	5	9	12	16	19	22	8	8
Vit. A (i.u.)					65 000			
Vit. D (i.u.)					6 500			
Vit. E (i.u.)					385			

Table 7 contd.
7.6 *Daily allowances for cows producing milk of 38 g fat and 34 g protein/kg, and weighing 650 kg* ($q_m = 0.65$)

Milk yield (kg/day)	0	5	10	15	20	25	30	Month of pregnancy	
								8	9
Liveweight change (kg/day)	0	+0.6	+0.4	+0.25	0	−0.2	−0.4		
DMI (kg)	10.5	13	15	17	17.5	17.5	18	14	14
ME (MJ)	64	117	134	153	167	187	206	85	102
RDP (g)	534	976	1 115	1 275	1 397	1 556	1 716	709	851
UDP (g)	31	63	150	236	331	410	496	0	0
E_{mp} (MJ)	47	79	90	102	111	122	134	50	52
APL	1.00	1.69	1.92	2.18	2.37	2.62	2.86	1.06	1.11
Ca (g)	24	37	51	64	77	90	103	39	39
P (g)	19	29	39	49	59	69	79	27	27
Mg (g)	12	16	20	24	28	31	35	15	15
Na (g)	5	9	12	16	19	22	26	8	8
Vit. A (i.u.)					65 000				
Vit. D (i.u.)					6 500				
Vit. E (i.u.)					385				

Table 7 contd.
7.7 Net energy (NE_{mp}) values of foods (MJ/kg DM) for lactating cattle

APL	M/D								
	9.0	9.5	10.0	10.5	11.0	11.5	12.0	12.5	13.0
1.00	6.1	6.5	6.9	7.4	7.8	8.3	8.8	9.3	9.8
1.10	6.0	6.4	6.8	7.3	7.7	8.2	8.6	9.1	9.6
1.20	5.9	6.3	6.7	7.2	7.6	8.1	8.5	9.0	9.5
1.30	5.8	6.2	6.6	7.1	7.5	8.0	8.4	8.9	9.4
1.40	5.7	6.1	6.6	7.0	7.4	7.9	8.3	8.8	9.3
1.50	5.7	6.1	6.5	6.9	7.4	7.8	8.3	8.7	9.2
1.75	5.5	5.9	6.3	6.8	7.2	7.6	8.1	8.5	9.0
2.00	5.4	5.8	6.2	6.6	7.1	7.5	7.9	8.4	8.9
2.25	5.3	5.7	6.1	6.5	7.0	7.4	7.8	8.3	8.7
2.50	5.2	5.6	6.0	6.4	6.8	7.3	7.7	8.2	8.6
2.75	5.2	5.5	5.9	6.3	6.8	7.2	7.6	8.0	8.5
3.00	5.1	5.5	5.9	6.3	6.7	7.1	7.5	8.0	8.4
3.25	5.0	5.4	5.8	6.2	6.6	7.0	7.4	7.9	8.3
3.50	5.0	5.3	5.7	6.1	6.5	6.9	7.3	7.8	8.2

TABLE 8 Feeding standards for growing cattle[2,3]

8.1 Daily allowances for early maturing heifers

W (kg)	q_m	Component		Liveweight gain (kg/day)					DMI (kg/day)
				0	0.5	0.75	1.00	1.25	
200	0.55	ME	(MJ)	28	44	55			5.0
		RDP	(g)	231	365	460			
		UDP	(g)	2	0	0			
	0.65	ME	(MJ)	26	40	49	60	75	6.5
		RDP	(g)	220	335	410	503	624	
		UDP	(g)	16	15	0	0	0	
		E_{mp}	(MJ)	19	27	31	36	41	
		APL		1.00	1.40	1.62	1.86	2.13	
		Ca	(g)	6.4	19.6	26.2	32.8	39.4	
		P	(g)	5.0	13.2	17.4	21.5	25.6	
		Mg	(g)	3.5	4.9	5.5	6.2	6.9	
		Na	(g)	1.6	2.5	2.9	3.3	3.7	
		Vit. A	(i.u.)		←	14 000	→		
		Vit. D	(i.u.)		←	1 200	→		
		Vit. E	(i.u.)		←	115	→		
400	0.55	ME	(MJ)	45	69	86			8.5
		RDP	(g)	373	576	719			
		UDP	(g)	18	0	0			
	0.65	ME	(MJ)	43	64	77	94	116	11.0
		RDP	(g)	355	530	644	784	964	
		UDP	(g)	31	0	0	0	0	
		E_{mp}	(MJ)	31	43	49	57	64	
		APL		1.00	1.38	1.59	1.82	2.07	
		Ca	(g)	14.2	26.6	32.7	38.9	45.1	
		P	(g)	9.9	18.2	22.3	26.5	30.6	
		Mg	(g)	7.0	8.4	9.1	9.7	10.4	
		Na	(g)	3.2	4.1	4.5	4.9	5.3	
		Vit. A	(i.u.)		←	28 000	→		
		Vit. D	(i.u.)		←	2 400	→		
		Vit. E	(i.u.)		←	195	→		

Table 8 contd.
8.2 *Daily allowances for steers of medium maturity*

W (kg)	q_m	Component		Liveweight gain (kg/day)					DMI (kg/day)
				0	0.5	0.75	1.0	1.25	
200	0.55	ME	(MJ)	28	40	48			5.0
		RDP	(g)	231	331	397			
		UDP	(g)	2	37	37			
	0.65	ME	(MJ)	26	37	43	51	61	6.5
		RDP	(g)	220	307	361	426	505	
		UDP	(g)	16	57	70	66	45	
		E_{mp}	(MJ)	19	25	28	32	36	
		APL		1.00	1.31	1.48	1.66	1.87	
		Ca	(g)	6.4	19.6	26.2	32.8	39.4	
		P	(g)	5.0	13.2	17.4	21.5	25.6	
		Mg	(g)	3.5	4.9	5.5	6.2	6.9	
		Na	(g)	1.6	2.5	2.9	3.3	3.7	
		Vit. A	(i.u.)	←		14 000		→	
		Vit. D	(i.u.)	←		1 200		→	
		Vit. E	(i.u.)	←		115		→	
400	0.55	ME	(MJ)	45	63	75	90		8.5
		RDP	(g)	373	525	625	750		
		UDP	(g)	18	1	0	0		
	0.65	ME	(MJ)	43	58	68	80	94	11.0
		RDP	(g)	355	487	570	668	787	
		UDP	(g)	31	35	15	0	0	
		E_{mp}	(MJ)	31	40	45	51	57	
		APL		1.00	1.29	1.45	1.63	1.83	
		Ca	(g)	14.2	26.6	32.7	38.9	45.1	
		P	(g)	9.9	18.2	22.3	26.5	30.6	
		Mg	(g)	7.0	8.4	9.1	9.7	10.4	
		Na	(g)	3.2	4.1	4.5	4.9	5.3	
		Vit. A	(i.u.)	←		28 000		→	
		Vit. D	(i.u.)	←		2 400		→	
		Vit. E	(i.u.)	←		195		→	

Table 8 contd.
8.3   *Daily allowances for bulls of late maturity*

W (kg)	q_m	Component		Liveweight gain (kg/day) 0	0.5	0.75	1.00	1.25	DMI (kg/day)
200	0.55	ME	(MJ)	32	40	45	51		5.0
		RDP	(g)	266	333	374	423		
		UDP	(g)	0	62	92	111		
	0.65	ME	(MJ)	30	37	42	47	52	6.5
		RDP	(g)	253	312	347	388	435	
		UDP	(g)	0	82	112	138	160	
		E_{mp}	(MJ)	22	26	29	31	34	
		APL		1.00	1.19	1.29	1.40	1.53	
		Ca	(g)	6.4	19.6	26.2	32.8	39.4	
		P	(g)	5.0	13.2	17.4	21.5	25.6	
		Mg	(g)	3.5	4.9	5.5	6.2	6.9	
		Na	(g)	1.6	2.5	2.9	3.3	3.7	
		Vit. A (i.u.)		←		14 000		→	
		Vit. D (i.u.)		←		1 200		→	
		Vit. E (i.u.)		←		115		→	
400	0.55	ME	(MJ)	51	64	71	80	91	8.5
		RDP	(g)	429	531	594	668	755	
		UDP	(g)	0	16	27	20	0	
	0.65	ME	(MJ)	49	60	66	74	82	11.0
		RDP	(g)	408	498	552	613	685	
		UDP	(g)	0	43	60	60	57	
		E_{mp}	(MJ)	36	42	46	49	54	
		APL		1.00	1.18	1.28	1.39	1.50	
		Ca	(g)	14.2	26.6	32.7	38.9	45.1	
		P	(g)	9.9	18.2	22.3	26.5	30.6	
		Mg	(g)	7.0	8.4	9.1	9.7	10.4	
		Na	(g)	3.2	4.1	4.5	4.9	5.3	
		Vit. A (i.u.)		←		28 000		→	
		Vit. D (i.u.)		←		2 400		→	
		Vit. E (i.u.)		←		195		→	

Table 8 contd.

8.4   *Net energy (NE$_{mp}$) values of foods (MJ/kg DM) for growing cattle*

APL	M/D												
	8.0	8.5	9.0	9.5	10.0	10.5	11.0	11.5	12.0	12.5	13.0	13.5	14.0
1.0	5.2	5.7	6.1	6.5	6.9	7.4	7.8	8.3	8.8	9.3	9.8	10.3	10.8
1.1	5.0	5.4	5.9	6.3	6.7	7.2	7.7	8.1	8.6	9.1	9.6	10.1	10.6
1.2	4.8	5.2	5.7	6.1	6.5	7.0	7.5	7.9	8.4	8.9	9.4	10.0	10.5
1.3	4.6	5.0	5.4	5.9	6.3	6.8	7.3	7.8	8.2	8.8	9.3	9.8	10.3
1.4	4.3	4.8	5.2	5.7	6.1	6.6	7.1	7.6	8.1	8.6	9.1	9.6	10.2
1.5	4.1	4.5	5.0	5.4	5.9	6.4	6.9	7.4	7.9	8.4	8.9	9.5	10.0
1.6	3.8	4.3	4.7	5.2	5.7	6.2	6.7	7.2	7.7	8.2	8.8	9.3	9.9
1.7	3.5	4.0	4.5	4.9	5.4	5.9	6.4	7.0	7.5	8.0	8.6	9.1	9.7
1.8	3.2	3.7	4.2	4.7	5.2	5.7	6.2	6.7	7.3	7.8	8.4	9.0	9.6
1.9	2.8	3.3	3.9	4.4	4.9	5.5	6.0	6.5	7.1	7.6	8.2	8.8	9.4
2.0	2.3	3.0	3.5	4.1	4.7	5.2	5.8	6.3	6.9	7.4	8.0	8.6	9.2
2.1	1.4	2.5	3.2	3.8	4.4	4.9	5.5	6.1	6.7	7.2	7.8	8.4	9.1
2.2		1.7	2.7	3.4	4.0	4.6	5.2	5.8	6.4	7.0	7.6	8.3	8.9
2.3			2.0	3.0	3.7	4.3	5.0	5.6	6.2	6.8	7.4	8.1	8.7
2.4				2.4	3.3	4.0	4.7	5.3	6.0	6.6	7.2	7.9	8.5
2.5					2.8	3.6	4.4	5.1	5.7	6.4	7.0	7.7	8.4
2.6					2.0	3.2	4.0	4.8	5.5	6.1	6.8	7.5	8.2

TABLE 9   Feeding standards for pregnant ewes[2,3]
9.1   *Daily allowances for ewes, assuming zero weight change*

Ewe weight (kg)	Component		Weeks before lambing							
			Single lamb				Twin lambs			
			8–7	6–5	4–3	2–1	8–7	6–5	4–3	2–1
55	DMI	(kg)	1.2	1.2	1.2	1.1	1.3	1.3	1.3	1.2
	ME	(MJ)	8.9	9.9	11.1	12.8	10.0	11.5	13.6	16.4*
	RDP	(g)	74	82	93	107	83	96	114	137
	UDP	(g)	22	23	27	36	21	23	26	33
	E_{mp}	(MJ)	5.4	5.5	5.6	5.9	5.5	5.7	6.0	6.3
	APL		1.04	1.07	1.10	1.14	1.07	1.11	1.16	1.23
	Ca	(g)	5.0	5.0	7.4	7.4	5.0	5.0	8.9	8.9
	P	(g)	4.0	4.0	5.1	5.1	4.0	4.0	5.7	5.7
	Mg	(g)	1.0	1.0	1.3	1.3	1.0	1.0	1.5	1.5
	Na	(g)	←	1.8	→		←	1.8	→	
	Vit. A	(i.u.)	←	3 500	→		←	3 500	→	
	Vit.D	(i.u.)	←	550	→		←	550	→	
	Vit.E	(i.u.)	←	35	→		←	35	→	
75	DMI	(kg)	1.5	1.5	1.5	1.4	1.6	1.6	1.6	1.5
	ME	(MJ)	11.3	12.5	14.1	16.2	12.7	14.6	17.3	20.7*
	RDP	(g)	94	104	118	135	106	122	144	173
	UDP	(g)	26	28	31	40	26	27	30	38
	E_{mp}	(MJ)	6.8	7.0	7.2	7.4	7.0	7.2	7.6	8.0
	APL		1.04	1.07	1.10	1.14	1.07	1.11	1.16	1.23
	Ca	(g)	6.7	6.7	9.8	9.8	6.7	6.7	11.7	11.7
	P	(g)	5.4	5.4	6.8	6.8	5.4	5.4	7.6	7.6
	Mg	(g)	1.3	1.3	1.7	1.7	1.3	1.3	2.0	2.0
	Na	(g)	←	2.5	→		←	2.5	→	
	Vit. A	(i.u.)	←	5 000	→		←	5 000	→	
	Vit. D	(i.u.)	←	750	→		←	750	→	
	Vit. E	(i.u.)	←	45	→		←	45	→	

* These energy levels are not attainable because of inadequate DMI.

Table 9 contd
9.2 *Daily allowances for 55 kg pregnant ewes losing 50 g liveweight/day, and 75 kg pregnant ewes losing 70 g liveweight/day*

Ewe weight (kg)	Component		Single lamb 8–7	6–5	4–3	2–1	Twin lambs. 8–7	6–5	4–3	2–1
55	DMI	(kg)	1.2	1.2	1.2	1.1	1.3	1.3	1.3	1.2
	ME	(MJ)	6.8	7.8	9.0	10.7	7.9	9.4	11.5	14.2
	RDP	(g)	57	65	75	89	66	79	96	119
	UDP	(g)	29	30	34	43	28	30	33	41
	E_{mp}	(MJ)	5.1	5.2	5.4	5.6	5.2	5.4	5.7	6.0
	APL		1.00	1.01	1.05	1.09	1.02	1.06	1.11	1.18
	Ca	(g)	5.0	5.0	7.4	7.4	5.0	5.0	8.9	8.9
	P	(g)	4.0	4.0	5.1	5.1	4.0	4.0	5.7	5.7
	Mg	(g)	1.0	1.0	1.3	1.3	1.0	1.0	1.5	1.5
	Na	(g)	←		1.8	→	←		1.8	→
	Vit. A	(i.u.)	←	3 500		→	←	3 500		→
	Vit. D	(i.u.)	←	550		→	←	550		→
	Vit. E	(i.u.)	←	35		→	←	35		→
75	DMI	(kg)	1.5	1.5	1.5	1.4	1.6	1.6	1.6	1.5
	ME	(MJ)	8.4	9.6	11.1	13.2	9.7	11.7	14.3	17.7
	RDP	(g)	70	80	93	110	81	97	119	148
	UDP	(g)	36	38	42	50	36	37	40	48
	E_{mp}	(MJ)	6.4	6.6	6.8	7.1	6.6	6.9	7.2	7.6
	APL		1.00	1.01	1.04	1.08	1.01	1.05	1.10	1.17
	Ca	(g)	6.7	6.7	9.8	9.8	6.7	6.7	11.7	11.7
	P	(g)	5.4	5.4	6.8	6.8	5.4	5.4	7.6	7.6
	Mg	(g)	1.3	1.3	1.7	1.7	1.3	1.3	2.0	2.0
	Na	(g)	←		2.5	→	←		2.5	→
	Vit. A	(i.u.)	←	5 000		→	←	5 000		→
	Vit. D.	(i.u.)	←	750		→	←	750		→
	Vit. E	(i.u.)	←	45		→	←	45		→

TABLE 10   Feeding standards for lactating ewes[2,3]
10.1   *Daily allowances for lactating ewes, assuming zero weight change*
$$(q_m = 0.625)$$

Ewe weight (kg)	Component		Week of lactation					
			Single lamb			Twin lambs		
			1–4	5–8	9–12	1–4	5–8	9–12
55	DMI	(kg)	1.5	1.7	1.6	1.6	1.8	1.7
	ME	(MJ)	16.6	15.7	13.0	22.3*	20.0	15.8
	RDP	(g)	139	131	108	186	167	132
	UDP	(g)	57	53	43	75	68	54
	E_{mp}	(MJ)	11.0	10.4	8.8	14.3	13.1	10.5
	APL		2.14	2.03	1.71	2.79	2.53	2.04
	Ca	(g)	8.1	7.7	6.7	10.3	9.4	7.8
	P	(g)	6.8	6.6	5.8	8.4	7.8	6.6
	Mg	(g)	2.2	2.1	1.7	2.9	2.6	2.1
	Na	(g)	2.1	2.0	1.9	2.4	2.3	2.0
	Vit. A	(i.u.)	←	5 500	→	←	5 500	→
	Vit. D	(i.u.)	←	550	→	←	550	→
	Vit. E	(i.u.)	←	40	→	←	45	→
75	DMI	(kg)	1.9	2.2	2.0	2.0	2.3	2.1
	ME	(MJ)	24.9	23.3	18.4	32.3*	27.4	21.3
	RDP	(g)	208	194	154	270	229	178
	UDP	(g)	84	78	62	108	92	72
	E_{mp}	(MJ)	16.2	15.2	12.3	20.6	17.7	14.1
	APL		2.48	2.33	1.89	3 14	2.71	2.15
	Ca	(g)	12.2	11.5	9.6	15.0	13.1	10.8
	P	(g)	10.1	9.7	8.3	12.1	10.8	9.1
	Mg	(g)	3.3	3.1	2.5	4.2	3.6	2.9
	Na	(g)	3.0	2.9	2.6	3.4	3.1	2.8
	Vit. A	(i.u.)	←	7 500	→	←	7 500	→
	Vit. D	(i.u.)	←	750	→	←	750	→
	Vit. E	(i.u.)	←	55	→	←	60	→

* These energy levels are not attainable because of inadequate DMI

Table 10 contd
10.2   *Daily allowances for 55 kg lactating ewes losing 50 g liveweight/day, and 75 kg lactating ewes losing 75 g liveweight/day*

$$(q_m = 0.625)$$

Ewe weight (kg)	Component		Week of lactation					
			Single lamb			Twin lambs		
			1–4	5–8	9–12	1–4	5–8	9–12
55	DMI	(kg)	1.5	1.7	1.6	1.6	1.8	1.7
	ME	(MJ)	14.7	13.8	11.1	20.3*	18.1	13.9
	RDP	(g)	123	115	93	169	151	116
	UDP	(g)	62	58	49	81	74	59
	E_{mp}	(MJ)	9.8	9.3	7.6	13.2	11.9	9.4
	APL		1.92	1.80	1.48	2.56	2.31	1.82
	Ca	(g)	8.1	7.7	6.7	10.3	9.4	7.8
	P	(g)	6.8	6.6	5.8	8.4	7.8	6.6
	Mg	(g)	2.2	2.1	1.7	2.9	2.6	2.1
	Na	(g)	2.1	2.0	1.9	2.4	2.3	2.0
	Vit. A	(i.u.)	←——— 5 500 ———→			←——— 5 500 ———→		
	Vit. D	(i.u.)	←——— 550 ———→			←——— 550 ———→		
	Vit. E	(i.u.)	←——— 40 ———→			←——— 45 ———→		
75	DMI	(kg)	1.9	2.2	2.0	2.0	2.3	2.1
	ME	(MJ)	22.0	20.4	15.6	29.4*	24.5	18.5
	RDP	(g)	184	170	130	245	205	154
	UDP	(g)	93	87	70	117	101	80
	E_{mp}	(MJ)	14.5	13.5	10.6	18.8	16.0	12.4
	APL		2.21	2.07	1.62	2.88	2.44	1.89
	Ca	(g)	12.2	11.5	9.6	15.0	13.1	10.8
	P	(g)	10.1	9.7	8.3	12.1	10.8	9.1
	Mg	(g)	3.3	3.1	2.5	4.2	3.6	2.9
	Na	(g)	3.0	2.9	2.6	3.4	3.1	2.8
	Vit. A	(i.u.)	←——— 7 500 ———→			←——— 7 500 ———→		
	Vit. D	(i.u.)	←——— 750 ———→			←——— 750 ———→		
	Vit. E	(i.u.)	←——— 55 ———→			←——— 60 ———→		

* These energy levels are not attainable because of inadequate DMI

Table 10 contd
10.3   *Net energy (Ne$_{mp}$) values of foods (MJ/kg DM) for lactating ewes*

APL	M/D								
	9.0	9.5	10.0	10.5	11.0	11.5	12.0	12.5	13.0
1.00	6.1	6.5	6.9	7.4	7.8	8.3	8.8	9.3	9.8
1.10	6.0	6.4	6.8	7.3	7.7	8.2	8.6	9.1	9.6
1.20	5.9	6.3	6.7	7.2	7.6	8.1	8.5	9.0	9.5
1.30	5.8	6.2	6.6	7.1	7.5	8.0	8.4	8.9	9.4
1.40	5.7	6.1	6.6	7.0	7.4	7.9	8.3	8.8	9.3
1.50	5.7	6.1	6.5	6.9	7.4	7.8	8.3	8.7	9.2
1.75	5.5	5.9	6.3	6.8	7.2	7.6	8.1	8.5	9.0
2.00	5.4	5.8	6.2	6.6	7.1	7.5	7.9	8.4	8.9
2.25	5.3	5.7	6.1	6.5	7.0	7.4	7.8	8.3	8.7
2.50	5.2	5.6	6.0	6.4	6.8	7.3	7.7	8.2	8.6
2.75	5.2	5.5	5.9	6.3	6.8	7.2	7.6	8.0	8.5
3.00	5.1	5.5	5.9	6.3	6.7	7.1	7.5	8.0	8.4
3.25	5.0	5.4	5.8	6.2	6.6	7.0	7.4	7.9	8.3
3.50	5.0	5.3	5.7	6.1	6.5	6.9	7.3	7.8	8.2

TABLE 11 Feeding standards for growing lambs[2,3]
11.1 *Daily allowances for castrated male lambs.*

Weight (kg)	q_m	Component		Liveweight gain (g/day)				DMI (kg/day)
				0	50	100	150	
20	0.55	ME	(MJ)	3.7	4.8			0.46
		RDP	(g)	31	40			
		UDP	(g)	14	18			
	0.65	ME	(MJ)	3.5	4.5	5.6	6.8	0.56
		RDP	(g)	29	38	47	57	
		UDP	(g)	16	20	23	26	
		E_{mp}	(MJ)	2.6	3.1	3.7	4.3	
		APL		1.00	1.22	1.44	1.66	
		Ca	(g)	1.5	2.3	3.1	3.9	
		P	(g)	1.0	1.2	1.5	1.8	
		Mg	(g)	0.35	0.47	0.59	0.71	
		Na	(g)	0.56	0.62	0.68	0.74	
		Vit. A	(i.u.)	←		660	→	
		Vit. D	(i.u.)	←		120	→	
		Vit. E	(i.u.)	←		21	→	
35	0.55	ME	(MJ)	5.7	7.3			0.77
		RDP	(g)	48	61			
		UDP	(g)	18	17			
	0.65	ME	(MJ)	5.4	6.8	8.4	10.1	0.92
		RDP	(g)	45	57	70	84	
		UDP	(g)	20	20	20	18	
		E_{mp}	(MJ)	4.0	4.8	5.6	6.4	
		APL		1.00	1.21	1.41	1.62	
		Ca	(g)	2.5	3.4	4.2	5.0	
		P	(g)	1.7	2.0	2.2	2.5	
		Mg	(g)	0.62	0.74	0.86	0.98	
		Na	(g)	0.98	1.0	1.1	1.2	
		Vit. A.	(i.u.)	←		1 200	→	
		Vit. D	(i.u.)	←		210	→	
		Vit. E	(i.u.)	←		25	→	

Table 11 contd.
11.2 *Daily allowances for entire male lambs*

Weight (kg)	q_m	Component		Liveweight gain (g/day)				DMI (kg/day)
				0	50	100	150	
20	0.55	ME	(MJ)	4.3				0.46
		RDP	(g)	36				
		UDP	(g)	10				
	0.65	ME	(MJ)	4.1	4.9	5.8	6.8	0.56
		RDP	(g)	34	41	49	57	
		UDP	(g)	12	17	22	26	
		E_{mp}	(MJ)	3.0	3.5	4.0	4.5	
		APL		1.00	1.17	1.34	1.51	
		Ca	(g)	1.5	2.3	3.1	3.9	
		P	(g)	1.0	1.2	1.5	1.8	
		Mg	(g)	0.35	0.47	0.59	0.71	
		Na	(g)	0.56	0.62	0.68	0.74	
		Vit. A	(i.u.)	←	660		→	
		Vit. D	(i.u.)	←	120		→	
		Vit. E	(i.u.)	←	21		→	
35	0.55	ME	(MJ)	6.6				0.77
		RDP	(g)	55				
		UDP	(g)	12				
	0.65	ME	(MJ)	6.2	7.6	9.0	10.5	0.92
		RDP	(g)	52	63	75	88	
		UDP	(g)	14	15	16	16	
		E_{mp}	(MJ)	4.6	5.3	6.1	6.9	
		APL		1.00	1.17	1.34	1.51	
		Ca	(g)	2.5	3.4	4.2	5.0	
		P	(g)	1.7	2.0	2.3	2.5	
		Mg	(g)	0.62	0.74	0.86	0.98	
		Na	(g)	0.98	1.0	1.1	1.2	
		Vit. A	(i.u.)	←	1 200		→	
		Vit. D	(i.u.)	←	210		→	
		Vit. E	(i.u.)	←	25		→	

Table 11 contd.

11.3 *Daily allowances for female lambs*

Weight (kg)	q_m	Component		Liveweight gain (g/day)				DMI (kg/day)
				0	50	100	150	
20	0.55	ME	(MJ)	3.7	4.9			0.46
		RDP	(g)	31	40			
		UDP	(g)	14	16			
	0.65	ME	(MJ)	3.5	4.5	5.7	6.9	0.56
		RDP	(g)	29	38	47	58	
		UDP	(g)	16	18	20	22	
		E_{mp}	(MJ)	2.6	3.2	3.7	4.3	
		APL		1.00	1.23	1.45	1.68	
		Ca	(g)	1.5	2.3	3.1	3.9	
		P	(g)	1.0	1.2	1.5	1.8	
		Mg	(g)	0.35	0.47	0.59	0.71	
		Na	(g)	0.56	0.62	0.68	0.74	
		Vit. A. (i.u.)		←		660	→	
		Vit. D (i.u.)		←		120	→	
		Vit. E (i.u.)		←		21	→	
35	0.55	ME	(MJ)	5.7	7.6			0.77
		RDP	(g)	48	63			
		UDP	(g)	18	13			
	0.65	ME	(MJ)	5.4	7.1	8.9	10.9	0.92
		RDP	(g)	45	59	74	91	
		UDP	(g)	20	17	13	8	
		E_{mp}	(MJ)	4.0	4.9	5.8	6.8	
		APL		1.00	1.24	1.47	1.71	
		Ca	(g)	2.5	3.4	4.2	5.0	
		P	(g)	1.7	2.0	2.3	2.5	
		Mg	(g)	0.62	0.74	0.85	1.0	
		Na	(g)	0.98	1.0	1.1	1.2	
		Vit. A (i.u.)		←		1 200	→	
		Vit. D (i.u.)		←		210	→	
		Vit. E (i.u.)		←		25	→	

Table 11 contd.

11.4   *Net energy (NE$_{mp}$) values of foods (MJ/kg DM) for growing lambs*

APL	M/D												
	8.0	8.5	9.0	9.5	10.0	10.5	11.0	11.5	12.0	12.5	13.0	13.5	14.0
1	5.2	5.7	6.1	6.5	6.9	7.4	7.8	8.3	8.8	9.3	9.8	10.3	10.8
1.1	5.0	5.4	5.9	6.3	6.7	7.2	7.7	8.1	8.6	9.1	9.6	10.1	10.6
1.2	4.8	5.2	5.7	6.1	6.5	7.0	7.5	7.9	8.4	8.9	9.4	10.0	10.5
1.3	4.6	5.0	5.4	5.9	6.3	6.8	7.3	7.8	8.2	8.8	9.3	9.8	10.3
1.4	4.3	4.8	5.2	5.7	6.1	6.6	7.1	7.6	8.1	8.6	9.1	9.6	10.2
1.5	4.1	4.5	5.0	5.4	5.9	6.4	6.9	7.4	7.9	8.4	8.9	9.5	10.0
1.6	3.8	4.3	4.7	5.2	5.7	6.2	6.7	7.2	7.7	9.2	8.8	9.3	9.9
1.7	3.5	4.0	4.5	4.9	5.4	5.9	6.4	7.0	7.5	8.0	8.6	9.1	9.7
1.8	3.2	3.7	4.2	4.7	5.2	5.7	6.2	6.7	7.3	7.8	8.4	9.0	9.6
1.9	2.8	3.3	3.9	4.4	4.9	5.5	6.0	6.5	7.1	7.6	8.2	8.8	9.4
2.0	2.3	3.0	3.5	4.1	4.7	5.2	5.8	6.3	6.9	7.4	8.0	8.6	9.2
2.1	1.4	2.5	3.2	3.8	4.4	4.9	5.5	6.1	6.7	7.2	7.8	8.4	9.1
2.2		1.7	2.7	3.4	4.0	4.6	5.2	5.8	6.4	7.0	7.6	8.3	8.9
2.3			2.0	3.0	3.7	4.3	5.0	5.6	6.2	6.8	7.4	8.1	8.7
2.4				2.4	3.3	4.0	4.7	5.3	6.0	6.6	7.2	7.9	8.5
2.5					2.8	3.6	4.4	5.1	5.7	6.4	7.0	7.7	8.4
2.6					2.0	3.2	4.0	4.8	5.5	6.1	6.8	7.5	8.2

TABLE 12   Dietary allowances (mg/kg DM) of trace elements for ruminants[2]

	Cattle		Sheep	
Copper	Pre-ruminant calf	2	Pre-ruminant lamb	1
	Others	12	Growing lambs	3
			Others	6
Iron	Before weaning	30	All classes	30
	After weaning	40		
	> 150 kg liveweight	30		
	Pregnant and lactating	40		
Iodine	Winter	0.5	Winter	0.5
	Summer	0.15	Summer	0.15
	Presence of goitrogens	2.00	Presence of goitrogens	2.00
Cobalt	All classes	0.11	All classes	0.11
Selenium	All classes	0.10	All classes	0.10
Zinc	All classes	40	All classes	40
Manganese	All classes	40	All classes	40

TABLE 13 Feeding standards for pigs[4,8]
13.1 *Typical dietary nutrient levels for growing pigs (fresh basis)**

Component	Liveweight (kg)		
	20	50	90
Feed (kg/day)	1.2	2.2	2.4
Digestible energy (MJ/kg)	14.0	13.5	13.0
Crude protein (g/kg)	220	180	140
Ideal protein (g/kg)	194	149	94
Lysine (g/kg)	13.6	10.4	6.6
Methionine + cystine (g/kg)	6.8	5.2	3.3
Threonine (g/kg)	8.2	6.3	4.0
Tryptophan (g/kg)	1.9	1.5	1.0
Calcium (g/kg)	9.8	8.1	7.8
Phosphorus (g/kg)	7.0	6.1	5.9
Salt (g/kg)	3.2	3.1	3.0
Iron (mg/kg)	62	59	57
Magnesium (mg/kg)	308	230	220
Zinc (mg/kg)	56	49	47
Copper (mg/kg)	5.6	5.4	5.2
Manganese (mg/kg)	11	11	11
Iodine (mg/kg)	0.15	0.15	0.15
Selenium (mg/kg)	0.15	0.15	0.15
Vitamin A (i.u./kg)	8 000	6 000	6 000
Vitamin D (i.u./kg)	1 000	750	750
Vitamin E (i.u./kg)	15	15	15
Thiamin (mg/kg)	2.0	1.5	1.5
Riboflavin (mg/kg)	3.0	3.0	3.0
Nicotinic acid (mg/kg)	15	15	15
Pantothenic acid (mg/kg)	10	10	10
Pyridoxine (mg/kg)	2.5	2.5	2.5
Choline (mg/kg)	1 000	1 000	1 000
Biotin (mg/kg)	0.2	0.2	0.2
Vitamin B_{12} (mg/kg)	0.01	0.01	0.01

* Growth rate 0.7 kg/day.

Table 13 contd.
13.2   *Typical dietary nutrient levels for breeding sows (fresh basis)*

Component	Pregnancy	Lactation			3 weeks' weaning
		5–8 weeks' weaning			
		11 piglets	9 piglets	7 piglets	All piglets
Feed (kg/day)	2.0	5.9	5.2	4.4	5.2
Digestible energy (MJ/kg)	13.0	13.0	13.0	13.0	13.0
Crude protein (g/kg)	130	160	160	160	170
Lysine (g/kg)	4.5	7.0	7.0	7.0	8.0
Methionine + cystine (g/kg)	3.1	3.8	3.8	3.8	4.4
Threonine (g/kg)	3.8	4.9	4.9	4.9	5.6
Isoleucine (g/kg)	3.9	4.9	4.9	4.9	5.6
Leucine (g/kg)	3.4	8.1	8.1	8.1	9.2
Tryptophan (g/kg)	0.7	1.3	1.3	1.3	1.5
Calcium (g/kg)	8.5	8.5	8.5	8.5	8.5
Phosphorus (g/kg)	6.5	6.5	6.5	6.5	6.5
Salt (g/kg)	3.0	3.0	3.0	3.0	3.0
Iron (mg/kg)	60	60	60	60	60
Zinc (mg/kg)	50	50	50	50	50
Copper (mg/kg)	6	6	6	6	6
Manganese (mg/kg)	16	16	16	16	16
Iodine (mg/kg)	0.5	0.5	0.5	0.5	0.5
Selenium (mg/kg)	0.15	0.15	0.15	0.15	0.15
Vitamin A (i.u./kg)	8 000	8 000	8 000	8 000	8 000
Vitamin D (i.u./kg)	1 000	1 000	1 000	1 000	1 000
Vitamin E (i.u./kg)	15	15	15	15	15
Thiamin (mg/kg)	1.5	1.5	1.5	1.5	1.5
Riboflavin (mg/kg)	3.0	3.0	3.0	3.0	3.0
Nicotinic acid (mg/kg)	15	15	15	15	15
Pantothenic acid (mg/kg)	10	10	10	10	10
Pyridoxine (mg/kg)	1.5	1.5	1.5	1.5	1.5
Choline (mg/kg)	1 500	1 500	1 500	1 500	1 500
Biotin (mg/kg)	0.30	0.30	0.30	0.30	0.30
Vitamin B_{12} (mg/kg)	0.015	0.015	0.015	0.015	0.015

TABLE 14 Feeding standards for poultry. Typical dietary nutrient levels (fresh basis)[1,6]

14.1 Chickens

	Growing chicks		Pullets 12–18 wks	Laying hens	Breeding hens	Broiler starter	Broiler finisher
	0–6 wks	6–12 wks					
ME (MJ/kg)	11.5	10.9	10.9	11.1	11.1	12.6	12.6
Crude protein (g/kg)	210	145	120	160	160	230	190
Amino acids (g/kg)							
Arginine	11	7.1	6.7	4.9	4.9	12.6	9.5
Glycine + serine	13.2	9.4	8.0	—	—	12.0	11.0
Histidine	5.1	3.3	2.4	1.6	1.6	5.0	5.0
Isoleucine	9	5.9	4.5	5.3	5.3	9.0	8.0
Leucine	14.7	9.9	8.4	6.6	6.6	16.0	13.0
Lysine	11	7.4	6.6	7.3	7.3	12.5	10.0
Methionine + cystine	9.2	6.2	4.5	5.5	4.6	9.2	8.0
Phenylalanine + tyrosine	15.8	10.8	8.0	7.0	7.0	15.8	14.0
Threonine	7.4	4.9	4.2	3.5	3.5	8.0	6.5
Tryptophan	2.0	1.4	1.2	1.4	1.4	2.3	1.9
Valine	10.4	6.6	5.3	5.3	5.3	10	9.0
Major minerals (g/kg)							
Calcium	12	10	8	35	33	12	10
Phosphorus (av.)	5	5	5	5	5	5	5
Magnesium	0.3*	0.3*	0.3*	0.3*	0.3*	0.3*	0.3*
Sodium	1.5	1.5	1.5	1.5	1.5	1.5	1.5
Potassium	3.0	—	—	—	—	3.0	3.0
Trace minerals (mg/kg)							
Copper	3.5*	3.5*	3.5*	3.5*	3.5*	3.5*	3.5*
Iodine	0.4*	0.4*	0.4*	0.4*	0.4*	0.4*	0.4*
Iron	80*	80*	80*	80*	80*	80*	45*
Manganese	100*	100*	100*	100*	100*	100*	100*
Zinc	50*	50*	50*	50*	50*	50*	50*
Selenium	0.15	—	—	—	—	0.15	0.15
Vitamins (i.u./kg)							
A	2 000*	2 000*	2 000*	6 000*	6 000*	2 000*	2 000*
D₃	600*	600*	600*	800*	800*	600*	600*
E	25*	25*	25*	25*	25*	25*	25*
Vitamins (mg/kg)							
K	1.3*	1.3*	1.3*	1.3*	1.3*	1.3*	1.3*
Thiamin	3	—	—	—	2	3	—
Riboflavin	4*	4*	4*	4*	4*	4*	4*
Nicotinic acid	28*	28*	28*	28*	28*	28*	28*
Pantothenic acid	10*	10*	10*	10*	10*	10*	10*
Choline	1 300	—	—	—	1 100	1 300	1 300
Vitamin B₁₂	—	—	—	—	0.01	—	

* Added as supplement

Table 14 contd.
14.2 *Turkeys*

	Growing turkeys			Breeding turkeys
	0–6 wks	*6–12 wks*	*12 + wks*	
ME (MJ/kg)	12.6	11.9	11.9	11.3
Crude protein (g/kg)	300	260	180	160
Amino acids (g/kg)				
Arginine	16	13	8	5
Glycine + serine	9	—	7	—
Histidine	6	5	3.5	2
Isoleucine	11	10	5.5	5.5
Leucine	19	15	8	7
Lysine	17	13	8	7.5
Methionine + cystine	10	8	6	5.5
Phenylalanine + Tyrosine	16	15	10	8
Threonine	10	9	5.5	4
Tryptophan	2.6	2.3	1.5	1.7
Valine	12	10	6	5
Major minerals (g/kg)				
Calcium	9	10	8	30
Phosphorus (av.)	4.5	5	4	5
Magnesium	0.36*	0.36*	0.36*	0.30*
Sodium	1.75	1.75	1.75	1.75
Potassium	—	—	—	—
Trace minerals (mg/kg)				
Copper	4.2*	4.2*	4.2*	3.5*
Iodine	0.48*	0.48*	0.48*	0.4*
Iron	96*	96*	96*	80*
Manganese	120*	120*	120*	100*
Zinc	60*	60*	60*	50*
Selenium	0.2	0.15	0.15	—
Vitamins (i.u./kg)				
A	12 000*	12 000*	12 000*	10 000*
D_3	1 800*	1 800*	1 800*	1 500*
E	36*	36*	36*	30*
Vitamins (mg/kg)				
K	4.8*	4.8*	4.8*	4.0*
Thiamin	4.8*	4.8*	4.8*	4.0*
Riboflavin	12*	12*	12*	10*
Nicotinic acid	60*	60*	60*	50*
Pantothenic acid	19.2*	19.2*	19.2*	16*
Pyridoxine	6*	6*	6*	5*
Biotin	0.12*	0.12*	0.12*	0.1*
Folic acid	2.4*	2.4*	2.4*	2.0*
Vitamin B_{12}	0.024*	0.024*	0.024*	0.02*
Choline	1 760	—	—	1 350

* Added as supplement

TABLE 15   Water allowances of farm animals[2,4]

15.1   *Cattle and sheep*

	kg water/kg DM intake		
	Environmental temperature (°C)		
	<16	*16–20*	*>20*
Cattle			
Calves, up to 6 weeks	7.0	8.0	9.0
Cattle, growing or adult, pregnant or non-pregnant	5.4	6.1	7.0
Sheep			
Lambs, up to 4 weeks	4.0	5.0	6.0
Sheep, growing or adult, non-pregnant	2.0	2.5	3.0
Ewes, mid pregnancy, twin bearing	3.3	4.1	4.9
Ewes, late pregnancy, twin bearing	4.4	5.5	6.6
Ewes, lactating, first month	4.0	5.0	6.0
Ewes, lactating, second/third month	3.0	3.7	4.5

15.2   *Lactating cows (600 kg liveweight)*

Milk yield (kg/day)	Daily water intake (kg/head)		
	Environmental temperature (°C)		
	<16	*16–20*	*> 20*
10	81	92	105
20	92	104	119
30	103	116	133
40	113	128	147

15.3   *Pigs*

	Daily water intake (kg/head)
Growing pigs	1.5–2.0 at 15 kg liveweight, increasing to 6.0 at 90 kg liveweight
Non-pregnant sows	5.0
Pregnant sows	5.0–8.0
Lactating sows	15.0–20.0

TABLE 16   Values for metabolic liveweight ($W^{0.75}$) for weights at 10 kg intervals to 590 kg

Base	Hundreds	Tens									
		0	10	20	30	40	50	60	70	80	90
$W^{0.75}$	0	0	5.6	9.5	12.8	15.9	18.8	21.6	24.2	26.8	29.2
	100	31.6	34.0	36.3	38.5	40.7	42.9	45.0	47.1	49.1	51.2
	200	53.2	55.2	57.1	59.1	61.0	62.9	64.8	66.6	68.4	70.3
	300	72.1	73.9	75.7	77.4	79.2	80.9	82.6	84.4	86.1	87.8
	400	89.4	91.1	92.8	94.4	96.1	97.7	99.3	100.9	102.6	104.2
	500	105.7	107.3	108.9	110.5	112.0	113.6	115.1	116.7	118.2	119.7

REFERENCES

1 Agricultural Research Council, London, 1975. *The Nutrient Requirements of Farm Livestock. No. 1, Poultry*. Commonwealth Agricultural Bureaux, Farnham Royal.

2 Agricultural Research Council, London, 1980. *The Nutrient Requirements of Ruminant Livestock*. Commonwealth Agricultural Bureaux, Farnham Royal.

3 Agricultural Research Council, London, 1984. *The Nutrient Requirements of Ruminant Livestock. Supplement No. 1*. Commonwealth Agricultural Bureaux, Farnham Royal.

4 Agricultural Research Council, London, 1981. *The Nutrient Requirements of Pigs*. Commonwealth Agricultural Bureaux, Farnham Royal.

5 Agricultural Research Council, London, 1976. *The Nutrient Requirements of Farm Livestock. No. 4. Composition of British Feedstuffs*. Commonwealth Agricultural Bureaux, Farnham Royal.

6 W. Bolton and R. Blair, 1977. *Poultry Nutrition*, 4th edn. MAFF Bull. No. 174. HMSO, London.

7 R. E. Evans, 1960. *Rations for Livestock*. MAFF Bull. No. 174. HMSO, London.

8 Ministry of Agriculture, Fisheries and Food, 1982. *Nutrient Allowances for Pigs*. Booklet 2089. MAFF (Publications), Alnwick.

9 Ministry of Agriculture, Fisheries and Food, Department of Agriculture and Fisheries for Scotland, Department of Agriculture for Northern Ireland, 1984. *Energy Allowances and Feeding Systems for Ruminants*. Reference book 433. HMSO, London.

10 National Academy of Sciences, National Research Council, 1984. *Nutrient Requirements of Beef Cattle*. National Academy Press, Washington, DC.

11 National Academy of Sciences, National Research Council, 1984. *Nutrient Requirements of Poultry*. Washington, DC.

12 Poultry Research Centre 1981. *Analytical Data of Poultry Feedstuffs, 1. General and Amino Acid Analyses, 1977–1980*. Occasional Publication No. 1. PRC, Roslin, Midlothian.

13 F. W. Wainman, 1975. *First Rept. of the Feedingstuffs Evaluation Unit*. Rowett Research Institute, Aberdeen.

14 F. W. Wainman. P. J. S. Dewey and A. W. Boyne, 1978. *Second Rept. of the Feedingstuffs Evaluation Unit*. Rowett Research Institute, Aberdeen.

15 F. W. Wainman, P. J. S. Dewey and A. W. Boyne, 1981. *Third Rept. of the Feedingstuffs Evaluation Unit*. Rowett, Research Institute, Aberdeen.

Index

abomasum, 131, 142, 147, 150
abortion, 110, 451
absorption, 130,139–42, 154
acetic acid
  in intake regulation, 381
  in large intestine, 137
  in rumen, 147–50, 348
  in silage, 405–14
  metabolism, 158, 159, 173–99, 337
  utilisation,
    for fattening, 236–9
    for maintenance, 234
    for milk production, 339, 348,
      361–3
acetoin, 406, 407
acetone, 16, 17, 337
acetyl coenzyme A, 75, 76, 118, 120,
  148, 164–99
acid detergent fibre (ADF), 6, 203
acidosis, 337
Ackerman factor, 473
activators, 121, 199
activity increment, 291, 292, 351, 368,
  370
acyl-carrier protein, 190
adenine, 55–7, 161, 187
adenosine, 55–7, 161, 185–9
adenosine diphosphate (ADP), 57,
  161–99
adenosine monophosphate (AMP),
  55, 56, 161, 164
adenosine triphosphate (ATP), 57,
  161–99, 223, 233, 234, 238,
  295
Adolph, 378

adrenal corticosteroids, 38, 337
aflatoxin, 459, 460
Agricultural and Food Research
  Council (AFRC), 247, 254–7, 277,
    278, 285, 286, 292, 300, 307,
    308, 347, 348
Agricultural Research Council (ARC)
  see Agricultural and Food
    Research Council
Agrostis spp. *see* bents
AIV silage, 413
alanine, 44, 51, 80, 180,183, 184, 407
albumins, 50, 339
aldaric acid, 15
aldehyde oxidase, 112
aldehydes, 10, 33
Alderman, G., 258
aldoses, 10
alfalfa *see* lucerne
alimentary canal 130–57
alkali disease, 113
alkaloids, 54, 434, 451
alkalosis, 98
allantoin, 53, 220, 476
allometric equation, 301–7
alopecia, 399
aluminium, 277, 341
American Net Energy Systems, 253–5
amides, 53, 152, 261, 389
amines, 52, 137, 152, 406, 407, 412
amino acids, 42–8, 51, 159, 260–83
  absorption of, 139–41
  as products of digestion, 135, 136,
    147, 150
  as source of energy, 179–83

available, 270, 271
catabolism of, 179–83, 296
chemical properties of, 46, 47
deamination of, 179, 407, 411, 412
deficiencies, 380
dispensable, 260–83
essential *see* indispensable amino
    acids
formulae, 42–7
in cereals, 439–51
in milk, 339
in protein concentrates, 455–83
in protein synthesis, 183–8
in silage, 406, 407
indispensable *see* indispensable
    amino acids
metabolism of, 159, 160
synthesis of, 183–9
aminoethylphosphoric acid, 279
aminopeptidases, 133, 136
amino sugars, 14, 15
ammonia,
    in metabolism, 124, 137, 159,
        180–2, 210
    in milk, 338
    in rumen, 150–3, 276–83, 479
    in silage, 409–14
    treatment of hay, 422
    treatment of straw, 212, 427, 428
amphipaths, 136
amygdalin, 17
amylase, 18, 132–5, 139, 146
amylopectin, 20, 21, 132, 440
amylose, 20, 440
anabolism, 158
anaemia, 71, 107
    copper deficiency, 104
    haemolytic, 402, 467
    iron deficiency, 102, 141, 336
androgens, 40
angustifolin, 469
Animal Production Level (APL), 250,
    252
animal protein concentrates, 470–5
Animal Protein Factor (APF), 471
anomers, 11
antioxidants, 34, 424, 461
apo-enzyme, 120
arabinas, 20
arabinose, 12, 17, 23, 24, 140
arachidic acid, 29
arachidonic acid, 29, 30
arginine, 45, 180, 184, 187 273, 389,
    406, 407, 439, 466, 474

as indispensable amino acid, 42,
    268, 314, 373
Armsby, H. P., 246
arsenic, 90, 116, 341
artificially dried forage, 423, 424
ascorbic acid *see* vitamin C
ash, 4, 207
asparagine, 44, 53, 183, 390, 432
aspartic acid, 44, 51, 53, 179–83
Aspergillus flavus, 459
associative effect of foods, 211, 242
Astragalus racemosus, 113
atherosclerosis, 39
atropine, 54
availability of minerals, 214, 215, 318,
    356, 365, 368
Avena sativa see oats
avidin, 84
Ayrshire, 100, 342–8

bacteria
    in digestive tract, 137, 143–57
    in silage, 404–15
bagasse, 399, 401
barley
    by-products, 443–5
    feed, 445
    grain, 438–42, 453
    straw, 425, 426
    whole, 400
basal metabolism, 287, 288
beans, 398, 467, 468, 478
bents, 394, 396
beri-beri, 58
berseem, 398
Beta vulgaris, 431
betaine, 52, 433
bile, 38, 40, 135, 140, 154
bilirubin, 135
biliverdin, 135
biological value (BV)
    in factorial calculations, 298, 310,
        315, 355
    of food proteins, 265–9, 373, 456,
        457, 471
    of microbial protein, 276
biotin, 74, 83, 84, 99, 341
biuret, 153, 478, 481
Blaxter, K. L., 247, 295, 353
blind staggers, 113
bloat, 149, 150, 399, 441, 479
blood, 376, 402
    dried, 473, 474

minerals in, 92, 95, 97, 100, 102
blood meal, 473, 474
body composition, estimation of, 232, 233
bomb calorimeter, 218
bone, 93–5, 100
bone flour, 94
boron, 341
bracken poisoning, 76
bran, 101, 433, 449, 450
Brassica campestris see turnip and rape
Brassica napus see swede
Brassica oleracea see kale and cabbage
Brassicas, 109, 401, 402, 431
brewers' grains, 433, 443, 444
brewers' yeast, 444
Brody, S., 289
bromine, 341
brown fat, 295
buffering capacity, 406, 409–12
butterfat, 31, 219, 339–74
butyric acid, 29, 137, 361–3
    in intake regulation, 381
    in milk, 340
    in rumen, 147–50
    in silage, 405, 410–5
    utilisation of, 158, 159, 172, 174, 182, 234, 237
Byerly, T. C., 327, 328

cabbage, 73, 100, 401
cadaverine, 52, 407
caeca, 138
caecum, 131, 156
calcium, 90–5, 440
    absorption, 141
    availability of, 214
    binding protein, 66
    deficiency symptoms, 93, 94
    in animal body, 91
    in eggs, 328–30
    in grass, 391, 396
    in milk, 340, 344
    interrelations with
        magnesium, 93
        manganese, 110
        phosphorus, 67, 93–5, 358
        vitamin D, 67
        zinc, 111
    requirements for
        egg production, 329

lactation, 356–8, 365, 368, 373
    maintenance and growth, 316–8
    reproduction, 326, 332
    sources of, 94
calcium phosphate, 94
calorie, 161
calorimeter
    animal 224–30, 290
    bomb, 218
    gradient layer, 225
calorimetry, 2244–33
calves
    digestion in, 142–57
    mineral deficiencies in, 90–116
    vitamin deficiencies in, 58–89, 319
cannibalism, 98, 330
cannula, 208
capric acid, 29
caproic acid, 29, 34, 415
caprylic acid, 29
carbamoyl phosphate, 180
carbohydrates, 3, 8–25, 51
    absorption, 140
    classification of, 8, 9
    determination of, 4–6
    digestion of, 132–9, 142–52
    in eggs, 328
    in grasses, 388, 389
    in milk, 338, 339, 343–7
    in silage, 406–15
    metabolism of, 158–78
    oxidation of, 165–9, 226
    synthesis of, 195–8
    utilisation of
        for maintenance, 234
        for production, 236, 237
carbon balance, 230, 231
carbonic anhydrase, 111
carboxyglutamic acid, 46, 73
carboxypeptidase, 133, 135
carcinogens, 460
carmine, 201
carnitine, 87, 190
carotene, 59, 61, 64, 141, 330
    in grasses, 390, 423
    in hay, 417, 422
    in milk, 341, 359
    in roots, 431, 434
    in silage, 408
    see also vitamin A
Carpenter, K. J., 271
carrots, 18, 61, 431, 434
caseins, 51, 219, 277, 316, 339, 474
cassava, 16, 432, 436

catabolism, 158–83
cattle
  digestibility trials with, 201–3
  digestion in, 142–57
  energy utilisation by, 233–41
  feeding standards for
    growth, 301–20
    lactation, 337–63
    maintenance, 284–301
    reproduction, 321–37
  intake regulation in, 381–6
  mineral deficiencies in, 90–116
  mineral requirements of, 316–8,
    356–8
  vitamin deficiencies in, 58–89
  vitamin requirements of, 318–20,
    358–60
  see also ruminants
cellobiose, 19, 146
cellulase, 206
cellulose, 19, 22, 209, 210, 219, 237,
  427
  digestion of
    in fowls, 139
    in horses, 156
    in pigs, 137
    in ruminants, 144–50, 209, 383
  in grass, 389
Centrosema pubescens (Centro), 397
cephalins, 37
cereal
  forage, 399, 404
  grains, 68, 76, 81, 84, 95, 212,
    438–54
  hay, 421
  processing, 452–4
  straws, 425
cerebrocortical necrosis, 76
cerebrosides, 35
ceruloplasmin, 103
cetyl alcohol, 38
chaconin, 434
chaff, 424, 428
chelates, 90, 215
chemical score, 269, 270, 273, 457
chemostatic theory of food
  intake regulation, 376
chick *see* poultry
chiral atoms, 10
chitin, 14, 23, 476
chlorine, 90, 98, 340, 343
chloroform, 222
chlorophyll, 41, 61, 90, 408, 434
choice feeding, 380, 381

cholecalciferol, 39, 64–8 *see also*
  vitamin D
cholecystokinin, 135, 376
cholesterol, 39, 51, 115, 135, 136, 314
choline, 36, 37, 52, 84, 85, 341
chondroitin, 24
chou moellier, 401
chromic oxide, 203
chromium, 90, 115
chromoproteins, 51
chylomicrons, 140, 160
chymosin, 133
chymotrypsin, 133, 135, 458
chymotrypsinogen, 135
citric acid, 118, 167–99, 220
citrulline, 180, 199
cloaca, 131, 138
clostridia, 52, 137, 404–15
clovers, 149, 397, 403, 407, 420
clubbed down, 77
cobalamins *see* vitamin B$_{12}$
cobalt, 90, 92, 106–154, 391
cocaine, 54
cocksfoot, 389, 392, 395
coconut meal, 462, 478
cod liver oil, 60, 65, 72, 362
codons, 186–9
coenzyme A, 81, 99, 118, 120, 164–99
coenzymes, 120, 160–99
coliform bacteria, 137, 405, 406
collagen, 43, 50, 185
colon, 131, 137, 156
colostrum, 63, 65, 341
comparative slaughter techniques, 232
competitive inhibition, 126
coniine, 54
cooking of foods, 212
copper, 90, 92, 103–6, 112, 215
  deficiency symptoms, 104, 335
  in animal body, 103, 104
  in eggs, 328
  in milk, 341
  in pasture, 391
  in wool production, 104, 316
  sources of, 105
  toxicity of, 105, 106
cortisol, 40, 337
cotton seed
  cake 101, 456
  meal, 457, 460–2, 478
coumestans, 403
cow *see* cattle and ruminants
crazy chick disease, 72
creatine, 164, 296, 338

creatinine, 220, 296, 297
crop (of fowl), 131, 138, 379
crude fibre, 4, 5, 202, 212
  determination of, 5
  effect on digestibility, 210
  effect on milk, 361
crude protein, 4
  as feeding standard, 260–2, 299,
    356
  components of, 4
  determination of, 4
  digestibility of, 202, 210
  in milk, 343–7
cryptoxanthin, 61, 445
curled toe paralysis, 77
cutin, 150
cyanogenetic glycosides, 16, 17, 109,
    436, 463, 467
cysteine, 44, 46, 99, 104, 134, 180
cystine, 46, 99, 269, 270, 273, 439
  in association with methionine, 268
  in oilseed meals, 457–66
  in wool, 315
cytidine, 185–9
cytochrome oxidase, 103
cytochromes, 51, 90, 102, 112, 119,
    163
cytosine, 54–7, 187

D value, 202
Dactylis glomerata see cocksfoot
Dasytricha, 145
Daucus carota see carrots
deamination, 179, 276, 406, 407, 411,
    412
decanoic acid *see* capric acid
degradable protein *see* rumen
    degradable protein
7-dehydrocholesterol, 39, 65, 141
dehydroretinol *see* vitamin A₂
deoxyadenosyl cobalamin, 86
deoxyribonuclease, 133, 136
deoxyribonucleic acid (DNA), 15, 50,
    56, 57, 66, 136, 186
deoxyribose, 15, 54–7
deoxy sugars, 15
deuterium, 232
dextrins, 21, 134, 136, 443
dhurrin, 17
diacylglycerols, 33, 136
diaminopimelic acid (DAPA), 279
dicoumarol, 74
digestibility, 200–16

and intake, 382–4
  factors affecting, 210–4
  in sacco, 206
  in vitro, 204–7, 393
  measurements, 201–7
  of grass, 391–3
  of hay, 421
  of proteins, 262
  of silage, 411
digestibility coefficient, 207
digestible crude protein, 262, 273,
    275–7
  calculation from crude protein, 202,
    275, 276
digestible energy, 219, 220, 224, 245,
    309, 371
digestion, 130–57
  in fowls, 138, 139, 156
  in horses, 156
  in pigs, 130–42
  in ruminants, 142–57
  work of, 223, 234, 239
digestive tract,
  of ruminant, 131, 142–57
  of fowl, 131, 138, 139
  of pig, 130–8
diglycerides *see* diacylglycerols
dihydroxyacetone phosphate, 166,
    173, 176, 192
dihydroxycholecalciferol, 66, 93
dilution rate, 155
dilution techniques, 232
dimethyl disulphide, 402
dipeptidases, 133, 136
disaccharides, 17–9
dispensable amino acids, 260–83
distillers' grains, 444, 445
distillers' solubles, 444, 445
diureidoisobutane, 481
diverticulum, 138
DNA *see* deoxyribonucleic acid
dodecanoic acid *see* lauric acid
draff, 443
dried grass, 423, 424
dulcitol, 15
duodenal juice, 134
duodenum, 131–6, 138, 208, 379
Dutch net energy system, 252, 254

East German net energy system, 247,
    251, 252, 254
Echinochloa crusgalli see millet

EFA *see* essential fatty acids
egg, 93, 328
egg production, 241, 242, 326–30
eicosanoic acid *see* arachidic acid
Eijkmann, 58
Eleusine coracana see millet
Embden and Meyerhof pathway, 165, 166, 171, 199
emulsification, 135
enantiomer, 10
encephalomalacia, 71, 72
endergonic reactions, 161
endogenous urinary nitrogen *see* nitrogen
endopeptidases, 135
energy
    content of foods, 217–59
    digestible *see* digestible energy
    feed unit, 252, 254
    gross *see* gross energy
    metabolisable *see* metabolisable energy
    metabolism, 160–83
    net *see* net energy
    productive values for poultry, 257
    requirements for
        egg production, 327–9
        growth, 307–9
        lactation, 347–54, 360–4, 368
        maintenance, 287–96
        pregnancy, 332–7
    retention, 224–42
    systems, 245–59
ensilage, 404–15
Enterobacteriaceae, 405
enterokinase, 198
enzootic ataxia, 104
enzymes, 117–29, 158–99
    digestive, 130–57
    inhibitors, 126, 212
    in plants, 417, 421, 443
    microbial, 142
    nature, 119, 124, 125
    nomenclature, 128
epimerases, 119
ergocalciferol, 39
    see also vitamin D
ergosterol, 39, 65, 141, 390, 419
ergot, 451
erucic acid, 32, 464
erythrocuprein, 103
Escherichia coli, 74, 82, 404
essential amino acids *see* indispensable amino acids

essential amino acid index (EAAI), 270
essential fatty acids, 30, 41, 362
ethanolamine, 36, 37
ether extract, 4, 214
ethylenediaminetetraacetic acid (EDTA), 5, 90
euglobulin, 339
exergonic reactions, 161
exogenous urinary nitrogen
    see nitrogen
exopeptidases, 135
exudative diathesis, 71, 72, 112

factorial method of estimation
    energy requirements, 347–54, 368, 371
    mineral requirements, 317, 356–8, 368, 373, 374
    protein requirements, 298, 299, 310–2, 354–6, 368, 372
faeces, 137, 155, 201–15, 219–22
    collection of, 201–4
falling disease, 104
fasting
    catabolism, 288
    metabolism, 249, 250, 287–96
fatal syncope, 72
fats, 3, 26–35
    absorption, 140
    brown, 295
    chemical properties of, 32–5
    composition of, 31, 32
    digestion of, 133, 135, 136, 153, 154
    energy value, 219
    hydrogenation, 34, 35
    hydrolysis of, 32, 33
    melting point, 27, 29
    metabolism of, 173–8, 337
    milk, 32, 339–74
    oxidation of, 33, 227
    synthesis of, 189–95
    storage of, 231, 302–7
    taints in, 33, 37
    utilisation of
        for maintenance, 234,
        for production, 237–42
fattening, 303
fatty acids, 27–41, 118, 136, 159–78
    absorption of, 140
    essential, 30
    in milk fat, 339, 340, 361
    oxidation of, 33, 176–8, 228

synthesis of, 190–5
see also volatile fatty acids
fatty liver and kidney syndrome, 84
favism, 467
feathers, 61, 103, 115
feedback inhibition, 199 ·
feeding standards for
  egg production, 326–30
  growth, 301–20
  lactation, 337–74
  maintenance, 284–301, 316–20
  pregnancy, 331–7
  reproduction, 321–37
  wool production, 314–6
ferric oxide, 201
ferritin, 88, 102, 103
fertilisers (nitrogenous), 101, 396,
  431, 447
Festuca spp., 394, 395
Fingerling, G., 257
fish factor, 87
fish meal, 470–2, 478
  as source of minerals, 94, 95,98,
    102, 109, 111, 471
  as source of vitamins, 67, 73, 471
fistulated animals, 204, 208
Flatt, W. P., 241
flavin adenine dinucleotide (FAD),
  77, 162, 163
flavin mononucleotide (FMN), 77, 120
flavoproteins, 51, 77, 102
flowmeal, 448
fluorine, 90, 92, 94, 113, 114, 341
flushing, 324
fodder beet, 430–2
  tops, 402
foetus, 331–7, 385
folacin, 82, 83, 341
folic acid, 74, 82, 83, 126
food
  digestibility of, 200–16
  digestion of, 130–57
  energy content of, 217–43
  energy value of, 244–59
  intake of, 360–3, 375–86, 397,
    411–4, 425
  protein evaluation, 260–83
Forbes, E. B., 348
formalin, 152, 316, 414, 475
formic acid, 148, 407, 413, 414
fowl *see* poultry
Fraps, G.S., 257
French net energy system, 253
Fries, 348

Friesian, 342–28
fructans, 22, 146, 388, 389, 395–7,
  417
fructose, 11–22, 136, 140, 146, 158,
  196, 388, 389, 406, 407
fructose phosphate, 165, 166, 170,
  173, 196
fructosides, 16
fucose, 15
fumaric acid, 148, 167, 180–2
fungi, 143
Funk, 58
furan, 11

galactans, 23
galatolipids, 35, 390
galactosamine, 14, 15, 24, 51
galactose, 11, 13, 15, 18, 19, 23, 24,
  35, 136, 140, 158, 196, 197, 339
galactosides, 16
galacturonic acid, 23
gall bladder, 135
gallic acid, 34
gangliosides, 36
gastrin, 134
gastric juice, 103, 134
gelatin, 59
genistein, 458
gizzard, 131, 138
gliadin, 448
globulins, 50, 138, 339
glucans, 20–2, 476
glucaric acid, 15
glucolipids, 26
gluconeogenesis, 236
gluconic acid, 15
glucosaminans, 23
glucosamine, 14, 15, 24, 51
glucose, 10–24, 219, 406, 407
  as product of digestion, 132, 136,
    139, 140, 146
  in intake regulation, 370, 381
  in pasture, 388, 389
  in pregnancy toxaemia, 336, 337
  metabolism of, 128, 144, 158–68,
    173, 176, 182, 195–8, 223,
    226, 227, 363
  utilisation of
    for maintenance, 234
    for production, 237
glucose-l-phosphate, 14, 196, 197
glucose-6-phosphate 14, 166, 196
glucosides, 16

glucosinolate, 464, 465
glucostatic theory of food intake
    regulation, 376, 381
glucuronates, 220
glucuronic acid, 15, 24, 88
glutamic acid, 44, 46, 51, 53, 82, 134,
    389, 447, 448
  metabolism of, 179–84
glutamine, 44, 53, 181, 183, 390
glutathione peroxidase, 69, 113
glutelin, 446, 448
gluten, 446, 448
glyceraldehyde, 10, 47
glycerol, 27–37
  in milk fat, 340
  metabolism of, 159, 160, 173, 176
  synthesis of, 193–5
glycine, 40, 43, 44, 51, 136, 140,
    180–4, 407
  requirement by chick, 268, 313,
    314, 329
glycocholic acid, 40, 135
glycogen, 11, 21, 158, 159
  as energy source, 168, 169
  digestion of, 132, 135
  synthesis of, 195, 196
glycolipids, 35, 36
glycolysis, 165, 166
glycoproteins, 13, 51
glycosides, 16, 17
goat, requirements of, 363–5
goitre, 108, 335, 402
goitrin, 109, 464
goitrogens, 109, 401, 402, 464, 465
gonadotrophic hormones, 324
gossypol, 461
gramineae, 394, 438–54
grass, 388–403
  dried, 423, 424
  silage, 404, 409–15
grass factor, 87
grinding of foods, 147, 155, 212, 222,
    238, 453
groats, 447, 448
gross energy, 218, 219, 224
  of milk, 347, 349, 370
  of silage, 411
gross protein value (GPV), 263, 264,
    457
groundnut meal, 457, 459, 460, 478
growth
  coefficient, 301–3
  energy requirements for, 307–9
  energy utilisation for, 235–41
  feeding standards for, 301–20

mineral and vitamin requirements
    for, 316–20
  protein requirements for, 309–14
guanine, 55–7, 185–9
Guernsey, 342–8
gulonolactone oxidase, 88
gums, 24

haem, 151, 119
haemoglobin, 51, 90, 101–3, 336, 390,
    402
haemolytic anaemia, 402, 467
haemosiderin, 102
hair, 103, 297, 355
halibut liver oil, 60, 65
Hammond, J., 306
hay, 416–22, 426
heat
  increment of foods, 222–42, 287,
    288, 377, 381
  loss from body, 217–42, 293–6
  of combustion, 219
  of fermentation, 223, 234
  production, 224–42, 287–96
heating of foods, 212, 471
Helianthus tuberosus see Jerusalem
    artichoke
hemicelluloses, 23, 210, 427
  digestion of, 137, 139, 145, 146
  in grass, 389, 391
  in silage, 406
hen *see* poultry
heptoses, 13, 14
herring meal, 115
heteroglycans, 10, 23, 24
hexanoic acid *see* caproic acid
hexadecanoic acid *see* palmitic acid
hexose-phosphate shunt *see* pentose
    phosphate pathway
hexoses, 8–17
hippuric acid, 220
histamine, 52, 407
histidine, 45, 47, 52, 82, 268, 273,
    373, 407
histones, 50
Hogg, J., 106
Holcus lanatus see Yorkshire fog
holo-enzymes, 120
holotrichs, 145
Holstein, 322, 342–8
homoglycans, 10, 20–3
Hopkins, F.G., 58
hops, 433, 443, 444
Hordeum sativum see barley

hormones, 99, 322, 324, 335, 340, 376
horse, 71, 76, 340
  digestion in, 156
Huxley, J. S., 301, 304
hyaluronic acid, 24
hydrogen cyanide, 16, 17, 463
hydrolases, 118
hydroperoxide, 33
hydroxybutyrate, 158–60, 172, 174,
    337, 339
hydroxycholecalciferol, 66
hydroxycobalamine, 86
hydroxylysine, 43
hydroxyproline, 43
hyparrhenia rufa see Jaragua grass
hypervitaminosis, 88, 89
hypocalcaemia, 66, 94, 432
hypoglycaemia, 337
hypomagnesaemia, 100, 101
hypomagnesaemic tetany, 100, 101,
    479
hypothalamus, 376–9
hypoxanthine, 181

Ideal Protein, 274, 275, 312
ileum, 131, 132, 208
ill thrift, 113
immunoglobulins, 140, 141
indicators (as markers), 201–4
indispensable amino acids, 47, 184
  importance in protein evaluation,
      260–83, 312–14
  requirements of pigs and poultry,
      312–14, 329, 372, 373
indole, 137
infertility, 63, 95, 404
infra red reflectance
    spectrophotometry, 6
inhibitors, 413, 414
inoculants, 413
inosinic acid, 181
inositol, 36, 87, 341
in sacco techniques, 206
insulin, 99, 376
intake of food *see* food intake
International Units, 62, 65, 69
intestine,
  small, 132–9, 147–54, 379
  large, 137, 156
intrinsic factor, 86
inulin, 22, 437
invertase *see* sucrase
in vitro techniques, 204–7, 393
iodine, 90, 92, 108, 109

absorption, 141
deficiency symptoms, 108, 109
in animal body, 108
in eggs, 328, 330
in milk, 341
requirements for,
  egg production, 330
  maintenance and growth, 316
  pregnancy, 335
Ipomoea batatas see sweet potato
iron, 90, 92, 101–3, 163
absorption of, 103, 141
availability of, 103, 214
deficiency symptoms, 102
in animal body, 101, 102
in eggs, 328, 330
in grass, 391
in milk, 341
in pregnancy, 336
in proteins, 120
requirements for
  egg production, 330
  maintenance and growth, 316
sources of, 102, 103
toxicity, 103
isobutylidene diurea, 153
isobutyric acid, 147, 150
isocitric acid, 162, 163
isoelectric point, 46
isoflavones, 403
isokestose, 19
isoleucine, 44, 47, 147, 268, 273, 329,
    373
isomerases, 119
isoprene, 40
isothiocyanates, 465
Isotrichidae, 145
Italian ryegrass, 389, 395

jacobine, 54
Jaragua grass, 394, 395
Java beans, 16
Jersey, 100, 342–8
Jerusalem artichoke, 432, 436, 437
joule, 161

kale, 73, 109, 401, 402
Kellner, 0,, 246, 247, 251–4, 348
keratin, 50, 315
kestose, 19
ketoglutaric acid, 162, 163, 167,
    179–84
ketones, 33, 228, 337

ketoses, 11
ketosis, 140, 228
kilocalorie, 161
kilojoule, 161
Kjeldahl, 4, 260, 262

lactalbumin, 339
lactase, 133, 136
lactation, 241, 242, 337–74, 379, 385
    tetany, 100
lactic acid, 18
    in metabolism, 118, 128, 138,
        167–70, 237
    in rumen, 144, 147, 148
    in silage, 405–15
lactic acid bacteria, 137, 143, 404,
        405, 409–14
lactoglobulin, 339, 474
lactose, 13, 18
    digestion of, 136, 138
    in milk, 338–74
    synthesis of, 195–8, 236
lanolin, 38
lard, 31, 32
large intestine *see* intestine
lathyrism, 467
Lathyrus sativus, 467
laurel oil, 28
lauric acid, 28, 29, 32, 38
lean body mass, 232
lecithinases, 133, 136
lecithins, 37, 51, 84, 136
legume, 397–9
    hay, 420
    seeds 467–70
    silage, 404, 407
    straw, 427
Leucaena leucocephala, 399
leucine, 44, 47, 135, 147, 180, 187,
        268, 273, 373, 407, 466, 474
levan, 22
level of feeding, 213, 214, 222, 304
ligases, 119
lignin, 24, 25, 137, 424, 425, 427
    effect on digestibility, 146, 150, 210
    in grass, 389, 391
limestone, 94
linamarin, 16, 17, 436, 463
Lind, J., 58
linoleic acid, 29–31, 35, 153, 339,
        363, 390, 440
linolenic acid 29–31, 35, 153, 339,
        363, 390
linseed, 30

cyanogenetic glycosides in, 16,
        17, 463
    meal, 101, 463, 478
    mucilage, 24, 463
lipase, 32, 33, 133–6, 447
lipids, 26–41, 160, 439
    classification, 26
    digestion of, 135, 136, 153, 154
    in eggs, 328
    in grass, 390
    nitrogenous, 36–8, 51
    see also fats
lipoic acid, 87
lipoproteins, 51, 160, 363
lipostatic theory of food intake,
        regulation, 377, 381
liver, 60
    fatty syndrome, 84
    importance in metabolism, 140, 151,
        158–60, 168, 170, 171
    necrosis, 460
    secretions, 135
Lolium multiflorum see Italian
        ryegrass
Lolium perenne see perennial ryegrass
lotaustralin, 17, 436
lucerne, 73, 397, 398, 403, 407, 410,
        423
    dried, 424
    hay, 420
lupin seed meal, 469
lupinin, 469
Lupinus spp., 469
lyases, 119
lysine, 45, 140, 180, 269–75, 407, 422
    as indispensable amino acid, 47,
        268, 313
    FDNB reactive, 271–3
    in cereal grains, 439, 446–8
    in grasses, 389
    in oilseed meals, 457–66
    requirement
        by poultry, 314, 329
        by pig, 314, 373
lysozyme, 132, 134

McHardy, F.V., 249
Macroptilium atropurpureum see
        Siratro
magnesium, 90, 93, 99–101, 185, 316
    absorption of, 101
    availability of, 214
    deficiency symptoms, 99–101
    in animal body, 100
    in eggs, 328

in grass, 391
in milk, 340, 365, 368
losses from body, 316
sources of, 101
tetany, 100, 101
Maillard reactions, 422
maintenance
    energy requirements for, 249–51,
        287–96, 351, 364, 368, 370
    feeding standards for, 284–301, 351
    mineral and vitamin requirements
        for, 316–8
    protein requirements for, 296–301
    utilisation of energy for, 233–5,
        247–56
maize
    by-products, 446
    grain, 79, 439, 445, 446, 453, 457
    silage, 409, 410
    straw, 425
malic acid, 148, 170
malonyl coenzyme A, 190, 192
malt coombs, 433, 443
maltase, 133, 136, 139, 146
maltose, 18, 19, 21, 443
    digestion of, 136, 138, 146
mammary gland, 168, 195–8, 332,
    337–74
manganese, 90, 109, 110
    and reproduction, 110, 326
    deficiency symptoms, 110
    in animal body, 109
    in eggs, 328, 330
    in enzyme action, 121, 190
    in grass, 391, 396
    requirements for egg production,
        330
    toxicity, 110
mangel,
    roots, 18, 430–2
    tops, 402
Manihot esculenta see cassava
manioc *see* cassava
mannans, 13, 23, 476
mannitol, 16, 407, 409
mannose, 11, 13, 15, 16, 23, 140, 196
margarine, 34
markers, 232, 279
mastitis, 343
meat and bonemeal, 94, 472, 473, 478
meatmeal, 98, 102, 111, 472, 473
Medicago sativa see lucerne
Megacalorie, 161
Megajoule, 161
melissic acid, 38

menadione, 73, 89
menaquinone, 73
metabolic faecal nitrogen *see* nitrogen
metabolic liveweight, 289, 379
metabolisability, 248, 249
metabolisable energy, 220–42,
        257–300
    of poultry diets, 220, 221
    requirements for,
        egg production, 327–9
        ruminants, 284–320, 332–7,
            347–54,364
    system of rationing ruminants,
        245–56
    utilisation for,
        fattening, 235–40, 411
        growth, 241, 411
        lactation, 241, 242, 347–54
        maintenance, 233–5, 247–51
metabolisable protein, 277, 278
metabolism, 158–99
methaemoglobin, 390
methane, 143, 147–57, 207, 220–2,
        228, 230, 411
methionine, 44, 82, 85, 99, 187,
        269–75
    as indispensable amino acid, 47,
        268
    association with cystine, 268
    association with glycine, 313
    available, 270, 271, 274
    in cereal grains, 439
    in oilseed meals, 457–66
    in wool, 315
    requirements by
        pig 373
        poultry, 313, 314, 329
methlycobalamin, 86
methylmalonyl coenzyme A, 86, 170
micelles, 136, 140, 154
micronisation, 212, 452
middlings, 449
milk
    by-products, 474, 475
    chemical composition of, 337–47,
        364, 366, 367
    digestion of, 142
    fat, 339–74
    minerals in, 94, 95, 104, 105, 336
    requirements for production of,
        337–74
    taints in, 431
    utilisation of energy for production
        of 241, 242, 251

vitamins in, 60, 65, 77, 83, 84, 336,
    358–60, 474
yields, 342–5, 353, 354, 360, 364,
    365, 370, 371
milk fever, 93, 94, 359
millet, 439, 451, 452
mimosine, 399
minerals, 90–116
    absorption of, 141
    availability of, 214, 215, 318, 356,
        365, 368
    deficiency symptoms, 90–116, 380
    essential, 90
    in cereal grains, 440–2
    in grass, 390, 391, 396
    in legumes, 397, 398
    in milk, 340, 341
    in straw, 426
    in tubers, 435
    requirements for
        egg production, 329, 330
        lactation, 356–8, 368
        maintenance and growth, 316–8
mitochondria, 162, 190, 192, 198
modified acid-detergent fibre (MADF),
    6
molasses, 19, 101, 399, 413, 432, 433,
    462, 480
molybdenum, 90, 92, 105, 112
    in animal body, 112
    in enzymes, 112
    in milk, 341
    in pasture, 391, 396
    toxicity of, 112
monensin, 222
monoacylglycerols, 33, 118, 135, 136,
    154
monoglycerides *see* monoacylglycerols
monopteroylglutamic acid, 82
monosaccharides, 9–17, 136
morphine, 54
Morrison, F. B., 246
mouth (digestion in), 132
mucilages, 24, 463
mucosal block theory, 103
mucopolysaccharides, 114
muscular dystrophy, 69–71
mycosterols, 39
mycotoxins, 417
myopathy, 71, 113
myrosinase, 465
myricyl alcohol, 38
myricyl palmitate, 38
myristic acid, 29
myxoxanthin, 61

naphthoquinones, 73
    see also vitamin K
necrosis of the liver, 460
Neptunia amplexicaulis, 113
net energy, 223, 224
    determination of, 223–33
    requirements for
        growth, 301–20
        lactation, 347–54, 364, 368, 370
        maintenance, 284–301
    systems for
        pigs and poultry, 256, 257
        ruminants, 245–56
    value of grass, 393, 397
    value of silage, 411
net protein retention (NPR), 263
net protein utilisation (NPU), 266
net protein value (NPV), 266
Neurath, 49
neutral-detergent-fibre (NDF), 5, 210
niacin, 78
nickel, 35, 90, 115
nicotinamide, 74, 78, 79
nicotinamide-adenine dinucleotide
    (NAD), 74, 79, 118, 120,
    159–99
nicotinamide-adenine-dinucleotide
    phosphate (NADP), 74, 79,
    120, 159–99
nicotine, 54.
nicotinic acid, 78, 320, 341
night blindness, 62, 63
nitrates, 53, 152, 390
nitrites, 53, 390, 472
nitrogen
    balance, 230, 231, 264, 299, 310,
        356
    degradable *see* rumen degradable
        protein
    endogenous urinary, 265, 266, 296,
        297
    exogenous urinary, 297
    faecal, 201, 262, 265, 266, 275
    metabolic faecal, 207, 211, 265,
        266, 275, 425
    undegradable *see* undegradable
        protein
    urinary, 227, 228, 296–9
    see also crude protein
nitrogen-free extractives, 4, 5
non-glucogenic ratio (NGR), 362
non-protein nitrogenous compounds,
    51–7, 151–3, 478–82
nucleic acids, 54–7, 136, 311, 476
nucleoproteins, 51

nucleosides, 55–7
nucleosidases, 133
nucleotides, 55–7, 185

oat
  by-products, 447, 448
  grain, 439, 440, 442, 447, 453, 457
  straw, 425
oat hay poisoning, 53
octadecanoic acid *see* stearic acid
octanoic acid *see* caprylic acid
oesophageal groove, 142
oesophagus, 131, 138, 379
oestrogens, 40, 325, 403
oil extraction, 455–7
oilseed cakes and meals, 455–67
oleic acid, 28, 29, 31, 34, 440
  in milk fat, 339
  synthesis of, 193
oligo-1, 6-glucosidase, 133, 136
oligosaccharides, 16–19, 132, 136
oligotrichs, 145
omasum, 131, 142, 147
Onobrychis sativa see sainfoin
Opaque-2, 446
Ophryscolecidae, 145
ornithine, 180, 199, 339, 406
orotic acid, 87
Oryza sativa see rice
osteomalacia, 67, 93, 95
ova, 321–6
overfeeding, 325
oxalacetic acid, 118, 148, 167–99, 337
oxalates, 466
oxalic acid, 402
oxidation, 33, 59, 61, 176–9, 226–8,
    408, 417–9
oxidative decarboxylation, 76, 165
oxidative phosphorylation, 162–99
oxidoreductases, 118

palatability, 378, 382
palmitic acid, 29, 38, 176–8, 190, 191,
    393, 390
palmitoleic acid, 29, 193
palm kernel meal, 462
pancreas, 135, 138
pancreatic juice, 135
pangamic acid, 87
Panicum miliaceum see millet
pantoic acid, 80
pantothenic acid, 74, 80–2, 335, 341
para-aminobenzoic acid, 126

parakeratosis, 111
parathyroid gland, 66, 93, 94
parotid glands, 132
parsnips, 434
Pastinaca sativa see parsnips
pasture, 388–403
  see also grass
peas, 398, 469
peat scours, 105
pectic acid, 23, 146
pectic substances, 23
pectin, 23, 146
Pediococcus acidilactici, 405, 413
pelleting of foods, 147, 222, 239, 372,
    423, 452, 453, 464
Pennisetum typhoideum see millet
pentanone, 34
pentose phosphate pathway, 168, 169
pentoses, 55–7, 146, 407
pepsin, 133, 134, 204, 206
pepsinogen, 134
peptide linkage, 48, 187
peptides, 48, 136, 150, 151
perennial ryegrass, 395, 419
perosis, 85, 110
peroxides, 69
phaeohphytin, 408
Phaseolus spp., 467, 468
phenol, 34, 137
phenylalanine, 45, 47, 134, 135, 180,
    187
  as indispensable amino acid, 47,
      268, 373
  association with tyrosine, 268, 313
phenylethylamine, 52
Phleum pratense see timothy
phosphatases, 133, 136
phosphatidic acid, 36, 194
phosphatidyl-ethanolamines, 37
phosphocreatine, 164, 165
phosphoenolpyruvate, 162, 166, 170,
    171, 183, 184
phosphogluconate oxidative pathway
    see pentose phosphate pathway
phosphoglycerate, 166, 395
phosphoglycerides, 36
phospholipids, 36, 94, 154
phosphoproteins, 51, 94
phosphoric acid,
  in carbohydrates, 14
  in lipids, 36, 37
  in metabolism, 161–99
  in nucleic acids, 54–7
  in proteins, 51
phosphorus, 51, 90–6, 279, 440

absorption, 141
association with carbohydrates, 14
availability of, 96, 214, 215
deficiency of, 95, 242
for reproduction, 332
in animal body, 94, 95
in eggs, 328, 330
in grass, 391
in metabolism, 143, 161–99
in milk, 340, 344
interrelation with
  calcium, 67, 93–5, 358
  magnesium, 96
  vitamin D, 67
  zinc, 111
requirements for
  egg production, 330
  lactation, 356–8, 365, 368, 373
  maintenance and growth, 318
  reproduction, 326, 332
phosphorylation, 161
phosphoserine, 183
phosvitin, 51
phylloquinone, 73
phytase, 215
phytates, 95, 96, 141, 330, 440
phytic acid, 90, 95, 96, 215, 330, 466
phytosterols, 39, 141
pica, 95, 107
pig
  composition of, 302–7
  critical temperature of, 294
  digestion and absorption in,
        130–42, 156
  energy
    requirements for, 308, 309, 370–2
    systems for, 256, 257
    utilisation by, 234, 237
  feeding standards for
    growth, 301–20
    lactation, 370–4
    maintenance, 284–301
    reproduction, 321–6, 331–7
  intake regulation in, 376–81
  mineral deficiencies in, 90–116
  mineral requirements of, 316–8,
        356–8, 373, 374
  protein evaluation for, 263–75
  protein requirements of, 312–4,
        372, 373
  vitamin deficiencies in, 58–89
  vitamin requirements of, 318–20,
        358–60, 374
piglet
  anaemia, 102

digestion in, 137, 138
pigments, in silage, 408
pining, 106
pinocytosis, 139
Pisum sativum see peas
polyneuritis, 76
polynucleotides, 56
polypeptides, 48, 189
polysaccharides, 20–4, 132–9, 142–50,
        195, 196
polysomes, 186
postabsorptive state, 288
pot ale, 445
potassium, 90, 96, 97, 215, 316, 328
  in animal body, 96, 233
  in grass, 391
  in milk, 340, 343
  losses from body, 315
potato, 156, 432, 434–6
poultry
  amino acid requirements of, 47,
        268, 312, 329
  digestibility measurements with, 201
  digestion in, 138, 139, 156
  energy
    requirements of, 293, 309
    systems for, 256, 257
    utilisation of, 241, 242
  feeding standards for, 293, 326–30
  intake regulation in, 376–81
  mineral deficiencies in, 90–116
  mineral requirements of, 316–8
  protein
    evaluation of, 263–75
    requirements of, 312–4
  vitamin deficiencies in, 58–89
  vitamin requirements for, 318–20
  waste, 153, 481, 482
pregnancy, 331–7, 379, 385
  anabolism, 333, 335
  toxaemia, 336, 337
procarboxypeptidase, 135
productive energy, 257
proline, 43, 45, 51, 140, 147, 184,
        187, 268
propionic acid, 137, 340, 422, 441,
        452, 453
  efficiency of utilisation
    for maintenance, 234
    for production, 237, 361–3
  in intake regulation, 381
  in rumen, 147–50
  in silage, 415
  metabolism of, 86, 107, 158, 159,
        169–71, 182

prostaglandins, 30, 41
prostanoic acid, 41
prosthetic groups, 51, 120
protamins,.50
proteases, 406, 417, 421, 458
protein concentrates, 455–83
protein efficiency ratio (PER), 263,
    457
protein equivalent (PE), 275
protein replacement value (PRV), 265
proteins, 42–54
  absorption, 140, 141
  classification, 50, 51
  degradable *see* rumen degradable
  digestibility of, 202, 204, 209
  digestion of, 133–5, 145, 150–2
  energy value of, 219
  evaluation of, 260–83
  in carcass, 232
  in cereals, 438–54
  in eggs, 328
  in grass, 389, 417
  in milk, 338, 339, 343–7
  in silage, 406
  metabolisable, 277, 278
  microbial, 276–83, 298
  oxidation of, 227, 228
  properties of, 49, 50
  requirements of,
    for egg production, 329
    growth, 309–16
    lactation, 354–6, 360–5, 368, 372,
      373
    maintenance, 296–301
    reproduction, 325, 332
    wool production, 314–6
  storage, 231, 301–4
  structure of, 47–9
  synthesis of, 183–9
  undegradable *see* undegradable
      protein
  utilisation of,
    for maintenance, 234, 235
    production, 235–42
protozoa, 143–5, 147, 149
proventriculus, 131
provitamins, 61, 65
proximate analysis of foods, 4, 5
pseudoglobulin, 339
Pseudomonas sp., 476, 477
Pteridium aquilina, 76
purines, 54–7, 82, 112
putrescine, 52, 407
pyran, 11
pyridoxal, 74, 79, 80

pyridoxamine, 79
pyridoxine, 79
pyrimidines, 54–7
pyruvic acid, 76, 118, 128, 147, 148,
    162, 165–8, 176, 180, 183, 184,
    407

quinine, 54
quinones, 34

radioactive isotopes, 92, 101, 215,
    232, 279
raffinose, 19, 388
rancidity, 33, 462
rape, 109, 401
rapeseed meal, 464, 465
rectum, 131
Reid, J. T., 306
rennin, 133, 134
reproduction, 321–37
respiration chamber, 220, 224, 225,
    229, 230
respiratory exchange, 229, 230
respiratory quotient, 227, 228, 288
reticulum, 131, 147
retinaldehyde, 62
retinol *see* vitamin A
rhamnose, 15, 24
rhodopsin, 62
riboflavin, 74, 76–8, 335, 341, 391
ribonuclease, 133, 136
ribonucleic acid (RNA), 13, 56, 57,
    66, 136, 185, 279
ribose, 12, 13, 54–7
ribosomes, 186–9
ribulose, 13, 169
ribulose–1, 5 biphosphate
    carboxylase, 389
rice
  by-products, 450
  grain, 439, 440, 450
  straw, 426
ricinine, 54
ricinoleic acid, 32
rickets, 65, 67, 93, 95, 330
roots, 430–4
rumen
  degradable protein (RDP), 151,
    209, 277–83, 298–301, 311,
    316, 354–6, 364, 476–8, 480
  digestion in, 131, 142–55
  protein breakdown in, 150–2,
    275–83

synthesis of vitamins in, 74, 75, 87, 107, 154
ruminants
  digestibility measurements with, 201–4
  digestion in, 142–57
  energy requirements for, 284–96, 307–9, 332–7
  intake regulation in, 381–6
  protein evaluation for, 275–83
  protein requirements of, 296–301, 309–16
  utilisation of energy by, 233–41
rumination, 142
rutin, 87
rye,
  grain, 439, 450, 451
  straw, 426
ryegrass, 393–5

sainfoin, 398
saliva, 132, 142, 152
Salmonella, 482
salt *see* sodium chloride,
saponification 32
Scandinavian feed unit system 253, 254
scurvy, 58, 88
seaweed, 109
sebaceous glands, 314
Secale cereale see rye
secretin, 135
sedoheptulose, 13, 14
selenium, 90, 113, 328, 391
  interrelationship with vitamin E, 68–72
  toxicity, 113, 463
Selenomonas ruminantium, 143
sensory appraisal, 377, 378, 381, 382
serine, 36, 44, 51, 82, 180, 183, 184, 187, 314, 406, 407
sesame meal, 466, 478
Setaria italica see millet,
Setaria sphecelata, 411
sheep
  digestibility trials with, 201–3
  digestion in, 142–57
  energy requirements of, 284–96, 307–9
  energy utilisation by, 233–41
  feeding standards for
    growth, 301–20
    lactation, 337–63, 365–9
    maintenance, 284–301

reproduction, 321–6, 331–7
intake regulation in, 381–6
mineral deficiencies in, 90–116
mineral requirements of, 316–8
protein requirements of, 296–301, 309–16
vitamin deficiencies in, 58–74
vitamin requirements of, 58–89, 318–20
wool production, 314–6
shorthorn, 345
silage, 349, 384, 404–15
silica, 114, 450
silicon, 90, 114, 341, 425
silicosis, 114
silos 404, 412, 415
single-cell protein, 475–7
Siratro, 397
skatole, 137
skim milk, 474, 475
small intestine *see* intestine
soap, 32
sodium, 90, 96–8, 139, 140, 215, 316, 317
  in animal body, 97
  in eggs, 328, 330
  in milk, 340, 343, 358
  losses from body 317
sodium chloride, 97, 109, 330, 470
  requirements, 358
  toxicity, 98
sodium hydroxide treatment
  of grain, 453, 454
  of straw, 212, 427, 428
solanidine, 434
solanine, 54, 434
Solanum tuberosum see potato
solids-not-fat, 344–74
sorbitol, 15
sorghum, 439, 452, 453
soya bean, 458
  meal, 457, 459, 478
specific gravity of animals, 233
spermaceti, 38
spermatozoa, 321–6
sphagnum moss, 433
sphingomyelins, 37, 38
sphingosine, 35, 37
spices, 378
splay-leg syndrome, 453
stachyose, 19, 388
stage of growth (of plants), 391–4, 419, 420
starch, ll, 20, 21, 219, 395, 397
  digestion of

by fowls, 139
by pigs, 132, 135
by ruminants, 144–50
in cereal grains, 440–52
in potatoes, 434, 435
utilisation for,
  maintenance, 234
  production, 237
starch equivalent, 247, 253
steaming up, 234
stearic acid, 28, 29, 34, 35, 136, 153,
  154, 193
sterioisomers, 10
sterility, 69–71, 325, 326
steroids, 38–40
sterols, 38–40, 65
Stickland reaction, 406, 407
stomach, 131, 134, 379
stravidin, 84
straw, 155, 424–8
  alkali treated, 212, 427, 428, 481
streptavidin, 84
streptococci, 18, 137, 144, 271, 404,
  405
strychnine, 54
sublingual glands, 132
submandibular glands, 132
submaxillary glands, 132
succinic acid, 147, 148, 167, 172
succinyl coenzyme A, 75, 86, 148,
  167, 170–2
sucrase, 18, 133, 136, 139
sucrose, 17, 18, 378, 388, 389, 431
  digestion of, 136, 138, 146
sudoriferous glands, 314
sugar acids, 15
sugar alcohols, 15, 16
sugar beet, 19, 52
  molasses, 433
  pulp, 432, 433
  tops, 402
sugar cane, 399, 401
sugars, 8–19
  absorption of, 139, 140
  digestion of, 136–9
  in grass, 388
  in roots, 430, 433
  in silage, 405, 406
suint, 314, 315
sulphanilamide, 126
sulphite oxidase, 112
sulphur, 42, 44, 90, 99, 105, 279, 391,
  480
sunflower seed meal, 465, 466, 478
surface area (of animals), 288, 289

swayback, 104, 335
swede, 401
  roots, 430–2
  tops, 402
sweet clover disease, 74
sweet potato, 423, 436

taints, 33, 431
taurocholic acid, 135
teart, 105, 112, 396
temperature, 127, 381, 422, 423
  body, 218, 293–6
  critical, 294–6
terpenes, 40
tetany, 100, 101, 479
tetradecanoic acid *see* myristic acid
tetrahymena, 271
tetrahymenol, 271
tetra-iodothyronine *see* thyroxine
tetrasaccharides, 19
tetroses, 8
thermal equivalent, 227, 228
thermostatic theory of food intake
    regulation, 377
thiaminase, 76
thiamin, 59, 74–6, 99, 319, 341, 449,
  460
thiamin diphosphate, 74, 165
thiocyanate, 401, 464
thioglucosides, 464
thiourea, 478
threonine, 44, 180, 268, 373
thromboxanes, 30
thumps, 102
thymine, 54–7
thyroglobulin, 43, 108
thyroid gland, 108, 399
thyroxine, 43, 92, 108
timothy, 391–5
tin, 90, 116, 341
tocopherols, 34, 68–72
  see also vitamin E
tocotrienols, 68
total digestible nutrients (TDN) 214,
  253, 254, 284
trace elements, 90, 91, 101–16
transamination, 47, 53, 179, 183
transferases, 99, 118
transferrin, 88
transmethylation, 84
trefoil, 17
triacylglycerols, 27–35, 118, 133–6,
  153, 154, 159, 160, 173, 194,
  195, 236, 339, 340, 363, 390

tricarboxylic acid cycle, 166–8, 170–3, 176, 179, 180
Trifolium spp. *see* clovers
triglycerides *see* triacyglycerols
tri-iodothyronine, 43, 108
trimethylamine, 53
triolein, 29
trioses, 8
tripalmitin, 182, 195, 227
trisaccharides, 19
triticale, 451
Triticum aestivum see wheat
tritium, 232
true metabolisable energy, 221
true protein, 51, 262, 311
trypsin, 121, 133, 135, 198
    inhibitors, 212, 458, 459, 468
trypsinogen, 121, 135, 198
tryptamine, 52, 407
tryptophan, 45, 47, 78, 134, 135, 180, 268–75, 329, 373
    in cereal grains, 439
tubers, 430, 434–7
turacin, 103
turkeys, 98, 313
turnip, 401
    roots, 430–2
    tops, 402
tussock grass, 394
twin lamb disease, 336
tyramine, 52
tyrosine, 43, 45, 109, 134, 135, 180, 268, 273

ubiquinone, 163
ultra-violet light, 39, 65
undegradable protein, (UDP), 151, 277–83, 299, 311, 316, 354–6, 365
underfeeding, 285, 323, 325, 334, 335, 369, 384
uracil, 54–7, 185–9
urea, 53, 99, 159, 220, 222, 275, 297, 384, 401, 427, 478–81
    in milk, 338
    in saliva, 152
    metabolism of, 124, 151–3, 180–4
urease, 124, 153, 427, 479
uric acid, 53, 181, 476, 482
uridine diphosphate galactose, 196, 197
uridine diphosphate glucose, 196, 197
urine, 151, 201, 296, 297

energy losses in, 220–2, 227, 228
minerals in, 317
uronic acids, 15, 146
U.S.A. net energy system, 253–5

valeric acid, 147, 362
valine, 44, 47, 75, 147, 150, 199, 268, 373
Van Es, E. J. H., 241, 348
vanadium, 90, 115
Van Soest, P. J., 5, 6
variable net energy system, 249
vernolic acid, 32
vetches, 398
Vicia faba, 398, 467
vicianin, 17
villi, 132, 140
vinquish, 106
vinylthioxazolidine, 109
Virtanen, A. I., 413
vitamin A, 59–64
    absorption of, 141
    chemical nature, 60
    deficiency, 62–4, 326, 335
    hypervitaminosis, 88, 89
    metabolism of, 62
    requirements for
        egg production, 330
        growth, 319
        lactation, 359, 374
vitamin A$_2$, 60
vitamin B complex, 51, 59, 74–87
    requirements for
        egg production, 330
        growth, 319
    synthesis of
        in intestine, 137
        in rumen, 154
vitamin B$_1$ *see* thiamin
vitamin B$_2$ *see* riboflavin
vitamin B$_6$, 59, 79, 80, 341
vitamin B$_{12}$, 59, 85–7, 90, 92, 106, 107, 142, 341
vitamin C, 59, 87, 88, 341
vitamin D, 59, 64–8, 93–5
    absorption of, 141
    and radiant energy, 65
    chemical nature, 39, 64
    deficiency, 67
    hypervitaminosis, 88, 89
    in roughage, 390, 324
    metabolism of, 66, 67
    potency of different forms, 67

requirements of,
  for egg production, 330
  lactation, 359, 374
vitamin E, 34, 59, 68–72, 424
  absorption of, 141
  chemical nature, 68
  deficiency of, 69–72, 326
  in cereals, 440, 449
  in grass, 68, 391
  interrelations with selenium, 68–72
  metabolism of, 69
vitamin K, 59, 72–4
  absorption of, 141
  chemical nature, 73
  deficiency, 74
  hypervitaminosis, 89
  metabolism of, 73
  synthesis of, 73, 154
vitamins, 3, 58–89
  absorption of, 141, 142
  and energy utilisation, 242
  in milk, 341, 474
  requirements of,
    for egg production, 330
    lactation, 358–60, 374
    maintenance and growth, 316–20
  synthesis of, 137, 154
volatile fatty acids, 137
  in intake regulation, 381
  in rumen, 143–57
  metabolism of, 169–73
  utilisation of,
    for maintenance, 234, 235
    production, 236–9, 348
voluntary intake of food, *see* food
    intake

wafering of foods, 212
water, 2
  in animal body, 301–4
  in foods, 416, 423
  in salt poisoning, 330
  metabolic, 2

requirements, 523
waxes, 38, 314
wheat,
  grain, 438, 448, 449, 453, 457
  by-products, 449, 450
  straw, 426
whey, 475
white-fish meal *see* fishmeal
Wood, P.D.P., 342
Woodman, H.E., 353
wool, 314–6
  minerals in, 99, 103, 113
  pigmentation in, 103
  production of, 314

xanthine, 181
xanthine oxidase, 112
xanthophylls, 61, 424
xerophthalmia, 63
xylans, 12, 20, 24
xylulose, 13, 146, 169
xylose, 12, 23, 24, 139, 140, 378

yeast, 67, 75–84, 101, 110, 111, 137,
    404, 409, 414, 443, 444, 476,
    477
Yorkshire fog, 394, 396

Zea mays see maize
zein, 446
zinc, 90, 111, 112
  absorption of, 141
  deficiency of, 111, 326
  in animal body, 111
  in eggs, 328, 330
  in grass, 391
  in milk, 341
  toxicity, 112
zoosterols, 39
zwitter ions, 46
zymogens, 121, 198